A Practical Introduction to Finite Element Analysis

A Practical Introduction to Finite Element Analysis

Y K CHEUNG

Ph.D., D.Sc., F.I.C.E., F.I.Struct.E., F.A.S.C.E.
Professor of Civil Engineering
University of Hong Kong

and

M F YEO

M.Sc., Ph.D., M.I.E.Aust.
Senior Lecturer in Civil Engineering
University of Adelaide
Australia

PITMAN

PITMAN PUBLISHING LIMITED
39 Parker Street, London WC2B 5PB

North American Editorial Office
1020 Plain Street, Marshfield, Massachusetts

Associated Companies
Fearon Pitman Publishers Inc, San Francisco
Copp Clark Pitman, Toronto
Pitman Publishing New Zealand Ltd, Wellington
Pitman Publishing Pty Ltd, Melbourne

LIBRARY OF CONGRESS CATALOGING IN PUBLICATION DATA
Cheung, Y.K.
A practical introduction to finite element analysis
Includes bibliographical references and index
1. Finite element method   2. Engineering data processing
I. Yeo, M.F., joint author   II. Title
TA347.F5C47   624.171   79-4262

ISBN 0 273 01083 2 hardcover
ISBN 0 273 01080 8 flexicover

Reproduced and printed by photolithography
in Great Britain at Biddles of Guildford

Preface

In this book the authors have attempted to present an introduction to the finite element method which has become the most powerful and versatile tool available to engineers. Unlike a number of other texts which dwell at length on the theoretical development and applications of the finite element method, the emphasis of this text is on various practical aspects of the method such as (i) the detailed description of various techniques used in the choice of displacement functions, stiffness formulation, numerical integration, assembly, solution and mesh generation; (ii) the detailed explanation of computer programs at various levels of development and comments for nearly all program statements; and (iii) intermediate output at all stages of program execution and final output through several case studies so that the readers can see what is happening at every step. The book is intended for final-year students, post-graduate students and for practising engineers. The material included is largely self-contained and the readers are required only to have some basic knowledge of matrix algebra and Fortran programming. To keep the book down to a reasonable size, only plane stress problems are discussed here, although obviously the basic numerical and programming techniques are also applicable to three-dimensional elasticity, to ideal flow and to other problems.

The theory of the finite element method is considered in Chapter 1, and considerable effort is devoted to the topic of displacement functions. In Chapter 2 a plane stress triangular element is derived and a simple but workable computer program is described in a step-by-step fashion. The isoparametric concept of formulating elements is introduced in Chapter 3 and numerical integration schemes and different techniques used in reducint the amount of efforts in calculating an element stiffness matrix are discussed. In Chapter 4, the front solution technique, which can be considered as one of the most (if not the most) effective solution schemes, is presented in great detail and simple illustrative examples, with all computation steps shown, are used in helping the readers to understand the manipulations. Chapter 5 gives the complete listing for an advance computer program and the sample input and output for an illustrative problem. Finally, two simple mesh generation techniques are intrcduced in Chapter 6 and their corresponding routines are also listed.

We are indebted to Mr G. Sved of the University of Adelaide for reading the manuscript and for making valuable suggestions for improving the presentation. Thanks are due to Mr S. Swaddiwudhipong and Mr C. Rigon, both doctoral students at the University of Adelaide, for checking parts of the manuscript, and to Mrs G. Stock, for her very skilful typing of the manuscript.

Finally we are deeply grateful to Mrs Yuk Bau Cheung for her constant support and encouragement in this effort.

August 1978

Y K Cheung
M F Yeo

Contents

1 The Finite Element Method

1.1 Introduction

It is a well-known fact that the finite element method, which by now has become the most powerful tool of analysis and is applicable to a wide range of problems, started as an extension of the stiffness or displacement method, in which a skeletal structure is assumed to be made up of an assemblage of one-dimensional elements such as bar elements (axial actions), beam elements (bending actions) and frame elements (axial, bending and torsional actions).

In the stiffness method for skeletal structures the elements of an actual structure are connected together at discrete joints, and equations of equilibrium involving external loads and member end forces are established at all joints to solve for joint displacements. The relationship between the end forces and end displacements of each member is represented by the stiffness matrix, which can be derived directly through the various energy theorems such as the principle of virtual work or principle of minimum total potential energy.

A similar situation exists for two- and three-dimensional elements for solid mechanics in which elements of some geometrical shapes such as triangles, quadrilaterals and hexahedra are connected at some artificially created points, and stiffness relationships between nodal displacements and equivalent nodal forces are derived in the same manner. In this way a continuum with infinite degrees of freedom can be discretized into an equivalent system with finite degrees of freedom.

In recent years the finite element method has been applied widely to non-structural problems, and the formulation is now mainly based on variational principles or weighted residual procedures. A detailed presentation can be found in a text by Zienkiewicz [1].

Throughout this book energy principles will be used for the derivation of the stiffness and load matrices. All elements and example problems presented later are in the field of two-dimensional elasticity, so as to keep the presentation as simple as possible. However, much of the material discussed herein is also directly applicable to three-dimensional elasticity and even to flow problems.

1.2 Principle of virtual work

The principle relates to two distinct and separate systems in which the first is a set of forces in equilibrium (with P and σ as the external forces and internal stresses respectively) and the second a set of geometrically compatible deformations (with Δ and ε as the displacements and strains, respectively). The principle states that for any system in equilibrium, the external virtual work must be equal to the internal virtual work, i.e.

System in equilibrium

$$\sum P.\Delta = \int_V \sigma.\varepsilon \, dv \tag{1.1}$$

Geometrically compatible deformations

Note that, in practice, one of the systems always relates to a real or actual structure in which some sort of solution is required, while the other is an imaginary or virtual system. Therefore, it is possible to have the option of establishing:

(i) Theorem of virtual forces in which a real system of displacements and strains is coupled to a virtual system of forces and stresses. Applying Eqn.(1.1), we have

$$\sum \text{(Virtual external forces)} \cdot \text{(Actual displacements)} = \int_V \text{(Virtual stresses)} \cdot \text{(Actual strains)} \, dv \tag{1.2}$$

(ii) Theorem of virtual displacements in which a real system of forces and stresses is coupled to a virtual system of displacements and strains. Again, using Eqn.(1.1).

$$\sum \text{(Real external forces)} \cdot \text{(Virtual displacements)} = \int_V \text{(Real stresses)} \cdot \text{(Virtual strains)} \, dv \tag{1.3}$$

Eqn.(1.2) is used to calculate displacements and leads to the unit load theorem while Eqn.(1.3) is used to calculate external forces and leads to the unit displacement theorem.

1.3 Principle of minimum total potential energy

The total potential energy of a system is defined as

$$\phi = U + W \tag{1.4}$$

in which W is the potential energy of the external forces in the deformed configuration and is defined as

$$W = -\sum P.\Delta \tag{1.5}$$

and U is the strain energy of the deformed structure and is given by

$$U = \int_V \left(\int \sigma d\varepsilon \right) dv \tag{1.6}$$

If the deformed system is now taken as the real or actual system, and a set of small geometrically compatible displacements $\delta\Delta$ as the virtual system, then, by virtue of Eqn.(1.3), it is possible to establish that

$$\sum P.\delta\Delta = \int_V (\sigma.\delta\varepsilon)\,dv \tag{1.7}$$

However, due to the imposed virtual displacements, the actual system will also undergo a change in total potential energy of

$$\begin{aligned}\delta\phi &= \delta U + \delta W \\ &= \int_V (\sigma.\delta\varepsilon)\,dv - \sum P.\delta\Delta\end{aligned} \tag{1.8}$$

By comparing Eqn.(1.7) with Eqn.(1.8), it can be concluded that

$$\delta\phi = 0. \tag{1.9}$$

In other words, the total potential energy of a system in equilibrium is stationary, and, because it can be further proved that the energy is always a minimum for stable structures, the statement is called the principle of stationary total potential energy or, more often, principle of minimum total potential energy. The principle can also be alternatively expressed as 'of all compatible displacements satisfying given boundary conditions, those which satisfy the equilibrium conditions make the total potential energy assume a stationary value'. A solution satisfying both equilibrium and compatibility is of course the correct solution for a linearly elastic problem. However, such solutions are difficult, if not impossible, to work out for the majority of cases and researchers have to resort to approximate solutions by assuming compatible displacements with undetermined parameters and determining the values of the parameters such that the total potential energy of the system is a minimum.

It follows that if a set of trial displacement functions with unknown parameters Δ_i is used to approximate the actual displacements of a system, then it is possible to determine such Δ_i by minimizing the total potential energy, i.e. by performing the operations

$$\frac{\partial\phi}{\partial\Delta_i} = 0 \tag{1.10}$$

Eqn.(1.10) has been used extensively in the derivation of finite element stiffness matrices.

1.4 Stiffness matrix of a bar member

Bar members are primarily used in trusses, and, by definition, they are assumed to be acted upon only by axial forces. The stiffness coefficients of a bar member can be derived directly from Hooke's Law, and referring to Fig.1.1(a) it is seen that if end 2 is fixed and end 1 is allowed to move, then

$$\begin{aligned}P_{x1} &= \frac{EA}{\ell}u_1 \\ P_{x2} &= -\frac{EA}{\ell}u_1\end{aligned} \tag{1.11}$$

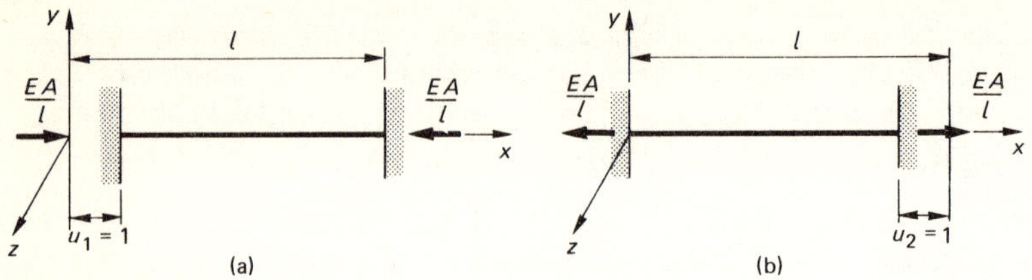

Fig.1.1 A bar member

Similarly, it can be concluded from Fig.1.1(b) that if end 1 is now fixed but end 2 allowed to move,

$$
\begin{aligned}
P_{x1} &= -\frac{EA}{\ell}u_2 \\
P_{x2} &= \frac{EA}{\ell}u_2
\end{aligned}
\tag{1.12}
$$

The complete stiffness relationship is now obtained by combining Eqn. (1.11) and Eqn.(1.12), and is given as

$$
\begin{Bmatrix} P_{x1} \\ P_{x2} \end{Bmatrix} = \frac{EA}{\ell}\begin{bmatrix} 1 & -1 \\ -1 & 1 \end{bmatrix}\begin{Bmatrix} u_1 \\ u_2 \end{Bmatrix}
\tag{1.13}
$$

1.5 Stiffness matrix of a beam member

A beam member is concerned with bending action only and its stiffness relationship is just a simple extension of the well-known slope-deflection equations, which are of the form

$$
\begin{aligned}
M_1 &= (6EI/\ell^2)v_1 + (4EI/\ell)\theta_1 - (6EI/\ell^2)v_2 + (2EI/\ell)\theta_2 \\
M_2 &= (6EI/\ell^2)v_1 + (2EI/\ell)\theta_1 - (6EI/\ell^2)v_2 + (4EI/\ell)\theta_2
\end{aligned}
\tag{1.14(a)}
$$

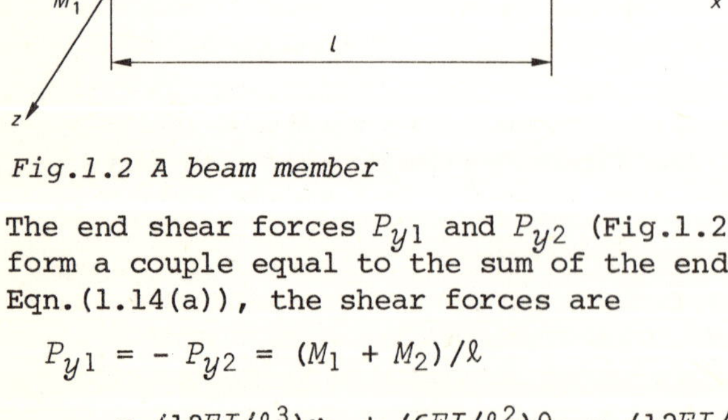

Fig.1.2 A beam member

The end shear forces P_{y1} and P_{y2} (Fig.1.2), which are equal and opposite, form a couple equal to the sum of the end moments. Therefore, from Eqn.(1.14(a)), the shear forces are

$$
\begin{aligned}
P_{y1} &= -P_{y2} = (M_1 + M_2)/\ell \\
&= (12EI/\ell^3)v_1 + (6EI/\ell^2)\theta_1 - (12EI/\ell^3)v_2 + (6EI/\ell^2)\theta_2
\end{aligned}
\tag{1.14(b)}
$$

Writing Eqn.(1.14) in matrix form, the stiffness relationships for a beam become

$$\begin{Bmatrix} P_{y1} \\ M_1 \\ P_{y2} \\ M_2 \end{Bmatrix} = \begin{bmatrix} 12EI/\ell^3 & 6EI/\ell^2 & -12EI/\ell^3 & 6EI/\ell^2 \\ 6EI/\ell^2 & 4EI/\ell & -6EI/\ell^2 & 2EI/\ell \\ -12EI/\ell^3 & -6EI/\ell^2 & 12EI/\ell^3 & -6EI/\ell^2 \\ 6EI/\ell^2 & 2EI/\ell & -6EI/\ell^2 & 4EI/\ell \end{bmatrix} \begin{Bmatrix} v_1 \\ \theta_1 \\ v_2 \\ \theta_2 \end{Bmatrix} \tag{1.15}$$

Note that the right-handed coordinate system is used here and will be used throughout the rest of the book.

1.6 Finite element procedure

Before we go any further into detailed derivations, it is perhaps worthwhile to point out that there are three types of commonly used elements. The displacement element using assumed displacement patterns form the overwhelming majority of all the known finite elements, while the equilibrium elements [2] (based on assumed stress patterns) and the hybrid element [3] (based on both assumed displacements and stresses) are used to a much lesser extent. In this book, only displacement elements will be discussed, and, since the assumed displacement patterns over the whole area of an element are described by the so-called displacement functions, the reader will note later that a great deal of attention is paid to the choice of suitable displacement functions for various finite elements.

As mentioned previously, the finite element method started as an extension of the stiffness method and was applied to two- and three-dimensional problems in structural mechanics. However, unlike skeletal structures, there are no well-defined joints where equilibrium of forces can be established and, therefore, the continuum must be discretized into a number of elements of arbitrary shapes and also artificial joints or nodes must be created. In this way the continuum is approximated by a system with a finite degree of freedom and a numerical solution can be achieved. A detailed description of the finite element procedure will be given below.

(i) The continuum is divided into two- or three-dimensional finite elements, which are separated by straight or curved lines (two-dimensional) or by flat or curved surfaces (three-dimensional).

Since each element is allowed to have its own shape, size, material properties and thickness, it is not surprising that the method is admirably suitable for problems involving non-homogeneous properties, irregular geometry, complex support conditions, and combination of various types of loadings.

In many cases only one type of element is used for one problem, but it is also possible to mix elements of different types such as a beam element connected to a plate bending element because of physical requirements, or higher-order elements connected to lower-order elements in stress concentration problems.

(ii) The elements are assumed to be inter-connected at a discrete number of nodal points situated at element corners or on element boundaries, although occasionally internal nodes (Fig.1.3(a)) which are not connected to any other element may also be present.

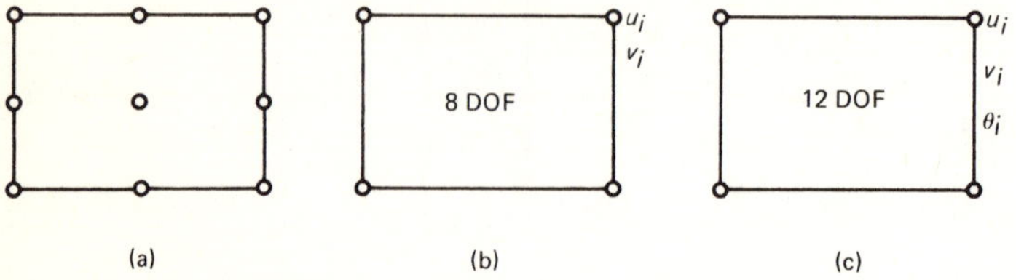

Fig.1.3 Rectangular plane stress elements

The nodal degrees of freedom (which can vary from node to node within each element), called nodal-displacement parameters, normally refer to the displacements and their first partial derivatives (rotations) at a node, but very often may include other terms such as stresses, strains and second, or even higher, partial derivatives. As an example, the rectangular plane stress element of Fig.1.3(b) (bilinear element) has eight degrees of freedom representing the translations u and v at the four corner nodes in the x and y directions, while the higher-order plane stress element (beam type element) of Fig.1.3(c) has four additional rotational nodal degrees of freedom.

(iii) A displacement function in terms of the coordinate variables x, y and the nodal displacement parameters (e.g. u_i, v_i in Fig.1.3) is chosen to represent the displacement variations within each element, and, by using the principle of virtual work or the principle of minimum total potential energy discussed previously, a stiffness matrix relating the nodal 'forces' to the nodal 'displacements' can be derived. Such a displacement function will try to approximate the actual displacement field over the whole element.

1.7 Displacement functions

From the above discussions, it is evident that the choice of suitable displacement functions is the most important part of the whole procedure. A good displacement function will lead to an element of high accuracy and with converging characteristics, and conversely a wrongly chosen displacement function yielding poor or non-converging results, and at times even worse, solutions which converge to an incorrect answer. The latter phenomenon was observed and discussed by Clough [4] but fortunately it is not a common occurrence.

A displacement function is either given as (i) a simple polynomial with undetermined coefficients which are subsequently transformed to become the relevant nodal displacement parameters, or (ii) directly in terms of shape functions, which are usually osculatory polynomials [5] having zero values at all other nodes of an element but unit values for the displacement or its partial derivatives at the node under consideration. Physically, the shape function associated with a nodal displacement parameter gives us the displacement field over an element when that particular nodal displacement parameter is given a unit value and all other nodal displacement parameters are given zero values. Some examples of shape functions are given in Fig.1.4. Thus a displacement function can either be given (from (i) above) as

$$f = A_1 + A_2x + A_3y + \ldots \qquad (1.16(a))$$

in which A_1, A_2 etc. are undetermined polynomial constants, or from (ii) as

$$f = N_1(x,y)f_1 + N_2(x,y)f_2 + \ldots \qquad (1.16(b))$$

in which f_1, f_2 etc. are the nodal displacement parameters and N_1, N_2 etc. the corresponding shape functions.

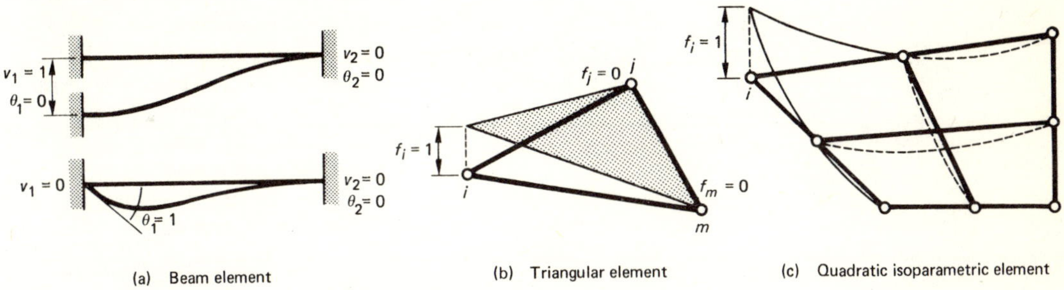

Fig.1.4 Examples of shape functions

At this stage it is appropriate to become familiar with the following concepts which will be helpful to us in the construction of displacement functions.

1.7.1 THE PASCAL TRIANGLE

For displacement functions of two-dimensional elements given in terms of simple polynomials, the Pascal triangle (Fig.1.5) is a useful aid for determining the combination of terms which should be used. The terminology used here is somewhat different from the one used in mathematics which deals with the coefficients of the binomial theorem.

$$\begin{array}{ccccccccccc}
 & & & & & 1 & & & & & \\
 & & & & x & & y & & & & \\
 & & & x^2 & & xy & & y^2 & & & \\
 & & x^3 & & x^2y & & xy^2 & & y^3 & & \\
 & x^4 & & x^3y & & x^2y^2 & & xy^3 & & y^4 & \\
x^5 & & - & & - & & - & & - & & y^5
\end{array}$$

Fig.1.5 Pascal Triangle

In Table 1.1 some examples on the use of the Pascal triangle are given. It should be noticed that for general purpose elements symmetry in the x, y terms is always preserved and that normally all the terms for a lower degree polynomial should be used before taking up more terms from a higher degree polynomial.

TABLE 1.1 *Typical finite elements with their degrees of freedom (DOF) and their displacement functions*

Element type	Pascal triangle	Remarks
Constant strain triangle u - 3 DOF v - 3 DOF	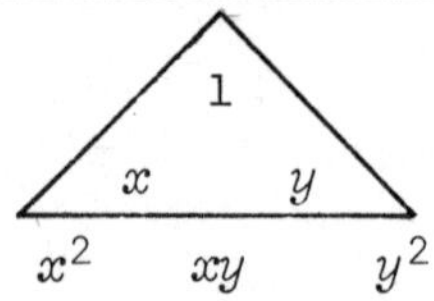	Linear variation of u and v
Plane stress rectangle u - 4 DOF v - 4 DOF	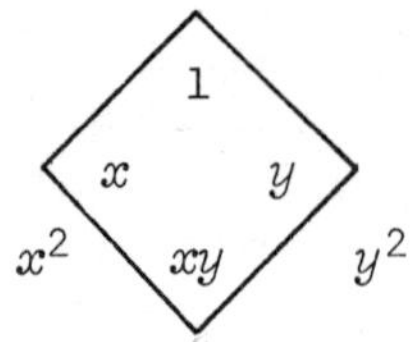	Linear variation of u and v along edges of element required. xy is the appropriate fourth term since it degenerates either to x or y only along the edges.
Bending rectangle w - 12 DOF	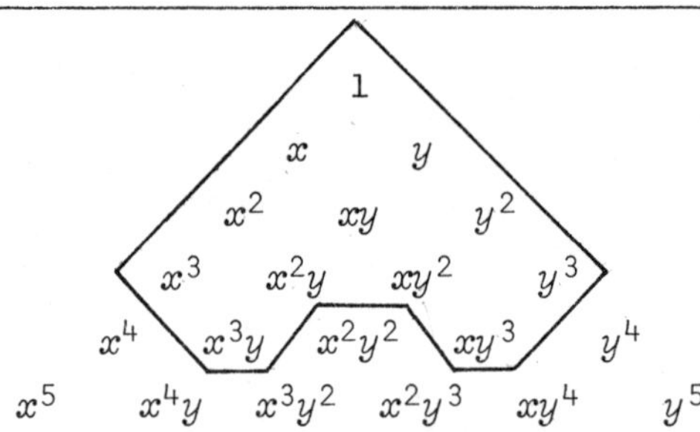	Cubic variation along edges of element required. x^3y and xy^3 degenerate to cubic functions of x and y only along the edges.
Rectangular beam type plane stress element u - 4 DOF v - 8 DOF	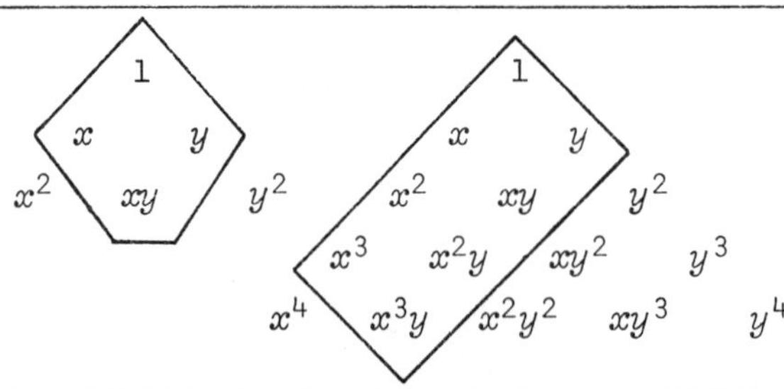	Linear variation of u along edges. Variation of v linear in y and cubic in x so as to simulate correctly beam behaviour.
Rectangular plane stress element with corner and midside nodes u - 8 DOF v - 8 DOF	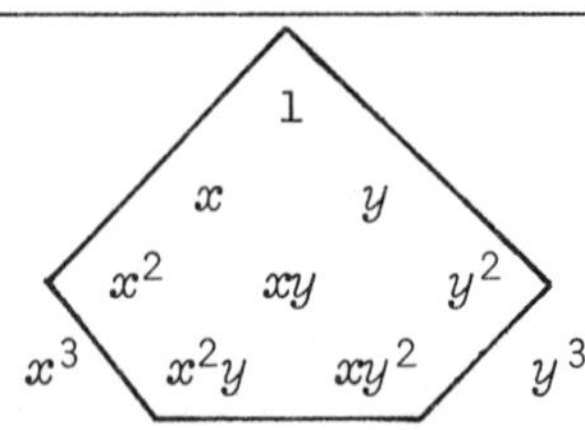	Parabolic variation of u and v along edges. x^3 and y^3 excluded.

1.7.2 LAGRANGE POLYNOMIAL

Lagrange polynomials (collocation polynomials) [5] are often used for the construction of shape functions of elements in which only function values but not derivatives are specified at the nodes. The basic form of the Lagrange polynomial in a single coordinate system with n nodes is

$$f(x) = \sum_{i=0}^{n} \mathcal{L}_i^n(x) f_i \tag{1.17}$$

where $\mathcal{L}_i^n(x)$ is called the Lagrange multiplier function and is given by

$$\mathcal{L}_i^n(x) = \frac{(x - x_0)(x - x_1) \ldots (x - x_{i-1})(x - x_{i+1}) \ldots (x - x_n)}{(x_i - x_0)(x_i - x_1) \ldots (x_i - x_{i-1})(x_i - x_{i+1}) \ldots (x_i - x_n)} \tag{1.18}$$

It is obvious that $\mathcal{L}_i^n(x)$ possesses the properties of

$$\mathcal{L}_i^n(x_k) \begin{cases} = 0 & k \neq i \\ = 1 & k = i \end{cases}$$

and thus fits in with the definition of a shape function. Some examples of Lagrange polynomials are given in Table 1.2.

It is also possible to apply the Lagrange polynomials to shape functions involving two or even three coordinates. Thus the shape functions for a two-dimensional problem would be

$$f(x,y) = \sum_{i=0}^{n} \sum_{j=0}^{m} \mathcal{L}_i^n(x) \mathcal{L}_j^m(y) f_{ij} \tag{1.19}$$

where n and m stand for the number of subdivisions or arguments in the x and y directions respectively.

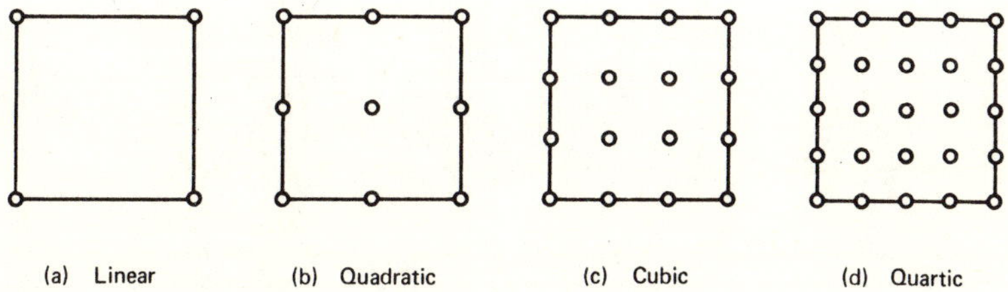

Fig.1.6 Family of Lagrange elements

The shape functions given in Eqn.(1.19) in fact form the basis of a family of plane stress elements given in Fig.1.6. However, the usefulness of this family of elements is severely limited by the presence of a large number of internal nodes, and in fact most research workers have restricted their attention to the linear element only.

TABLE 1.2 *Lagrange polynomials (single coordinate only)*

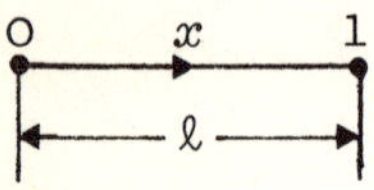

Linear

$$\mathscr{L}_0^1(x) = \frac{x - x_1}{0 - x_1} = 1 - \frac{x}{\ell}$$

$$\mathscr{L}_1^1(x) = \frac{x}{x_1} = \frac{x}{\ell}$$

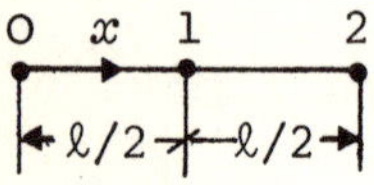

Parabolic

$$\mathscr{L}_0^2(x) = \frac{(x - x_1)(x - x_2)}{(0 - x_1)(0 - x_2)} = \frac{(x - x_1)(x - x_2)}{x_1 x_2}$$

$$= \left(2\frac{x}{\ell} - 1\right)\left(\frac{x}{\ell} - 1\right) = \left(1 - 2\mathscr{L}_1^1\right)\mathscr{L}_0^1$$

$$\mathscr{L}_1^2(x) = \frac{(x - 0)(x - x_2)}{(x_1 - 0)(x_1 - x_2)} = \frac{x(x - x_2)}{x_1(x_1 - x_2)}$$

$$= 4\frac{x}{\ell}\left(1 - \frac{x}{\ell}\right) = 4\mathscr{L}_0^1\mathscr{L}_1^1$$

$$\mathscr{L}_2^2(x) = \frac{(x - 0)(x - x_1)}{(x_2 - 0)(x_2 - x_1)} = \frac{x(x - x_1)}{x_2(x_2 - x_1)}$$

$$= \frac{x}{\ell}\left(2\frac{x}{\ell} - 1\right) = \mathscr{L}_1^1\left(2\mathscr{L}_1^1 - 1\right)$$

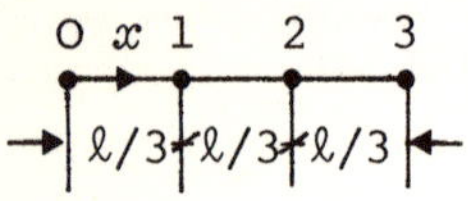

Cubic

$$\mathscr{L}_0^3(x) = \frac{(x - x_1)(x - x_2)(x - x_3)}{(0 - x_1)(0 - x_2)(0 - x_3)}$$

$$= \frac{(x_1 - x)(x_2 - x)(x_3 - x)}{x_1 x_2 x_3}$$

$$= \frac{1}{2}\left(1 - \frac{3x}{\ell}\right)\left(2 - \frac{3x}{\ell}\right)\left(1 - \frac{x}{\ell}\right)$$

$$= \frac{1}{2}\left(1 - 3\mathscr{L}_1^1\right)\left(2 - 3\mathscr{L}_1^1\right)\mathscr{L}_0^1$$

$$\mathscr{L}_1^3(x) = \frac{(x - 0)(x - x_2)(x - x_3)}{(x_1 - 0)(x_1 - x_2)(x_1 - x_3)} = \frac{x(x - x_2)(x - x_3)}{x_1(x_1 - x_2)(x_1 - x_3)}$$

$$= \frac{9}{2}\frac{x}{\ell}\left(3\frac{x}{\ell} - 2\right)\left(\frac{x}{\ell} - 1\right) = \frac{9}{2}\mathscr{L}_1^1\left(2 - 3\mathscr{L}_1^1\right)\mathscr{L}_0^1$$

$$\mathscr{L}_2^3(x) = \frac{(x - 0)(x - x_1)(x - x_3)}{(x_2 - 0)(x_2 - x_1)(x_2 - x_3)} = \frac{x(x - x_1)(x - x_3)}{x_2(x_2 - x_1)(x_2 - x_3)}$$

$$= \frac{9}{2}\frac{x}{\ell}\left(3\frac{x}{\ell} - 1\right)\left(1 - \frac{x}{\ell}\right) = \frac{9}{2}\mathscr{L}_1^1\left(3\mathscr{L}_1^1 - 1\right)\mathscr{L}_0^1$$

$$\mathscr{L}_3^3(x) = \frac{(x - 0)(x - x_1)(x - x_2)}{(x_3 - 0)(x_3 - x_1)(x_3 - x_2)} = \frac{x(x - x_1)(x - x_2)}{x_3(x_3 - x_1)(x_3 - x_2)}$$

$$= \frac{1}{2}\frac{x}{\ell}\left(3\frac{x}{\ell} - 1\right)\left(3\frac{x}{\ell} - 2\right) = \frac{1}{2}\mathscr{L}_1^1\left(3\mathscr{L}_1^1 - 1\right)\left(3\mathscr{L}_1^1 - 2\right)$$

1.7.3 HERMITIAN POLYNOMIAL

The Hermitian or osculatory polynomials [5] not only agree in value with a given function at specified locations, as is the case for collocation polynomials, but their derivatives will also match with the derivatives of the given functions up to any given order at the said locations. With only a first-order osculation, the Hermitian polynomial is the simplest one and has the form

$$f(x) = \sum_{i=0}^{n} U_i(x) f_i + \sum_{i=0}^{n} V_i(x) f_i' \tag{1.20}$$

where f_i and f_i' are the function value and the first derivative respectively at point i, and

$$U_i(x) = \left[1 - 2\frac{d}{dx}\left(\mathscr{L}_i^n(x)\right)_{x=x_i} (x - x_i)\right]\left[\mathscr{L}_i^n(x)\right]^2$$

$$V_i(x) = (x - x_i)\left[\mathscr{L}_i^n(x)\right]^2 \tag{1.21}$$

The most commonly used Hermitian polynomial is the so-called beam function involving two points only, and taking the distance between the two points as ℓ, then we have

$$\mathscr{L}_0^n(x) = \mathscr{L}_0^1(x), \quad \mathscr{L}_1^n(x) = \mathscr{L}_1^1(x)$$

(see Table 1.2). From Eqn.(1.21),

$$U_0(x) = \left(1 + \frac{2x}{x_1}\right)\left(\frac{x^2 - 2xx_1 + x_1^2}{x_1^2}\right)$$

$$= 1 - \frac{3x^2}{x_1^2} + \frac{2x^3}{x_1^3}$$

$$= 1 - \frac{3x^2}{\ell^2} + \frac{2x^3}{\ell^3}$$

$$U_1(x) = \left(1 - \frac{2x}{x_1} + 2\right)\left(\frac{x}{x_1}\right)^2 = \frac{3x^2}{x_1^2} - \frac{2x^3}{x_1^3} = \frac{3x^2}{\ell^2} - \frac{2x^3}{\ell^3} \tag{1.22(a)}$$

Similarly,

$$V_0(x) = x\left(1 - 2\frac{x}{\ell} + \frac{x^2}{\ell^2}\right) \tag{1.22(b)}$$

$$V_1(x) = x\left(\frac{x^2}{\ell^2} - \frac{x}{\ell}\right)$$

Another commonly used Hermitian polynomial is the one involving three equally spaced points with $x_0 = 0$, $x_1 = \ell/2$ and $x_2 = \ell$, and, in this case, $\mathcal{L}_0^2(x)$, $\mathcal{L}_1^2(x)$ and $\mathcal{L}_2^2(x)$ (*see* Table 1.2) will be the appropriate terms for $\mathcal{L}_i^n(x)$ in Eqn.(1.21).

$$U_0(x) = \left(1 + 6\frac{x}{\ell}\right)\left(4\frac{x^4}{\ell^4} - 12\frac{x^3}{\ell^3} + 13\frac{x^2}{\ell^2} - 6\frac{x}{\ell} + 1\right)$$

$$= 24\frac{x^5}{\ell^5} - 68\frac{x^4}{\ell^4} + 66\frac{x^3}{\ell^3} - 23\frac{x^2}{\ell^2} + 1$$

$$V_0(x) = x\left(4\frac{x^4}{\ell^4} - 12\frac{x^3}{\ell^3} + 13\frac{x^2}{\ell^2} - 6\frac{x}{\ell} + 1\right) \qquad (1.23(a))$$

Similarly, by taking $i = 1$ and 2 in Eqn.(1.21), it is possible to arrive at

$$U_1(x) = 16\frac{x^4}{\ell^4} - 32\frac{x^3}{\ell^3} + 16\frac{x^2}{\ell^2}$$

$$V_1(x) = x\left(16\frac{x^4}{\ell^4} - 40\frac{x^3}{\ell^3} + 32\frac{x^2}{\ell^2} - 8\frac{x}{\ell}\right)$$

$$U_2(x) = -24\frac{x^5}{\ell^5} + 52\frac{x^4}{\ell^4} - 34\frac{x^3}{\ell^3} + 7\frac{x^2}{\ell^2} \qquad (1.23(b))$$

$$V_2(x) = x\left(4\frac{x^4}{\ell^4} - 8\frac{x^3}{\ell^3} + 5\frac{x^2}{\ell^2} - \frac{x}{\ell}\right)$$

The above functions have been used in bending elements (finite strips) with mid-side nodal lines [6]. The use of such higher-order functions tends to improve the accuracy of the finite elements.

1.7.4 AREA COORDINATES

All the previous discussions are concerned with Cartesian (x,y) coordinates, and the examples given on two-dimensional problems are valid only for rectangular elements. For general quadrilateral elements with straight or curved sides similar functions can be used, but they should be in curvilinear (ξ,η) coordinates, and a detailed discussion will be given later in Chapter 3. The Cartesian coordinates are also not very convenient for triangular elements, and a special type of coordinate system called area coordinates should be used.

Referring to Fig.1.7 it is seen that the internal point ℓ will divide the triangle ijm into three smaller triangles, and depending on the position of the point ℓ, the area of each one of the triangles ℓim, ℓmj and ℓji can vary from zero to Δ, which is the area of the triangle ijm. In other words, the ratios A_i/Δ, A_j/Δ and A_m/Δ will take up any value between zero and unity in the same way as a first-order Lagrange polynomial. These ratios are called area coordinates, and they are defined by

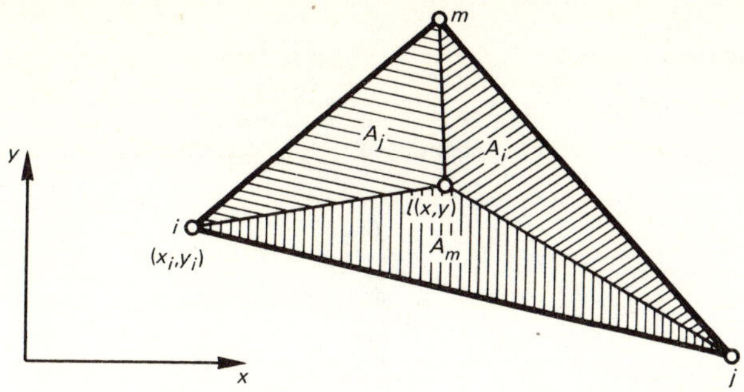

Fig.1.7 A triangle and the area coordinate system

$$\left.\begin{aligned}\mathcal{L}_i &= A_i/\Delta = (a_i + b_i x + c_i y)/2\Delta\\ \mathcal{L}_j &= A_j/\Delta = (a_j + b_j x + c_j y)/2\Delta\\ \mathcal{L}_m &= A_m/\Delta = (a_m + b_m x + c_m y)/2\Delta\end{aligned}\right\} \qquad (1.24)$$

in which

$$\left.\begin{aligned}a_i &= x_j y_m - x_m y_j\\ b_i &= y_j - y_m\\ c_i &= x_m - x_j\end{aligned}\right\} \qquad (1.25)$$

$$2\Delta = \det\begin{bmatrix}1 & x_i & y_i\\ 1 & x_j & y_j\\ 1 & x_m & y_m\end{bmatrix} = 2(\text{area of triangle } ijm),$$

x_i, y_i, etc. are the nodal coordinates and a_j, b_j, c_j, etc. can be computed through a cyclic permutation of the subscripts.

It is interesting to note that some other useful information can be derived from Eqn.(1.24), which in matrix form is

$$\begin{Bmatrix}\mathcal{L}_i\\ \mathcal{L}_j\\ \mathcal{L}_m\end{Bmatrix} = \frac{1}{2\Delta}\begin{bmatrix}a_i & b_i & c_i\\ a_j & b_j & c_j\\ a_m & b_m & c_m\end{bmatrix}\begin{Bmatrix}1\\ x\\ y\end{Bmatrix} \qquad (1.26)$$

Solving for 1, x, y,

$$\begin{Bmatrix}1\\ x\\ y\end{Bmatrix} = \begin{bmatrix}1 & 1 & 1\\ x_i & x_j & x_m\\ y_i & y_j & y_m\end{bmatrix}\begin{Bmatrix}\mathcal{L}_i\\ \mathcal{L}_j\\ \mathcal{L}_m\end{Bmatrix} \qquad (1.27(a))$$

TABLE 1.3 *Shape functions for triangular elements*

	Lagrange polynomial (Table 1.2)	Triangular elements in terms of area coordinates
Linear (Fig.1.8(a))	$\mathcal{L}_0^1(x)$ $\mathcal{L}_1^1(x)$	$N_1 = \mathcal{L}_1$ $N_2 = \mathcal{L}_2$ $N_3 = \mathcal{L}_3$
Quadratic (Fig.1.8(b))	$\mathcal{L}_0^2(x) = \left(1 - 2\mathcal{L}_1^1\right)\mathcal{L}_0^1$ $\mathcal{L}_1^2(x) = 4\mathcal{L}_0^1\mathcal{L}_1^1$ $\mathcal{L}_2^2(x) = \left(2\mathcal{L}_1^1 - 1\right)\mathcal{L}_1^1$	Corner nodes $N_1 = \left(2\mathcal{L}_1 - 1\right)\mathcal{L}_1$ $N_2 = \left(2\mathcal{L}_2 - 1\right)\mathcal{L}_2$ $N_3 = \left(2\mathcal{L}_3 - 1\right)\mathcal{L}_3$ Mid-side nodes $N_4 = 4\mathcal{L}_1\mathcal{L}_2$ $N_5 = 4\mathcal{L}_2\mathcal{L}_3$ $N_6 = 4\mathcal{L}_3\mathcal{L}_1$
Cubic (Fig.1.8(c))	$\mathcal{L}_0^3(x) = \frac{1}{2}\left(3\mathcal{L}_1^1 - 1\right)\left(3\mathcal{L}_1^1 - 2\right)\mathcal{L}_0^1$ $\mathcal{L}_1^3(x) = \frac{9}{2}\mathcal{L}_1^1\left(3\mathcal{L}_0^1 - 1\right)\mathcal{L}_0^1$ $\mathcal{L}_2^3(x) = \frac{9}{2}\mathcal{L}_1^1\left(3\mathcal{L}_1^1 - 1\right)\mathcal{L}_0^1$ $\mathcal{L}_3^3(x) = \frac{1}{2}\mathcal{L}_1^1\left(3\mathcal{L}_1^1 - 1\right)\left(3\mathcal{L}_1^1 - 2\right)$	Corner nodes $N_1 = \frac{1}{2}\left(3\mathcal{L}_1 - 1\right)\left(3\mathcal{L}_1 - 2\right)\mathcal{L}_1$ $N_2 = \frac{1}{2}\left(3\mathcal{L}_2 - 1\right)\left(3\mathcal{L}_2 - 2\right)\mathcal{L}_2$ $N_3 = \frac{1}{2}\left(3\mathcal{L}_3 - 1\right)\left(3\mathcal{L}_3 - 2\right)\mathcal{L}_3$ Edge nodes $N_4 = \frac{9}{2}\mathcal{L}_1\mathcal{L}_2\left(3\mathcal{L}_1 - 1\right)$ $N_5 = \frac{9}{2}\mathcal{L}_1\mathcal{L}_2\left(3\mathcal{L}_2 - 1\right)$ $N_6 = \frac{9}{2}\mathcal{L}_2\mathcal{L}_3\left(3\mathcal{L}_2 - 1\right)$ $N_7 = \frac{9}{2}\mathcal{L}_2\mathcal{L}_3\left(3\mathcal{L}_3 - 1\right)$ $N_8 = \frac{9}{2}\mathcal{L}_3\mathcal{L}_1\left(3\mathcal{L}_3 - 1\right)$ $N_9 = \frac{9}{2}\mathcal{L}_3\mathcal{L}_1\left(3\mathcal{L}_1 - 1\right)$ Internal node $N_{10} = 27\mathcal{L}_1\mathcal{L}_2\mathcal{L}_3$ †

† No direct comparison possible with Lagrange polynomial.

which when written out in 'long hand' yields

$$\left.\begin{aligned} &\mathcal{L}_i + \mathcal{L}_j + \mathcal{L}_m = 1 \\ &x = \mathcal{L}_i x_i + \mathcal{L}_j x_j + \mathcal{L}_m x_m \\ &y = \mathcal{L}_i y_i + \mathcal{L}_j y_j + \mathcal{L}_m y_m \end{aligned}\right\} \qquad (1.27(b))$$

The first equation in Eqn.(1.27(b)) shows that only two of the variables are in fact independent, while the second and the third equations give the relationship between the Cartesian coordinates and the area coordinates.

With the help of the area coordinates, it is now possible to establish a whole family of triangular plane stress elements, some of which are shown in Fig.1.8. Unlike the Lagrange elements, this family of triangular elements are used much more often in practice, especially for the linear and the quadratic elements.

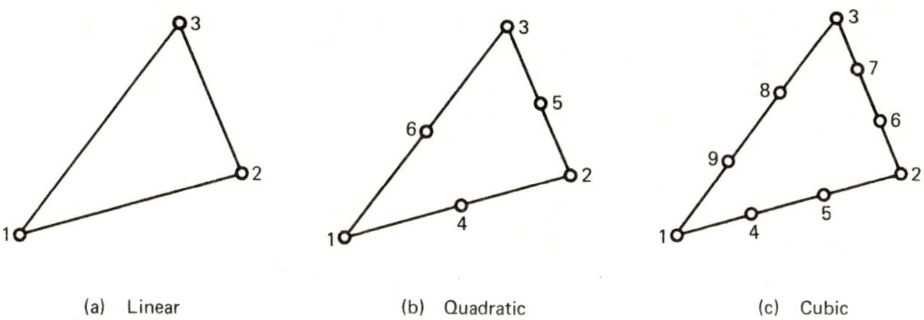

Fig.1.8 Family of triangular elements

Based on our knowledge of Lagrange polynomials, it is now a simple matter to construct the shape functions for the elements given in Fig. 1.8 and the process is shown in Table 1.3.

1.8 Choice of displacement functions

In general, a good displacement function should satisfy the following criteria.

(1) The displacement function, if given in the form of a simple polynomial, must have the same number of polynomial constants as the total number of degrees of freedom of the element. Care should be exercised in deciding on the degrees of freedom at a node, since on the one hand they must not be allowed to go below a minimum number, which ensures that the relevant deformation pattern will be adequately represented, and on the other hand they should not be so large as to make the computational procedure complex or cumbersome, even though increasing the number of degrees of freedom will usually mean improving the accuracy of an element. For example, while it is only necessary to have nodal displacements as degrees of freedom for two- or three-dimensional elasticity problems, it is, however, also necessary to use rotations (first derivative of the displacements) as additional degrees of freedom for beam and plate elements which resist external loads through bending. Degrees of freedom representing nodal strains can also be used if so desired.

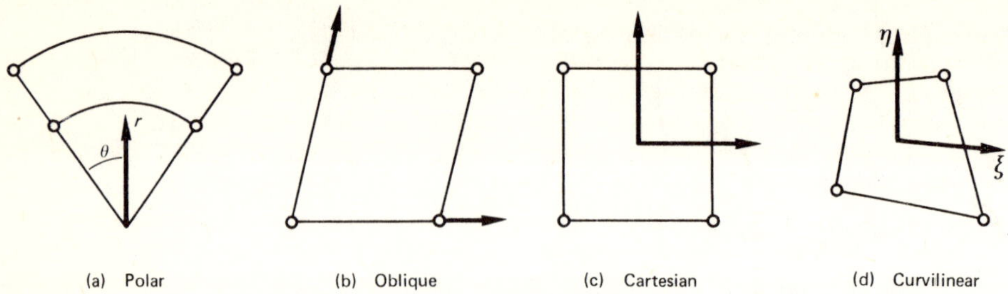

Fig.1.9 Plane elements of different shapes

A point worth noting is that different elements require the use of different coordinate systems. For example, the displacement functions for the sector element (Fig.1.9(a)) and parallelogrammic element (Fig. 1.9(b)) should be given in polar coordinates and oblique coordinates respectively, and not in Cartesian coordinates as for the rectangular element in Fig.1.9(c)). A much more difficult case is the quadrilateral element shown in Fig.1.9(d), where one is tempted to apply the same function $A_1 + A_2x + A_3y + A_4xy$. Unfortunately, while this function is a good one for the rectangular element, it is certainly not suitable for the quadrilateral and the correct displacement function should be $A_1 + A_2\xi + A_3\eta + A_4\xi\eta$ with ξ and η being a set of curvilinear coordinate axes.

It is not always an easy task to select the correct polynomial terms for a given number of degrees of freedom, and one of the best examples can be found in the case of a triangular bending element (Fig.1.10(a)) with nine degrees of freedom although there are altogether ten terms for a complete cubic polynomial (Fig.1.10(b)). The choice here is not at all obvious and in fact various researchers have demonstrated that there is no satisfactory solution available. The first successful nine-degrees-of-freedom triangular element was formulated through the use of area coordinates and not directly as a simple polynomial function.

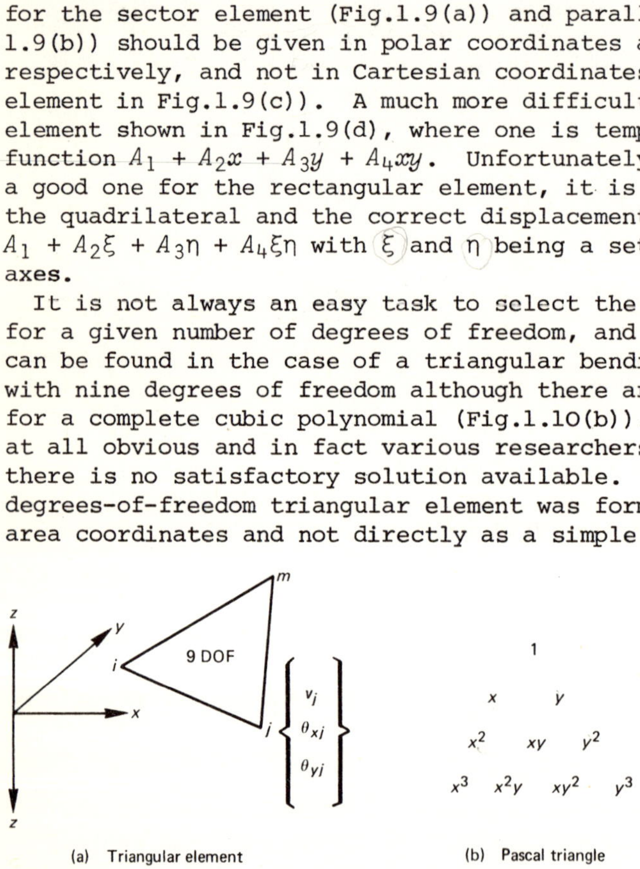

Fig.1.10 A triangular bending element

(2) In most cases, the displacement function should be balanced with respect to all coordinate axes, since most elements are classified as general purpose elements applicable to all types of problems. In this context, it will not be advisable to delete either the x^2y or xy^2 term for the triangular element shown in Fig.1.10(a) in order to satisfy criterion (1) discussed above, since such an element will yield solutions with accuracy biased in one direction.

It is important to note, however, that the statement in this criterion in no way excludes the use of special purpose elements with unbalanced displacement functions, and a good example of such unbalanced elements is the rectangular beam type plane stress element listed in Table 1.1. As can be seen, the variation of u (axial displacement of beam) and v (vertical displacement of beam) are linear and cubic respectively along the axial direction, and this in fact agrees well with the actual behaviour of a beam. Other examples having wide applications are the finite strips developed by Cheung [6] and others, in which a polynomial in one direction is used in conjunction with transcendental functions in another direction.

(3) The displacement function must allow the element to undergo rigid body movements without at the same time causing any internal strain.

For displacement functions given in terms of simple polynomials, it is only necessary to make certain that the first few terms at the top of the Pascal triangle have been included. For example, the terms, ℓ, x, y in Fig.1.9(b) would represent adequately the rigid body movements of the triangular bending element shown in Fig.1.10(a), since they would allow the element to take up another position in space on a flat inclined plane.

For displacements given in terms of shape functions it is fairly difficult to identify the rigid body terms directly, and the following two procedures are used.

(i) For elements with only scalar degrees of freedom, such as the constant strain triangle and the isoparametric elements which will be presented in detail in later chapters, the element will undergo translation or rotation during rigid body movements. For translation in any one direction all the nodes should have the same displacement f_r. This means that the displacement function given in Eqn.(1.16(b)) becomes

$$f_r = \left(\sum N_i\right) f_r \quad \text{or}$$

$$\sum N_i = 1 \tag{1.28(a)}$$

For a rigid body rotation θ around the origin of the coordinate axes means that the u displacement at any point will be equal to θy and, therefore, we have $u_i = \theta y$. Eqn.(1.16(b)) can be modified to

$$f = u = \theta y = \sum N_i \theta y_i$$

or

$$y = \sum N_i y_i \tag{1.28(b)}$$

Similarly, it can be shown that, by examining the v displacements,

$$x = \sum N_i x_i \tag{1.28(c)}$$

Thus, criterion (3) will be satisfied if all the nodal shape functions of an element add up to unity and if the x and y coordinates can be expressed as the sum of the products of the shape functions and the relevant nodal coordinates.

As an example, the shape functions for the quadratic plane stress triangular element of Fig.1.8(b) are

$$[N] = \left[(2\mathcal{L}_1 - 1)\mathcal{L}_1,\ (2\mathcal{L}_2 - 1)\mathcal{L}_2,\ (2\mathcal{L}_3 - 1)\mathcal{L}_3,\ 4\mathcal{L}_1\mathcal{L}_2,\ 4\mathcal{L}_2\mathcal{L}_3,\ 4\mathcal{L}_3\mathcal{L}_1\right] \qquad (1.28(d))$$

and

$$\sum N_i = (2\mathcal{L}_1 - 1)\mathcal{L}_1 + (2\mathcal{L}_2 - 1)\mathcal{L}_2 + (2\mathcal{L}_3 - 1)\mathcal{L}_3 + 4\mathcal{L}_1\mathcal{L}_2 + 4\mathcal{L}_2\mathcal{L}_3 + 4\mathcal{L}_3\mathcal{L}_1$$

Taking into account that $\mathcal{L}_1 + \mathcal{L}_2 + \mathcal{L}_3 = 1$ (*see* Eqn.(1.27(b)), and using the relationship to eliminate $\mathcal{L}_1$

$$\begin{aligned}\sum N_i &= (1 - 2\mathcal{L}_2 - 2\mathcal{L}_3)(1 - \mathcal{L}_2 - \mathcal{L}_3) + (2\mathcal{L}_2 - 1)\mathcal{L}_2 \\ &\quad + (2\mathcal{L}_3 - 1)\mathcal{L}_3 + 4(1 - \mathcal{L}_2 - \mathcal{L}_3)\mathcal{L}_2 + 4\mathcal{L}_2\mathcal{L}_3 \\ &\quad + 4\mathcal{L}_3(1 - \mathcal{L}_2 - \mathcal{L}_3) \\ &= 1\end{aligned}$$

To prove that Eqn.(1.28(b)) and Eqn.(1.28(c)) are also valid we take into account that $x_4 = (x_1 + x_2)/2$, $x_5 = (x_2 + x_3)/2$ and $x_6 = (x_3 + x_1)/2$. Therefore,

$$\begin{aligned}\sum N_i x_i &= (2\mathcal{L}_1 - 1)\mathcal{L}_1 x_1 + (2\mathcal{L}_2 - 1)\mathcal{L}_2 x_2 + (2\mathcal{L}_3 - 1)\mathcal{L}_3 x_3 \\ &\quad + 4\mathcal{L}_1\mathcal{L}_2(x_1 + x_2)/2 + 4\mathcal{L}_2\mathcal{L}_3(x_2 + x_3)/2 \\ &\quad + 4\mathcal{L}_2\mathcal{L}_3(x_3 + x_1)/2 \\ &= x_1\mathcal{L}_1(2\mathcal{L}_1 + 2\mathcal{L}_2 + 2\mathcal{L}_3 - 1) + x_2\mathcal{L}_2(2\mathcal{L}_1 + 2\mathcal{L}_2 + 2\mathcal{L}_3 - 1) \\ &\quad + x_3\mathcal{L}_3(2\mathcal{L}_1 + 2\mathcal{L}_2 + 2\mathcal{L}_3 - 1) \\ &= x_1\mathcal{L}_1 + x_2\mathcal{L}_2 + x_3\mathcal{L}_3\end{aligned}$$

Thus, by virtue of Eqn.(1.27(b)),

$$x = \sum N_i x_i$$

Similarly,

$$y = \sum N_i y_i$$

(ii) For elements with additional degrees of freedom such as rotations, an indirect test procedure is adopted.

(a) Give an element some sort of rigid body movements and identity the corresponding nodal displacements.

(b) Substitute these nodal displacements into the displacement function and find out whether the function will in fact be reduced to the assumed rigid body terms.

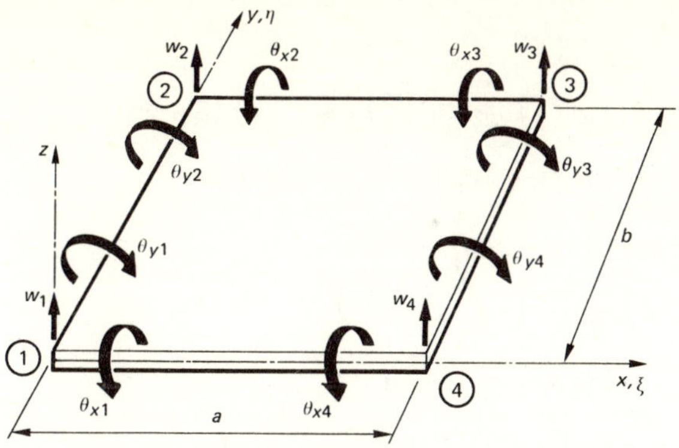

Fig.1.11 Rectangular plate element

For example, the displacement function of a non-conforming rectangular bending element [7] (Fig.1.11) is

$$w = [N]\{\delta\} \tag{1.29(a)}$$

in which the shape functions are

$$\begin{aligned}[N] = \big[&\big(1-\xi\eta-(3-2\xi)\xi^2(1-\eta)(1-\xi)(3-2\eta)\eta^2\big),\ (1-\xi)\eta(1-\eta)^2 b,\\ &-\xi(1-\xi)^2(1-\eta)a,\ \big((1-\xi)(3-2\eta)\eta^2+\xi(1-\xi)(1-2\xi)\eta\big),\\ &-(1-\xi)(1-\eta)\eta^2 b,\ -\xi(1-\xi)^2\eta a,\ (3-2\xi)\xi^2\eta-\xi\eta(1-\eta)(1-2\eta),\\ &-\xi(1-\eta)\eta^2 b,\ (1-\xi)\xi^2\eta a,\ \big((3-2\xi)\xi^2(1-\eta)+\xi\eta(1-\eta)(1-2\eta)\big),\\ &\xi\eta(1-\eta)^2 b,\ (1-\xi)\xi^2(1-\eta)a\big]\end{aligned} \tag{1.29(b)}$$

and the displacement parameters are

$$\{\delta\} = [w_1,\ \theta_{x1},\ \theta_{y1},\ w_2\ .\ .\ .\ \theta_{y4}]^T \tag{1.29(c)}$$

in which $\xi = x/a$ and $\eta = y/b$

For a rigid body movement we would expect the element to take up the position of an inclined plane in space, i.e.

$$w = A_1 + A_2x + A_3y$$

in which case the nodal displacements would be

$$w_1 = A_1$$

$$w_2 = A_1 + A_3b$$

$$w_3 = A_1 + A_2a + A_3b$$

$$w_4 = A_1 + A_2a$$

$$\theta_{x1} = \theta_{x2} = \theta_{x3} = \theta_{x4} = \frac{\partial w}{\partial y} = A_3$$

$$\theta_{y1} = \theta_{y2} = \theta_{y3} = \theta_{y4} = \frac{\partial w}{\partial x} = -A_2$$

As a result of this Eqn.(1.29) is reduced to

$$w = A_1\Big\{\big(1 - \xi\eta - (3 - 2\xi)\xi^2(1 - \eta) - (1 - \xi)(3 - 2\eta)\eta^2\big) + \big((1 - \xi)(3 - 2\eta)\eta^2 + \xi(1 - \xi)(1 - 2\xi)\eta\big) + \big((3 - 2\xi)\xi^2\eta - \xi\eta(1 - \eta)(1 - 2\eta)\big) + \big((3 - 2\xi)\xi^2(1 - \eta) + \xi\eta(1 - \eta)(1 - 2\eta)\big)\Big\}$$

$$+ A_2\Big\{\big((3 - 2\xi)\xi^2\eta - \xi\eta(1 - \eta)(1 - 2\eta)\big)a + \big((3 - 2\xi)\xi^2(1 - \eta) + \xi\eta(1 - \eta)(1 - 2\eta)\big)a - \big(-\xi(1 - \xi)^2(1 - \eta)a\big) - \big(-\xi(1 - \xi)^2\eta a\big) - \big((1 - \xi)\xi^2\eta a\big) - \big((1 - \xi)\xi^2(1 - \eta)a\big)\Big\}$$

$$+ A_3\Big\{\big((1 - \xi)(3 - 2\eta)\eta^2 + \xi(1 - \xi)(1 - 2\xi)\eta\big)b + \big((3 - 2\xi)\xi^2(1 - \eta) + \xi\eta(1 - \eta)(1 - 2\eta)\big)b + \big((1 - \xi)\eta(1 - \eta)^2 b\big) + \big(-(1 - \xi)(1 - \eta)\eta^2 b\big) + \big(-\xi(1 - \eta)\eta^2 b\big) + \xi\eta(1 - \eta)^2 b\Big\}$$

$$= A_1 + A_2\xi a + A_3\eta b$$

$$= A_1 + A_2 x + A_3 y$$

Therefore it is concluded that rigid body displacement terms have been included in the shape functions.

(4) The displacement function must be able to represent a state of constant strain since this is the expected outcome if the elements are made smaller and smaller so that any smooth strain curve will be represented by a series of steps in the limit.

For displacement functions given in terms of simple polynomials, the solution is again quite simple, and it is only necessary to retain all the terms which are connected with strains or curvatures. For example, it is wrong to eliminate the xy term in the displacement function given in Fig.1.10(b) which represents the constant term for the twist $(\partial^2 w/\partial x\partial y)$ expression. Without such a constant term, the twist will always be zero at the origin where $x = 0$ and $y = 0$.

For displacements given in terms of shape functions, the same procedure outlined in criterion (3(ii)) is used with the modification that the set of nodal displacements should correspond to some constant strain modes such as pure bending or direct compression.

As an example, consider the biaxial compression mode on the rectangular plane stress element considered previously. The corresponding nodal displacements are $u_i = -u_j = -u_k = u_m = c$ and $-v_i = v_j = v_k = v_m = d$ and, substituting these values into Eqn.(1.28), we have

$$u = \frac{1}{4}\{(1 - \xi)(1 - \eta) - (1 + \xi)(1 - \eta) - (1 + \xi)(1 - \eta) + (1 - \xi)(1 + \eta)\}c$$

$$= -cx$$

$$v = \frac{1}{4}\{-(1 - \xi)(1 - \eta) - (1 + \xi)(1 - \eta) + (1 + \xi)(1 - \eta) + (1 - \xi)(1 + \eta)\}d$$

$$= -dy$$

Since $\varepsilon_x = \partial u/\partial x = -c$ and $\varepsilon_y = \partial v/\partial y = -d$, we can satisfy ourselves that the constant strain mode is indeed present in the shape functions.

A more complicated example involves the quadratic plane stress triangular element discussed previously, in which the displacement functions are

$$u = \left(2\mathscr{L}_1 - 1\right)\mathscr{L}_1 u_1 + \left(2\mathscr{L}_2 - 1\right)\mathscr{L}_2 u_2 + \left(2\mathscr{L}_3 - 1\right)\mathscr{L}_3 u_3$$

$$+ 4\mathscr{L}_1\mathscr{L}_2 u_4 + 4\mathscr{L}_2\mathscr{L}_3 u_5 + 4\mathscr{L}_3\mathscr{L}_1 u_6$$

$$v = \left(2\mathscr{L}_1 - 1\right)\mathscr{L}_1 v_1 + \left(2\mathscr{L}_2 - 1\right)\mathscr{L}_2 v_2 + \left(2\mathscr{L}_3 - 1\right)\mathscr{L}_3 v_3$$

$$+ 4\mathscr{L}_1\mathscr{L}_2 v_4 + 4\mathscr{L}_2\mathscr{L}_3 v_5 + 4\mathscr{L}_3\mathscr{L}_1 v_6 \qquad (1.30(a))$$

For a general constant strain state, the displacement field must be linear and, therefore,

$$u = B_1 + B_2x + B_3y$$
$$v = B_4 + B_5x + B_6y \qquad (1.30(b))$$

and the corresponding nodal displacements are obtained by substituting into Eqn. (1.30) the appropriate nodal coordinates x_1, x_2, x_3, $x_4 = (x_1 + x_2)/2$, $x_5 = (x_2 + x_3)/2$, $x_6 = (x_3 + x_1)/2$, etc. Eqn. (1.30(a)) can now be written as (*vide* Eqn. (1.27(b)))

$$u = \left(2\mathscr{L}_1 - 1\right)\mathscr{L}_1(B_1 + B_2x_1 + B_3y_1) + \left(2\mathscr{L}_2 - 1\right)\mathscr{L}_2(B_1 + B_2x_2 + B_3y_2)$$

$$+ \left(2\mathscr{L}_3 - 1\right)\mathscr{L}_3(B_1 + B_2x_3 + B_3y_3) + 4\mathscr{L}_1\mathscr{L}_2\left(B_1 + B_3\frac{x_1 + x_2}{2} + B_3\frac{y_1 + y_2}{2}\right)$$

$$+ 4\mathscr{L}_2\mathscr{L}_3\left(B_1 + B_2\frac{x_2 + x_3}{2} + B_3\frac{y_2 + y_3}{2}\right) + 4\mathscr{L}_3\mathscr{L}_1\left(B_1 + B_2\frac{x_3 + x_1}{2} + B_3\frac{y_3 + y_1}{2}\right)$$

$$= \left\{2\left(\mathscr{L}_1 + \mathscr{L}_2 + \mathscr{L}_3\right)^2 - \left(\mathscr{L}_1 + \mathscr{L}_2 + \mathscr{L}_3\right)\right\}B_1$$

$$+ \left\{\left(2x_1\mathscr{L}_1\left(\mathscr{L}_1 + \mathscr{L}_2 + \mathscr{L}_3\right) - x_1\mathscr{L}_1\right) + \left(2x_2\mathscr{L}_2\left(\mathscr{L}_1 + \mathscr{L}_2 + \mathscr{L}_3\right) - x_2\mathscr{L}_2\right)\right.$$

$$\left. + \left(2x_3\mathscr{L}_3\left(\mathscr{L}_1 + \mathscr{L}_2 + \mathscr{L}_3\right) - x_3\mathscr{L}_3\right)\right\}B_2 + \left\{\left(2y_1\mathscr{L}_1\left(\mathscr{L}_1 + \mathscr{L}_2 + \mathscr{L}_3\right) - y_1\mathscr{L}_1\right)\right.$$

$$\left. + \left(2y_2\mathscr{L}_2\left(\mathscr{L}_1 + \mathscr{L}_2 + \mathscr{L}_3\right) - y_2\mathscr{L}_2\right) + \left(2y_3\mathscr{L}_3\left(\mathscr{L}_1 + \mathscr{L}_2 + \mathscr{L}_3\right) - y_3\mathscr{L}_3\right)\right\}B_3$$

$$= B_1 + \left(x_1\mathscr{L}_1 + x_2\mathscr{L}_2 + x_3\mathscr{L}_3\right)B_2 + \left(y_1\mathscr{L}_1 + y_2\mathscr{L}_2 + y_3\mathscr{L}_3\right)B_3$$

$$= B_1 + xB_2 + yB_3$$

and

$$v = B_4 + xB_5 + yB_6$$

Once again the constant strain states are found to be present in the shape functions.

(5) The displacement functions should satisfy the compatibility conditions along common boundaries between adjacent elements. To be more specific, there should be at least compatibility of displacements for two- and three-dimensional elasticity problems, and for plate bending problems these would be an additional requirement in the compatibility of the first partial derivatives of the displacements.

The satisfaction of criteria (4) and (5) will guarantee the convergence of the strain energy of the system to the correct level. However, in practice criterion (5) has often been partially relaxed to produce non-conforming elements which could be quite well-behaved, but with no monotonic convergence characteristics.

1.9 Formulation of stiffness matrix and load matrix

Once the displacement function has been determined, it is a fairly straightforward matter to obtain all the strains and stresses within the element and to formulate the stiffness matrix and consistent load matrices and in this section the principle of minimum total potential energy will be used in the formulation. It should be noted that a consistent load matrix contains equivalent nodal forces which represent the action of external distributed loads or body forces, and the magnitude of these nodal forces will vary with the displacement functions used for the element. Naturally the resultant of all the forces will remain unaltered.

1.9.1 DISPLACEMENT FUNCTIONS

A displacement function given in terms of simple polynomials can be written in matrix form as

$$\{f\} = [R]\{A\} \tag{1.31}$$

in which $\{f\}$ may have one or more components (e.g. $[u,v]^T$ for plane stress element and $\{w\}$ for bending element), $[R]$ is a function of the coordinates and $\{A\}$ is a vector of polynomial constants.

It is necessary to express the displacement function in terms of nodal displacement parameters $\{\delta\}$ and the relationships between the parameters and the polynomial constants are obtained by simply substituting into Eqn.(1.31) the nodal coordinates of the elements. Thus

$$\{\delta\} = [C]\{A\} \tag{1.32}$$

from which

$$\{A\} = [C]^{-1}\{\delta\} \tag{1.33}$$

and finally

$$\begin{aligned}\{f\} &= [R][C]^{-1}\{\delta\} \\ &= [N]\{\delta\} = \sum [N_i]\{\delta_i\}\end{aligned} \tag{1.34}$$

which is equivalent to saying that the displacements are now given in shape function form after carrying out the transformation process.

1.9.2 STRAINS

The strains are obtained through making appropriate differentiations of the displacement function with respect to the relevant coordinate variables x or y. Thus,

$$\{\varepsilon\} = [B]\{\delta\} \tag{1.35}$$

in which $[B]$ is called the strain matrix.

As an example, in the case of a plane stress element,

$$\{\varepsilon\} = \begin{Bmatrix} \varepsilon_x \\ \varepsilon_y \\ \gamma_{xy} \end{Bmatrix} = \begin{Bmatrix} \dfrac{\partial u}{\partial x} \\ \dfrac{\partial v}{\partial y} \\ \dfrac{\partial u}{\partial y} + \dfrac{\partial v}{\partial x} \end{Bmatrix} \tag{1.36}$$

and, in the case of a bending element,

$$\{\varepsilon\} = \begin{Bmatrix} \chi_x \\ \chi_y \\ 2\chi_{xy} \end{Bmatrix} = \begin{Bmatrix} -\dfrac{\partial^2 w}{\partial x^2} \\ -\dfrac{\partial^2 w}{\partial y^2} \\ 2\dfrac{\partial^2 w}{\partial x \partial y} \end{Bmatrix} \tag{1.37}$$

1.9.3 STRESSES

The stresses are related to the strains by

$$\{\sigma\} = [D]\{\varepsilon\}$$

$$= [D][B]\{\delta\} \tag{1.38}$$

The matrix $[D]$ is often referred to as the elasticity or property matrix. For a plane stress element with isotropic properties

$$[D] = \frac{E}{1 - \nu^2}\begin{bmatrix} 1 & \nu & 0 \\ \nu & 1 & 0 \\ 0 & 0 & \dfrac{1-\nu}{2} \end{bmatrix} \tag{1.39(a)}$$

while for orthotropic properties

$$[D] = \begin{bmatrix} \frac{E_x}{1-\nu_x\nu_y} & \frac{\nu_x E_y}{1-\nu_x\nu_y} & 0 \\ \frac{\nu_x E_y}{1-\nu_x\nu_y} & \frac{E_y}{1-\nu_x\nu_y} & 0 \\ 0 & 0 & G \end{bmatrix} \tag{1.39(b)}$$

1.9.4 MINIMIZATION OF TOTAL POTENTIAL ENERGY

(a) Strain energy
The strain energy of a linear elastic body is given by

$$U = \frac{1}{2}\int\{\varepsilon\}^T\{\sigma\}\mathrm{d(vol)} \tag{1.40(a)}$$

which after substituting for $\{\varepsilon\}$ and $\{\sigma\}$ from Eqn.(1.35) and Eqn.(1.38) becomes

$$U = \frac{1}{2}\int\{\delta\}^T[B]^T[D][B]\{\delta\}\mathrm{d(vol)} \tag{1.40(b)}$$

(b) Potential energy
The potential energy due to distributed loads $\{q\}$ can be written as

$$\begin{aligned} W &= -\int\{f\}^T\{q\}\mathrm{d(vol)} \\ &= -\int\{\delta\}^T[N]^T\{q\}\mathrm{d(vol)} \end{aligned} \tag{1.41}$$

Eqn.(1.41) can be degenerated into an area integral or line integral in case of surface loads or line loads. For a concentrated load, integration is no longer necessary and the expression is reduced to that of a load multiplied by the corresponding displacement.

(c) Total potential energy
From Eqn.(1.4) the total potential energy is given by

$$\begin{aligned} \phi &= U + W \\ &= \frac{1}{2}\int\{\delta\}^T[B]^T[D][B]\{\delta\}\mathrm{d(vol)} - \int\{\delta\}^T[N]^T\{q\}\mathrm{d(vol)} \end{aligned} \tag{1.42}$$

(d) Minimization procedure
Referring to Eqn.(1.10), the minimization procedure requires that

$$\left\{\frac{\partial\phi}{\partial\{\delta\}}\right\} = \{0\} \tag{1.43}$$

Therefore, if Eqn.(1.42) is substituted into Eqn.(1.43) and the partial differentiation is carried out, then

$$\left\{\frac{\partial\phi}{\partial\{\delta\}}\right\} = \int[B]^T[D][B]\{\delta\}\mathrm{d(vol)} - \int[N]^T\{q\}\mathrm{d(vol)} = \{0\} \tag{1.44}$$

or

$$[k]\{\delta\} - \{P\} = 0 \tag{1.45}$$

in which the stiffness matrix

$$[k] = \int[B]^T[D][B]\mathrm{d(vol)} \tag{1.46}$$

and the consistent load matrix

$$\{P\} = \int[N]^T\{q\}\mathrm{d(vol)} \tag{1.47}$$

A great deal of integration work has to be performed in Eqns.(1.46) and (1.47) and, while it is possible to carry out the integration in closed form for simple cases, it is advisable to use numerical integration procedure such as Gaussian quadrature for the majority of cases.

1.10 Bar element stiffness matrix rederived

Consider the bar element of Fig.1.1 which has two nodes and one degree of freedom at each node. With the total number of degrees of freedom equal to two, in accordance with criterion (1) in Section 1.8 there should only be two polynomial constants. The stiffness matrix, which has been given in Eqn.(1.13), will be rederived in accordance with the finite element formulation procedure.

(i) *Displacement function in polynomial form*

$$\{f\} = \{u\} = A_1 + A_2x$$

Therefore,

$$[R] = [1 \quad x]$$

$$\{A\} = \begin{Bmatrix} A_1 \\ A_2 \end{Bmatrix} \tag{1.48}$$

(ii) *Nodal displacement parameter-polynomial constant relationship*
Substituting the nodal coordinates $x = 0$ and $x = \ell$ into Eqn.(1.48) one after the other, we have

$$u_1 = A_1 + A_2(0)$$

$$u_2 = A_1 + A_2(\ell)$$

Therefore,

$$[C] = \begin{bmatrix} 1 & 0 \\ 1 & \ell \end{bmatrix} \qquad [C]^{-1} = \frac{1}{\ell}\begin{bmatrix} \ell & 0 \\ -1 & 1 \end{bmatrix}$$

(iii) *Displacement function in shape function form*
Referring to Eqn.(1.34), the shape functions

$$[N] = [R][C]^{-1}$$

$$= [1 \quad x]\frac{1}{\ell}\begin{bmatrix} \ell & 0 \\ -1 & 1 \end{bmatrix}$$

$$= \left[\left(1 - \frac{x}{\ell}\right) \quad \frac{x}{\ell}\right] = [N_1 \quad N_2] \tag{1.49}$$

The above expression in fact corresponds to the linear Lagrange polynomial given in Table 1.2. Thus it is quite easy to formulate the displacements in shape function directly in this case.

(iv) *Strain-nodal displacement parameter relationship*
Only axial strain is present in this case and

$$\{\varepsilon\} = \varepsilon_x = \frac{du}{dx} = \begin{bmatrix} \frac{dN_1}{dx} & \frac{dN_2}{dx} \end{bmatrix}\begin{Bmatrix} u_1 \\ u_2 \end{Bmatrix}$$

$$= \begin{bmatrix} -\frac{1}{\ell} & \frac{1}{\ell} \end{bmatrix}\begin{Bmatrix} u_1 \\ u_2 \end{Bmatrix}$$

Therefore,

$$[B] = \begin{bmatrix} -\frac{1}{\ell} & \frac{1}{\ell} \end{bmatrix} \tag{1.50}$$

(v) *Stress-strain relationship*
The relationship is simply given by Hooke's Law, and

$$\{\sigma\} = \sigma_x = E\varepsilon_x$$

Therefore

$$[D] = [E] \tag{1.51}$$

(vi) *Stiffness matrix*
From Eqn.(1.46), the stiffness matrix is

$$[k] = \int [B]^T[D][B]\,d(\text{vol})$$

$$= \int_A d(\text{area}) \int_0^\ell [B]^T[D][B]\,dx$$

$$= A\int_0^\ell \begin{Bmatrix} \frac{-1}{\ell} \\ \frac{1}{\ell} \end{Bmatrix}[E]\begin{bmatrix} \frac{-1}{\ell} & \frac{1}{\ell} \end{bmatrix}dx$$

$$= \frac{EA}{\ell}\begin{bmatrix} 1 & -1 \\ -1 & 1 \end{bmatrix} \tag{1.52}$$

A quick check will show that the two matrices given in Eqns.(1.13) and (1.52) are identical, and this demonstrates the validity of the finite element formulation.

1.11 Beam element stiffness matrix rederived

Consider now the beam element shown in Fig.1.2. The two nodal deflections will obviously only provide a rigid body translation for the element, and in order for bending action to occur, additional rotational degrees of freedom must be incorporated. Thus there are altogether four degrees of freedom which would define a cubic polynomial uniquely. The stiffness matrix of the beam element, which was given in Eqn.(1.15), will also be rederived in accordance with the finite element procedure.

(i) *Displacement function in polynomial form*

$$\{f\} = \{v\} = A_1 + A_2x + A_3x^2 + A_4x^3$$

i.e.

$$[R] = [1 \quad x \quad x^2 \quad x^3]$$

$$\{A\} = [A_1 \quad A_2 \quad A_3 \quad A_4]^T \tag{1.53}$$

(ii) *Nodal displacement parameter-polynomial constant relationship*
Before proceeding any further it is necessary first of all to obtain an expression for the rotation θ from the deflection v.

$$\theta = \frac{dv}{dx} = A_2 + 2A_3x + 3A_4x^2 \tag{1.54}$$

If the nodal coordinates are now substituted into Eqn.s(1.53) and (1.54), it is possible to establish the following equations (*vide* Eqn.(1.32)).

$$v_1 = A_1 + A_2(0) + A_3(0)^2 + A_4(0)^3$$

$$\theta_1 = A_2 + 2A_3(0) + 3A_4(0)^2$$

$$v_2 = A_1 + A_2(\ell) + A_3(\ell)^2 + A_4(\ell)^3$$

$$\theta_2 = A_2 + 2A_3(\ell) + 3A_4(\ell)^2$$

Therefore,

$$[C] = \begin{bmatrix} 1 & 0 & 0 & 0 \\ 0 & 1 & 0 & 0 \\ 1 & \ell & \ell^2 & \ell^3 \\ 0 & 1 & 2\ell & 3\ell^2 \end{bmatrix} \qquad [C]^{-1} = \begin{bmatrix} 1 & 0 & 0 & 0 \\ 0 & 1 & 0 & 0 \\ -3/\ell^2 & -2/\ell & 3/\ell^2 & -1/\ell \\ 2/\ell^3 & 1/\ell^2 & -2/\ell^3 & 1/\ell^2 \end{bmatrix}$$

(iii) *Displacement function in shape function form*

$$[N] = [R][C]^{-1} = [1 \quad x \quad x^2 \quad x^3]\begin{bmatrix} 1 & 0 & 0 & 0 \\ 0 & 1 & 0 & 0 \\ -3/\ell^2 & -2/\ell & 3/\ell^2 & -1/\ell \\ 2/\ell^3 & 1/\ell^2 & -2/\ell^3 & 1/\ell^2 \end{bmatrix}$$

$$= \left[\left(1 - \frac{3x^2}{\ell^2} + \frac{2x^3}{\ell^3}\right),\ x\left(1 - \frac{2x}{\ell} + \frac{x^2}{\ell^2}\right),\ \left(\frac{3x^2}{\ell^2} - \frac{2x^3}{\ell^3}\right),\ x\left(\frac{x^2}{\ell^2} - \frac{x}{\ell}\right)\right] \tag{1.55}$$

The expression in Eqn.(1.55) is in fact the same as the first-order Hermitian polynomial given in Eqn.(1.22)

(iv) *Strain-nodal displacement parameter relationship*
For a beam element, the strain corresponds to the curvature and, therefore,

$$\{\varepsilon\} = -\frac{d^2v}{dx^2} = \left[\left(\frac{6}{\ell^2} - \frac{12x}{\ell^3}\right),\ \left(\frac{4}{\ell} - \frac{6x}{\ell^2}\right),\ \left(\frac{-6}{\ell^2} + \frac{12x}{\ell^3}\right),\ \left(\frac{2}{\ell} - \frac{6x}{\ell^2}\right)\right]\{\delta\}$$

or

$$[B] = \left[\left(\frac{6}{\ell^2} - \frac{12x}{\ell^3}\right),\ \left(\frac{4}{\ell} - \frac{6x}{\ell^2}\right),\ \left(\frac{-6}{\ell^2} + \frac{12x}{\ell^3}\right),\ \left(\frac{2}{\ell} - \frac{6x}{\ell^2}\right)\right] \tag{1.56}$$

(v) *Stress-strain relationship*
The stress-strain relationship corresponds to moment-curvature relationship in the case of a beam and we have

$$\{\sigma\} \equiv M = EI\left(-\frac{d^2v}{dx^2}\right)$$

Therefore,

$$[D] \equiv EI$$

(vi) *Stiffness matrix*
With $[B]$ and $[D]$ known, it is a simple matter to work out the stiffness matrix in accordance with Eqn.(1.46), bearing in mind that the volume integral can now be reduced to a line integral.

$$[k] = \int_0^{\ell} [B]^T[D][B]\,dx$$

$$= EI\int_0^{\ell}\begin{bmatrix} \frac{36}{\ell^4}-\frac{144x}{\ell^5}+\frac{144x^2}{\ell^6} & & & \text{Symmetrical} \\ \frac{24}{\ell^3}-\frac{84x}{\ell^4}+\frac{72x^2}{\ell^5} & \frac{16}{\ell^2}-\frac{48x}{\ell^3}+\frac{36x^2}{\ell^4} & & \\ \frac{-36}{\ell^4}+\frac{144x}{\ell^5}-\frac{144x^2}{\ell^6} & \frac{-24}{\ell^3}+\frac{84x}{\ell^4}-\frac{72x^2}{\ell^5} & \frac{36}{\ell^4}-\frac{144x}{\ell^5}+\frac{144x^2}{\ell^6} & \\ \frac{12}{\ell^3}-\frac{60x}{\ell^4}+\frac{72x^2}{\ell^5} & \frac{8}{\ell^2}-\frac{36x}{\ell^3}+\frac{36x^2}{\ell^4} & \frac{-12}{\ell^3}+\frac{60x}{\ell^4}-\frac{72x^2}{\ell^5} & \frac{4}{\ell^2}-\frac{24x}{\ell^3}+\frac{36x^2}{\ell^4} \end{bmatrix}dx$$

$$= EI\begin{bmatrix} \frac{12}{\ell^3} & \text{Symmetrical} & & \\ \frac{6}{\ell^2} & \frac{4}{\ell} & & \\ \frac{-12}{\ell^3} & \frac{-6}{\ell^2} & \frac{12}{\ell^3} & \\ \frac{6}{\ell^2} & \frac{2}{\ell} & \frac{-6}{\ell^2} & \frac{4}{\ell} \end{bmatrix} \tag{1.58}$$

As expected, the stiffness matrices given in Eqns.(1.15) and (1.58) are identical, although the formulation procedure is different.

(vi) *Consistent load matrix for uniformly distributed loading*
For a beam element, Eqn.(1.47) should be reduced to

$$\{P\} = \int_0^{\ell}[N]^T\{q\}dx$$

$$= \int_0^{\ell}\begin{Bmatrix} \left(1-\frac{3x^2}{\ell^2}+\frac{2x^3}{\ell^3}\right) \\ x\left(1-\frac{2x}{\ell}+\frac{x^2}{\ell^2}\right) \\ \left(3\frac{x^2}{\ell^2}-\frac{2x^3}{\ell^3}\right) \\ x\left(-\frac{x}{\ell}+\frac{x^2}{\ell^2}\right) \end{Bmatrix} q\,dx \tag{1.59(a)}$$

In the case of uniformly distributed load, q is constant and, therefore,

$$\{P\} = q\left\{\begin{matrix} x - \dfrac{x^3}{\ell^2} + \dfrac{x^4}{2\ell^3} \\ \\ \dfrac{x^2}{2} - \dfrac{2}{3}\dfrac{x^3}{\ell} + \dfrac{1}{4}\dfrac{x^4}{\ell^2} \\ \\ \dfrac{x^3}{\ell^2} - \dfrac{1}{2}\dfrac{x^4}{\ell^3} \\ \\ \dfrac{-x^3}{3\ell} + \dfrac{1}{4}\dfrac{x^4}{\ell^2} \end{matrix}\right\}_0^\ell = \left\{\begin{matrix} \dfrac{q\ell}{2} \\ \\ \dfrac{q\ell^2}{12} \\ \\ \dfrac{q\ell}{2} \\ \\ \dfrac{-q\ell^2}{12} \end{matrix}\right\} \qquad (1.59(b))$$

Eqn.(1.59(b)) coincides with the nodal forces obtained by reversing the fixed end forces acting on a clamped beam under uniformly distributed load.

References

1. O.C. Zienkiewicz. *Finite Element Methods in Engineering Science*. New York, McGraw-Hill Book Co., 1971.
2. B. Fraeijs de Veubeke. Upper and lower bounds in matrix structural analysis. In *Matrix Methods of Structural Analysis*. AGARD, Vol.72, 1964, pp.165-201.
3. T.H.H. Pian. Derivation of element stiffness matrices. *Journal of the American Institute of Aeronautics and Astronautics*, 2, 576-7, 1964.
4. R.W. Clough. The finite element method in structural mechanics. In *Stress Analysis* (O.C. Zienkiewicz and G.S. Holister, eds). New York, John Wiley, 1965.
5. F. Schied. *Theory and Problems of Numerical Analysis*. New York, McGraw-Hill Book Co., 1968.
6. Y.K. Cheung. *Finite Strip Method in Structural Analysis*. Oxford, Pergamon Press, 1976.
7. J.S. Przemieniecki. *Theory of Matrix Structural Analysis*. New York, McGraw-Hill Book Co., 1968.

2 Triangular Element for Plane Elasticity and a Simple Computer Program

2.1 Introduction

With the basis of the finite element formulation established in Chapter 1 and its application to one-dimensional elements fully demonstrated, it is appropriate at this stage to attend to continuum problems, and in particular to two-dimensional elasticity problems [1] which were the first successful examples of the application of the finite element method.

In this chapter, the simple triangular element listed in Table 1.1 will be derived in detail, and a simple but workable computer program will be described in a step by step fashion. Because of its simplicity the triangular element is the logical element to start off with in the learning process, although in no way should it be construed as being the best element available, since the isoparametric eight-node element to be discussed in the next chapter would be a very much better element from the point of computational effort and accuracy. The simple computer program is comprehensive and automatic in as much as all the finite element analysis steps of data input, stiffness assembly, node fixing, solution, stress computation and output have been included; the program is not, however, sophisticated because of its low level schemes of assembly and solution. This is done on purpose in order to keep the coding easy to understand, and the readers should refer to the next chapter for an advanced computer program.

It is possible to include both plane stress ($\sigma_z = 0$) and plane strain ($\varepsilon_z = 0$) problems into the analysis and this is discussed in the derivation of the stiffness matrix. However, to avoid any unnecessary confusion the simple program has been written only for plane stress analysis. For plane strain analysis modified elastic constants will have to be used as data input.

2.2 Constant strain triangular element

The simple triangular element referred to in Table 1.1 and shown in Fig. 2.1 is very often called a constant strain triangle, since the strains can be proved to be constant at every point within the element. The stiffness and consistent load matrices of this element will now be derived by applying the standard equations given in Section 1.9.

It is obvious that at each node there must be at least two degrees

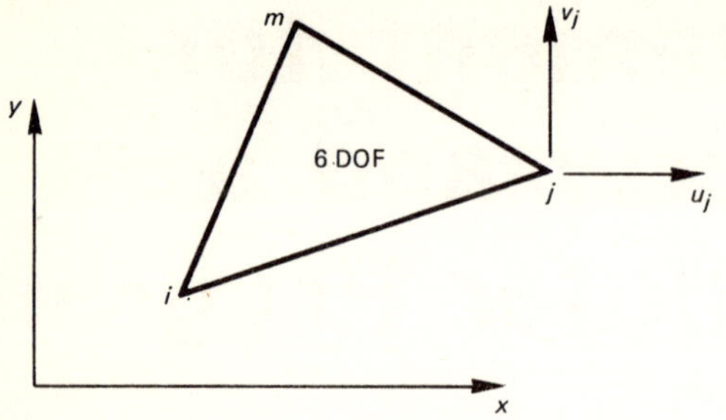

Fig.2.1 Constant strain triangular element

of freedom, a u displacement in the x direction and a v displacement in the y direction. Hence there would be a total of three degrees of freedom for u and v respectively, and consequently three polynomial constants for each displacement function.

2.2.1 DISPLACEMENT FUNCTION IN POLYNOMIAL FORM FROM THE PASCAL TRIANGLE

The choice of the polynomial function presents no difficulty and has already been shown in Table 1.1.

$$u = A_1 + A_2x + A_3y$$

$$= [1 \quad x \quad y][A_1 \quad A_2 \quad A_3]^T$$

$$= [R]\{A\} \tag{2.1a}$$

$$v = B_1 + B_2x + B_3y$$

$$= [R]\{B\} \tag{2.1b}$$

Therefore

$$\{f\} = \begin{Bmatrix} u \\ v \end{Bmatrix} = \begin{bmatrix} [R] & [0] \\ [0] & [R] \end{bmatrix} \begin{Bmatrix} \{A\} \\ \{B\} \end{Bmatrix} \tag{2.1c}$$

2.2.2 NODAL DISPLACEMENT PARAMETER-POLYNOMIAL CONSTANT RELATIONSHIP

Since both u and v have the same shape function it is only necessary to find the relationship for one of them. Substituting the x and y coordinates of the nodes into Eqn.(2.1a) one after the other, we obtain

$$u_i = A_1 + x_iA_2 + y_iA_3 \tag{2.2}$$

$$u_j = A_1 + x_jA_2 + y_jA_3 \tag{2.3}$$

$$u_m = A_1 + x_mA_2 + y_mA_3 \tag{2.4}$$

The polynomial constants can be expressed in terms of the nodal displacements by carrying out Gaussian elimination as follows.

(i) Eliminate A_1 by performing Eqn.(2.3) - Eqn.(2.2) and Eqn.(2.4) - Eqn.(2.2),

$$u_j - u_i = (x_j - x_i)A_2 + (y_j - y_i)A_3 \tag{2.5}$$

$$u_m - u_i = (x_m - x_i)A_2 + (y_m - y_i)A_3 \tag{2.6}$$

(ii) Eliminate A_2 by performing the calculations of Eqn.(2.6) × $(x_j - x_i)$ - Eqn.(2.5) × $(x_m - x_i)$,

$$(x_m - x_j)u_i + (x_i - x_m)u_j + (x_j - x_i)u_m$$
$$= \left\{(x_j - x_i)(y_m - y_i) - (x_m - x_i)(y_j - y_i)\right\}A_3$$

or

$$A_3 = \frac{1}{2\Delta}(x_{mj}u_i + x_{im}u_j + x_{ji}u_m) \tag{2.7}$$

where $x_{mj} = (x_m - x_j)$ and $y_{jm} = (y_j - y_m)$, etc. and Δ is in fact the area of triangle ijm and can be computed, *vide* Eqn.(1.25b).

(iii) Compute A_2 by substituting Eqn.(2.7) into Eqn.(2.5),

$$A_2 = \frac{1}{2\Delta}(y_{jm}u_i + y_{mi}u_j + y_{ij}u_m) \tag{2.8}$$

(iv) Compute A_1 by substituting Eqn.(2.7) and Eqn.(2.8) into Eqn.(2.2),

$$A_1 = \frac{1}{2\Delta}\left[(x_jy_m - x_my_j)u_i + (x_my_i - x_iy_m)u_j + (x_iy_j - x_jy_i)u_m\right] \tag{2.9}$$

Writing Eqns.(2.9), (2.8) and (2.7) in matrix form,

$$\begin{Bmatrix} A_1 \\ A_2 \\ A_3 \end{Bmatrix} = \frac{1}{2\Delta}\begin{bmatrix} x_jy_m - x_my_j & x_my_i - x_iy_m & x_iy_j - x_jy_i \\ y_{jm} & y_{mi} & y_{ij} \\ x_{mj} & x_{im} & x_{ji} \end{bmatrix}\begin{Bmatrix} u_i \\ u_j \\ u_m \end{Bmatrix} \tag{2.10}$$

The expressions in Eqn.(2.10) can be simplified considerably by introducing the notations given in Eqn.(1.25) and Eqn.(2.10) now becomes

$$\begin{Bmatrix} A_1 \\ A_2 \\ A_3 \end{Bmatrix} = \frac{1}{2\Delta}\begin{bmatrix} a_i & a_j & a_m \\ b_i & b_j & b_m \\ c_i & c_j & c_m \end{bmatrix}\begin{Bmatrix} u_i \\ u_j \\ u_m \end{Bmatrix} \tag{2.11}$$

and a similar equation can be written for the v displacements:

$$\begin{Bmatrix} B_1 \\ B_2 \\ B_3 \end{Bmatrix} = \frac{1}{2\Delta}\begin{bmatrix} a_i & a_j & a_m \\ b_i & b_j & b_m \\ c_i & c_j & c_m \end{bmatrix}\begin{Bmatrix} v_i \\ v_j \\ v_m \end{Bmatrix} \tag{2.12}$$

In finite element analysis, the displacement parameters at each node are always grouped together (for example u_i, v_i, w_i, θ_{xi}, θ_{yi}, θ_{zi}), for ease of assembly, and therefore the displacement parameter vector would be

$$\{\delta\} = [u_i,\ v_i,\ u_j,\ v_j,\ u_m,\ v_m] \tag{2.13}$$

and the $[C]^{-1}$ matrix, unlike in previous cases, must now be constructed from Eqns.(2.11) and (2.12).

$$[C]^{-1} = \frac{1}{2\Delta}\begin{bmatrix} a_i & 0 & a_j & 0 & a_m & 0 \\ b_i & 0 & b_j & 0 & b_m & 0 \\ c_i & 0 & c_j & 0 & c_m & 0 \\ 0 & a_i & 0 & a_j & 0 & a_m \\ 0 & b_i & 0 & b_j & 0 & b_m \\ 0 & c_i & 0 & c_j & 0 & c_m \end{bmatrix} \tag{2.14}$$

2.2.3 DISPLACEMENT FUNCTIONS IN SHAPE FUNCTION FORM

From Eqn.(1.34) and Eqn.(2.1c),

$$\{f\} = \begin{Bmatrix} u \\ v \end{Bmatrix} = \begin{bmatrix} [R] & [0] \\ [0] & [R] \end{bmatrix} [C]^{-1}\{\delta\} = [N]\{\delta\} \tag{2.15}$$

Therefore the shape functions

$$[N] = \frac{1}{2\Delta}\left[\begin{array}{c|c|c|c|c|c} a_i + b_ix + c_iy & 0 & a_j + b_jx + c_jy & 0 & a_m + b_mx + c_my & 0 \\ 0 & a_i + b_ix + c_iy & 0 & a_i + b_jx + c_jy & 0 & a_m + b_mx + c_my \end{array}\right]$$

$$= \begin{bmatrix} \mathscr{L}_i & 0 & \mathscr{L}_j & 0 & \mathscr{L}_m & 0 \\ 0 & \mathscr{L}_i & 0 & \mathscr{L}_j & 0 & \mathscr{L}_m \end{bmatrix}$$

in which $\mathscr{L}_i$, $\mathscr{L}_j$, $\mathscr{L}_m$ are the area coordinates and thus the shape functions are really nothing but the linear functions shown in Table 1.3, which can be written down directly in the first place.

2.2.4 STRAIN-NODAL DISPLACEMENT PARAMETER RELATIONSHIP

The strain components for a two-dimensional elasticity problem are two direct strains ε_x, ε_y and shear strain γ_{xy} and they are given by

$$\varepsilon_x = \frac{\partial u}{\partial x}, \qquad \varepsilon_y = \frac{\partial v}{\partial y}, \qquad \gamma_{xy} = \frac{\partial u}{\partial y} + \frac{\partial v}{\partial x} \tag{2.16}$$

Substituting the displacement functions given in Eqn.(2.15) and carrying out the appropriate partial differentiations required in Eqn.(2.16),

$$\varepsilon_x = \frac{1}{2\Delta}(b_i u_i + b_j u_j + b_m u_m)$$

$$\varepsilon_y = \frac{1}{2\Delta}(c_i v_i + c_j v_j + c_m v_m)$$

$$\gamma_{xy} = \frac{1}{2\Delta}(c_i u_i + b_i v_i + c_j u_j + b_j v_j + c_m u_m + b_m v_m)$$

and the strain matrix $[B]$ is established as

$$[B] = \frac{1}{2\Delta}\begin{bmatrix} b_i & 0 & b_j & 0 & b_m & 0 \\ 0 & c_i & 0 & c_j & 0 & c_m \\ c_i & b_i & c_j & b_j & c_m & b_m \end{bmatrix} \tag{2.17}$$

2.2.5 STRESS-STRAIN RELATIONSHIP

The two cases of plane stress and plane strain will be discussed separately since the stress-strain relationships are different for the two cases. Only isotropic materials will be considered here.

(a) Plane stress

In plane stress problems only the three stress components (σ_x, σ_y, τ_{xy}) within the x-y plane are present, the other three components (σ_z, τ_{yz}, τ_{zx}) being equal to zero. The stress-strain equations for three-dimensional elasticity are thus reduced to

$$\sigma_x = \frac{E}{(1+\nu)(1-2\nu)}\left[(1-\nu)\varepsilon_x + \nu\varepsilon_y + \nu\varepsilon_z\right]$$

$$\sigma_y = \frac{E}{(1+\nu)(1-2\nu)}\left[\nu\varepsilon_x + (1-\nu)\varepsilon_y + \nu\varepsilon_z\right]$$

$$\sigma_z = 0 = \frac{E}{(1+\nu)(1-2\nu)}\left[\nu\varepsilon_x + \nu\varepsilon_y + (1-\nu)\varepsilon_z\right]$$

$$\tau_{xy} = \frac{E}{2(1+\nu)}\,\gamma_{xy}$$

which can be further reduced to

$$\left.\begin{aligned}\sigma_x &= \frac{E}{1-\nu^2}(\varepsilon_x + \nu\varepsilon_y)\\ \sigma_y &= \frac{E}{1-\nu^2}(\nu\varepsilon_x + \varepsilon_y)\\ \tau_{xy} &= \frac{E}{2(1+\nu)}\gamma_{xy}\end{aligned}\right\} \quad (2.18)$$

In matrix form, Eqn.(2.18) becomes

$$\begin{Bmatrix}\sigma_x\\ \sigma_y\\ \tau_{xy}\end{Bmatrix} = \begin{bmatrix}\frac{E}{1-\nu^2} & \frac{\nu E}{1-\nu^2} & 0\\ \frac{\nu E}{1-\nu^2} & \frac{E}{1-\nu^2} & 0\\ 0 & 0 & \frac{E}{2(1+\nu)}\end{bmatrix}\begin{Bmatrix}\varepsilon_x\\ \varepsilon_y\\ \gamma_{xy}\end{Bmatrix} \quad (2.19)$$

and thus the elasticity matrix $[D]$ has been established.

(b) Plane strain

Unlike plane stress it is ε_z and not σ_z which is equal to zero in plane strain. Because of this, it is more convenient to express the strains in terms of the stresses, i.e.

$$\varepsilon_x = \frac{1}{E}(\sigma_x - \nu\sigma_y - \nu\sigma_z)$$

$$\varepsilon_y = \frac{1}{E}(-\nu\sigma_x + \sigma_y - \nu\sigma_z)$$

$$\varepsilon_z = 0 = \frac{1}{E}(-\nu\sigma_x - \nu\sigma_y + \sigma_z)$$

$$\gamma_{xy} = \frac{2(1+\nu)}{E}\tau_{xy}$$

from which it can be established that

$$\sigma_z = \nu(\sigma_x + \sigma_y)$$

and that

$$\begin{Bmatrix}\varepsilon_x\\ \varepsilon_y\\ \gamma_{xy}\end{Bmatrix} = \frac{1}{E}\begin{bmatrix}(1-\nu^2) & -\nu(1+\nu) & 0\\ -\nu(1+\nu) & (1-\nu^2) & 0\\ 0 & 0 & 2(1+\nu)\end{bmatrix}\begin{Bmatrix}\sigma_x\\ \sigma_y\\ \tau_{xy}\end{Bmatrix} \quad (2.20)$$

Finally the stress-strain relationship is arrived at by solving Eqn. (2.20) so that

$$\begin{Bmatrix} \sigma_x \\ \sigma_y \\ \tau_{xy} \end{Bmatrix} = \frac{E}{1+\nu}\begin{bmatrix} \frac{1-\nu}{1-2\nu} & \frac{\nu}{1-2\nu} & 0 \\ \frac{\nu}{1-2\nu} & \frac{1-\nu}{1-2\nu} & 0 \\ 0 & 0 & \frac{1}{2} \end{bmatrix}\begin{Bmatrix} \varepsilon_x \\ \varepsilon_y \\ \gamma_{xy} \end{Bmatrix} \tag{2.21}$$

The two elasticity matrices given in Eqns.(2.19) and (2.21) have the same form and it is more convenient to represent them by the following common elasticity matrix

$$[D] = \begin{bmatrix} C_1 & C_1C_2 & 0 \\ C_1C_2 & C_1 & 0 \\ 0 & 0 & C_{12} \end{bmatrix} \tag{2.22}$$

in which

$C_1 = E/(1 - \nu^2)$ and $C_2 = \nu$ for plane stress,

$C_1 = \dfrac{E(1-\nu)}{(1+\nu)(1-2\nu)}$ and $C_2 = \dfrac{\nu}{1-\nu}$ for plane strain,

and

$C_{12} = C_1(1 - C_2)/2$ for both cases.

The element stresses are given by Eqn.(1.38), which after substituting in Eqns.(2.22) and (2.17), becomes

$$\begin{Bmatrix} \sigma_x \\ \sigma_y \\ \tau_{xy} \end{Bmatrix} = \frac{1}{2\Delta}\begin{bmatrix} C_1 & C_1C_2 & 0 \\ C_1C_2 & C_1 & 0 \\ 0 & 0 & C_{12} \end{bmatrix}\begin{bmatrix} b_i & 0 & b_j & 0 & b_m & 0 \\ 0 & c_i & 0 & c_j & 0 & c_m \\ c_i & b_i & c_j & b_j & c_m & b_m \end{bmatrix}\begin{Bmatrix} u_i \\ v_i \\ u_j \\ v_j \\ u_m \\ v_m \end{Bmatrix}$$

$$= \frac{1}{2\Delta}\begin{bmatrix} C_1 b_i & C_1 C_2 c_i & C_1 b_j & C_1 C_2 c_j & C_1 b_m & C_1 C_2 c_m \\ C_1 C_2 b_i & C_1 c_i & C_1 C_2 b_j & C_1 c_j & C_1 C_2 b_m & C_1 c_m \\ C_{12} c_i & C_{12} b_i & C_{12} c_j & C_{12} b_j & C_{12} c_m & C_{12} b_m \end{bmatrix} \begin{Bmatrix} u_i \\ v_i \\ u_j \\ v_j \\ u_m \\ v_m \end{Bmatrix} \tag{2.23}$$

Since none of the coefficients in the stress matrix of Eqn.(2.23) are functions of x and y, it is concluded that the stresses are constant over the whole element resulting in stress discontinuities between adjacent elements. In practice, the values of the stresses are assigned to the centroid of each triangular element so that a smooth flow of stresses can be achieved throughout the whole domain under consideration.

TABLE 2.1 *Stiffness matrix of constant strain triangle*

$$[k] = \frac{t}{4\Delta}\left[\begin{array}{c|c|c|c|c|c} C_1 b_i^2 + C_{12} c_i^2 & & & & & \text{symmetrical} \\ \hline C_1 C_2 b_i c_i + C_{12} b_i c_i & C_1 c_i^2 + C_{12} b_i^2 & & & & \\ \hline C_1 b_i b_j + C_{12} c_i c_j & C_1 C_2 b_j c_i + C_{12} b_i c_j & C_1 b_j^2 + C_{12} c_j^2 & & & \\ \hline C_1 C_2 b_i c_j + C_{12} b_j c_i & C_1 c_i c_j + C_{12} b_i b_j & C_1 C_2 b_j c_j + C_{12} b_j c_j & C_1 c_j^2 + C_{12} b_j^2 & & \\ \hline C_1 b_i b_m + C_{12} c_i c_m & C_1 C_2 b_m c_i + C_{12} b_i c_m & C_1 b_j b_m + C_{12} c_j c_m & C_1 C_2 b_m c_j + C_{12} b_j c_m & C_1 b_m^2 + C_{12} c_m^2 & \\ \hline C_1 C_2 b_i c_m + C_{12} b_m c_i & C_1 c_i c_m + C_{12} b_i b_m & C_1 C_2 b_j c_m + C_{12} b_m c_j & C_1 c_j c_m + C_{12} b_j b_m & C_1 C_2 b_m c_m + C_{12} b_m c_m & C_1 c_m^2 + C_{12} b_m^2 \end{array}\right]$$

2.2.6 STIFFNESS MATRIX

The stiffness matrix is given by Eqn. (1.46) as

$$[k] = \int [B]^T [D][B] \mathrm{d(vol)}$$

and, as indicated by Eqns. (2.17) and (2.22), both $[B]$ and $[D]$ matrices are independent of the variables x and y. Thus the stiffness matrix reduces to the following simple form

$$[k] = t\Delta [B]^T [D][B] \tag{2.24}$$

in which t is the thickness of the element and is assumed to be constant. It is a fairly simple matter to carry out the matrix multiplications given in Eqn. (2.24) and the resulting stiffness matrix is listed in Table 2.1.

In the previous discussions on a beam element it was found that the consistent load matrix corresponds to the fixed end forces (reversed in direction) of a clamped end beam under the same load. For two-dimensional finite elements under distributed loadings such fixed end forces at the nodes cannot be determined directly from classical structural mechanics and Eqn. (1.47) must be used for the computation of equivalent nodal forces. To take the simplest case of uniformly distributed forces q_x and q_y per unit area over an element, the consistent load matrix can be obtained as

$$\{P\} = \int \frac{1}{2\Delta} \begin{bmatrix} a_i + b_i x + c_i y & 0 & a_j + b_j x + c_j y & 0 & a_m + b_m x + c_m y & 0 \\ 0 & a_i + b_i x + c_i y & 0 & a_j + b_j x + c_j y & 0 & a_m + b_m x + c_m y \end{bmatrix}^T \begin{Bmatrix} q_x \\ q_y \end{Bmatrix} \mathrm{d}x\mathrm{d}y \tag{2.25}$$

The three integrals involved are $\int \mathrm{d}x\mathrm{d}y = \Delta$, $\int x\, \mathrm{d}x\mathrm{d}y$ and $\int y\, \mathrm{d}x\mathrm{d}y$. By choosing the centroid of the triangle as the origin of the coordinate axes, the last two integrals are equal to zero, and Eqn. (2.25) is reduced to

$$\{P\} = \frac{1}{2} \begin{Bmatrix} a_i q_x \\ a_i q_y \\ a_j q_x \\ a_j q_y \\ a_m q_x \\ a_m q_y \end{Bmatrix} \tag{2.26}$$

A further simplification is possible by noting that if the point ℓ in Fig.1.7 coincides with the origin then $x = y = 0$, and Eqn.(1.24) will take up the form of

$$\mathcal{L}_i = A_i/\Delta = a_i/2\Delta$$

$$\mathcal{L}_j = A_j/\Delta = a_j/2\Delta$$

$$\mathcal{L}_m = A_m/\Delta = a_m/2\Delta$$

By virtue of the fact that $A_i = A_j = A_m$, it is finally established that

$$a_i = a_j = a_m = \frac{2}{3}\Delta$$

and

$$\{P\} = \frac{\Delta}{3}\begin{Bmatrix} q_x \\ q_y \\ q_x \\ q_y \\ q_x \\ q_y \end{Bmatrix} \qquad (2.27)$$

2.3 Example problem for all subsequent computations

As mentioned previously, a simple computer program using the constant strain triangular element will be presented, in which the various steps of data input, stiffness generation and assembly, introduction of boundary conditions, solution, stress computation and output of results will be discussed in detail and illustrated by carrying out the numerical analysis of a simple problem. The problem chosen is that of a thick cylinder (Fig.2.2) subjected to internal pressure and zero axial force, for which an analytical solution [2] is available and the stresses are given as

Radial stress, $$\sigma_r = \frac{a^2 p}{b^2 - a^2}\left(1 - \frac{b^2}{r^2}\right) \qquad (2.28a)$$

Tangential stress, $$\sigma_\theta = \frac{a^2 p}{b^2 - a^2}\left(1 + \frac{b^2}{r^2}\right) \qquad (2.28b)$$

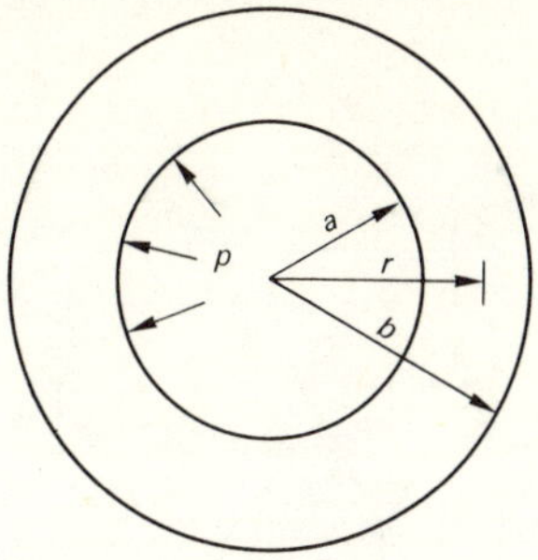

Fig.2.2 Thick cylinder under internal pressure

Before any analysis can be made it is necessary for the engineer to make two decisions, the type of elements and the mesh configuration to be used.

The first of these decisions is very often restricted by the availability of element subroutines, and engineers are forced to use lower-order elements (simple elements with few degrees of freedom and using lower-order polynomial functions) which are more readily found in program packages, in spite of the fact that some higher-order element (element of higher accuracy but with many degrees of freedom and using higher-order polynomial functions) might be preferable in view of its accuracy and ease of fitting the geometry.

The second of these decisions is linked with the first in that the order of the element used in the analysis is in general inversely proportional to the number of elements used to represent the structure, but given a certain element type and a particular problem, more elements will be required to model complex geometrical shapes and areas in which the stress gradient is steep. As there are no hard and fast rules for establishing the meshes, experience plays a very important role in the decision making. For new or important problems the standard approach is to repeat the analysis several times with successively refined meshes, and when acceptable convergence has occurred the last mesh is the standard one for all future computations for similar types of loadings.

For a number of elements, and in particular the constant strain triangular element, the accuracy of the results deteriorates rapidly with increased aspect ratio, and in general, long and narrow elements should be avoided.

To demonstrate the effect of mesh refinement on the convergence of results consider the illustrative problem of the thick cylinder given in Fig.2.2. Because of axisymmetry it is normally only required to analyse a small segment of the cylinder but for our simple program a 90° segment (Fig.2.3) is used here to take advantage of the symmetry conditions along the x and y axes.

A number of different mesh divisions are shown in Fig.2.3. Mesh (a) will be seen to yield bad results because of the extreme coarseness while meshes (b) and (c) will still give bad results because of insufficient refinement of mesh in the radial direction. From the results plotted in Fig.2.4 it can be seen that the results for meshes (d) and (e) are quite reasonable since the radial stress results are coincident with the analytical curve while for tangential stresses the finite element results straddle the exact curve.

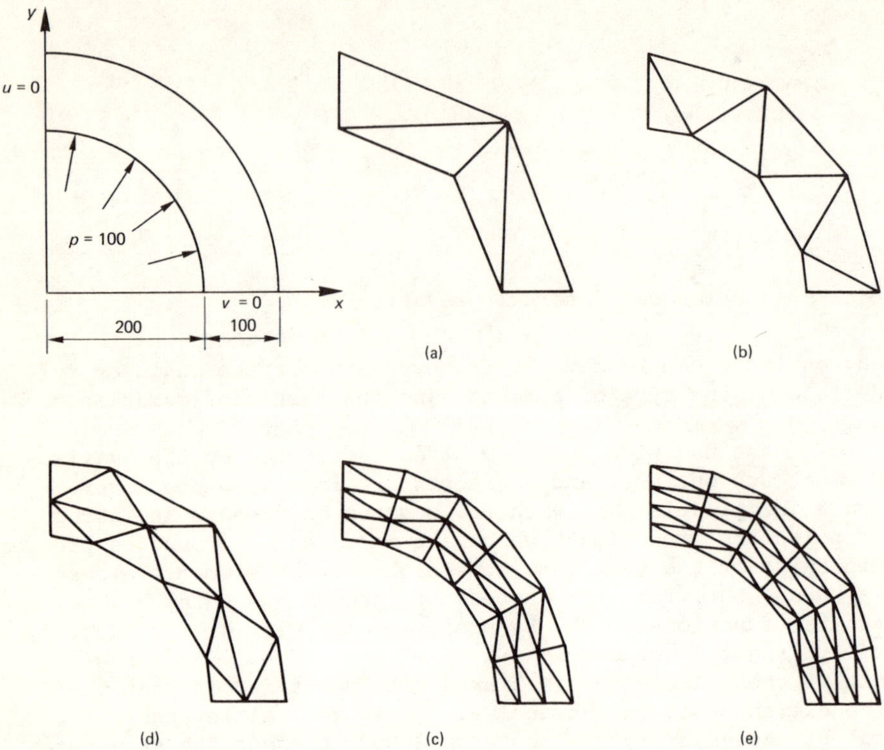

Fig.2.3 Different mesh divisions for the example problem

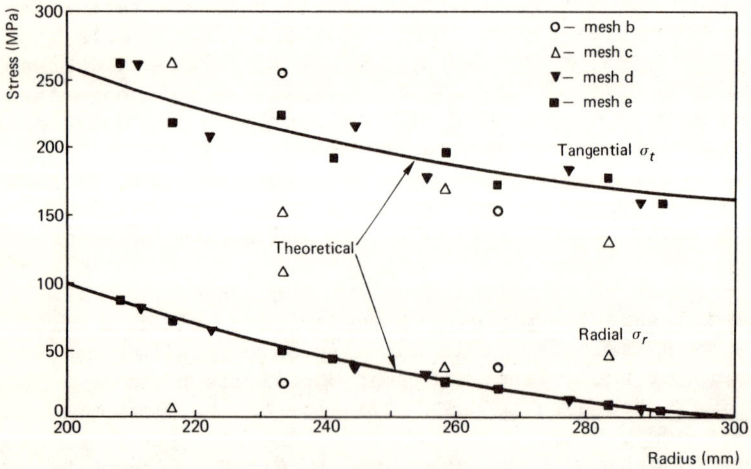

Fig.2.4 Stresses in the thick cylinder

However, in order to demonstrate explicitly all the computational steps in the execution of the finite element program mesh (a) will be used because of the comparatively small amount of computation involved.

2.4 Data preparation

After having decided on the element type and mesh division it is now possible to prepare the data for input into the program. Such data is extracted directly from the drawing of the structure in which all information concerning node and element numbering, material properties, load and boundary conditions are available (Fig.2.5). These data will be grouped under a number of headings.

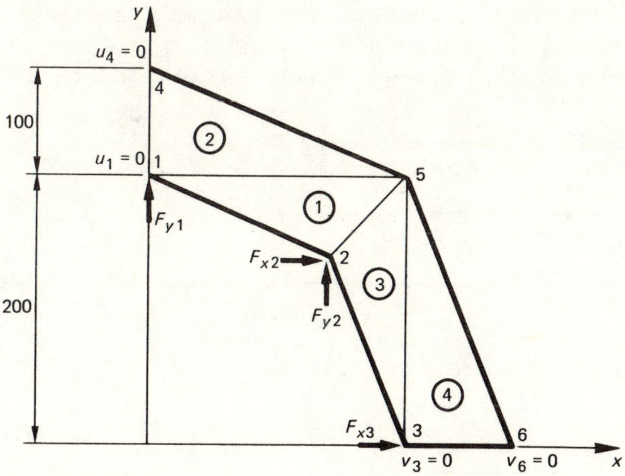

Fig.2.5 Details of example problem

2.4.1 CONTROL DATA

To make the program general and not problem dependent, it is necessary to define the main control parameters for the structure to be analysed, and this is achieved through the input of one or several control data cards at the start of the program. The control data consists of four variables and their values for the sample problem are listed below.

MAXNEL (number of elements in the structure) = 4

MAXNOD (number of nodes in the structure) = 6

MAXLOD (number of nodes with external applied forces) = 3

MAXFIX (number of nodes with constraints) = 4

These variables are used to control the reading of the main body of data and for determining the limits of the main loops inside the assembly and solution part of the program. Once a program has been compiled and stored on permanent file, it can be used time after time to analyse different problems by reading in a new set of control variables without going through any recoding process.

2.4.2 ELEMENT DEFINITION

The variable MAXNEL controls the input of geometric and material properties data for all the elements. Each of the elements is defined by three nodes and the nodes must be defined in the same order as the one used in formulation, which is anticlockwise for the set of axes given in Fig.2.1. Failure to comply with this will result in a non-positive definite stiffness [3] for the element and incorrect results for the whole structure.

In the element definition any one of the three nodes can be used as the first node and the element number will in fact correspond to its position in the data set, i.e. the first data card will correspond to element number one, etc.

Referring to Fig.2.5, the element definitions for the illustrative problem are

Element number	Node i	Node j	Node m
1	1	2	5
2	1	5	4
3	2	3	5
4	3	6	5

In accordance with the discussion given previously, it would be equally proper to define element one as 2, 5, 1.

2.4.3 NODAL COORDINATES

The variable MAXNOD controls the input for all nodal coordinates, which are required for the computation of stiffness, stress and load matrices and at times for determining the orientation of an element in space. For the illustrative problem the nodal coordinates are defined as

Node	x coordinate	y coordinate
1	0.00	200.00
2	141.42	141.42
3	200.00	0.00
4	0.00	300.00
5	212.13	212.13
6	300.00	0.00

These nodal coordinates are relative to the centre of the cylinder with the x-y plane orientated as shown in Fig.2.5. It is allowable, however, to use any other point as origin, and provided that there is no rotation of the x-y plane, the various matrices will remain unchanged.

2.4.4 APPLIED LOADS

The loads which are applied to a structure must be defined in the data. For many problems the majority of nodes do not have any loads applied to them, and the variable MAXLOD needs only to define the number of loaded nodes and to read in a corresponding number of data cards. By initializing the load array to zero the zero values at all unloaded nodes would be provided for automatically.

The x and y components of the loads have been calculated manually here, although for more advanced programs they are very often computed automatically by simply specifying the pressure values and the lines or surfaces over which they are acting.

The load data for the sample problem is listed below:

Node	x component	y component
1	2929	7071
2	10 000	10 000
3	7071	2929

Note that the loaded nodes do not have to be entered in a sequential order since their locations in the load array are defined by the node number preceding the load components.

2.4.5 NODAL FIXITY

The variable MAXFIX indicates the number of nodes with constraints. Such constraints are necessary to eliminate the rigid body modes which are associated with the element stiffness matrix as well as the stiffness matrix of the whole structure. From statics it is well known that for a plane problem at least three independent displacements must be constrained.

The nodal fixity is indicated to the program by a two digit fixity code. The first digit refers to the u displacement and the second digit to the v displacement. A digit set to zero indicates that the associated displacement is free to move while for the case of a digit set to unity the associated displacement is constrained.

Due to symmetry, nodes 1, 4 and 3, 6 of the sample problem are not allowed to move in the x and y directions respectively. The corresponding data input is therefore as follows.

Node	u-displacement	v-displacement
6	0	1
1	1	0
3	0	1
4	1	0

Note that once again the order of the constrained nodes in the data set is immaterial.

2.4.6 ELEMENT PROPERTIES

Before element stiffness can be calculated element material properties must be available. For this elementary program it is assumed that all the elements have the same isotropic properties and thickness. For more advanced programs it is usual to allow the material properties and element thickness to vary from one element to the next and also to allow anisotropic material properties for each element.

For the illustrative problem the following properties were chosen: Young's modulus = 0.1×10^6 MPa, Poisson's ratio = 0.25, thickness = 1.0 mm.

2.4.7 DATA ECHO

In the majority of finite element programs a data echo is included and the input data are printed as a matter of course, although sometimes it is not strictly an 'echo' because of small discrepancies in the input and output formats. The inclusion of such a data echo will allow the checking of input data easily and quickly in case of an error, and also will serve as a means of identification if the results are stored for future reference.

Data input coding

```
      PROGRAM SIMPLE(INPUT,OUTPUT,TAPE5=INPUT,TAPE6=OUTPUT)

      DIMENSION LNODS(65,3),COORDS(50,2),ALOAD(50,2),NFIX(50,2)
      DIMENSION ELSTIF(6,6),SS(90,90)
      DIMENSION SL(90),D(3,3),B(3,6),DB(3,6),DISPL(90),DISP(6)        (1)
      DIMENSION STRESS(65,3)
      DIMENSION REACT(2)

C     NVABZ IS THE NUMBER OF VARIABLES PER NODE
C     NNODZ IS THE NUMBER OF NODES PER ELEMENT

      NVABZ=2                                                         (2)
      NNODZ=3

C     READ IN DATA

      READ(5,1001) MAXNEL,MAXNOD,MAXLOD,MAXFIX                        (3)
      WRITE(6,1002) MAXNEL,MAXNOD,MAXLOD,MAXFIX

C     INITIALIZE VECTORS AND MATRICES

      MAXVAR=MAXNOD*NVABZ
      DO 2 I=1,MAXNOD
      DO 1 J=1,2
```

```
      ALOAD(I,J)=0.0
      NFIX(I,J)=0.0
    1 CONTINUE
    2 CONTINUE
      DO 4 I=1,MAXVAR
      SL(I)=0.0
      DISPL(I)=0.0                                                    (4)
      DO 3 J=1,MAXVAR
      SS(I,J)=0.0
    3 CONTINUE
    4 CONTINUE
      DO 6 I=1,MAXNEL
      DO 5 J=1,3
      STRESS(I,J)=0.0
    5 CONTINUE
    6 CONTINUE

C     READ IN ELEMENT DEFINITIONS

      WRITE(6,1003)
      DO 11 NEL=1,MAXNEL
      READ(5,1004) (LNODS(NEL,I),I=1,NNODZ)                           (5)
      WRITE(6,1005) NEL,(LNODS(NEL,I),I=1,NNODZ)
   11 CONTINUE

C     READ IN NODAL CO-ORDINATES

      WRITE(6,1006)
      DO 12 NOD=1,MAXNOD
      READ(5,1007) NIC,(COORDS(NIC,I),I=1,NVABZ)                      (6)
      WRITE(6,1008) NIC,(COORDS(NIC,I),I=1,NVABZ)
   12 CONTINUE

C     READ IN APPLIED LOADS AND FIXITIES

      DO 14 J=1,MAXLOD
      READ(5,1010) NIC,(ALOAD(NIC,I),I=1,NVABZ)                       (7)
   14 CONTINUE
      DO 15 J=1,MAXFIX
      READ(5,1011) NIC,(NFIX(NIC,I),I=1,NVABZ)                        (8)
   15 CONTINUE

      WRITE(6,1009)
      DO 16 I=1,MAXNOD
      WRITE(6,1012) I,(ALOAD(I,J),J=1,NVABZ),(NFIX(I,J),J=1,NVABZ)
   16 CONTINUE

C     READ IN ELASTIC PROPERTIES

      READ(5,1013) YM,PR,T                                            (9)
      WRITE(6,1014) YM,PR,T
```

(1) Dimension the program.
(2) Define the number of variables per node and the number of nodes per element.
(3) Read in the control data (number of elements, nodes, loaded nodes and fixed nodes).
(4) Initialize vectors and matrices.
(5) Read in element definitions.
(6) Read in nodal coordinates.
(7) Read in applied loads.
(8) Read in restrained nodes.
(9) Read in element properties.

Data echo of illustrative problem

```
 TOTAL NUMBER OF ELEMENTS =     4
                    NODES =     6
                    LOADS =     3
                 FIXITIES =     4


ELEMENT DEFINITIONS
ELEMENT   NODE1   NODE2   NODE3

    1        1       2       5
    2        1       5       4
    3        2       3       5
    4        3       6       5


NODAL CO-ORDINATES
NODE        X           Y

   1       0.000     200.000
   2     141.420     141.420
   3     200.000       0.000
   4       0.000     300.000
   5     212.130     212.130
   6     300.000       0.000


APPLIED LOADS AND FIXITIES
NODE      APPLIED LOADS          FIXITIES

   1      .293E+04   .707E+04     1     0
   2      .100E+05   .100E+05     0     0
   3      .707E+04   .293E+04     0     1
   4     0.         0.            1     0
   5     0.         0.            0     0
   6     0.         0.            0     1


MATERIAL PROPERTIES

YOUNGS MODULUS FOR MATERIAL     =   .100E+06
POISSONS RATIO FOR MATERIAL     =   .250E+00
CONSTANT ELEMENT THICKNESS      =   .100E+01

DATA COMPLETE
```

2.5 Generation of individual stiffness matrices

The structural stiffness matrix is obtained by adding together, or assembling as it is called, all the element stiffness matrices, which must therefore be computed beforehand. The simple nature of the constant strain triangle allows the element stiffness coefficients to be determined explicitly and from Table 2.1 it can be seen that for the computation of the coefficients the following data is required.

(i) The node numbers forming the element.

(ii) The coordinates of these nodes.

(iii) The element material properties and thickness.

In the program the three nodal numbers are retrieved from the matrix of element definitions and are called NIC1, NIC2 and NIC3. The associated nodal coordinates are X1, Y1; X2, Y2 and X3, Y3 respectively. The material properties and element thickness are input as data and the constants b_i, c_i, b_j, etc. can be calculated. It is then a simple matter to program the coefficients given in Table 2.1 and to establish the element stiffness matrix ELSTIF.

The flow chart for calculating the element stiffness is shown in Fig. 2.6 and the numerical values of the element stiffnesses for the illustrative problems are given in Section 2.6.

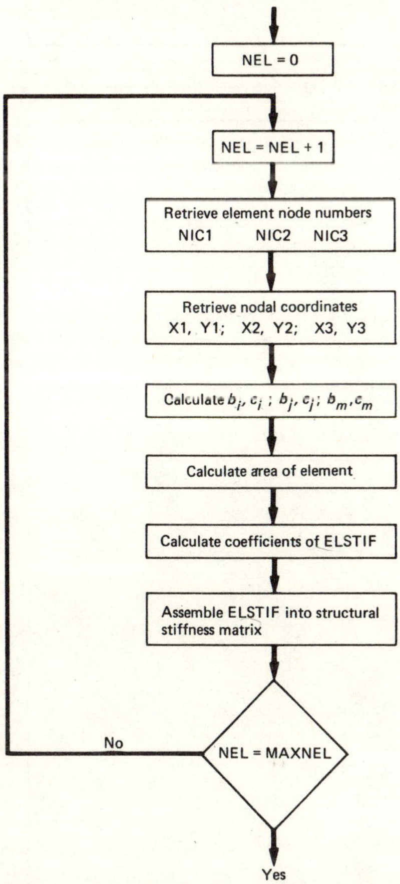

Fig.2.6 Flow chart for stiffness generation (NEL = element number)

Coding for generating element stiffness

```
C       SET UP STRUCTURAL STIFFNESS MATRIX

        EC1=YM/(1-PR*PR)
        EC2=PR                                                        (1)
        EC12=EC1*(1-EC2)/2.0

        DO 100 NEL=1,MAXNEL                                           (2)

C       SET UP ELEMENT GEOMETRY

        NIC1=LNODS(NEL,1)
        NIC2=LNODS(NEL,2)                                             (3)
        NIC3=LNODS(NEL,3)

        X1=COORDS(NIC1,1)
        Y1=COORDS(NIC1,2)
        X2=COORDS(NIC2,1)                                             (4)
        Y2=COORDS(NIC2,2)
        X3=COORDS(NIC3,1)
        Y3=COORDS(NIC3,2)

        AREA=(X2*Y3-X3*Y2-X1*(Y3-Y2)+Y1*(X3-X2))/2.0                  (5)

        BI=Y2-Y3
        CI=X3-X2
        BJ=Y3-Y1                                                      (6)
        CJ=X1-X3
        BM=Y1-Y2
        CM=X2-X1

C       SET UP ELEMENT STIFFNESS MATRIX

        ELSTIF(1,1)=EC1*BI**2.0+EC12*CI**2.0
        ELSTIF(2,1)=(EC1*EC2+EC12)*BI*CI
        ELSTIF(2,2)=EC1*CI**2.0+EC12*BI**2.0
        ELSTIF(3,1)=EC1*BI*BJ+EC12*CI*CJ
        ELSTIF(3,2)=EC1*EC2*BJ*CI+EC12*BI*CJ
        ELSTIF(3,3)=EC1*BJ**2.0+EC12*CJ**2.0
        ELSTIF(4,1)=EC1*EC2*BI*CJ+EC12*BJ*CI
        ELSTIF(4,2)=EC1*CI*CJ+EC12*BI*BJ
        ELSTIF(4,3)=(EC1*EC2+EC12)*BJ*CJ
        ELSTIF(4,4)=EC1*CJ**2.0+EC12*BJ**2.0
        ELSTIF(5,1)=EC1*BI*BM+EC12*CI*CM                              (7)
        ELSTIF(5,2)=EC1*EC2*BM*CI+EC12*BI*CM
        ELSTIF(5,3)=EC1*BJ*BM+EC12*CJ*CM
        ELSTIF(5,4)=EC1*EC2*BM*CJ+EC12*BJ*CM
        ELSTIF(5,5)=EC1*BM**2.0+EC12*CM**2.0
        ELSTIF(6,1)=EC1*EC2*BI*CM+EC12*BM*CI
        ELSTIF(6,2)=EC1*CI*CM+EC12*BI*BM
        ELSTIF(6,3)=EC1*EC2*BJ*CM+EC12*BM*CJ
        ELSTIF(6,4)=EC1*CJ*CM+EC12*BJ*BM
        ELSTIF(6,5)=(EC1*EC2+EC12)*BM*CM
        ELSTIF(6,6)=EC1*CM**2.0+EC12*BM**2.0
```

```
      DO 40 I=1,5
      IP1=I+1
      DO 39 J=IP1,6
      ELSTIF(I,J)=ELSTIF(J,I)                                  (8)
   39 CONTINUE
   40 CONTINUE
      DO 42 I=1,6
      DO 41 J=1,6
      ELSTIF(I,J)=ELSTIF(I,J)*T/(4.0*AREA)                      (9)
   41 CONTINUE
   42 CONTINUE
```

(1) Calculate coefficients of the material property matrix.
(2) Element loop.
(3) Determine the node numbers forming the element.
(4) Nodal coordinates of the element nodes.
(5) Area of the element.
(6) Calculate variables b_i, c_i, b_j, c_j, b_m, c_m.
(7) Calculate the values of the coefficients in the lower triangle of the stiffness matrix using Table 2.1 - with the common multiplier T/(4*AREA) missing.
(8) Generate the upper triangle of the element stiffness using symmetry.
(9) Multiply by Thickness/(4*AREA) to obtain the correct element stiffness.

2.6 Assemblage of the structural stiffness matrix

The stiffness relationship for an element has been given by Eqn.(1.45) as

$$[k]\{\delta\} = \{P\}$$

For the triangular element in question, we have

$$\begin{bmatrix} k_{ii} & k_{ij} & k_{im} \\ k_{ji} & k_{jj} & k_{jm} \\ k_{mi} & k_{mj} & k_{mm} \end{bmatrix} \begin{Bmatrix} \delta_i \\ \delta_j \\ \delta_m \end{Bmatrix} = \begin{Bmatrix} P_i \\ P_j \\ P_m \end{Bmatrix}$$

in which all the terms are in fact sub-matrices.

Since a structure is made up of many elements, it follows that the structural stiffness matrix is also made up of a corresponding number of element stiffness matrices, and is given as

$$[K]\{\delta\} = \{F\}$$

where $[K]$ is the structural stiffness matrix, $\{F\}$ is a structural load matrix with a number of load cases, $\{\delta\}$ is a matrix of all nodal displacements.

The process of combining the element stiffness matrices together is called assembly, and the theory behind the assembly process will be explained by considering the structure shown in Fig.2.7(a), in which the triangular element definitions are:

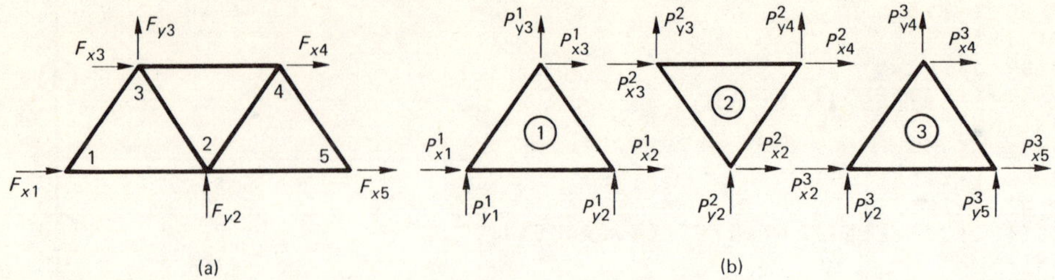

Fig.2.7 Structure and its component elements

Element number	Node i	Node j	Node m
1	1	2	3
2	2	4	3
3	2	5	4

The forces (including reactions) applied to the structure are F_{x1}, F_{y2}, F_{x3}, F_{y3}, F_{x4} and F_{x5}. The forces are in equilibrium and therefore although all of them have been shown acting in a positive sense, some of them are in fact negative. For simplication, the constraints are not shown in the figure.

If the structure is split up into its component elements and the external forces are divided between the appropriate elements (Fig.2.7(b)) such that equilibrium and compatibility are maintained at nodal, element and structural levels, then it is obvious that from joint equilibrium

$$\begin{aligned}
&\text{Node 1}\begin{cases} p^1_{x1} = F_{x1} \\ p^1_{y1} = 0 \end{cases}\\
&\text{Node 2}\begin{cases} p^1_{x2} + p^2_{x2} + p^3_{x2} = 0 \\ p^1_{y2} + p^2_{y2} + p^3_{y2} = F_{y2} \end{cases}\\
&\text{Node 3}\begin{cases} p^1_{x3} + p^2_{x3} = F_{x3} \\ p^1_{y3} + p^2_{y3} = F_{y3} \end{cases}\\
&\text{Node 4}\begin{cases} p^2_{x4} + p^3_{x4} = F_{x4} \\ p^2_{y4} + p^3_{y4} = 0 \end{cases}\\
&\text{Node 5}\begin{cases} p^3_{x5} = F_{x5} \\ p^3_{y5} = 0 \end{cases}
\end{aligned} \tag{2.29}$$

From element equilibrium and displacement compatibility condition we have

$$\left.\begin{array}{ll}
\begin{bmatrix} k^1_{ii} & k^1_{ij} & k^1_{im} \\ k^1_{ji} & k^1_{jj} & k^1_{jm} \\ k^1_{mi} & k^1_{mj} & k^1_{mm} \end{bmatrix} \begin{Bmatrix} \delta_1 \\ \delta_2 \\ \delta_3 \end{Bmatrix} = \begin{Bmatrix} P^1_1 \\ P^1_2 \\ P^1_3 \end{Bmatrix} & \text{for element 1} \\[2ex]
\begin{bmatrix} k^2_{ii} & k^2_{ij} & k^2_{im} \\ k^2_{ji} & k^2_{jj} & k^2_{jm} \\ k^2_{mi} & k^2_{mj} & k^2_{mm} \end{bmatrix} \begin{Bmatrix} \delta_2 \\ \delta_4 \\ \delta_3 \end{Bmatrix} = \begin{Bmatrix} P^2_2 \\ P^2_4 \\ P^2_3 \end{Bmatrix} & \text{for element 2} \\[2ex]
\begin{bmatrix} k^3_{ii} & k^3_{ij} & k^3_{im} \\ k^3_{ji} & k^3_{jj} & k^3_{jm} \\ k^3_{mi} & k^3_{mj} & k^3_{mm} \end{bmatrix} \begin{Bmatrix} \delta_2 \\ \delta_5 \\ \delta_4 \end{Bmatrix} = \begin{Bmatrix} P^3_2 \\ P^3_5 \\ P^3_4 \end{Bmatrix} & \text{for element 3}
\end{array}\right] \quad (2.30)$$

in which the superscripts refer to the element numbers, and k^1_{ii}, δ_1, P^1_1, etc. are submatrices involving x and y components. Substituting Eqn. (2.30) into Eqn.(2.29) gives

$$\left.\begin{array}{l}
k^1_{ii}\delta_1 + k^1_{ij}\delta_2 + k^1_{im}\delta_3 = P^1_1 = \begin{Bmatrix} F_{x1} \\ 0 \end{Bmatrix} \\[2ex]
k^1_{ji}\delta_1 + (k^1_{jj} + k^2_{ii} + k^3_{ii})\delta_2 + (k^1_{jm} + k^2_{im})\delta_3 + (k^2_{ij} + k^3_{im})\delta_4 \\
\quad + k^3_{ij}\delta_5 = P^1_2 + P^2_2 + P^3_2 = \begin{Bmatrix} 0 \\ F_{y2} \end{Bmatrix} \\[2ex]
k^1_{mi}\delta_1 + (k^1_{mj} + k^2_{mi})\delta_2 + (k^1_{mm} + k^2_{mm})\delta_3 + k^2_{mj}\delta_4 = P^1_3 + P^2_3 = \begin{Bmatrix} F_{x3} \\ F_{y3} \end{Bmatrix} \\[2ex]
(k^2_{ji} + k^3_{mi})\delta_2 + k^2_{jm}\delta_3 + (k^2_{jj} + k^3_{mm})\delta_4 + k^3_{mj}\delta_5 = P^2_4 + P^3_4 = \begin{Bmatrix} F_{x4} \\ 0 \end{Bmatrix} \\[2ex]
k^3_{ji}\delta_2 + k^3_{jm}\delta_4 + k^3_{jj}\delta_5 = P^3_5 = \begin{Bmatrix} F_{x5} \\ 0 \end{Bmatrix}
\end{array}\right] \quad (2.31)$$

Rewriting Eqn.(2.31) in matrix form

$$\begin{bmatrix} k^1_{ii} & k^1_{ij} & k^1_{im} & & \\ k^1_{ji} & k^1_{jj}+k^2_{ii}+k^3_{ii} & k^1_{jm}+k^2_{im} & k^2_{ij}+k^3_{im} & k^3_{ij} \\ k^1_{mi} & k^1_{mj}+k^2_{mi} & k^1_{mm}+k^2_{mm} & k^2_{mj} & \\ & k^2_{ji}+k^3_{mi} & k^2_{jm} & k^2_{jj}+k^3_{mm} & k^3_{mj} \\ & k^3_{ji} & & k^3_{jm} & k^3_{jj} \end{bmatrix} \begin{Bmatrix} \delta_1 \\ \delta_2 \\ \delta_3 \\ \delta_4 \\ \delta_5 \end{Bmatrix} = \begin{Bmatrix} F_1 \\ F_2 \\ F_3 \\ F_4 \\ F_5 \end{Bmatrix} \qquad (2.32)$$

Eqn.(2.31) is of course the expanded form of Eqn.(2.32) for the example structure.

The above process described the theoretical basis of combining the individual element equations to form the overall structural equations. For the computer an automatic assembly process yielding the same end results will be used. To effect this automatic assembling an element destination vector is established. For example the element definition vector for element 2 is (2, 4, 3) and for the simple solution subroutine to be presented in Section 2.9, this will become also the destination vector which will determine the order in which the element displacements are stored in the element displacement vector, i.e. the displacements are stored as $[\delta_2 \quad \delta_4 \quad \delta_3]^T$.

When the front solution technique is described in Chapter 3 the reader will observe that this destination vector is obtained through a more complicated procedure.

For the purposes of explanation only, the operation of the destination vector can be seen most easily if the vector is placed above and to the left of the element stiffness and alongside the element load vector. Thus for element 2 we have

$$\begin{array}{c} \\ 2 \\ 4 \\ 3 \end{array} \begin{array}{c} \begin{array}{ccc} 2 & 4 & 3 \end{array} \\ \begin{bmatrix} k^2_{ii} & k^2_{ij} & k^2_{im} \\ k^2_{ji} & k^2_{jj} & k^2_{jm} \\ k^2_{mi} & k^2_{mj} & k^2_{mm} \end{bmatrix} \end{array} \begin{Bmatrix} \delta_2 \\ \delta_4 \\ \delta_3 \end{Bmatrix} = \begin{Bmatrix} P^2_2 \\ P^2_4 \\ P^2_3 \end{Bmatrix} \begin{array}{c} 2 \\ 4 \\ 3 \end{array}$$

The structural destination vector is simply a vector of the order in which the displacements are stored in the structural displacement vector. Usually it is an array in ascending order and for this example it is (1, 2, 3, 4, 5). If this structural destination vector is now placed above and to the left of the structural stiffness matrix and alongside the structural load vector the required automatic assembly process is obtained by adding any one of the 2 × 2 submatrices in the element stiffness into the structural stiffness according to the location defined by the destination vectors above and to the left of the element stiffness. For example k^2_{jm} will be assembled into the 2 × 2 submatrix area defined by location (4,3) in the structural stiffness matrix. The row position, 4, is obtained from the vertical destination vector to the left of the element stiffness and the column position, 3, from the horizontal destination vector above the element stiffness.

In a similar manner the load vector is also assembled. For example P^2_4 has alongside it the coefficient 4 from the destination vector, and is therefore added into the structural load vector into the 2 × 1 subvector area that has alongside it a 4 from the structural destination vector.

The structural stiffness after assembly of element 2 only is shown in Eqn.(2.33).

$$
\begin{matrix} & 1 & 2 & 3 & 4 & 5 \\ 1 \\ 2 \\ 3 \\ 4 \\ 5 \end{matrix}
\begin{bmatrix} & & & & \\ & k^2_{ii} & k^2_{im} & k^2_{ij} & \\ & k^2_{mi} & k^2_{mm} & k^2_{mj} & \\ & k^2_{ji} & k^2_{jm} & k^2_{jj} & \\ & & & & \end{bmatrix}
\begin{Bmatrix} \delta_1 \\ \delta_2 \\ \delta_3 \\ \delta_4 \\ \delta_5 \end{Bmatrix}
=
\begin{Bmatrix} \\ P^2_2 \\ P^2_3 \\ P^2_4 \\ \\ \end{Bmatrix}
\begin{matrix} 1 \\ 2 \\ 3 \\ 4 \\ 5 \end{matrix}
\qquad (2.33)
$$

The reader can now check Eqn.(2.32) and confirm that the coefficients of element 2 are assembled into the structural coefficients as shown in Eqn.(2.33).

If elements 1 and 3 are assembled in the same manner Eqn.(2.32) would be reproduced exactly.

Coding for assembling structural stiffness matrix

```
C     ADD IN ELEMENT STIFFNESS INTO STRUCTURAL MATRIX

      DO 50 I=1,NNODZ
      DO 49 II=1,NVABZ
      ISTRST=(LNODS(NEL,I)-1)*NVABZ+II  ---------------- (1)
      IELEMT=(I-1)*NVABZ+II  --------------------------- (2)
      DO 48 J=1,NNODZ
      DO 47 JJ=1,NVABZ
      JSTRST=(LNODS(NEL,J)-1)*NVABZ+JJ  ---------------- (3)
      JELEMT=(J-1)*NVABZ+JJ  --------------------------- (4)
      SS(ISTRST,JSTRST)=SS(ISTRST,JSTRST)+ELSTIF(IELEMT,JELEMT)
   47 CONTINUE
   48 CONTINUE
   49 CONTINUE
   50 CONTINUE
  100 CONTINUE
```

(1),(3) Destination in structural stiffness matrix (ISTRST, JSTRST) of coefficient.

(2),(4) (IELEMT, JELEMT) from the element stiffness.

Element stiffness no. 1

Destination vector

	1		2		3	
1	.259E+05	-.118E+05	-.244E+05	.154E+05	-.148E+04	-.357E+04
	-.118E+05	.259E+05	.220E+05	-.578E+05	-.102E+05	.319E+05
2	-.244E+05	.220E+05	.642E+05	-.606E+04	-.397E+05	-.160E+05
	.154E+05	-.578E+05	-.606E+04	.170E+06	-.929E+04	-.112E+06
3	-.148E+04	-.102E+05	-.397E+05	-.929E+04	.412E+05	.195E+05
	-.357E+04	.319E+05	-.160E+05	-.112E+06	.195E+05	.803E+05

Structural stiffness matrix after assembly of element no. 1

Structural destination vector

	1		2		3		4		5		6	
1	.259E+05	-.118E+05	-.244E+05	.154E+05	0.	0.	0.	0.	-.148E+04	-.357E+04	0.	0.
	-.118E+05	.259E+05	.220E+05	-.578E+05	0.	0.	0.	0.	-.102E+05	.319E+05	0.	0.
2	-.244E+05	.220E+05	.642E+05	-.606E+04	0.	0.	0.	0.	-.397E+05	-.160E+05	0.	0.
	.154E+05	-.578E+05	-.606E+04	.170E+06	0.	0.	0.	0.	-.929E+04	-.112E+06	0.	0.
3	0.	0.	0.	0.	0.	0.	0.	0.	0.	0.	0.	0.
	0.	0.	0.	0.	0.	0.	0.	0.	0.	0.	0.	0.
4	0.	0.	0.	0.	0.	0.	0.	0.	0.	0.	0.	0.
	0.	0.	0.	0.	0.	0.	0.	0.	0.	0.	0.	0.
5	-.148E+04	-.102E+05	-.397E+05	-.929E+04	0.	0.	0.	0.	.412E+05	.195E+05	0.	0.
	-.357E+04	.319E+05	-.160E+05	-.112E+06	0.	0.	0.	0.	.195E+05	.803E+05	0.	0.
6	0.	0.	0.	0.	0.	0.	0.	0.	0.	0.	0.	0.
	0.	0.	0.	0.	0.	0.	0.	0.	0.	0.	0.	0.

Element stiffness no. 4

Destination vector

	3		6		5	
3	.120E+06	.293E+05	-.112E+06	-.160E+05	-.828E+04	-.133E+05
	.293E+05	.618E+05	-.929E+04	-.397E+05	-.200E+05	-.221E+05
6	-.112E+06	-.929E+04	.113E+06	-.404E+04	-.114E+04	.133E+05
	-.160E+05	-.397E+05	-.404E+04	.428E+05	.200E+05	-.305E+04
5	-.828E+04	-.200E+05	-.114E+04	.200E+05	.943E+04	0.
	-.133E+05	-.221E+05	.133E+05	-.305E+04	0.	.251E+05

Structural stiffness matrix after assembly of all four elements

Structural destination vector

	1		2		3		4		5		6	
1	.878E+05	.175E+05	-.244E+05	.154E+05	0.	0.	-.397E+05	-.929E+04	-.236E+05	-.236E+05	0.	0.
	.175E+05	.146E+06	.220E+05	-.578E+05	0.	0.	-.160E+05	-.112E+06	-.236E+05	.236E+05	0.	0.
2	-.244E+05	.220E+05	.234E+06	-.121E+05	-.578E+05	.154E+05	0.	0.	-.152E+06	-.252E+05	0.	0.
	.154E+05	-.578E+05	-.121E+05	.234E+06	.220E+05	-.244E+05	0.	0.	-.252E+05	-.152E+06	0.	0.
3	0.	0.	-.578E+05	.220E+05	.146E+06	.175E+05	0.	0.	.236E+05	-.236E+05	-.112E+06	-.160E+05
	0.	0.	.154E+05	-.244E+05	.175E+05	.878E+05	0.	0.	-.236E+05	-.236E+05	-.929E+04	-.397E+05
4	-.397E+05	-.160E+05	0.	0.	0.	0.	.428E+05	-.404E+04	-.305E+04	.200E+05	0.	0.
	-.929E+04	-.112E+06	0.	0.	0.	0.	-.404E+04	.113E+06	.133E+05	-.114E+04	0.	0.
5	-.236E+05	-.236E+05	-.152E+06	-.252E+05	.236E+05	-.236E+05	-.305E+04	.133E+05	.156E+06	.391E+05	-.114E+04	.200E+05
	-.236E+05	.236E+05	-.252E+05	-.152E+06	-.236E+05	-.236E+05	.200E+05	-.114E+04	.391E+05	.156E+06	.133E+05	-.305E+04
6	0.	0.	0.	0.	-.112E+06	-.929E+04	0.	0.	-.114E+04	.133E+05	.113E+06	-.404E+04
	0.	0.	0.	0.	-.160E+05	-.397E+05	0.	0.	.200E+05	-.305E+04	-.404E+04	.428E+05

2.7 Introduction of prescribed displacement conditions

The problem of nodal fixity was discussed briefly in Section 2.4. Such fixities include prescribed displacements which very often are zero displacements at rigid supports. Non-zero displacements are prescribed for support settlements and also for boundary points of local fine mesh analysis in which the displacements were first obtained through a coarse mesh analysis of a much larger domain.

There are a number of methods for prescribing displacements, and four of them will be described below. Method 4 is the method used throughout this text because of its simplicity in the computation of reactions, which should be computed and summed for important problems so as to ensure that equilibrium between external loads and reactions is satisfied.

The coding for calculating nodal reactions will be listed at the end of Section 2.9, since reactions can be calculated only after the determination of displacements.

For the purpose of discussion it will be assumed here that a matrix of size $N \times N$ is of the form

$$\begin{bmatrix} a_{11} & a_{12} & \cdots & a_{1n} & \cdots & a_{1N} \\ a_{21} & a_{22} & \cdots & a_{2n} & \cdots & a_{2N} \\ \vdots & \vdots & & \vdots & & \vdots \\ a_{n1} & a_{n2} & \cdots & a_{nn} & \cdots & a_{nN} \\ \vdots & \vdots & & \vdots & & \vdots \\ a_{N1} & a_{N2} & \cdots & a_{Nn} & \cdots & a_{NN} \end{bmatrix} \begin{Bmatrix} x_1 \\ x_2 \\ \vdots \\ x_n \\ \vdots \\ x_N \end{Bmatrix} = \begin{Bmatrix} b_1 \\ b_2 \\ \vdots \\ b_n \\ \vdots \\ b_N \end{Bmatrix} \tag{2.34}$$

in which x_i is a nodal displacement and b_i a nodal load, and that furthermore the displacement x_n has been prescribed to be equal to α, which can be a zero or non-zero value.

2.7.1 METHOD 1

In this method, the row vector corresponding to the prescribed variable is deleted while the column vector is multiplied by α and transferred to the right-hand side. The matrix is reduced in size to $(N-1) \times (N-1)$ and takes up the form of

$$\begin{bmatrix} a_{11} & a_{12} & \cdots & a_{1,n-1} & a_{1,n+1} & \cdots & a_{1N} \\ a_{21} & a_{22} & \cdots & a_{2,n-1} & a_{2,n+1} & \cdots & a_{2N} \\ \vdots & \vdots & & \vdots & \vdots & & \vdots \\ a_{n-1,1} & a_{n-1,2} & \cdots & a_{n-1,n-1} & a_{n-1,n+1} & \cdots & a_{n-1,N} \\ a_{n+1,1} & a_{n+1,2} & \cdots & a_{n+1,n-1} & a_{n+1,n+1} & \cdots & a_{n+1,N} \\ \vdots & \vdots & & \vdots & \vdots & & \vdots \\ a_{N1} & a_{N2} & \cdots & a_{N,n-1} & a_{N,n+1} & \cdots & a_{NN} \end{bmatrix} \begin{Bmatrix} x_1 \\ x_2 \\ \vdots \\ x_{n-1} \\ x_{n+1} \\ \vdots \\ x_N \end{Bmatrix} = \begin{Bmatrix} b_1 \\ b_2 \\ \vdots \\ b_{n-1} \\ b_{n+1} \\ \vdots \\ b_N \end{Bmatrix} - \alpha \begin{Bmatrix} a_{1n} \\ a_{2n} \\ \vdots \\ a_{n-1,n} \\ a_{n+1,n} \\ \vdots \\ a_{N,n} \end{Bmatrix} \tag{2.35}$$

This method is only used in manual computation because of the advantage of having a reduced matrix. For automatic computation the efforts involved in rearranging the coefficients far outweigh the gain achieved through the reduction in matrix size except perhaps for eigenvalue problems in which core storage is critical.

For the computation of reaction corresponding to the displacement x_n, the original row vector must be stored and then multiplied with the computed displacements, i.e. the reaction is given by

$$R_n = \sum_{i=1}^{N} a_{ni}x_i - b_n \tag{2.36}$$

For the more common case of $\alpha = 0$ it is of course not necessary to carry out the modification of the right-hand side as indicated in Eqn.(2.35).

2.7.2 METHOD 2

This method is simple to understand and is easier to implement when compared with Method 1. The diagonal coefficient corresponding to x_n is made unity and the rest of the coefficients in the row are then set equal to zero. To maintain symmetry in the matrix, the coefficients in the corresponding column are multiplied by α and transferred to the right-hand side, and after this the column (apart from the diagonal coefficient) is also set to zero. Lastly the coefficient b_n is replaced by α and the final form of the equations is shown below.

$$\begin{bmatrix} a_{11} & a_{12} & \cdots & a_{1,n-1} & 0 & a_{1,n+1} & \cdots & a_{1N} \\ a_{21} & a_{22} & \cdots & a_{2,n-1} & 0 & a_{2,n+1} & \cdots & a_{2N} \\ \vdots & \vdots & & \vdots & \vdots & \vdots & & \vdots \\ a_{n-1,1} & a_{n-1,2} & \cdots & a_{n-1,n-1} & 1 & a_{n-1,n+1} & \cdots & a_{n-1,N} \\ 0 & 0 & & 0 & 0 & 0 & & 0 \\ a_{n+1,1} & a_{n+1,2} & \cdots & a_{n+1,n-1} & 0 & a_{n+1,n+1} & \cdots & a_{n+1,N} \\ a_{N1} & a_{N2} & \cdots & a_{N,n-1} & 0 & a_{N,n+1} & \cdots & a_{NN} \end{bmatrix} \begin{Bmatrix} x_1 \\ x_2 \\ \vdots \\ x_{n-1} \\ x_n \\ x_{n+1} \\ x_N \end{Bmatrix} = \begin{Bmatrix} b_1 \\ b_2 \\ \vdots \\ b_{n-1} \\ \alpha \\ b_{n+1} \\ b_N \end{Bmatrix} - \alpha \begin{Bmatrix} a_{1n} \\ a_{2n} \\ \vdots \\ a_{n-1,n} \\ 0 \\ a_{n+1,n} \\ a_{N,n} \end{Bmatrix} \tag{2.37}$$

The reaction corresponding to x_n is obtained in the same way as indicated by Eqn.(2.36).

2.7.3 METHOD 3

Method 3 requires very few operations and is preferable to Method 4 for non-zero displacements and when reactions are not required. In this method the diagonal coefficient corresponding to x_n is multiplied by a very large number, say 10^{12}, and the load term b_n is replaced by $\alpha \times a_{nn} \times 10^{12}$. The modified equations are shown below.

$$\begin{bmatrix} a_{11} & a_{12} \cdots & a_{1n} & \cdots & a_{1N} \\ a_{21} & a_{22} & a_{2n} & & a_{2N} \\ \vdots & \vdots & \vdots & & \vdots \\ a_{n1} & a_{n2} & a_{nn}\times 10^{12} & \cdots & a_{nN} \\ \vdots & \vdots & \vdots & & \vdots \\ a_{N1} & a_{N2} & a_{Nn} & \cdots & a_{NN} \end{bmatrix} \begin{Bmatrix} x_1 \\ x_2 \\ \vdots \\ x_n \\ \vdots \\ x_N \end{Bmatrix} = \begin{Bmatrix} b_1 \\ b_2 \\ \vdots \\ \alpha\times a_{nn}\times 10^{12} \\ \vdots \\ b_N \end{Bmatrix} \tag{2.38}$$

Allowing for the fact that the coefficients in a row are approximately of the same order of magnitude, it is quite simple to see that the equation corresponding to x_n is in fact very nearly equivalent to $x_n = \alpha$. The reactions are obtained as before by storing the original equation and performing the calculations as given in Eqn.(2.36).

2.7.4 METHOD 4

This is the method that has been implemented in the simple program since the reaction can be obtained directly from the modified equation.

The method consists of adding a very large number, say 10^{50}, to the diagonal coefficients which physically corresponds to 'earthing' the structure with a very stiff spring. For a rigid support we obtain a very small displacement instead of an absolute zero and the reaction for that support can be computed directly as

Reaction = - (Big spring stiffness) × (Very small displacement)

For non-zero prescribed displacement it is only necessary to modify the right-hand side as is indicated in Method 2. The small displacement obtained in the solution must be replaced by the prescribed displacement before the displacement vector is used to calculate the element stresses.

This method will fail if the big spring stiffness is not significantly larger than the stiffness coefficients of the structure. However, with a value of 10^{50} this is unlikely to occur in practice.

The modified equations are given below for completeness.

$$\begin{bmatrix} a_{11} & a_{12} \cdots & a_{1n} & \cdots & a_{1N} \\ a_{21} & a_{22} \cdots & a_{2n} & \cdots & a_{2N} \\ \vdots & \vdots & \vdots & & \vdots \\ a_{n1} & a_{n2} \cdots & a_{nn}+10^{50} & \cdots & a_{nN} \\ \vdots & \vdots & \vdots & & \vdots \\ a_{N1} & a_{N2} \cdots & a_{Nn} & \cdots & a_{NN} \end{bmatrix} \begin{Bmatrix} x_1 \\ x_2 \\ \vdots \\ x_n \\ \vdots \\ x_N \end{Bmatrix} = \begin{Bmatrix} b_1 \\ b_2 \\ \vdots \\ b_n \\ \vdots \\ b_N \end{Bmatrix} - \alpha \begin{Bmatrix} a_{1n} \\ a_{2n} \\ \vdots \\ a_{nn} \\ \vdots \\ a_{Nn} \end{Bmatrix} \tag{2.39}$$

Coding for application of nodal fixities

```
C     APPLY FIXITIES


      I=0
      DO 200 NIC=1,MAXNOD
      DO 199 NVB=1,NVABZ
      I=I+1
      IF(NFIX(NIC,NVB).EQ.0) GO TO 199 ————— (1)
      SS(I,I)=1.0E+50+SS(I,I) ————— (2)
  199 CONTINUE
  200 CONTINUE
```

(1) Check if displacement prescribed.
(2) Add big spring stiffness to impose prescribed zero displacement.

Structural stiffness matrix after fixity imposed

.100E+51	.175E+05	-.244E+05	.154E+05	0.	0.	-.397E+05	-.929E+04	-.236E+05	-.236E+05	0.	0.	.293E+04
.175E+05	.146E+06	.220E+05	-.578E+05	0.	0.	-.160E+05	-.112E+06	-.236E+05	.236E+05	0.	0.	.141E+05
-.244E+05	.220E+05	.234E+06	-.121E+05	-.578E+05	.154E+05	0.	0.	-.152E+06	-.252E+05	0.	0.	.171E+05
.154E+05	-.578E+05	-.121E+05	.234E+06	.220E+05	-.244E+05	0.	0.	-.252E+05	-.152E+06	0.	0.	.171E+05
0.	0.	-.578E+05	.220E+05	.146E+06	.175E+05	0.	0.	.236E+05	-.236E+05	-.112E+06	-.160E+05	.141E+05
0.	0.	.154E+05	-.244E+05	.175E+05	.100E+51	0.	0.	-.236E+05	-.236E+05	-.929E+04	-.397E+05	.293E+04
-.397E+05	-.160E+05	0.	0.	0.	0.	.100E+51	-.404E+04	-.305E+04	.200E+05	0.	0.	0.
-.929E+04	-.112E+06	0.	0.	0.	0.	-.404E+04	.113E+06	.133E+05	-.114E+04	0.	0.	0.
-.236E+05	-.236E+05	-.152E+06	-.252E+05	.236E+05	-.236E+05	-.305E+04	.133E+05	.156E+06	.391E+05	-.114E+04	.200E+05	0.
-.236E+05	.236E+05	-.252E+05	-.152E+06	-.236E+05	-.236E+05	.200E+05	-.114E+04	.391E+05	.156E+06	.133E+05	-.305E+04	0.
0.	0.	0.	0.	-.112E+06	-.929E+04	0.	0.	-.114E+04	.133E+05	.113E+06	-.404E+04	0.
0.	0.	0.	0.	-.160E+05	-.397E+05	0.	0.	.200E+05	-.305E+04	-.404E+04	.100E+51	0.

2.8 Load matrix

Only two types of loading will be considered for this initial elementary program:

(i) concentrated loads applied at nodes;

(ii) uniformly distributed loads over an element edge.

Other types of loading such as gravity load, centrifugal loads, temperature loads, non-uniformly distributed loads, etc., will be described in Chapter 3 relating to isoparametric elements.

2.8.1 CONCENTRATED LOADS (NODAL POINT LOADS)

These are the easiest type of applied load to code in any finite element program. In the program described in this section the solution of equations is not carried out until all elements have been assembled into the structural stiffness matrix. The concentrated nodal loads can therefore be added into the structural load vector, SL, immediately after assembly of all the elements. Because there are two variables per node the load components X_N and Y_N applied to node N are added, or assembled, into the coefficients of SL associated with node N as listed below.

$X_N \rightarrow$ SL(N*2-1)

$Y_N \rightarrow$ SL(N*2)

X_N and Y_N are of course inputted as data.

2.8.2 UNIFORMLY DISTRIBUTED EDGE LOADS

This type of loading is somewhat more complicated to program than nodal point loads because it is usual in finite element programs to define only the magnitude of the distributed load per unit length, and the edge over which it is acting. The program then has to calculate the equivalent nodal forces, to determine the relative orientation of the edge to the global axes, and then transform the nodal forces from a direction perpendicular to the edge into components parallel to the global axes.

To facilitate this process, it is usual to define a locally orientated set of axes x'-y' which are parallel and perpendicular to the loaded edge respectively. For example if an element defined by nodes i, j, m in which edge i-j is loaded then the locally defined x' axis will be directed from i to j, as shown in Fig.2.8.

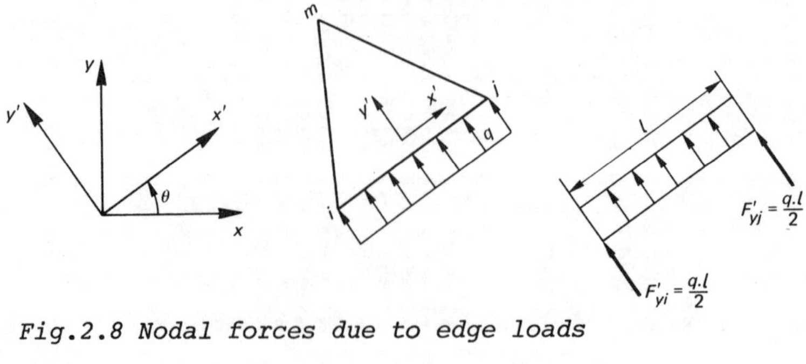

Fig.2.8 Nodal forces due to edge loads

To conform to the right-handed rule adopted in this text the y' axis must be in the direction shown. The distributed load can be either positive or negative depending on its direction.

The forces in the two systems are related to each other by

$$\begin{Bmatrix} F_x \\ F_y \end{Bmatrix} = \begin{bmatrix} \cos\theta & -\sin\theta \\ \sin\theta & \cos\theta \end{bmatrix} \begin{Bmatrix} F'_x \\ F'_y \end{Bmatrix}$$

or

$$\{F\} = [L]\{F'\}$$

in which $[L]$ is the transformation matrix between the two systems, and θ is positive as indicated in Fig.2.8.

The direction cosines are easily computed from geometrical considerations:

$$\cos\theta = (x_j - x_i)/\ell$$

$$\sin\theta = (y_j - y_i)/\ell$$

Thus the equivalent global components of forces at a node for the present example are

$$\begin{Bmatrix} F_x \\ F_y \end{Bmatrix} = \begin{bmatrix} \cos\theta & -\sin\theta \\ \sin\theta & \cos\theta \end{bmatrix} \begin{Bmatrix} 0 \\ q\ell/2 \end{Bmatrix} = (q\ell/2) \begin{Bmatrix} -\sin\theta \\ \cos\theta \end{Bmatrix}$$

Coding for assembly of applied nodal loads

```
C     ADD IN APPLIED LOADS

      DO 181 I=1,MAXNOD
      DO 180 J=1,NVABZ
      SL(2*I-2+J)=ALOAD(I,J)+SL(2*I-2+J)  ——————— (1)
  180 CONTINUE
  181 CONTINUE
```

(1) Add the nodal loads from the data into the load vector.

Load vector after assembly

```
 .293E+04  ┐
 .707E+04  ┘ (1)
 .100E+05  ┐
 .100E+05  ┘ (2)
 .707E+04
 .293E+04
0.
0.
0.
0.
0.
0.
```

(1) F_x and F_y for node 1.
(2) F_x and F_y for node 2.

2.9 Solution of simultaneous equations

The most frequently used method of solving a set of simultaneous equations is Gaussian elimination [4]. This method consists of eliminating the equations one at a time so that there is a corresponding reduction in the size of the modified matrix until finally the matrix is reduced to one equation containing only one variable.

The set of eliminated equations forms a triangular matrix and is used for backsubstitution purposes. By starting at the last equation and working backwards to the first equation, one variable will be determined at each step by using the variables which are already known.

2.9.1 GAUSSIAN ELIMINATION USING A FULL MATRIX

To demonstrate the procedure for this solution technique consider a set of four simultaneous equations represented by

$$[A]\{x\} = \{b\} \tag{2.40}$$

The reduction or triangularisation of $[A]$ is carried out by computing

$$\begin{aligned} a^*_{ij} &= a_{ij} - a_{is}a_{sj}/a_{ss} \\ b^*_i &= b_i - (a_{is}/a_{ss})b_s \end{aligned} \tag{2.41}$$

in which s refers to the equation being reduced and $i,j > s$.

The modified form of Eqn.(2.40) after the first step is

$$\begin{bmatrix} a_{11} & a_{12} & a_{13} & a_{14} \\ 0 & a_{22}-\dfrac{a_{21}}{a_{11}}a_{12} & a_{23}-\dfrac{a_{21}}{a_{11}}a_{13} & a_{24}-\dfrac{a_{21}}{a_{11}}a_{14} \\ 0 & a_{32}-\dfrac{a_{31}}{a_{11}}a_{12} & a_{33}-\dfrac{a_{31}}{a_{11}}a_{13} & a_{34}-\dfrac{a_{31}}{a_{11}}a_{14} \\ 0 & a_{42}-\dfrac{a_{41}}{a_{11}}a_{12} & a_{43}-\dfrac{a_{41}}{a_{11}}a_{13} & a_{44}-\dfrac{a_{41}}{a_{11}}a_{14} \end{bmatrix} \begin{Bmatrix} x_1 \\ x_2 \\ x_3 \\ x_4 \end{Bmatrix} = \begin{Bmatrix} b_1 \\ b_2-\dfrac{a_{21}}{a_{11}}b_1 \\ b_2-\dfrac{a_{31}}{a_{11}}b_1 \\ b_4-\dfrac{a_{41}}{a_{11}}b_1 \end{Bmatrix} \tag{2.42}$$

and the final form after the completion of all reduction steps is

$$\begin{bmatrix} a_{11} & a_{12} & a_{13} & a_{14} \\ 0 & a^*_{22} & a^*_{23} & a^*_{24} \\ 0 & 0 & a^{**}_{33} & a^{**}_{34} \\ 0 & 0 & 0 & a^{***}_{44} \end{bmatrix} \begin{Bmatrix} x_1 \\ x_2 \\ x_3 \\ x_4 \end{Bmatrix} = \begin{Bmatrix} b_1 \\ b^*_2 \\ b^{**}_3 \\ b^{***}_4 \end{Bmatrix} \tag{2.43}$$

where the asterisks are used to indicate the number of times each coefficient has been modified. The backsubstitution process is obvious and requires no comment. The computer coding for this technique is listed later on in this section. The whole matrix is stored in core and the coding is straightforward and easy to understand. However, the subroutine is inefficient in terms of core storage and computer time because no advantage has been taken of the symmetric and banded nature of a structural stiffness matrix.

2.9.2 GAUSSIAN ELIMINATION FOR A BANDED MATRIX OF CONSTANT WIDTH

The procedure described above uses a full matrix and is therefore inefficient both in terms of core requirement and the number of arithmetic operations in achieving a solution. Fortunately in many matrices generally met in finite element problems the non-zero coefficients are clustered near the diagonal, and outside this band of non-zero coefficients all the coefficients are of zero value. Such matrices are called banded matrices and for most structural problems in which the nodes have been numbered sensibly, the matrices are banded.

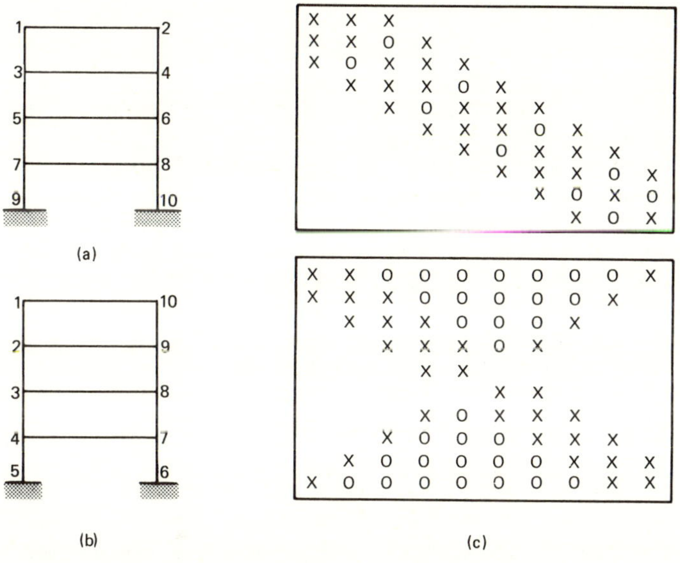

Fig.2.9 Typical frame and stiffness matrix (X = non-zero coefficient)

In Fig.2.9(a) the nodes of a typical frame structure have been numbered sensibly and the structural stiffness matrix can be seen to be a fairly compact band matrix. On the other hand, by numbering the nodes incorrectly as shown in Fig.2.9(b), the matrix is no longer banded and the entire stiffness matrix must be used in the solution process.

The symmetry and banded nature of the matrix can be utilized in developing a more efficient band matrix solution routine, in which only the upper half (or lower half) of the band is stored as a rectangular array, as is shown in Fig.2.10. It is obvious that since the core requirement is directly proportional to the half band width which is in turn related to the maximum difference between the nodal numbers of an

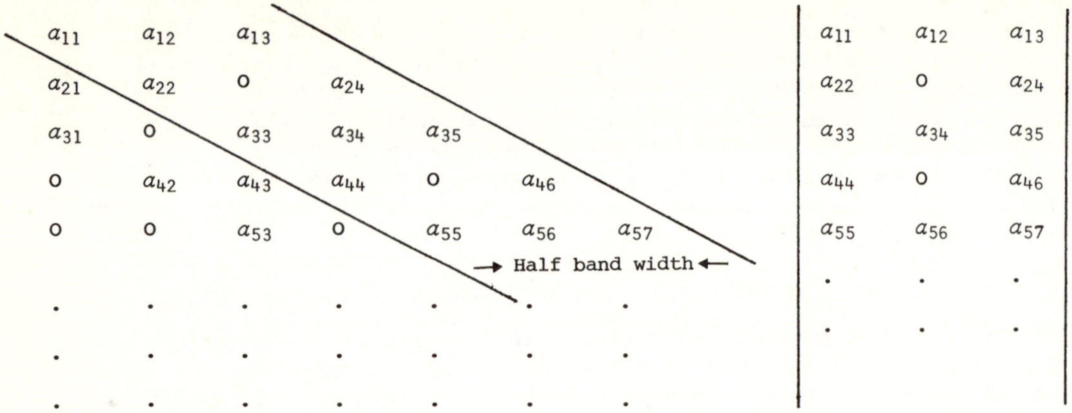

(a) Original band matrix

(b) Rectangular array storage of upper half of band

Fig.2.10 Storage scheme for a banded matrix

element, great care must be exercised in the numbering of nodes so that the difference between connecting nodes can be minimised.

Using the band solver routine, the storage requirement is reduced drastically since the half band width is very often only about 30 per cent of the matrix size. The execution time is also reduced because all the zero coefficients outside the band will not be operated upon.

This extra efficiency is gained at the expense of a slightly more complicated computer coding due to the shifting of the stiffness coefficients. The equivalent equation to Eqn.(2.41) is

$$a^*_{ij} = a_{ij} - \left(a_{s,(i-s+1)} / a_{s1}\right) a_{s,(j+1-s)} \tag{2.44}$$

with the limits

$$(s + 1) \leqslant i \leqslant (s + \text{HBW} - 1)$$

$$1 \leqslant j \leqslant \left(\text{HBW} - (i - s)\right)$$

where HBW = the half band width of the matrix. The computer listing of the band solver is given at the end of this chapter and includes modifications to the assembly, reduction and backsubstitution.

Coding for Gaussian reduction and backsubstitution

```
C       CARRY OUT GAUSSIAN REDUCTION

        M1=MAXNOD*NVABZ ———————————————— (1)
        M2=M1-1
        DO 300 I=1,M2 ————————————————— (2)
        II=I+1
        DO 290 K=II,M1 ———————————————— (3)
        FACT=SS(K,I)/SS(I,I) ——————————— (4)
```

```
      DO 280 J=II,M1
      SS(K,J)=SS(K,J)-FACT*SS(I,J)              (5)
  280 CONTINUE
      SS(K,I)=0.0
      SL(K)=SL(K)-FACT*SL(I)                    (6)
  290 CONTINUE
  300 CONTINUE

C     BACK SUBSTITUTE

      DO 400 I=1,M1                             (7)
      II=M1-I+1
      PIVOT=SS(II,II)                           (8)
      SS(II,II)=0.0
      DO 350 J=II,M1
      SL(II)=SL(II)-SS(II,J)*DISPL(J)           (9)
  350 CONTINUE
      DISPL(II)=SL(II)/PIVOT                    (10)
  400 CONTINUE

      WRITE(6,1018)
      DO 401 I=1,MAXNOD
      WRITE(6,1019) I,DISPL(2*I-1),DISPL(2*I)   (11)
  401 CONTINUE
```

Reduction

(1) M1 is the number of equations to be solved.

(2) Loop through each equation I in turn from 1 to M1-1 using each equation to reduce all equations below it.

(3) Loop to reduce all equations K below the equation I being used for reduction.

(4) Calculate the common multiplier a_{io}/a_{ss} from Eqn.(2.41) to avoid repetition of the same calculation.

(5) Modify all coefficients of equation K, the equation being reduced.

(6) Reduce the load term of equation K.

Backsubstitution

(7) Loop through all the reduced equations in turn from M1 back to 1 using each equation to calculate one displacement.

(8) For each equation, II, store the value of the diagonal coefficient, PIVOT, and then set the pivot to zero in the stored equation.

(9) Subtract from the load term the sum of the products of the displacements multiplied by the equation coefficients.

(10) Divide the result of (9) by the value of the pivot to calculate the displacement.

(11) Print out the displacements.

Structural stiffness matrix and load vector during the reduction procedure

There is no observable change in the stiffness matrix, other than in column 1, after reduction by the first equation because of the large diagonal coefficient in this equation. After reduction by the second equation the structural stiffness matrix and load vector contain

.100E+51	.175E+05	-.244E+05	.154E+05	0.	0.	-.397E+05	-.929E+04	-.236E+05	-.236E+05	0.	0.	.293E+04
0.	.146E+06	.220E+05	-.578E+05	0.	0.	-.160E+05	-.112E+06	-.236E+05	.236E+05	0.	0.	.707E+04
0.	0.	.231E+06	-.344E+04	-.578E+05	.154E+05	.240E+04	.169E+05	-.148E+06	-.288E+05	0.	0.	.894E+04
0.	0.	-.344E+04	.211E+06	.220E+05	-.244E+05	-.630E+04	-.443E+05	-.346E+05	-.143E+06	0.	0.	.128E+05
0.	0.	-.578E+05	.220E+05	.146E+06	.175E+05	0.	0.	.236E+05	-.236E+05	-.112E+06	-.160E+05	.707E+04
0.	0.	.154E+05	-.244E+05	.175E+05	.100E+51	0.	0.	-.236E+05	-.236E+05	-.929E+04	-.397E+05	.293E+04
0.	0.	.240E+04	-.630E+04	0.	0.	.100E+51	-.163E+05	-.562E+04	.226E+05	0.	0.	.771E+03
0.	0.	.169E+05	-.443E+05	0.	0.	-.163E+05	.274E+05	-.473E+04	.169E+05	0.	0.	.542E+04
0.	0.	-.148E+06	-.346E+05	.236E+05	-.236E+05	-.562E+04	-.473E+04	.152E+06	.428E+05	-.114E+04	.200E+05	.114E+04
0.	0.	-.288E+05	-.143E+06	-.236E+05	-.236E+05	.226E+05	.169E+05	.428E+05	.152E+06	.133E+05	-.305E+04	-.114E+04
0.	0.	0.	0.	-.112E+06	-.929E+04	0.	0.	-.114E+04	.133E+05	.113E+06	-.404E+04	0.
0.	0.	0.	0.	-.160E+05	-.397E+05	0.	0.	.200E+05	-.305E+04	-.404E+04	.100E+51	0.

After complete triangularisation the result is:

.100E+51	.175E+05	-.244E+05	.154E+05	0.	0.	-.397E+05	-.929E+04	-.236E+05	-.236E+05	0.	0.	.293E+04
0.	.146E+06	.220E+05	-.578E+05	0.	0.	-.160E+05	-.112E+06	-.236E+05	.236E+05	0.	0.	.707E+04
0.	0.	.231E+06	-.344E+04	-.578E+05	.154E+05	.240E+04	.169E+05	-.148E+06	-.288E+05	0.	0.	.894E+04
0.	0.	0.	.211E+06	.212E+05	-.242E+05	-.626E+04	-.440E+05	-.368E+05	-.143E+06	0.	0.	.129E+05
0.	0.	0.	0.	.130E+06	.238E+05	.123E+04	.863E+04	-.988E+04	-.165E+05	-.112E+06	-.160E+05	.801E+04
0.	0.	0.	0.	0.	.100E+51	-.110E+04	-.775E+04	-.161E+05	-.350E+05	.113E+05	-.368E+05	.235E+04
0.	0.	0.	0.	0.	0.	.100E+51	-.178E+05	-.507E+04	.188E+05	.106E+04	.151E+03	.985E+03
0.	0.	0.	0.	0.	0.	0.	.164E+05	-.885E+03	-.968E+04	.746E+04	.106E+04	.692E+04
0.	0.	0.	0.	0.	0.	0.	0.	.497E+05	-.232E+04	-.928E+04	.188E+05	.101E+05
0.	0.	0.	0.	0.	0.	0.	0.	0.	.440E+05	.310E+04	-.357E+04	.143E+05
0.	0.	0.	0.	0.	0.	0.	0.	0.	0.	.110E+05	-.145E+05	.465E+04
0.	0.	0.	0.	0.	0.	0.	0.	0.	0.	0.	.100E+51	.400E+04

Program output for nodal displacements

```
NODAL DISPLACEMENTS

              X COMP            Y COMP

   1          .00000            .45815
   2          .35490            .35490
   3          .45815            .00000
   4          .00000            .42169
   5          .29581            .29581
   6          .42169            .00000
```

Once the displacements have been determined the reactions at fixed nodes can be determined by multiplying the small displacement in the direction of the fixity by the big spring stiffness.

Coding for calculating nodal reactions

```
      WRITE(6,1017)
      DO 450 I=1,MAXNOD                          (1)
      II=0
      DO 440 J=1,NVABZ
      REACT(J)=0.0
      IF(NFIX(I,J).EQ.0) GO TO 440               (2)
      III=(I-1)*NVABZ+J                          (3)
      REACT(J)=-DISPL(III)*1.0E50                (4)
      II=1
  440 CONTINUE
      IF(II.EQ.1) WRITE(6,1025) I,REACT          (5)
  450 CONTINUE
```

(1) Loop through each of the nodes in turn.
(2) Check each fixity code for earthed displacement.
(3) Locate the associated displacement when an earthed displacement is located.
(4) Calculate the reaction.
(5) Print out reaction.

Program output for nodal reactions

```
NODAL REACTIONS

NODE       X COMP        Y COMP

   1    -.160E+05    0.
   3    0.           -.160E+05
   4    -.400E+04    0.
   6    0.           -.400E+04
```

2.10 Calculation of element stresses

Once the displacements have been obtained it is possible to compute the stresses of each element in turn by using Eqn.(1.38)

$$\{\sigma\} = [D][B]\{\delta\}$$

The matrices $[D]$ and $[B]$ for the constant strain element are given by Eqn.(2.22) and Eqn.(2.17) respectively and the computer coding for the whole process is listed below. The stress output for the example problem is given at the end of the computer listing.

Coding for calculating element stress

```
      CALCULATE STRESSES

      WRITE(6,1020)

      DO 602 I=1,3                                              ─┐
      DO 601 J=1,3                                               │
      D(I,J)=0.0                                                 (1)
  601 CONTINUE                                                   │
  602 CONTINUE                                                  ─┤
                                                                 │
      D(1,1)=EC1                                                 │
      D(1,2)=EC1*EC2                                             │
      D(2,1)=EC1*EC2                                             (2)
      D(2,2)=EC1                                                 │
      D(3,3)=EC12                                               ─┘

      DO 700 NEL=1,MAXNEL

      DO 604 I=1,3                                              ─┐
      DO 603 J=1,6                                               │
      B(I,J)=0.0                                                 (3)
  603 CONTINUE                                                   │
  604 CONTINUE                                                  ─┤
                                                                 │
      NIC1=LNODS(NEL,1)                                          │
      NIC2=LNODS(NEL,2)                                          (4)
      NIC3=LNODS(NEL,3)                                          │
                                                                ─┤
      X1=COORDS(NIC1,1)                                          │
      Y1=COORDS(NIC1,2)                                          │
      X2=COORDS(NIC2,1)                                          (5)
      Y2=COORDS(NIC2,2)                                          │
      X3=COORDS(NIC3,1)                                          │
      Y3=COORDS(NIC3,2)                                         ─┘

      A=(X2*Y3-X3*Y2-X1*(Y3-Y2)+Y1*(X3-X2))/2.0 ──────────────── (6)
```

```
      B(1,1)=(Y2-Y3)/(2.0*A)
      B(1,3)=(Y3-Y1)/(2.0*A)
      B(2,4)=(X1-X3)/(2.0*A)                                  (7)
      B(2,2)=(X3-X2)/(2.0*A)
      B(1,5)=(Y1-Y2)/(2.0*A)
      B(2,6)=(X2-X1)/(2.0*A)
      DO 605 I=1,3
      B(3,2*I-1)=B(2,2*I)                                     (8)
      B(3,2*I)=B(1,2*I-1)
  605 CONTINUE

      DO 608 I=1,3
      DO 607 J=1,6
      DB(I,J)=0.0
      DO 606 K=1,3                                            (9)
      DB(I,J)=DB(I,J)+D(I,K)*B(K,J)
  606 CONTINUE
  607 CONTINUE
  608 CONTINUE

      DO 609 I=1,3
      II=2*LNODS(NEL,I)-1
      JJ=2*LNODS(NEL,I)                                       (10)
      DISP(2*I-1)=DISPL(II)
      DISP(2*I)=DISPL(JJ)
  609 CONTINUE

      DO 620 I=1,3
      DO 619 J=1,6                                            (11)
      STRESS(NEL,I)=STRESS(NEL,I)+DB(I,J)*DISP(J)
  619 CONTINUE
  620 CONTINUE
      WRITE(6,1021) NEL,(STRESS(NEL,J),J=1,3)
  700 CONTINUE
```

(1) Initialize the material property matrix [D].
(2) Set up the [D] matrix.
(3) Initialize the strain matrix [B].
(4) Determine the node numbers forming the element.
(5) Nodal coordinates of the element nodes.
(6) Area of the element.
(7) Set up the first two rows of the [B] matrix.
(8) Set up the third row of [B] using the values of the derivatives calculated in (7).
(9) Multiply [D][B].
(10) Set up the element displacement vector DISP( ).
(11) Calculate the element stresses by multiplying the result of (9) by DISP.

Program output for element stresses

ELEMENT STRESSES

	SIGMA X-X	SIGMA Y-Y	TAU X-Y
1	161.18303	32.83678	-125.05258
2	139.02280	-1.70267	-29.77643
3	32.83678	161.18303	-125.05258
4	-1.70267	139.02280	-29.77643

Stress calculation for element no. 1

$$[D] = \begin{bmatrix} .107E+06 & .267E+05 & 0. \\ .267E+05 & .107E+06 & 0. \\ 0. & 0. & .400E+05 \end{bmatrix}$$

$$[B] = \begin{bmatrix} -.500E-02 & 0. & .858E-03 & 0. & .414E-02 & 0. \\ 0. & .500E-02 & 0. & -.150E-01 & 0. & .100E-01 \\ .500E-02 & -.500E-02 & -.150E-01 & .858E-03 & .100E-01 & .414E-02 \end{bmatrix}$$

$$[D][B] = \begin{bmatrix} -.533E+03 & .133E+03 & .915E+02 & -.400E+03 & .442E+03 & .267E+03 \\ -.133E+03 & .533E+03 & .229E+02 & -.160E+04 & .110E+03 & .107E+04 \\ .200E+03 & -.200E+03 & -.600E+03 & .343E+02 & .400E+03 & .166E+03 \end{bmatrix}$$

$$\{\delta\} = \begin{Bmatrix} .160E-45 \\ .458E+00 \\ .355E+00 \\ .355E+00 \\ .296E+00 \\ .296E+00 \end{Bmatrix}$$

$$\{\delta\} = [D][B]\{\delta\} = \begin{Bmatrix} 161.18303 \\ 32.83678 \\ -125.05258 \end{Bmatrix}$$

The theoretical stress vector is $\begin{Bmatrix} 143.14 \\ 16.86 \\ -136.28 \end{Bmatrix}$. The discrepancy is due to the coarseness of the mesh.

Format statements

```
1001 FORMAT(4I5)
1002 FORMAT(1H1,5X,
    1*TOTAL NUMBER OF ELEMENTS =*,I5,/,25X,*NODES =*,I5,
    1/,25X,*LOADS =*,I5,/,22X,*FIXITIES =*,I5)
1003 FORMAT(///,5X,*ELEMENT DEFINITIONS*,/,5X,*ELEMENT*,2X,*NODE1*,2X,
    1*NODE2*,2X,*NODE3*,/)
1004 FORMAT(3I5)
1005 FORMAT(5X,I5,3(2X,I5))
1006 FORMAT(///,5X,*NODAL CO-ORDINATES*,/,5X,*NODE*,6X,*X*,9X,*Y*,/)
1007 FORMAT(I5,2(5X,F10.6))
1008 FORMAT(5X,I4,2F10.3)
1010 FORMAT(I5,2(5X,F10.3))
1011 FORMAT(3I5)
1009 FORMAT(///,5X,*APPLIED LOADS AND FIXITIES*,/,5X,*NODE*,5X,
    1*APPLIED LOADS*,8X,*FIXITIES*,/)
1012 FORMAT(5X,I4,3X,2E10.3,2I5)
1013 FORMAT(3E10.3)
1014 FORMAT(///,5X,*MATERIAL PROPERTIES*,//,5X,*YOUNGS MODULUS FOR *, *
    1MATERIAL  =*,E10.3,/,5X,*POISSONS RATIO FOR MATERIAL  =*,
    1E10.3,/,5X,*CONSTANT ELEMENT THICKNESS    =*,E10.3,
    1//,5X,*DATA COMPLETE*)
1017 FORMAT(1H1,5X,*NODAL REACTIONS*,///,
    1         6X,4HNODE,6X,6HX COMP,6X,6HY COMP,/)
1018 FORMAT(1H1,///,5X,*NODAL DISPLACEMENTS*,///,
    1              19X,6HX COMP,9X,6HY COMP,//)
1019 FORMAT(5X,I5,5X,F10.5,5X,F10.5)
1020 FORMAT(1H1,///,5X,*ELEMENT STRESSES*,///,21X,*SIGMA X-X*,11X,
    1*SIGMA Y-Y*,13X,*TAU X-Y*,/)
1021 FORMAT(5X,I5,5X,3(F15.5,5X),/)
1025 FORMAT(I10,2(2X,E10.3))


      STOP
      END
```

Complete listing of the program with the band solver implemented

```
      PROGRAM SIMPLE(INPUT,OUTPUT,TAPE5=INPUT,TAPE6=OUTPUT)


      DIMENSION LNODS(65,3),COORDS(50,2),ALOAD(50,2),NFIX(50,2)
      DIMENSION ELSTIF(6,6),SS(90,90)
      DIMENSION SL(90),D(3,3),B(3,6),DB(3,6),DISPL(90),DISP(6)
      DIMENSION STRESS(65,3)
      DIMENSION REACT(2)
      INTEGER HBW


C     NVABZ IS THE NUMBER OF VARIABLES PER NODE
C     NNODZ IS THE NUMBER OF NODES PER ELEMENT


      HBW=0
      NVABZ=2
      NNODZ=3



C     READ IN DATA


      READ(5,1001) MAXNEL,MAXNOD,MAXLOD,MAXFIX
      WRITE(6,1002) MAXNEL,MAXNOD,MAXLOD,MAXFIX
```

```
C       INITIALIZE VECTORS AND MATRICES

        MAXVAR=MAXNOD*NVABZ
        DO 2 I=1,MAXNOD
        DO 1 J=1,2
        ALOAD(I,J)=0.0
        NFIX(I,J)=0.0
      1 CONTINUE
      2 CONTINUE
        DO 4 I=1,MAXVAR
        SL(I)=0.0
        DISPL(I)=0.0
        DO 3 J=1,MAXVAR
        SS(I,J)=0.0
      3 CONTINUE
      4 CONTINUE
        DO 6 I=1,MAXNEL
        DO 5 J=1,3
        STRESS(I,J)=0.0
      5 CONTINUE
      6 CONTINUE

C       READ IN ELEMENT DEFINITIONS

        WRITE(6,1003)
        DO 11 NEL=1,MAXNEL
        READ(5,1004) (LNODS(NEL,I),I=1,NNODZ)
        WRITE(6,1005) NEL,(LNODS(NEL,I),I=1,NNODZ)
        NIC1=LNODS(NEL,1)
        NIC2=LNODS(NEL,2)
        NIC3=LNODS(NEL,3)
        IF(IABS(NIC1-NIC2).GT.HBW) HBW=IABS(NIC1-NIC2)
        IF(IABS(NIC2-NIC3).GT.HBW) HBW=IABS(NIC2-NIC3)
        IF(IABS(NIC3-NIC1).GT.HBW) HBW=IABS(NIC3-NIC1)
     11 CONTINUE
        HBW=(HBW+1)*NVABZ

C       READ IN NODAL CO-ORDINATES

        WRITE(6,1006)
        DO 12 NOD=1,MAXNOD
        READ(5,1007) NIC,(COORDS(NIC,I),I=1,NVABZ)
        WRITE(6,1008) NIC,(COORDS(NIC,I),I=1,NVABZ)
     12 CONTINUE

C       READ IN APPLIED LOADS AND FIXITIES

        DO 14 J=1,MAXLOD
        READ(5,1010) NIC,(ALOAD(NIC,I),I=1,NVABZ)
     14 CONTINUE
        DO 15 J=1,MAXFIX
        READ(5,1011) NIC,(NFIX(NIC,I),I=1,NVABZ)
     15 CONTINUE

        WRITE(6,1009)
        DO 16 I=1,MAXNOD
        WRITE(6,1012) I,(ALOAD(I,J),J=1,NVABZ),(NFIX(I,J),J=1,NVABZ)
     16 CONTINUE

C       READ IN ELASTIC PROPERTIES

        READ(5,1013) YM,PR,T
        WRITE(6,1014) YM,PR,T
```

```
C     SET UP STRUCTURAL STIFFNESS MATRIX

      EC1=YM/(1-PR*PR)
      EC2=PR
      EC12=EC1*(1-EC2)/2.0


      DO 100 NEL=1,MAXNEL

C     SET UP ELEMENT GEOMETRY

      NIC1=LNODS(NEL,1)
      NIC2=LNODS(NEL,2)
      NIC3=LNODS(NEL,3)

      X1=COORDS(NIC1,1)
      Y1=COORDS(NIC1,2)
      X2=COORDS(NIC2,1)
      Y2=COORDS(NIC2,2)
      X3=COORDS(NIC3,1)
      Y3=COORDS(NIC3,2)

      AREA=(X2*Y3-X3*Y2-X1*(Y3-Y2)+Y1*(X3-X2))/2.0

      BI=Y2-Y3
      CI=X3-X2
      BJ=Y3-Y1
      CJ=X1-X3
      BM=Y1-Y2
      CM=X2-X1


C     SET UP ELEMENT STIFFNESS MATRIX

      ELSTIF(1,1)=EC1*BI**2.0+EC12*CI**2.0
      ELSTIF(2,1)=(EC1*EC2+EC12)*BI*CI
      ELSTIF(2,2)=EC1*CI**2.0+EC12*BI**2.0
      ELSTIF(3,1)=EC1*BI*BJ+EC12*CI*CJ
      ELSTIF(3,2)=EC1*EC2*BJ*CI+EC12*BI*CJ
      ELSTIF(3,3)=EC1*BJ**2.0+EC12*CJ**2.0
      ELSTIF(4,1)=EC1*EC2*BI*CJ+EC12*BJ*CI
      ELSTIF(4,2)=EC1*CI*CJ+EC12*BI*BJ
      ELSTIF(4,3)=(EC1*EC2+EC12)*BJ*CJ
      ELSTIF(4,4)=EC1*CJ**2.0+EC12*BJ**2.0
      ELSTIF(5,1)=EC1*BI*BM+EC12*CI*CM
      ELSTIF(5,2)=EC1*EC2*BM*CI+EC12*BI*CM
      ELSTIF(5,3)=EC1*BJ*BM+EC12*CJ*CM
      ELSTIF(5,4)=EC1*EC2*BM*CJ+EC12*BJ*CM
      ELSTIF(5,5)=EC1*BM**2.0+EC12*CM**2.0
      ELSTIF(6,1)=EC1*EC2*BI*CM+EC12*BM*CI
      ELSTIF(6,2)=EC1*CI*CM+EC12*BI*BM
      ELSTIF(6,3)=EC1*EC2*BJ*CM+EC12*BM*CJ
      ELSTIF(6,4)=EC1*CJ*CM+EC12*BJ*BM
      ELSTIF(6,5)=(EC1*EC2+EC12)*BM*CM
      ELSTIF(6,6)=EC1*CM**2.0+EC12*BM**2.0

      DO 40 I=1,5
      IP1=I+1
      DO 39 J=IP1,6
      ELSTIF(I,J)=ELSTIF(J,I)
   39 CONTINUE
   40 CONTINUE
```

```
      DO 42 I=1,6
      DO 41 J=1,6
      ELSTIF(I,J)=ELSTIF(I,J)*T/(4.0*AREA)
   41 CONTINUE
   42 CONTINUE

C     ADD IN ELEMENT STIFFNESS INTO STRUCTURAL MATRIX

      DO 50 I=1,NNODZ
      DO 49 II=1,NVABZ
      ISTRST=(LNODS(NEL,I)-1)*NVABZ+II
      IELEMT=(I-1)*NVABZ+II
      DO 48 J=1,NNODZ
      DO 47 JJ=1,NVABZ
      JSTRST=(LNODS(NEL,J)-1)*NVABZ+JJ
      JELEMT=(J-1)*NVABZ+JJ
      IF(ISTRST.GT.JSTRST) GO TO 47
      JSTRST=JSTRST-ISTRST+1
      SS(ISTRST,JSTRST)=SS(ISTRST,JSTRST)+ELSTIF(IELEMT,JELEMT)
   47 CONTINUE
   48 CONTINUE
   49 CONTINUE
   50 CONTINUE
  100 CONTINUE

C     ADD IN APPLIED LOADS

      DO 181 I=1,MAXNOD
      DO 180 J=1,NVABZ
      SL(2*I-2+J)=ALOAD(I,J)+SL(2*I-2+J)
  180 CONTINUE
  181 CONTINUE

C     APPLY FIXITIES

      I=0
      DO 200 NIC=1,MAXNOD
      DO 199 NVB=1,NVABZ
      I=I+1
      IF(NFIX(NIC,NVB).EQ.0) GO TO 199
      SS(I,1)=SS(I,1)+1.0E+50
  199 CONTINUE
  200 CONTINUE

C     CARRY OUT GAUSSIAN REDUCTION

      M1=MAXNOD*NVABZ
      M2=M1-1
      DO 300 I=1,M2
      I1=I+1
      I2=I+HBW-1
```

```
      IF(I2.GT.M1) I2=M1
      DO 290 K=I1,I2
      FACT=SS(I,K-I+1)/SS(I,1)
      NSF1=HBW-K+I
      DO 280 J=1,NSF1
      SS(K,J)=SS(K,J)-FACT*SS(I,J+K-I)
  280 CONTINUE
      SL(K)=SL(K)-FACT*SL(I)
  290 CONTINUE
  300 CONTINUE


C     BACK SUBSTITUTE

      DO 400 I=1,M1
      II=M1-I+1
      PIVOT=SS(II,1)
      SS(II,1)=0.0
      DO 350 J=1,HBW
      K=II+J-1
      IF(K.GT.M1) GO TO 350
      SL(II)=SL(II)-SS(II,J)*DISPL(K)
  350 CONTINUE
      DISPL(II)=SL(II)/PIVOT
  400 CONTINUE

      WRITE(6,1018)
      DO 401 I=1,MAXNOD
      WRITE(6,1019) I,DISPL(2*I-1),DISPL(2*I)
  401 CONTINUE

      WRITE(6,1017)
      DO 450 I=1,MAXNOD
      II=0
      DO 440 J=1,NVABZ
      REACT(J)=0.0
      IF(NFIX(I,J).EQ.0) GO TO 440
      III=(I-1)*NVABZ+J
      REACT(J)=-DISPL(III)*1.0E50
      II=1
  440 CONTINUE
      IF(II.EQ.1) WRITE(6,1025) I,REACT
  450 CONTINUE

C     CALAULATE STRESSES

      WRITE(6,1020)


      DO 602 I=1,3
      DO 601 J=1,3
      D(I,J)=0.0
  601 CONTINUE
  602 CONTINUE

      D(1,1)=EC1
      D(1,2)=EC1*EC2
      D(2,1)=EC1*EC2
      D(2,2)=EC1
      D(3,3)=EC12

      DO 700 NEL=1,MAXNEL

      DO 604 I=1,3
      DO 603 J=1,6
      B(I,J)=0.0
  603 CONTINUE
  604 CONTINUE
```

```
      NIC1=LNODS(NEL,1)
      NIC2=LNODS(NEL,2)
      NIC3=LNODS(NEL,3)

      X1=COORDS(NIC1,1)
      Y1=COORDS(NIC1,2)
      X2=COORDS(NIC2,1)
      Y2=COORDS(NIC2,2)
      X3=COORDS(NIC3,1)
      Y3=COORDS(NIC3,2)

      A=(X2*Y3-X3*Y2-X1*(Y3-Y2)+Y1*(X3-X2))/2.0

      B(1,1)=(Y2-Y3)/(2.0*A)
      B(1,3)=(Y3-Y1)/(2.0*A)
      B(2,4)=(X1-X3)/(2.0*A)
      B(2,2)=(X3-X2)/(2.0*A)
      B(1,5)=(Y1-Y2)/(2.0*A)
      B(2,6)=(X2-X1)/(2.0*A)
      DO 605 I=1,3
      B(3,2*I-1)=B(2,2*I)
      B(3,2*I)=B(1,2*I-1)
  605 CONTINUE

      DO 608 I=1,3
      DO 607 J=1,6
      DB(I,J)=0.0
      DO 606 K=1,3
      DB(I,J)=DB(I,J)+D(I,K)*B(K,J)
  606 CONTINUE
  607 CONTINUE
  608 CONTINUE

      DO 609 I=1,3
      II=2*LNODS(NEL,I)-1
      JJ=2*LNODS(NEL,I)
      DISP(2*I-1)=DISPL(II)
      DISP(2*I)=DISPL(JJ)
  609 CONTINUE

      DO 620 I=1,3
      DO 619 J=1,6
      STRESS(NEL,I)=STRESS(NEL,I)+DB(I,J)*DISP(J)
  619 CONTINUE
  620 CONTINUE
      WRITE(6,1021) NEL,(STRESS(NEL,J),J=1,3)
  700 CONTINUE

 1001 FORMAT(4I5)
 1002 FORMAT(1H1,5X,
     1*TOTAL NUMBER OF ELEMENTS =*,I5,/,25X,*NODES =*,I5,
     1/,25X,*LOADS =*,I5,/,22X,*FIXITIES =*,I5)
 1003 FORMAT(///,5X,*ELEMENT DEFINITIONS*,/,5X,*ELEMENT*,2X,*NODE1*,2X,
     1*NODE2*,2X,*NODE3*,/)
 1004 FORMAT(3I5)
 1005 FORMAT(5X,I5,3(2X,I5))
 1006 FORMAT(///,5X,*NODAL CO-ORDINATES*,/,5X,*NODE*,6X,*X*,9X,*Y*,/)
 1007 FORMAT(I5,2(5X,F10.6))
 1008 FORMAT(5X,I4,2F10.3)
 1010 FORMAT(I5,2(5X,F10.3))
 1011 FORMAT(3I5)
 1009 FORMAT(///,5X,*APPLIED LOADS AND FIXITIES*,/,5X,*NODE*,5X,
     1*APPLIED LOADS*,8X,*FIXITIES*,/)
 1012 FORMAT(5X,I4,3X,2E10.3,2I5)
```

```
1013 FORMAT(3E10.3)
1014 FORMAT(///,5X,*MATERIAL PROPERTIES*,//,5X,*YOUNGS MODULUS FOR *,
    1MATERIAL  =*,E10.3,/,5X,*POISSONS RATIO FOR MATERIAL  =*,
    1E10.3,/,5X,*CONSTANT ELEMENT THICKNESS   =*,E10.3,
    1//,5X,*DATA COMPLETE*)
1017 FORMAT(1H1,5X,*NODAL REACTIONS*,///,
    1         6X,4HNODE,6X,6HX COMP,6X,6HY COMP,/)
1018 FORMAT(1H1,///,5X,*NODAL DISPLACEMENTS*,///,
    1             19X,6HX COMP,9X,6HY COMP,//)
1019 FORMAT(5X,I5,5X,F10.5,5X,F10.5)
1020 FORMAT(1H1,///,5X,*ELEMENT STRESSES*,///,21X,*SIGMA X-X*,11X,
    1*SIGMA Y-Y*,13X,*TAU X-Y*,/)
1021 FORMAT(5X,I5,5X,3(F15.5,5X),/)
1025 FORMAT(I10,2(2X,E10.3))


      STOP
      END
```

References

1. M.J. Turner, R.W. Clough, H.C. Martin and L.J. Topp. Stiffness and deflection analysis of complex structures. *Journal of Aeronautical Science*, 23, 805-823, 1956.
2. E.P. Popov. *Mechanics of Materials*, 2nd ed. Englewood Cliffs, N.J., Prentice-Hall, 1976.
3. S.H. Crandall. *Engineering Analysis*. New York, McGraw-Hill Book Co., 1956.
4. V.N. Faddeeva. *Computational Methods of Linear Algebra*. New York, Dover Publications, 1956.

3 Quadratic Isoparametric Element for Plane Elasticity

3.1 Introduction

The constant strain element presented in Chapter 2 is the lowest order element available, and an element with somewhat better characteristics is a quadrilateral element with four nodes. At first sight it seems possible to apply the displacement functions of a rectangular element (Table 1.1) directly to the quadrilateral element, such that

$$\begin{aligned} u &= A_1 + A_2x + A_3y + A_4xy \\ v &= B_1 + B_2x + B_3y + B_4xy \end{aligned} \qquad (3.1a)$$

However, upon closer examination it will be found that this is unsuitable since the displacements will vary parabolically along the edges which are in general not parallel to the x and y axes. With only two nodes along an edge it is no longer possible to guarantee the compatibility of displacements between adjacent elements.

Eqn.(3.1a) can be rewritten in non-dimensional form as

$$u = A_1 + A_2x/a + A_3y/b + A_4xy/ab$$

or

$$\begin{aligned} u &= A_1 + A_2\xi + A_3\eta + A_4\xi\eta \\ v &= B_1 + B_2\xi + B_3\eta + B_4\xi\eta \end{aligned} \qquad (3.1b)$$

The coordinates in ξ and η in Eqn.(3.1b) vary between -1 and +1, as shown in Fig.3.1(a). It can be seen that it is possible to use another set of curvilinear coordinates for quadrilateral element such that ξ and η, while not simply equal to x/a and y/b respectively as in Eqn. (3.1b) and is to be explained in detail subsequently, will nevertheless take up unit values at the inclined edges (Fig.3.1(b)). It should be noted that the ξ and η axes will bisect the edges in each case.

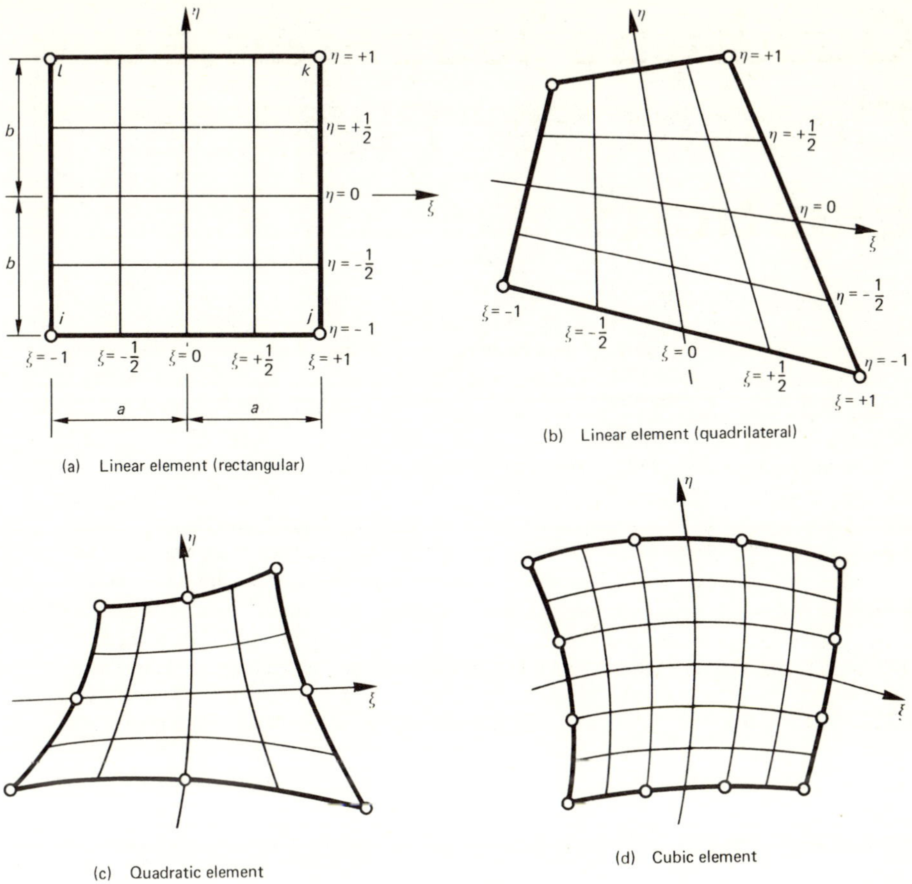

Fig.3.1 Cartesian and curvilinear coordinates

3.2 Isoparametric concept [1]

The displacement field for a two-dimensional element of general shape is given by

$$\begin{aligned} u &= N_1 u_1 + N_2 u_2 + N_3 u_3 + \ldots + N_r u_r \\ &= \sum N_i u_i \\ v &= N_1 v_1 + N_2 v_2 + N_3 v_3 + \ldots + N_r v_r \\ &= \sum N_i v_i \end{aligned} \tag{3.2}$$

in which N_i is a function of the curvilinear coordinates ξ and η, and r is the number of nodes.

The coordinates x and y inside the element domain can be described in a similar way by

$$x = M_1x_1 + M_2x_2 + M_3x_3 + \ldots + M_rx_r$$
$$= \sum M_ix_i$$
$$y = M_1y_1 + M_2y_2 + M_3y_3 + \ldots + M_ry_r \qquad (3.3)$$
$$= \sum M_iy_i$$

in which M_i is also a function of ξ and η.

For the particular case in which N_i and M_i are identical, i.e. the shape functions defining the displacement fields and geometry are the same, the element is termed isoparametric. Eqn.(3.3) provides the relationship between the Cartesian and the curvilinear coordinate systems, so that appropriate transformations can be carried out in the stiffness formulation.

There are a number of interesting points concerning the isoparametric concept and they are discussed below.

(i) The isoparametric concept allows any arbitrary geometry to be closely approximated, thereby minimizing any error associated with modelling the geometry and without resorting to the use of a fine mesh along the boundaries. This is because the geometry of the edges of an element will vary in the same way as the displacement function, i.e. a linear element will have straight edges all round, a quadratic element will have edges varying as parabolas, a cubic element will have edges varying as a third-degree polynomial curve, etc., as shown in Fig.3.1. For curved boundaries it is thus more desirable to use a higher order isoparametric element.

(ii) The rigid body displacement criterion discussed in Section 1.8 is satisfied for isoparametric elements. As we can see, Eqn.(1.28b) and (1.28c) are satisfied automatically since

$$x = \sum N_ix_i \qquad \text{and} \qquad y = \sum N_iy_i$$

by definition, while Eqn.(1.28a) requires that $\sum N_i = 1$ is also valid for all isoparametric elements because the shape functions are symmetric with respect to the coordinates ξ, η, and will cancel out during the summation process. Incidentally, this symmetry may also be an advantage from the programming point of view, and for the integration of the stiffness coefficients only a quarter of the function and derivative values need to be calculated while the rest are obtained by simple cyclic permutation.

(iii) The constant strain criterion is satisfied. This can be proved by proceeding as described in Section 1.8 and assigning a 'constant strain' field involving only constants and linear variables as given below.

$$f = A_1 + A_2x + A_3y \qquad (3.4)$$

At any node i the following equality exists,

$$f_i = A_1 + A_2x_i + A_3y_i \qquad (3.5a)$$

and, substituting into Eqn.(3.2),

$$f = A_1\sum N_i + A_2\sum N_i x_i + A_3\sum N_i y_i \qquad (3.5b)$$

By comparing Eqn.(3.5b) with Eqn.(3.4), it is seen that the constant strain criterion is satisfied, provided that

$$\sum N_i = 1 \qquad (3.6a)$$

$$\sum N_i x_i = x \qquad (3.6b)$$

$$\sum N_i y_i = y \qquad (3.6c)$$

Eqn.(3.6) is in fact a restatement of the rigid body displacement criterion discussed above and in Section 1.8.

(iv) Unlike the Lagrange elements (*see* Section 1.7), the commonly used isoparametric elements (up to the cubic element) have no internal nodes and are therefore more efficient from the computational point of view.

3.3 Family of isoparametric element

A whole family of two-dimensional isoparametric elements can be formulated but only three of them are shown in Fig.3.1, and of these three, the quadratic element is considered to be the best and will be used in the example program. Elements of an order higher than three are not really practical owing to the excessive computer time required for the generation of stiffness matrices.

At this stage it is worthwhile for us to re-examine the Lagrange polynomials given in Table 1.2. Obviously they can be utilized in the construction of shape functions for isoparametric elements although it is necessary to shift the coordinate origin from the left end to the centre and to change the variables from x/ℓ to $(1+\xi)/2$ in the expressions. The modified results are given in Table 3.1.

3.3.1 LINEAR ELEMENT (Fig.3.1(b))

This element is simply a more general form of the linear Lagrange element given in Fig.1.6(a). The displacement functions in simple polynomial form have been given in Eqn.(3.1b) while the shape functions can be constructed directly as products of the linear Lagrange polynomials given in Table 3.1, and are of the form

$$N = \mathscr{L}^1(\xi)\mathscr{L}^1(\eta)$$

More specifically, we have

$$\text{at node } i,\quad \xi_i = -1,\quad \eta_i = -1 \quad \text{and} \quad N_i = \tfrac{1}{4}(1-\xi)(1-\eta)$$

$$\text{at node } j,\quad \xi_j = +1,\quad \eta_j = -1 \quad \text{and} \quad N_j = \tfrac{1}{4}(1+\xi)(1-\eta)$$

$$\text{at node } k,\quad \xi_k = +1,\quad \eta_k = +1 \quad \text{and} \quad N_k = \tfrac{1}{4}(1+\xi)(1+\eta)$$

$$\text{at node } \ell,\quad \xi_\ell = -1,\quad \eta_\ell = +1 \quad \text{and} \quad N_\ell = \tfrac{1}{4}(1-\xi)(1+\eta)$$

TABLE 3.1 *Lagrange polynomials in dimensionless form*

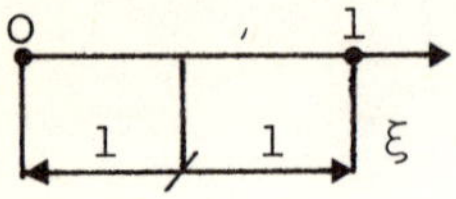

$$\mathcal{L}_0^1(\xi) = \frac{1}{2}(1 - \xi)$$

$$\mathcal{L}_1^1(\xi) = \frac{1}{2}(1 + \xi)$$

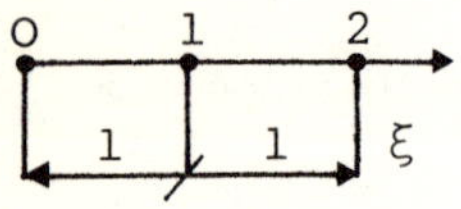

$$\mathcal{L}_0^2(\xi) = \frac{1}{2}\xi(\xi - 1)$$

$$\mathcal{L}_1^2(\xi) = (1 - \xi^2)$$

$$\mathcal{L}_2^2(\xi) = \frac{1}{2}(1 + \xi)\xi$$

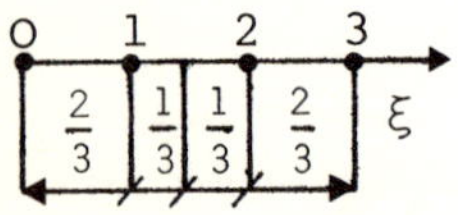

$$\mathcal{L}_0^3(\xi) = \frac{1}{16}(3\xi + 1)(3\xi - 1)(1 - \xi)$$

$$\mathcal{L}_1^3(\xi) = \frac{9}{16}(1 - \xi^2)(1 - 3\xi)$$

$$\mathcal{L}_2^3(\xi) = \frac{9}{16}(1 - \xi^2)(1 + 3\xi)$$

$$\mathcal{L}_3^3(\xi) = \frac{1}{16}(1 + \xi)(3\xi + 1)(3\xi - 1)$$

or, using i as the general subscript for all nodes,

$$N_i = \frac{1}{4}(1 + \xi\xi_i)(1 + \eta\eta_i) \tag{3.7}$$

in which ξ_i and η_i are the coordinate values of node i.

3.3.2 QUADRATIC ELEMENT (Fig.3.1(c))

There are altogether eight nodes for this element and with only three nodes along one edge, the displacement variation should be parabolic to satisfy compatibility. The correct displacement functions are

$$\begin{aligned} u &= A_1 + A_2\xi + A_3\eta + A_4\xi^2 + A_5\xi\eta + A_6\eta^2 + A_7\xi^2\eta + A_8\xi\eta^2 \\ v &= B_1 + B_2\xi + B_3\eta + B_4\xi^2 + B_5\xi\eta + B_6\eta^2 + B_7\xi^2\eta + B_8\xi\eta^2 \end{aligned} \tag{3.8}$$

As far as shape functions are concerned, for corner nodes they should vary as parabolas in the ξ and η directions and by definition, they should always have zero values at the midside nodes (Fig.3.2(c)). For

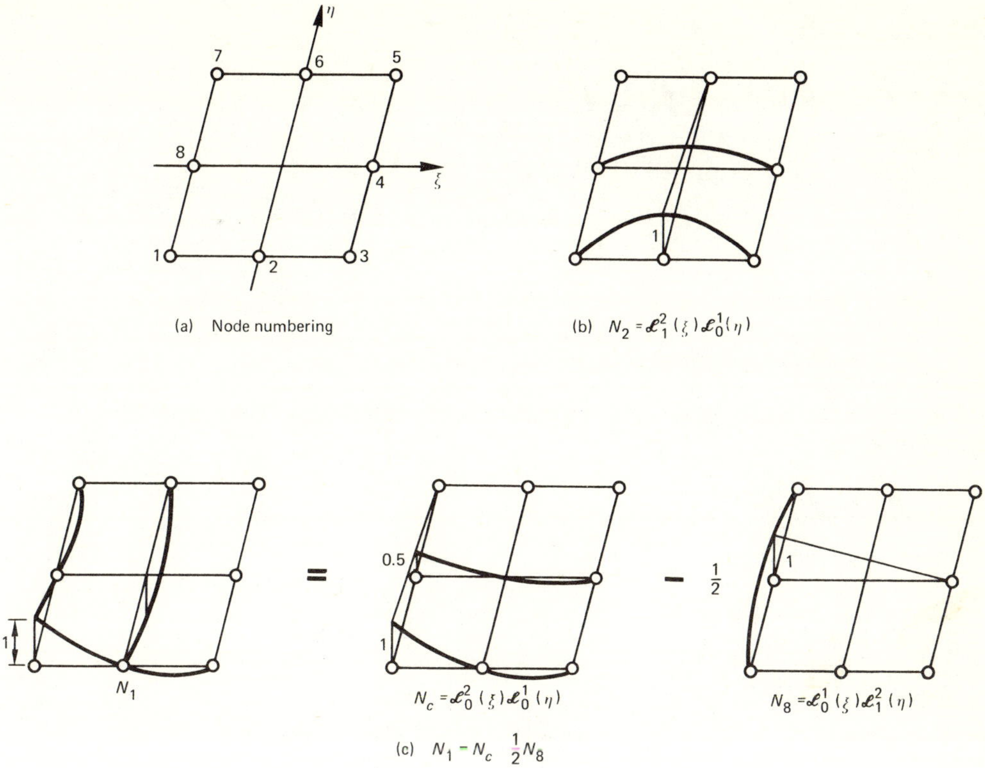

Fig.3.2 Construction of quadratic shape functions

midside nodes the construction is fairly straightforward and using node 2 in Fig.3.2(a) as an example we have

$$N_2 = \mathscr{L}_1^2(\xi)\mathscr{L}_0^1(\eta)$$
$$-\frac{1}{2}(1 - \xi)^2(1 \quad \eta)$$

In general for midside nodes with $\xi_i = 0$,

$$N_i = \frac{1}{2}(1 - \xi^2)(1 + \eta\eta_i) \qquad (3.9a)$$

and for midside nodes with $\eta_i = 0$

$$N_i = \frac{1}{2}(1 + \xi\xi_i)(1 - \eta^2) \qquad (3.9b)$$

The construction of the shape function for a corner node is more complex. Referring to Fig.3.2(c) it is seen that a shape function for node 1

$$N_c = \mathscr{L}_0^2(\xi)\mathscr{L}_0^1(\eta)$$

will satisfy the requirements along the edges $\eta = \pm 1$, and $\xi = +1$, although at node 8 on the edge $\xi = -1$, the function is equal to 0.5 instead of zero. This value can be corrected by subtracting Eqn.(3.9b) from N_c in the manner shown in Fig.3.2(c), i.e.

$$N_1 = N_c - \frac{1}{2}N_8$$

$$= \mathcal{L}_0^2(\xi)\mathcal{L}_0^1(\eta) - \frac{1}{2}\mathcal{L}_0^1(\xi)\mathcal{L}_1^2(\eta)$$

$$= \frac{1}{4}\xi(\xi - 1)(1 - \eta) - \frac{1}{4}(1 - \xi)(1 - \eta^2)$$

$$= \frac{1}{4}(1 - \xi)(1 - \eta)(- \xi - 1 - \eta)$$

The general equation for shape functions at all corner nodes is

$$N_i = \frac{1}{4}(1 + \xi\xi_i)(1 + \eta\eta_i)(\xi\xi_i + \eta\eta_i - 1) \qquad (3.9c)$$

Note that it is also possible to construct the shape function by starting with

$$N_c = \mathcal{L}_0^1(\xi)\mathcal{L}_0^1(\eta)$$

and then writing

$$N_1 = N_c - \frac{1}{2}N_2 - \frac{1}{2}N_8$$

3.3.3 CUBIC ELEMENT (Fig.3.1(d))

The displacement functions in simple polynomial form are

$$\begin{aligned} u &= A_1 + A_2\xi + A_3\eta + A_4\xi^2 + A_5\xi\eta + A_6\eta^2 + A_7\xi^3 + A_8\xi^2\eta \\ &\quad + A_9\xi\eta^2 + A_{10}\eta^3 + A_{11}\xi^3\eta + A_{12}\xi\eta^3 \\ v &= B_1 + B_2\xi + B_3\eta + B_4\xi^2 + B_5\xi\eta + B_6\eta^2 + B_7\xi^3 + B_8\xi^2\eta \\ &\quad + B_9\xi\eta^2 + B_{10}\eta^3 + B_{11}\xi^3\eta + B_{12}\xi\eta^3 \end{aligned} \qquad (3.10)$$

which conform to the requirements of having twelve nodes and cubic variations of displacements along the edges.

The construction of the shape functions along the same line as that indicated for the quadratic element and as an example, for node 2 with $\xi = -1$ and $\eta = -1/3$,

$$N_2 = \mathcal{L}_1^3(\xi)\mathcal{L}_0^1(\eta)$$

$$= \frac{9}{32}(1 - \xi^2)(1 - \eta)(1 - 3\xi)$$

or in general, for all edge nodes with $\xi_i = \pm 1/3$,

$$N_i = \frac{9}{32}(1 - \xi^2)(1 + \eta\eta_i)(1 + 9\xi\xi_i) \qquad (3.11a)$$

For node 12 with $\xi = -1$ and $\eta = -1/3$

$$N_{12} = \mathcal{L}_0^1(\xi)\mathcal{L}_1^3(\eta)$$

$$= \frac{9}{32}(1 - \xi)(1 - \eta^2)(1 - 3\eta)$$

or in general, for all edge nodes with $\eta_i = \pm 1/3$,

$$N_i = \frac{9}{32}(1 + \xi\xi_i)(1 - \eta^2)(1 + 9\eta\eta_i) \tag{3.11b}$$

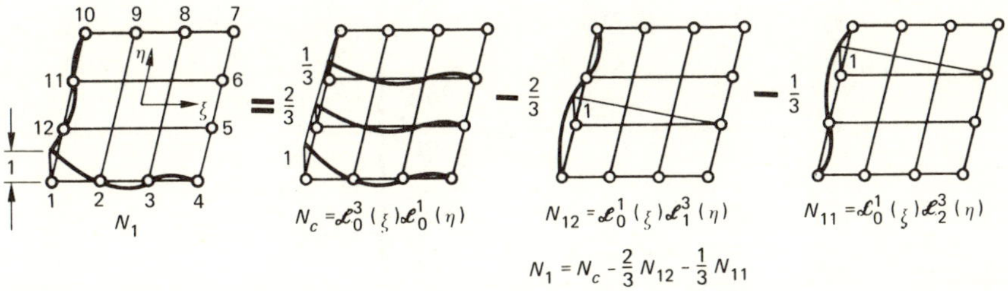

Fig.3.3 Construction of a cubic shape function for corner nodes

For corner nodes, the same procedure of correction as previously demonstrated for the quadratic element has to be made. From Fig.3.3 it is seen that a shape function N_c for node 1 will yield incorrect values at nodes 11 and 12, which should be corrected by the values $1/3$ and $2/3$ respectively. Thus the proper shape function for node 1 is

$$N_1 = N_c - \frac{2}{3}N_{12} - \frac{1}{3}N_{11}$$

$$= \mathcal{L}_0^3(\xi)\mathcal{L}_0^1(\eta) - \frac{2}{3}\mathcal{L}_0^1(\xi)\mathcal{L}_1^3(\eta) - \frac{1}{3}\mathcal{L}_0^1(\xi)\mathcal{L}_2^3(\eta)$$

$$= \frac{1}{32}(3\xi + 1)(3\xi - 1)(1 - \xi)(1 - \eta)$$

$$- \frac{6}{32}(1 - \xi)(1 - \eta^2)(1 - 3\eta) - \frac{3}{32}(1 - \xi)(1 - \eta^2)(1 + 3\eta)$$

$$= \frac{1}{32}(1 - \xi)(1 - \eta)\left[- 10 + 9(\xi^2 + \eta^2)\right]$$

The general shape function for all corner nodes is

$$N_i = \frac{1}{32}(1 + \xi\xi_i)(1 + \eta\eta_i)\left[- 10 + 9(\xi^2 + \eta^2)\right] \tag{3.11c}$$

3.4 Stiffness matrix formulation

It has already been shown in Chapter 1 that the stiffness matrix of an element can be derived through either a total potential energy or a virtual work approach. The expression (1.46) for element stiffness has been used a number of times already and is of the general form

$$[k] = \int [B]^T[D][B]\,d(\text{vol})$$

For two-dimensional problems,

$$[k] = t\int [B]^T[D][B]\,\mathrm{d(area)} \tag{3.12}$$

The $[B]$ matrix given in Eqn.(3.12) expresses strains ε_x, ε_y, γ_{xy} in terms of the nodal displacements and therefore contains first derivatives of the shape functions with respect to the global axes x and y. For the quadratic element the strain matrix $[B]$ is

$$[B] = \begin{bmatrix} \frac{\partial N_1}{\partial x} & 0 & \frac{\partial N_2}{\partial x} & 0 & \cdots & \frac{\partial N_8}{\partial x} & 0 \\ 0 & \frac{\partial N_1}{\partial y} & 0 & \frac{\partial N_2}{\partial y} & \cdots & 0 & \frac{\partial N_8}{\partial y} \\ \frac{\partial N_1}{\partial y} & \frac{\partial N_1}{\partial x} & \frac{\partial N_2}{\partial y} & \frac{\partial N_2}{\partial x} & \cdots & \frac{\partial N_8}{\partial y} & \frac{\partial N_8}{\partial x} \end{bmatrix}$$

and a typical submatrix is given by

$$[B_i] = \begin{bmatrix} \frac{\partial N_i}{\partial x} & 0 \\ 0 & \frac{\partial N_i}{\partial y} \\ \frac{\partial N_i}{\partial y} & \frac{\partial N_i}{\partial x} \end{bmatrix} \tag{3.13}$$

However, the shape functions for isoparametric elements (*see* Section 3.3) are defined with respect to the curvilinear coordinates ξ and η and therefore they cannot be differentiated directly with respect to the global x, y axes.

In order to overcome this difficulty it is necessary to obtain a relationship between the derivatives of the two sets of coordinates and this can be achieved through the normal chain rule of partial differentiation. For a two-dimensional problem the derivatives are related by

$$\left.\begin{aligned} \frac{\partial N}{\partial \xi} &= \frac{\partial N}{\partial x}\frac{\partial x}{\partial \xi} + \frac{\partial N}{\partial y}\frac{\partial y}{\partial \xi} \\ \frac{\partial N}{\partial \eta} &= \frac{\partial N}{\partial x}\frac{\partial x}{\partial \eta} + \frac{\partial N}{\partial y}\frac{\partial y}{\partial \eta} \end{aligned}\right\} \tag{3.14}$$

which in matrix form gives

$$\begin{Bmatrix} \frac{\partial N}{\partial \xi} \\ \frac{\partial N}{\partial \eta} \end{Bmatrix} = \begin{bmatrix} \frac{\partial x}{\partial \xi} & \frac{\partial y}{\partial \xi} \\ \frac{\partial x}{\partial \eta} & \frac{\partial y}{\partial \eta} \end{bmatrix} \begin{Bmatrix} \frac{\partial N}{\partial x} \\ \frac{\partial N}{\partial y} \end{Bmatrix} = [J] \begin{Bmatrix} \frac{\partial N}{\partial x} \\ \frac{\partial N}{\partial y} \end{Bmatrix} \tag{3.15}$$

ORDER NUMBER

PO NYAATP08

BOOK NUMBER

4311713

BAY NUMBER

6063-2

AUTHOR	Cheung, Y. K.
TITLE	A Practical Introduction to Finite Element Analysis

DESCRIPTION

This is an ex-library book and may have the usual library/used-book markings inside.This book has soft covers. In poor condition, suitable as a reading copy.

The matrix $[J]$ relating the derivatives of the two systems is called the Jacobian matrix and its coefficients can be obtained by differentiating Eqn.(3.3) with respect to the ξ and η coordinates.

For the quadratic element,

$$[J] = \begin{bmatrix} \frac{\partial N_1}{\partial \xi} & \frac{\partial N_2}{\partial \xi} & \cdots & \frac{\partial N_8}{\partial \xi} \\ \frac{\partial N_1}{\partial \eta} & \frac{\partial N_2}{\partial \eta} & \cdots & \frac{\partial N_8}{\partial \eta} \end{bmatrix} \begin{bmatrix} x_1 & y_1 \\ x_2 & y_2 \\ \vdots & \vdots \\ x & y \end{bmatrix} \tag{3.16}$$

$$= \begin{bmatrix} \frac{\partial \sum N_i x_i}{\partial \xi} & \frac{\partial \sum N_i y_i}{\partial \xi} \\ \frac{\partial \sum N_i x_i}{\partial \eta} & \frac{\partial \sum N_i y_i}{\partial \eta} \end{bmatrix}$$

With $[J]$ determined it is a straightforward matter to express $\partial N/\partial x$, $\partial N/\partial y$ with respect to $\partial N/\partial \xi$, $\partial N/\partial \eta$ through the inverse of $[J]$.

$$\begin{Bmatrix} \frac{\partial N}{\partial x} \\ \frac{\partial N}{\partial y} \end{Bmatrix} = [J]^{-1} \begin{Bmatrix} \frac{\partial N}{\partial \xi} \\ \frac{\partial N}{\partial \eta} \end{Bmatrix} \tag{3.17}$$

To complete the transformation between the two systems it is also necessary to express the elemental area in Eqn.(3.12) in the form of $d\xi\ d\eta$.

Consider an elemental parallelogram enclosed by two vectors

$$\vec{d\xi} = \begin{Bmatrix} \frac{\partial x}{\partial \xi} \\ \frac{\partial y}{\partial \xi} \end{Bmatrix} d\xi \quad \text{and} \quad \vec{d\eta} = \begin{Bmatrix} \frac{\partial x}{\partial \eta} \\ \frac{\partial y}{\partial \eta} \end{Bmatrix} d\eta \tag{3.18}$$

with the vectors defined from the relationship between the x,y and ξ,η coordinate systems and directed tangentially to the ξ = constant and η = constant contours respectively. The cross-product of the two vectors is equal to the area of the elemental parallelogram concerned and thus

$$d(\text{area}) = \vec{d\xi} \times \vec{d\eta}$$

$$= \det\begin{bmatrix} \dfrac{\partial x}{\partial \xi} & \dfrac{\partial x}{\partial \eta} \\ \\ \dfrac{\partial y}{\partial \xi} & \dfrac{\partial y}{\partial \eta} \end{bmatrix} (d\xi \quad d\eta)$$

$$= \det[J]d\xi d\eta \tag{3.19}$$

It follows that the integration limits should be changed to ± 1 and Eqn.(3.12) can now be written as

$$[K] = t\int_{-1}^{+1}\int_{-1}^{+1}[B]^T[D][B]\det[J]d\xi d\eta \tag{3.20}$$

with the understanding that $[B]^T[D][B]$ is a function of ξ and η only.

A necessary requirement for working out Eqn.(3.17) to obtain $\partial N/\partial x$ and $\partial N/\partial y$ is that $[J]$ can be inverted. This inverse exists if there is no excessive distortion of the element such that lines of constant ξ or η intersect inside or on the element boundaries, as shown in Fig.3.4.

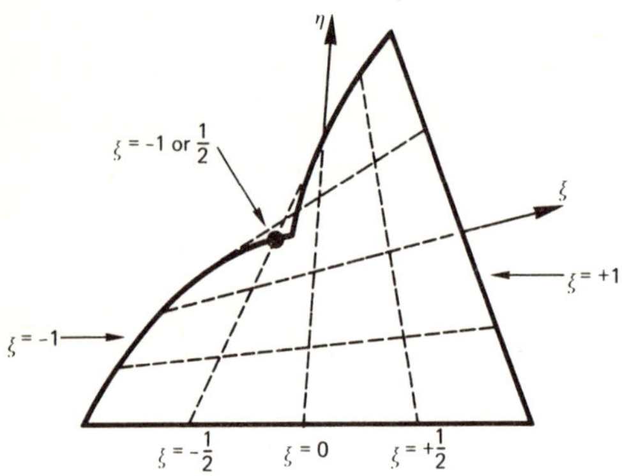

Fig.3.4 Excessive distortion of an isoparametric element

When the element is degenerated into a triangle by increasing an internal angle to 180° then $[J]$ is singular at that corner. It is, however, possible to obtain the element stiffness because $[J]$ is still unique at the Gauss points, but should the stresses be determined at that corner they will be indeterminate. A similar situation occurs when two adjacent corner nodes are made coincident to produce a triangular element. Therefore to ensure that $[J]$ can be inverted, any internal angle of each corner node of the element should be less than 180°, and, as an internal angle approaches 180° there is a loss in accuracy in the element stress, particularly at that corner.

In the majority of cases, the complexity of the expressions for the stiffness coefficients will make the explicit integration of Eqn.(3.20) exceedingly difficult, if not impossible, and thus numerical integration is mandatory for nearly all cases.

3.5 Numerical integration

For complex functions which cannot be integrated explicitly, a numerical integration can be carried out by first of all approximating the function with a polynomial. For example, if a linear approximation is made in place of the function shown in Fig.3.5(a), the integral, which represents the area under the curve, will be approximated by the area of a trapesium.

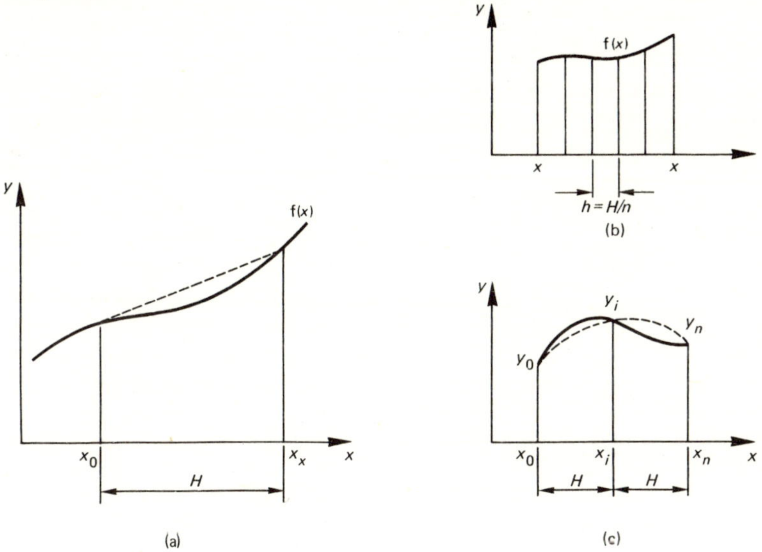

Fig.3.5 Linear approximation of a curve

$$I = \int_{x_0}^{x_n} f(x)\,\mathrm{d}x \simeq \frac{f(x_0) + f(x_n)}{2} H$$

$$\simeq \frac{y_0 + y_n}{2} H \qquad (3.21a)$$

Should the function depart appreciably from linearity then a significant error is likely to be introduced. This error can be reduced, however, by further subdividing the interval between x_0 and x_n into n divisions, as shown in Fig.3.5(b). Thus

$$I \simeq \frac{y_0 + y_1}{2} h + \frac{y_1 + y_2}{2} h + \ldots + \frac{y_{n-1} + y_n}{2} h$$

$$\simeq \frac{h}{2}(y_0 + 2y_1 + 2y_2 + \ldots + 2y_{n-1} + y_n) \qquad (3.21b)$$

Eqn.(3.21) is the well-known trapezoidal rule, but it is rarely used for finite element work because of its inefficiency.

An alternative is to approximate the function by a higher-order polynomial, this time in the form of a parabola through three points, as shown in Fig.3.5(c).

The area under the parabola is given by

$$I \simeq \frac{H}{6}(y_0 + 4y_i + y_n) \tag{3.22a}$$

and if a further subdivision is made we have Simpson's rule and the integral is

$$I \simeq \frac{h}{6}\Big[y_0 + y_n + 4(y_1 + y_3 + \ldots + y_{n-1}) + 2(y_2 + y_4 + \ldots + y_{n-2})\Big] \tag{3.22b}$$

Even higher-order polynomials can be used to approximate the function so as to improve the accuracy, and the general form of the integral is

$$I = \sum_{i=0}^{n} W_i f(x_i) \tag{3.23}$$

where $n + 1$ is the number of sample points for the polynomial. In Eqn. (3.23) the weighting coefficients are constant for equally spaced sampling points (as in Eqn.(3.22a)), and as a result for N points there are only N variables to fit a polynomial of order N-1. However, if unequally spaced sampling points are allowed and they are located so as to give the best approximation of the polynomial, there will be $2N$ variables to fit a polynomial of order $2N$-1. This is obviously a more efficient approach and is known as Gaussian quadrature[2].

For reasons of generality the coordinates of the sampling or Gauss points and the weighting coefficients are usually defined for limits of integration between -1 and +1, i.e.

$$\int_{-1}^{+1} f(\xi)\,d\xi = \sum_{i=1}^{N} W_i f(\xi_i) \tag{3.24}$$

which fits partly with the requirements of Eqn.(3.20). The values of W_i and ξ_i for various values of N are given in Table 3.2.

To extend Eqn.(3.24) to two dimensions so as to meet fully the requirement of Eqn.(3.20), it is required to integrate first with respect to one variable and then with respect to the other. Thus

$$\begin{aligned}\int_{-1}^{+1}\int_{-1}^{+1} f(\xi,\eta)\,d\xi d\eta &= \int_{-1}^{+1}\left[\int_{-1}^{+1} f(\xi,\eta)\,d\xi\right]d\eta \\ &= \int_{-1}^{1}\left[\sum_{i=1}^{N} W_i f(\xi_i,\eta)\right]d\eta \\ &= \int_{-1}^{+1}\left[\sum_{i=1}^{N} W_i g_i(\eta)\right]d\eta \\ &= \sum_{j=1}^{M}\sum_{i=1}^{N} W_i W_j g(\eta_j) \\ &= \sum_{j=1}^{M}\sum_{i=1}^{N} W_i W_j f(\xi_i,\eta_j)\end{aligned} \tag{3.25}$$

in which M and N can be equal or different.

TABLE 3.2 *Abscissae and weight coefficients of the Gaussian quadrature formula*

$$\int_{-1}^{1} f(x)\,dx = \sum_{i=1}^{n} W_i f(a_i)$$

± *a*				*W*		
			$n = 2$			
0.577 35	026 91	896 26		1.000 00	000 00	000 00
			$n = 3$			
0.774 59	666 92	414 83		0.555 55	555 55	555 55
0.000 00	000 00	000 00		0.888 88	888 88	888 89
			$n = 4$			
0.861 13	631 15	940 53		0.347 85	484 51	374 54
0.339 98	104 35	848 56		0.652 14	515 48	625 46
			$n = 5$			
0.906 17	984 59	386 64		0.236 92	688 50	561 89
0.538 46	931 01	056 83		0.478 62	867 04	993 66
0.000 00	000 00	000 00		0.568 88	888 88	888 89
			$n = 6$			
0.932 46	951 42	031 52		0.171 32	449 23	791 70
0.661 20	938 64	662 65		0.360 76	157 30	481 39
0.238 61	918 60	831 97		0.467 91	393 45	726 91
			$n = 7$			
0.949 10	791 23	427 59		0.129 48	496 61	688 70
0.741 53	118 55	993 94		0.279 70	539 14	892 77
0.405 84	515 13	773 97		0.381 83	005 05	051 19
0.000 00	000 00	000 00		0.417 95	918 36	734 69
			$n = 8$			
0.960 28	985 64	975 36		0.101 22	853 62	903 76
0.796 66	647 74	136 27		0.222 38	103 44	533 74
0.525 53	240 99	163 29		0.313 70	664 58	778 87
0.183 43	464 24	956 50		0.362 68	378 33	783 62
			$n = 9$			
0.968 16	023 95	076 26		0.081 27	438 83	615 74
0.836 03	110 73	266 36		0.180 64	816 06	948 57
0.613 37	143 27	005 90		0.260 61	069 64	029 35
0.324 25	342 34	038 09		0.312 34	707 70	400 03
0.000 00	000 00	000 00		0.330 23	935 50	012 60
			$n = 10$			
0.973 90	652 85	717 72		0.066 67	134 43	086 88
0.865 06	336 66	889 85		0.149 45	134 91	505 81
0.679 40	956 82	990 24		0.219 08	636 25	159 82
0.433 39	539 41	292 47		0.269 26	671 93	099 96
0.148 87	433 89	816 31		0.295 52	422 47	147 53

In the isoparámetric formulation of the element stiffness of a rectangular eight-nodal element the $[B]^T[D][B]$ to be integrated is a fourth-order function in both directions. To integrate this exactly requires more Gauss points than are in a 2 × 2 mesh but less than the number of points present in a 3 × 3 mesh. Zienkiewicz [3] argues that in the limit the element will be subjected to constant stress and hence the integral required is the lowest order necessary to evaluate accurately the area of the element, i.e. 3 × 3. On the other hand, the displacement formulation of the finite element analysis yields a lower bound on the strain energy and thereby overestimating the stiffness, while by not integrating the coefficients exactly a lower stiffness is obtained. Therefore by using a reduction in the integration order that is required for exact integration a better solution is often obtained because the two effects tend to cancel out each other. This is known as the 'reduced integration technique'.

Bathe and Wilson [4] indicated that the best integrating order for a rectangular eight-noded isoparametric element is 2 × 2. For an eight-noded quadrilateral element det$[J]$ is no longer a constant and has to be included in the integration as an extra variable and therefore a higher-order integration has to be used. For exact integration a 4 × 4 Gauss point mesh should be used but better displacements will actually result from using a 3 × 3 mesh due to reduced integration.

3.6 Calculation of element stiffness

In a typical finite element program using numerically integrated isoparametric elements and an advanced solution routine such as the front solver, the computer time spent in calculating the element stiffness can vary from 30 to 80 per cent of the total time required for a complete analysis, and therefore the algorithm used to compute the element stiffness has a significant effect on the overall efficiency of the program. With the algorithms in common use, the computer time required to calculate an element stiffness matrix varies by a factor of nine between the best and the least efficient.

For the quadratic element, the stiffness matrix is of the form

$$[K] = \left[\begin{array}{cc|cc|cc|cc}
k_{11} & k_{12} & k_{13} & k_{14} & k_{15} & k_{16} & k_{17} & k_{18} \\
 & k_{22} & k_{23} & k_{24} & k_{25} & k_{26} & k_{27} & k_{28} \\
\hline
 & & k_{33} & k_{34} & k_{35} & k_{36} & k_{37} & k_{38} \\
 & & & k_{44} & k_{45} & k_{46} & k_{47} & k_{48} \\
\hline
 & & & & k_{55} & k_{56} & k_{57} & k_{58} \\
\text{symmetrical} & & & & & k_{66} & k_{67} & k_{68} \\
\hline
 & & & & & & k_{77} & k_{78} \\
 & & & & & & & k_{88}
\end{array}\right] \qquad (3.26)$$

For efficiency of core storage, only the upper triangular matrix is used and the coefficients are stored in a one-dimensional array as shown below

$$\begin{bmatrix} k_{11} & k_{12} & k_{22} & | & k_{13} & k_{14} & k_{23} & k_{24} & | & k_{15} \; . \; . \; . \; . \; k_{28} & | \\ k_{33} & k_{34} & k_{44} & | & k_{35} \; . \; . \; . \; k_{88} \end{bmatrix} \tag{3.27}$$

The broken lines in Eqn.(3.26) indicate a $[k]_{ij}$ submatrix which results from Eqns.(3.20) and (3.25) with the value of t in Eqn.(3.20) being taken to be unity.

$$[k]_{ij} = \sum_{p=1}^{N}\sum_{q=1}^{N} W_p W_q \begin{bmatrix} \frac{\partial N_i}{\partial x} & 0 & \frac{\partial N_i}{\partial y} \\ 0 & \frac{\partial N_i}{\partial y} & \frac{\partial N_i}{\partial x} \end{bmatrix} \begin{bmatrix} D_1 & D_2 & 0 \\ D_2 & D_4 & 0 \\ 0 & 0 & D_5 \end{bmatrix} \begin{bmatrix} \frac{\partial N_j}{\partial x} & 0 \\ 0 & \frac{\partial N_j}{\partial y} \\ \frac{\partial N_j}{\partial y} & \frac{\partial N_j}{\partial x} \end{bmatrix} \det[J]$$

$$= \begin{bmatrix} k_1 & k_2 \\ k_3 & k_4 \end{bmatrix}_{ij} \tag{3.28}$$

Note that in Eqn.(3.28) only three coefficients need to be calculated when $i = j$ because of symmetry, and this corresponds to the storage arrangement given in Eqn.(3.27).

There are several methods of evaluating the expression in Eqn.(3.28). The most inefficient one is to carry out the full matrix multiplication of the triple product, and then to multiply it by the scalar quantity $W_p W_q \det[J]$ to give the required matrix, and finally to sum the matrices for all the Gauss points.

A far more efficient method is to multiply the matrices by hand to obtain explicit expressions for k_{1ij}, k_{2ij}, k_{3ij} and k_{4ij}. Due to the presence of a number of zero coefficients in both $[B]$ and $[D]$ matrices, the resulting expressions are relatively simple. Writing $W = W_p W_q \det[J]$ we have

$$\begin{aligned} k_{1ij} &= \sum\sum W\left(D_1 \frac{\partial N_i}{\partial x}\frac{\partial N_j}{\partial x} + D_5 \frac{\partial N_i}{\partial y}\frac{\partial N_j}{\partial y}\right) \\ k_{2ij} &= \sum\sum W\left(D_2 \frac{\partial N_i}{\partial x}\frac{\partial N_j}{\partial x} + D_5 \frac{\partial N_i}{\partial y}\frac{\partial N_j}{\partial y}\right) \\ k_{3ij} &= \sum\sum W\left(D_2 \frac{\partial N_i}{\partial y}\frac{\partial N_j}{\partial x} + D_5 \frac{\partial N_i}{\partial x}\frac{\partial N_j}{\partial y}\right) \\ k_{4ij} &= \sum\sum W\left(D_4 \frac{\partial N_i}{\partial y}\frac{\partial N_j}{\partial y} + D_5 \frac{\partial N_i}{\partial x}\frac{\partial N_j}{\partial x}\right) \end{aligned} \tag{3.29}$$

By using these explicit expressions instead of carrying out the full matrix multiplication for the triple product, it is possible to reduce the element stiffness formulation time by a factor of four.

A close inspection of Eqn.(3.29) shows that all the terms are of the form (constant) × (material property coefficient) × (product of shape function derivatives). Therefore, providing that the material properties are constant throughout the element, further economies can be introduced into the computation by storing separately in an array $[S]_{ij}$ the sum of the products of the shape function derivatives multiplied by W, such that

$$[S]_{ij} = \begin{bmatrix} \sum\sum W \frac{\partial N_i}{\partial x}\frac{\partial N_j}{\partial x} & \sum\sum W \frac{\partial N_i}{\partial x}\frac{\partial N_j}{\partial y} \\ \sum\sum W \frac{\partial N_i}{\partial y}\frac{\partial N_j}{\partial y} & \sum\sum W \frac{\partial N_i}{\partial y}\frac{\partial N_j}{\partial x} \end{bmatrix} \qquad (3.30)$$

and then to work out Eqn.(3.29) by incorporating the relevant material constants.

This method, which was first reported by Gupta [5], will cut the computing time for element stiffness formulation by a factor of nine over the full matrix multiplication method, and is the method implemented in the illustrative program to be shown later. For problems involving variable material properties within the elements, however, the Gupta algorithm cannot be used.

In the illustrative program the element stiffness formulation is controlled by subroutine STIFN which sets up the geometry of each element in turn and then carries out the numerical integration.

Inside the innermost loop of the numerical integration there is a call to subroutine JACOB which, for the Gauss point under consideration at that time, calculates the values of the shape function, the shape function derivatives and det$[J]$.

Immediately after the call to JACOB there is a call to MULT1 which assembles, in the element stiffness, the product of the shape functions as per Eqn.(3.30). After the Gauss loops have been completed MULT2 is called which carries out the rearrangement to produce the element stiffness as per Eqn.(3.29).

Subroutine STIFN

```
C     INSERT COMMON BLOCK


      REWIND 4                                          (1)
      IF(MAXMAT.EQ.1) CALL DMAT(1)                      (2a)
      DO 900 NEL =1,MAXNEL
      NCHK=0
      DO 1 I=1,16
      RS(I)=0.0
    1 CONTINUE                                          (3)
      DO 2 I=1,136
      S(I)=0.0
    2 CONTINUE
      IF(MAXMAT.EQ.1) GO TO 3
      I1=MAT(NEL)                                       (2b)
      CALL DMAT(I1)
```

```
    3 DO 10 J=1,NNODZ
      NIC=IABS(LDEF(NEL,J))
      NELDES(J)=LDEST(NIC)
      NELDEF(J)=NIC                                              (4)
      X(J)=CORD(NIC,1)
      Y(J)=CORD(NIC,2)
   10 CONTINUE
      CALL INTCRD                                                (5)
      IF(MAXRCT.NE.0) CALL PLOAD                                 (6)
      DO 100 JA=1,NRULE
      XL=VECTLC(JA)                                              (7)
      WX=WTFUN(JA)
      DO 90 JB=1,NRULE
      YL=VECTLC(JB)                                              (8)
      WY=WTFUN(JB)
      CALL JACOB(0)                                              (9)
      CALL MULT1                                                 (10)
      IF(GRAVX.NE.0.0.OR.GRAVY.NE.0.0) CALL GRAV                 (11)
      IF(WIRL.NE.0) CALL ROTAT                                   (12)
   80 CONTINUE
   90 CONTINUE
  100 CONTINUE
      CALL MULT2                                                 (13)
      IF(MAXPRS.NE.0) CALL PRES1                                 (14)
      IF(MAXTNS.NE.0) CALL TRANSF                                (15)
      WRITE(4) S,RS,(LDEF(NEL,J),J=1,NNODZ),NELDES               (16)
  900 CONTINUE
      IF(NSTOP.EQ.0) RETURN
      WRITE(6,1000)                                              (17)
      STOP
 1000 FORMAT( 50H1ILLCONDITIONING OR GEOMETRY OR DEFINITION ERRORS.,/,
     1 22H EXECUTION TERMINATED.)
      END
```

(1) Tape 4 is reset to receive the element stiffness and load vectors for reading back in subroutine FRONT.

(2) Subroutine DMAT sets up the material property matrix for the element. If all elements are made up of the same material DMAT is called once only (2a) or once for each element (2b).

(3) The element load vector RS and the element stiffness S are initialized.

(4) The element destination vector NELDES, the element definition vector NELDEF and the nodal coordinate vectors X and Y are set up.

(5) Subroutine INTCRD checks for any mid-side nodal coordinates that have not been defined. These coordinates are then defined by a linear interpretation between the appropriate corner nodes.

(6) If there are any nodal point loads subroutine PLOAD is called to check for any such loads in the element under consideration.

(7) XL is the ξ coordinate of each Gauss point in turn and WX the associated weighting function.

(8) YL is the η coordinate of each Gauss point in turn and WY the associated weighting function.

(9) Subroutine JACOB is called to calculate the shape function and the shape function derivatives.

(10) Subroutine MULT1 assembles into the element stiffness vector the products of the shape function derivatives and the weighting function as per equations (3.30).

(11) Subroutine GRAV is called if there are any gravity loads.

(12) Subroutine ROTAT is called if there are any rotational loads.

(13) Subroutine MULT2 is called to calculate the actual element stiffness from the product of shape function derivatives etc. as set up by MULT1.
(14) Subroutine PRES1 is called if there are any pressure loads.
(15) Subroutine TRANSF is called if there are any nodes which have to have their displacements defined relative to a local set of axes and not the global axis.
(16) The element stiffness, load vector, definition vector and destination vector are written onto tape 4.
(17) If badly distorted elements have been encountered, indicated by DETJ<O, then execution stops.

Subroutine MULT1

```
C     INSERT COMMON BLOCK

      N=0
      WAIT=WX*WY*DETJ                     (1)
      DO 20 I=1,NNODZ
      DXI=DX(I)*WAIT                      (2)
      DYI=DY(I)*WAIT
      DO 15 J=I,NNODZ
      DXJ=DX(J)                           (3)
      DYJ=DY(J)
      S(N+1)=S(N+1)+DXI*DXJ
      S(N+2)=S(N+2)+DXI*DYJ               (4)
      S(N+3)=S(N+3)+DYI*DYJ
      N=N+3
      IF(I.EQ.J) GO TO 15
      S(N+1)=S(N+1)+DYI*DXJ               (5)
      N=N+1
   15 CONTINUE
   20 CONTINUE
      RETURN
      END
```

(1) WAIT = $W_i * W_j * \det[J]$

(2) DX(I) = $\dfrac{\partial N_i}{\partial x}$

DY(I) = $\dfrac{\partial N_i}{\partial y}$

(3) DX(J) = $\dfrac{\partial N_j}{\partial x}$

DY(J) = $\dfrac{\partial N_j}{\partial y}$

(4) Setting up k_1, k_2 and k_3 as per Eqn.(3.30)

(5) If I $\neq$ J then k_4 is set up.

Subroutine MULT2

```
C     INSERT COMMON BLOCK


      N=0
      DO 40 I=1,NNODZ
      DO 30 J=I,NNODZ
      S1=S(N+1)
      S2=S(N+2)
      S3=S(N+3)
      IF(I.EQ.J) GO TO 20
      S4=S(N+4)
      S(N+1)=S1*D(1)+S3*D(5)
      S(N+2)=S2*D(2)+S4*D(5)
      S(N+3)=S4*D(3)+S2*D(5)
      S(N+4)=S3*D(4)+S1*D(5)
      N=N+4
      GO TO 30
   20 S(N+1)=S1*D(1)+S3*D(5)
      S(N+2)=S2*(D(3)+D(5))
      S(N+3)=S3*D(4)+S1*D(5)
      N=N+3
   30 CONTINUE
   40 CONTINUE
      RETURN
      END
```

(1) The products of the shape function derivatives and weighting functions are retrieved from the element stiffness.
(2) If I $\neq$ J then the 2 × 2 submatrix element stiffness is calculated by multiplying the products of the shape functions etc. by coefficients of the material property matrix.
(3) If I = J then the upper triangle of the 2 × 2 submatrix element stiffness is calculated.

Subroutine DMAT(I1)

```
C     INSERT COMMON BLOCK


      YMOD=YM(I1)
      PRAT=PR(I1)
      WEIGHT=WT(I1)
      D(1)=YMOD/(1.0-PRAT*PRAT)
      D(2)=D(1)*PRAT
      D(3)=D(2)
      D(4)=D(1)
      D(5)=YMOD/(2.0*(1.0+PRAT))
      RETURN
      END
```

(1) Young's modulus.
(2) Poisson's ratio.
(3) Mass/unit area of material forming the element.
(4) The coefficients of the material property matrix are calculated.

Subroutine JACOB (I1)

```
C     INSERT COMMON BLOCK

      DO 5 I=1,4                                                   (1)
      CW(I)=0.0
    5 CONTINUE
      DO 100 I=1,NNODZ                                             (2)
      XXI=XX(I)
      YYI=YY(I)
      GO TO (10,20,10,30,10,20,10,30),I                            (3)
   10 A1=XL*XXI
      A2=YL*YYI                                                    (4)
      DX(I)=0.25*XXI*(1.0+A2)*(2.0*A1+A2)
      DY(I)=0.25*YYI*(1.0+A1)*(2.0*A2+A1)
      SHP(I)=0.25*(1.0+A1)*(1.0+A2)*(A1+A2-1.0)
      GO TO 90
   20 DX(I)=0.5*XXI*(1.0-YL*YL)                                    (5)
      DY(I)=-YL*(1.0+XL*XXI)
      SHP(I)=0.5*(1.0+XL*XXI)*(1.0-YL*YL)
      GO TO 90
   30 DX(I)=-XL*(1.0+YL*YYI)                                       (6)
      DY(I)=0.5*YYI*(1.0-XL*XL)
      SHP(I)=0.5*(1.0+YL*YYI)*(1.0-XL*XL)
   90 CW(1)=CW(1)+DX(I)*X(I)                                       (7)
      CW(2)=CW(2)+DX(I)*Y(I)
      CW(3)=CW(3)+DY(I)*X(I)
      CW(4)=CW(4)+DY(I)*Y(I)
  100 CONTINUE
      DETJ=CW(1)*CW(4)-CW(2)*CW(3)                                 (8)
      IF(DETJ.GT.0.0) GO TO 110
      WRITE(6,1000) NEL,DETJ
      NSTOP=1
  110 AW(1)=CW(4)                                                  (9)
      AW(2)=-CW(2)
      AW(3)=-CW(3)
      AW(4)=CW(1)
      IF(I1.EQ.1) RETURN
      DO 120 I=1,4
      AW(I)=AW(I)/DETJ
  120 CONTINUE
      DO 200 I=1,NNODZ
      DXI=DX(I)
      DYI=DY(I)
      DX(I)=AW(1)*DXI+AW(2)*DYI                                    (10)
      DY(I)=AW(3)*DXI+AW(4)*DYI
  200 CONTINUE
      RETURN
 1000 FORMAT(//,5X,28HNEGATIVE OR ZERO DETERMINENT,/,5X,8HELEMENT ,I5,
     1       12H DETERMINENT,2X,E10.3)
      END
```

(1) The vector CW that will eventually contain the Jacobian matrix is initialized.

(2) (XXI, YYI) the local coordinates of each of the nodes in turn.

(3) The equation to be used to calculate the shape function and its derivatives with respect to ξ,η depends upon which node is being considered.

(4) Corner nodes $\quad DX(I) = \dfrac{\partial N_i}{\partial \xi} \qquad DY(I) = \dfrac{\partial N_i}{\partial \eta}$

(5) Midside nodes $\eta = 0$.
(6) Midside nodes $\xi = 0$.
(7) The Jacobian matrix is assembled as

$$[J] = \begin{bmatrix} CW(1) & CW(2) \\ CW(3) & CW(4) \end{bmatrix}$$

(8) $\det[J]$.

(9)
$$[J]^{-1} = \begin{bmatrix} AW(1) & AW(2) \\ AW(3) & AW(4) \end{bmatrix}$$

(10) Derivatives of shape function with respect to x,y:

$$DX(I) = \frac{\partial N_i}{\partial x}$$

$$DY(I) = \frac{\partial N_i}{\partial y}$$

3.7 Stress matrix

The calculation of the stresses in an element is a simple matter once the nodal displacements have been computed. Eqn.(1.38) is used with the displacement vector containing the nodal displacements for the element considered.

The explicit form of Eqn.(1.38) for the quadratic element is

$$\begin{Bmatrix} \sigma_x \\ \sigma_y \\ \tau_{xy} \end{Bmatrix} = \begin{bmatrix} D_1 & D_2 & 0 \\ D_2 & D_4 & 0 \\ 0 & 0 & D_5 \end{bmatrix} \begin{bmatrix} \frac{\partial N_1}{\partial x} & 0 & \frac{\partial N_2}{\partial x} & 0 & \cdots & 0 \\ 0 & \frac{\partial N_1}{\partial y} & 0 & \frac{\partial N_2}{\partial y} & \cdots & \frac{\partial N_8}{\partial y} \\ \frac{\partial N_1}{\partial y} & \frac{\partial N_1}{\partial x} & \frac{\partial N_2}{\partial y} & \frac{\partial N_2}{\partial x} & \cdots & \frac{\partial N_8}{\partial x} \end{bmatrix} \begin{Bmatrix} u_1 \\ v_1 \\ u_2 \\ \cdot \\ \cdot \\ \cdot \\ v_8 \end{Bmatrix} \qquad (3.31a)$$

or in longhand form,

$$\begin{aligned} \sigma_x &= \left[D_1 \sum_{i=1}^{8} \frac{\partial N_i}{\partial x} u_i + D_2 \sum_{i=1}^{8} \frac{\partial N_i}{\partial y} v_i\right] \\ \sigma_y &= \left[D_2 \sum_{i=1}^{8} \frac{\partial N_i}{\partial x} u_i + D_4 \sum_{i=1}^{8} \frac{\partial N_i}{\partial y} v_i\right] \\ \tau_{yx} &= D_5 \left[\sum_{i=1}^{8} \left(\frac{\partial N_i}{\partial y} u_i + \frac{\partial N_i}{\partial x} v_i\right)\right] \end{aligned} \qquad (3.31b)$$

For computational efficiency it is preferable to use Eqn.(3.31b). In the illustrative program subroutine STRESS controls the stress calculation. It sets up the geometry of each element in turn and loops through each of the element nodes. For each node it sets up the ξ,η coordinates and calls JACOB to determine the shape function derivatives. The calculation of stresses as per Eqn.(3.31b) is carried out by subroutine MULT3.

In a number of programs the stresses are determined at the nodes, since the nodal positions are readily located and it is convenient to output the displacements and stresses at the same points. It has been found that nodal stresses from a quadratic element are usually incorrect, however if the stresses of all elements meeting at a node are averaged much closer answers are obtained.

A better alternative is to calculate the stresses at the Gauss points, in which it will be found that because of the superior accuracy no averaging is necessary, or indeed, is possible. The stresses can be presented as stress contours; or a least square smoothing technique is used to yield a surface function from which accurate nodal stresses are obtained.

One explanation for the Gauss point stresses being more accurate than the nodal stresses is that the element stiffness is calculated by sampling at the Gauss points and it is therefore reasonable to expect the most accurate stresses and strains at the same points.

Subroutine STRESS

```
C     INSERT COMMON BLOCK

      WRITE(6,1000)
      DO 10 I=1,MAXNOD
      WRITE(6,1001) I,(DISPL(I,J),J=1,NVABZ)                  (1)
   10 CONTINUE
      IF(NPUT.EQ.0) GO TO 20                                  (2)
      N1=MAXNOD*4
      IF(N1.LE.MAXSS) GO TO 11
      NPUT=0                                                  (3)
      WRITE(6,2000)
      GO TO 20
   11 DO 12 I=1,N1
      SS(I)=0.0                                               (4)
   12 CONTINUE
   20 DO 100 NEL=1,MAXNEL
      IF(MAXMAT.EQ.1) GO TO 25
      N2=MAT(NEL)                                             (5)
      CALL DMAT(N2)
   25 WRITE(6,1002) NEL
      DO 30 NOD=1,NNODZ
      N3=IABS(LDEF(NEL,NOD))
      NELDEF(NOD)=N3
      X(NOD)=CORD(N3,1)
      Y(NOD)=CORD(N3,2)                                       (6)
      U(NOD)=DISPL(N3,1)
      V(NOD)=DISPL(N3,2)
   30 CONTINUE
      DO 50 NOD=1,NNODZ
      XL=XX(NOD)                                              (7)
      YL=YY(NOD)
```

```
      CALL JACOB(0) ---------------------------------------- (8)
      CALL MULT3 ------------------------------------------- (9)
      NIC=NELDEF(NOD)
      WRITE(6,1003) NIC,SIGMA ------------------------------ (10)
      IF(NPUT.EQ.0) GO TO 50
      N4=(NIC-1)*4
      DO 40 I=1,3                                         --+
      SS(N4+I)=SS(N4+I)+SIGMA(I)                             (11)
   40 CONTINUE                                            --+
      SS(N4+4)=SS(N4+4)+1.0 -------------------------------- (12)
   50 CONTINUE
  100 CONTINUE
      IF(NPUT.EQ.0) RETURN
      N5=-4
      WRITE(6,1004)                                       --+
      DO 150 NOD=1,MAXNOD                                   |
      N5=N5+4                                               |
      DIV=SS(N5+4)                                          |
      IF(DIV.EQ.0.0) GO TO 150                              |
      DO 140 I=1,3                                         (13)
      SIGMA(I)=SS(N5+I)/DIV                                 |
  140 CONTINUE                                              |
      WRITE(6,1003) NOD,SIGMA                               |
  150 CONTINUE                                            --+
      RETURN

 1000 FORMAT(20H1NODAL DISPLACEMENTS,/,
     1          7X,4HNODE,3X,6HX-COMP,9X,6HY-COMP)
 1001 FORMAT(6X,I5,3(5X,F10.7))
 1002 FORMAT(/////,17H STRESSES ELEMENT,I4,//,
     1          7X,4HNODE,6X,9HSIGMA X-X,3X,9HSIGMA Y-Y,5X,7HTAU X-Y)
 1003 FORMAT(/,1X,I10,3X,6(2X,F10.1))
 1004 FORMAT(23H1AVERAGE NODAL STRESSES,//,
     1          7X,4HNODE,6X,9HSIGMA X-X,3X,9HSIGMA Y-Y,5X,7HTAU X-Y)
 2000 FORMAT( 65H1INSUFFICIENT SPACE IN SS VECTOR TO ALLOW NODAL STRESS
     1AVERAGING.,/, 21H AVERAGING CANCELLED.)
       END
```

(1) The nodal displacements as calculated by the solution subroutines are printed out.

(2) If nodal stress averaging is requested, NPUT=1, then the stresses and the counter, used to indicate the number of elements connected to each node, are stored in the structural stiffness vector SS.

(3) If there is insufficient room in SS then nodal stress averaging is cancelled.

(4) The SS vector is initialized if nodal stress averaging is requested.

(5) The material property matrix for the element is set up.

(6) The element definition vector NELDEF, the nodal global coordinate vectors, X,Y and the nodal displacement vectors U,V are set up.

(7) The local coordinates of each of the nodes in turn XL = ξ, YL = η.

(8) Subroutine JACOB calculates the shape function derivatives for the node defined by (XL,YL).

(9) MULT3 determines the nodal stress as per Eqn.(3.31b).

(10) The nodal stresses are printed out.

(11) If nodal stress averaging requested the nodal stresses are stored in the SS vector.

(12) The counter is incremented.

(13) The average nodal stresses are calculated and printed out.

Subroutine MULT3

```
C     INSERT COMMON BLOCK

      DO 1 I=1,4
      CW(I)=0.0                                    (1)
    1 CONTINUE
      DO 10 I=1,NNODZ
      CW(1)=CW(1)+DX(I)*U(I)
      CW(2)=CW(2)+DX(I)*V(I)
      CW(3)=CW(3)+DY(I)*U(I)                       (2)
      CW(4)=CW(4)+DY(I)*V(I)
   10 CONTINUE
      SIGMA(1)=CW(1)*D(1)+CW(4)*D(2)
      SIGMA(2)=CW(1)*D(3)+CW(4)*D(4)               (3)
      SIGMA(3)=D(5)*(CW(2)+CW(3))
      RETURN
      END
```

(1) The CW vector is initialized.

(2) The products of the shape function derivatives and the nodal displacements are stored in CW.

(3) The nodal stresses are calculated as per Eqn.(3.31b).

3.8 Some consistent load matrices

The types of loading that are applied to a two-dimensional element can be divided into two groups: edge loads such as pressure and nodal points loads, the body forces such as those due to gravity and centrifugal loads.

Unlike the triangular element, in which all loads can be assigned to nodes intuitively or by statics, for a quadratic isoparametric the nodal loads due to distributed loads must be computed in accordance with Eqn. (1.47). The equivalent nodal forces are added, element by element, into a load vector to be used in the solution procedure. The treatment of the different type of loads is discussed in detail in the following subsections.

3.8.1 NODAL POINT LOADS

It is usually assumed that point loads will always be applied at nodes and not at an arbitrary point on an element boundary. This requirement is easily accommodated by putting in nodes at the points where the concentrated loads are acting.

For the front solver to be described in Chapter 4, the assembly is independent of the numbering of nodes, and at any one time, only the stiffness and the load coefficients connected with the front are stored. It is therefore more convenient to treat a nodal point load as a concentrated load acting on only one of the elements which are connected to the node in question, and then the nodal load vectors for all types of loading will be calculated on an element basis and assembled by the subroutine ASMBLE. Subroutine PLOAD determines the nodal point loads.

3.8.2 EDGE PRESSURE

It is usual to disregard the actual pressure variation along an element edge but simply to assume a parabolic distribution defined by the pressure values at each of the three nodes along that edge, and all intermediate pressure values can be calculated using the shape functions. For convenience in coding, all the nodes of the element are used in the computation so that there is no need to sort out the appropriate shape functions for the three nodes with given pressure values, which can include an input value of zero. Thus

$$P = \sum_{k=1}^{8} N_k P_k \tag{3.32}$$

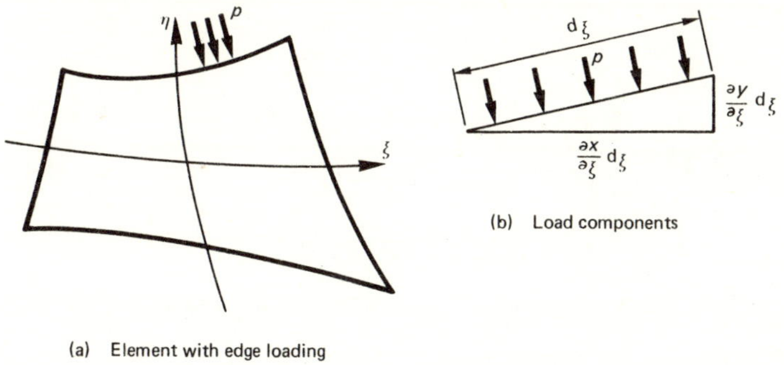

Fig.3.6 Edge pressure for an isoparametric element

For example, if we consider a pressure distributed in some manner along the $\eta = +1$ edge of an element (Fig.3.6), the components of the force $p\,d\xi$ acting upon an elemental length $d\xi$ are

$$\begin{Bmatrix} P_x \\ P_y \end{Bmatrix} = P \begin{Bmatrix} \dfrac{\partial y}{\partial \xi} \\ -\dfrac{\partial x}{\partial \xi} \end{Bmatrix} d\xi \tag{3.33}$$

p is specified in force per unit length since $d\xi$ (*see* Eqn.(3.18) is in fact defined with respect to x,y and not to the dimensionless ξ,η coordinates.

The consistent load for p is given by a modified form of Eqn.(1.47), in which the volume integral has been reduced to a line integral.

$$\{P\} = \int_{-1}^{+1} P[N]^T \begin{Bmatrix} \dfrac{\partial y}{\partial \xi} \\ -\dfrac{\partial x}{\partial \xi} \end{Bmatrix} d\xi \tag{3.34a}$$

where

$$[N] = \begin{bmatrix} N_1 & 0 & N_2 & 0 & \dots & 0 \\ 0 & N_1 & 0 & N_2 & \dots & N_8 \end{bmatrix} \tag{3.34b}$$

The above equation is usually integrated numerically and hence needs to be rewritten as

$$\{P\} = \sum_{i=1}^{n} W_i [N]^T p_i \begin{Bmatrix} \dfrac{\partial y}{\partial \xi} \\ -\dfrac{\partial x}{\partial \xi} \end{Bmatrix}_i \tag{3.34c}$$

where p_i is the pressure at the Gauss point i along the $\eta = +1$ edge and is computed by Eqn.(3.32).

If the same pressure acts on the edge $\eta = -1$ the sign of the load vector $\{P\}$ would be reversed since a positive pressure is assumed to act towards the centre of an element. A more general equation applicable to the two edges $\eta = \pm 1$ is

$$\{P\} = \sum_{i=1}^{n} W_i W_\eta [N]^T \left(\sum_{k=1}^{8} N_k p_k \right) \begin{Bmatrix} \dfrac{\partial y}{\partial \xi} \\ -\dfrac{\partial x}{\partial \xi} \end{Bmatrix} \tag{3.35a}$$

where W_η takes up the value of the η coordinate for the loaded edge. A similar expression for pressure loads on the $\xi = \pm 1$ edges is

$$\{P\} = \sum_{i=1}^{n} W_i W_\xi [N]^T \left(\sum_{k=1}^{8} N_k p_k \right) \begin{Bmatrix} -\dfrac{\partial y}{\partial \eta} \\ \dfrac{\partial x}{\partial \eta} \end{Bmatrix} \tag{3.35b}$$

Since the location of the Gauss points along the edge are not the same as those used to calculate the element stiffness it is usual to call the pressure subroutines from STIFN (Section 3.6) after the computation of element stiffness matrix. PRES1 determines whether an element has a pressure loading and also identifies the edge on which the pressure acts. An edge is only considered to be pressure loaded if the pressure values (zero or non-zero) for all three nodes are defined. PRES1 then calls PRES2 to determine the value of W_η or W_ξ, which in turn will call JACOB to evaluate $[N]$ and $[J]$ for each of the Gauss points, and then PRES3 to determine p and to compute the load vector from Eqn.(3.35).

3.8.3 CENTRIFUGAL FORCES

For ease of programming it is assumed that the centre of rotation coincides with the origin of the x,y axes.

The radially outward body force p_r on an elemental area dA is

$$p_r = r\Omega^2 m \tag{3.36}$$

in which

Ω is the angular velocity in radians per second;
m is the mass per unit area of the material;
r is the radial distance from the origin to the centroid of the elemental area.

p_r can be resolved into components parallel to x and y axes such that

$$\begin{Bmatrix} P_x \\ P_y \end{Bmatrix} = \Omega^2 m \begin{Bmatrix} x \\ y \end{Bmatrix} \tag{3.37}$$

where x and y correspond to the coordinates of the centroid of the elemental area.

By virtue of Eqn.(1.47), the equivalent nodal forces are

$$\{P\} = \int \Omega^2 m [N]^T \begin{Bmatrix} x \\ y \end{Bmatrix} dxdy \tag{3.38}$$

Once again the integration is carried out numerically and Eqn.(3.38) takes up the form of

$$\{P\} - \int_{-1}^{+1}\int_{-1}^{+1} \Omega^2 m [N]^T \begin{Bmatrix} x \\ y \end{Bmatrix} \det[J] d\xi d\eta$$

$$= \sum_{i=1}^{n}\sum_{j=1}^{n} W_i W_j \Omega^2 m [N]^T \begin{Bmatrix} x_{ij} \\ y_{ij} \end{Bmatrix} \det[J] \tag{3.39}$$

x_{ij} and y_{ij} are the x and y coordinates of each of the Gauss points respectively, and are easily computed with the help of Eqns.(3.6b) and (3.6c):

$$\begin{Bmatrix} x_{ij} \\ y_{ij} \end{Bmatrix} = \sum_{k=1}^{8} \begin{Bmatrix} N_k(\xi_i, \eta_j) x_k \\ N_k(\xi_i, \eta_j) y_k \end{Bmatrix} \tag{3.40}$$

It is natural to use the same Gauss points for computing the stiffness matrix and the equivalent load vector, and subroutine ROTAT is called inside the inner loop of STIFN for each Gauss point in turn to work out Eqn.(3.39).

3.8.4 BODY FORCES

The usual types of body forces are those due to gravity (g_y) and those due to earthquake loading (g_x). The formulation is identical to that for centrifugal forces and we have

$$\{P\} = \int_{-1}^{+1}\int_{-1}^{+1} m[N]^T \begin{Bmatrix} g_x \\ g_y \end{Bmatrix} \det[J]\, d\xi\, d\eta$$

$$= \sum_{i=1}^{n}\sum_{j=1}^{n} W_i W_j m[N]^T \begin{Bmatrix} g_x \\ g_y \end{Bmatrix} \det[J] \qquad (3.41)$$

where m is the mass per unit area and g_x and g_y the accelerations in the x and y directions respectively. The subroutine GRAV for computing Eqn. (3.41) is also called inside the inner loop of STIFN.

Subroutine PLOAD

```
C     INSERT COMMON BLOCK


      DO 50 NOD=1,NNODZ
      NIC=NELDEF(NOD) ---------------------------------------- (1)
      DO 49 I=1,MAXRCT
      IF(NIC.NE.NREACT(I)) GO TO 49 -------------------------- (2)
      II=(NOD-1)*NVABZ
      DO 45 J=1,NVABZ
      RS(II+J)=RS(II+J)+REACT(I,J) --------------------------- (3)
   45 CONTINUE
      NREACT(I)=0 -------------------------------------------- (4)
   49 CONTINUE
   50 CONTINUE
      RETURN
      END
```

(1) Each node in the element is taken in turn and
(2) compared with each node that has a point load applied.
(3) If a node has an applied point load then that load is assembled into the element load vector.
(4) To ensure that the point loads at the node are not assembled into another element load vector, of an element containing that node, the node number in the loaded node list NREACT is reset to zero.

Subroutine PRES1

```
C     INSERT COMMON BLOCK


      DIMENSION IP(8)
      LOGICAL IP
      N=0
      DO 1 I=1,NNODZ                          --+
      P(I)=0.0                                  | (1)
      IP(I)=.FALSE.                           --+
    1 CONTINUE
      DO 50 J=1,MAXPRS
      NOD=NPRES(J) ------------------------------------------- (2)
      DO 40 I=1,NNODZ
      IF(NELDEF(I).NE.NOD) GO TO 40 -------------------------- (3)
```

```
      P(I)=PRES(J)
      N=N+1                                              (4)
      IP(I)=.TRUE.
      GO TO 50
   40 CONTINUE
   50 CONTINUE
      IF(N.LT.3) RETURN                                  (5)
      IF(IP(1).AND.IP(2).AND.IP(3)) CALL PRES2(1)
      IF(IP(5).AND.IP(6).AND.IP(7)) CALL PRES2(2)        (6)
      IF(IP(3).AND.IP(4).AND.IP(5)) CALL PRES2(3)
      IF(IP(7).AND.IP(8).AND.IP(1)) CALL PRES2(4)
      RETURN
      END
```

(1) IP and P are initialized. IP indicates if a node has a pressure applied. P indicates the pressure at the node.

(2) Each node in the pressure loaded node list is taken in turn and

(3) compared with each of the nodes forming the element.

(4) If an element node is pressure loaded then the pressure is stored in P and the indicator IP is set to true.

(5) If there are less than three nodes loaded no edge has a pressure load.

(6) Each of the element edges is considered in turn and PRES2 called if an edge is pressure loaded.

Subroutine PRES2(I1)

```
C     INSERT COMMON BLOCK

      GO TO(50,52,100,102),I1                            (1)
   50 WX=1.0
      XL=1.0                                             (2)
      GO TO 53
   52 WX=-1.0                                            (3)
      XL=-1.0
   53 DO 60 JA=1,NRULE
      YL=VECTLC(JA)                                      (4)
      WY=WTFUN(JA)
      CALL JACOB(1)                                      (5)
      CALL PRES3(1,3)                                    (6)
   60 CONTINUE
      RETURN
  100 WY=1.0                                             (7)
      YL=1.0
      GO TO 103
  102 WY=-1.0                                            (8)
      YL=-1.0
  103 DO 120 JA=1,NRULE
      XL=VECTLC(JA)                                      (9)
      WX=WTFUN(JA)
      CALL JACOB(1)                                      (10)
      CALL PRES3(2,4)                                    (11)
  120 CONTINUE
      RETURN
      END
```

(1) I1 is the number of the edge loaded defined by PRES1.
(2) Edge 1: WX = W_ξ, XL = ξ = 1.
(3) Edge 2: WX = W_ξ, XL = ξ = - 1.
(4) YL = η coordinate of Gauss point in turn and WY = weighting function.
(5) JACOB calculates the shape function and shape function derivatives for point (XL,YL).
(6) PRES3 calculates the equivalent nodal force to the edge pressure.
(7)-(11) For edges 3 and 4, η = 1 and - 1 respectively, procedures are the same as (1)-(6) above.

Subroutine PRES3(I1, I2)

```
C     INSERT COMMON BLOCK

      APRES=0.0
      DO 10 I=1,NNODZ
      APRES=APRES+P(I)*SHP(I)                    (1)
   10 CONTINUE
      WAIT=WX*WY*APRES                           (2)
      M=0
      DO 20 I=1,NNODZ
      SHPI=SHP(I)*WAIT                           (3)
      RS(M+1)=RS(M+1)-SHPI*AW(I1)                (4)
      RS(M+2)=RS(M+2)-SHPI*AW(I2)
      M=M+NVABZ
   20 CONTINUE
      RETURN
      END
```

(1) APRES = $\sum N_i P_i$

(2) WAIT = W_i * (W_ξ) or (W_η) * $\left(\sum N_i P_i\right)$ * det$[J]$

(3) SHPI = N_i * W_i * (W_ξ) or (W_η) * $\left(\sum N_i P_i\right)$ * det$[J]$

(4) The equivalent nodal loads are assembled as per Eqns.(3.35a) or (3.35b) depending upon which edge is loaded.

Subroutine GRAV

```
C     INSERT COMMON BLOCK

      M=0
      WAIT=WEIGHT*DETJ*WX*WY                     (1)
      DO 100 I=1,NNODZ
      SHPI=SHP(I)*WAIT                           (2)
      RS(M+1)=RS(M+1)+SHPI*GRAVX                 (3)
      RS(M+2)=RS(M+2)+SHPI*GRAVY
      M=M+NVABZ
  100 CONTINUE
      RETURN
      END
```

(1) WAIT = M * det$[J]$ * W_i * W_j

(2) SHPI = N_i * M * det$[J]$ * W_i * W_j

(3) The equivalent nodal loads are assembled as per Eqn.(3.41).

Subroutine ROTAT

```
C     INSERT COMMON BLOCK

      WAIT=WEIGHT*DETJ*WX*WY*WIRL*WIRL  ---------------- (1)
      CX=0.0
      CY=0.0
      DO 10 I=1,NNODZ                        --+
      CX=CX+SHP(I)*X(I)                        | (2)
      CY=CY+SHP(I)*Y(I)                      --+
   10 CONTINUE
      M=0
      DO 100 I=1,NNODZ
      SHPI=SHP(I)*WAIT                       --+
      RS(M+1)=RS(M+1)+SHPI*CX                  | (3)
      RS(M+2)=RS(M+2)+SHPI*CY                --+
      M=M+NVABZ
  100 CONTINUE
      RETURN
      END
```

(1) WAIT = $W_i W_j \Omega^2 m \det[J]$

(2) CX, CX } Global coordinates of the Gauss point.

(3) Equivalent nodal loads calculated as per Eqn.(3.39) and assembled into the element load vector.

3.8.5 EXAMPLES OF LOADING

For a rectangular isoparametric element with edges parallel to the x,y axes it is possible to integrate the equations for the equivalent nodal loads explicitly. For example, consider the element of Fig.3.7 with a uniform pressure load along the top edge.

Using Eqns.(3.9) and (3.34a) and noting that for edge $\eta = 1$, $N_1 = N_2 = N_3 = N_4 = N_8 = 0$ and for the rectangular element, $\partial y/\partial\xi = 0$ and $\partial x/\partial\xi = a$, we find that

$$\{P\} = \begin{Bmatrix} P_{y5} \\ P_{y6} \\ P_{y7} \end{Bmatrix} = \int_{-1}^{1} p \begin{Bmatrix} N_5 \\ N_6 \\ N_7 \end{Bmatrix} (-a)\,d\xi$$

Writing $p(2a) = W$, the total pressure load, then

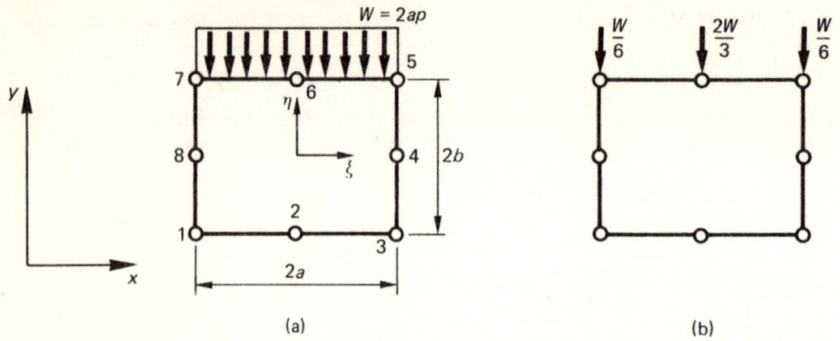

Fig.3.7 Edge loading and equivalent nodal forces

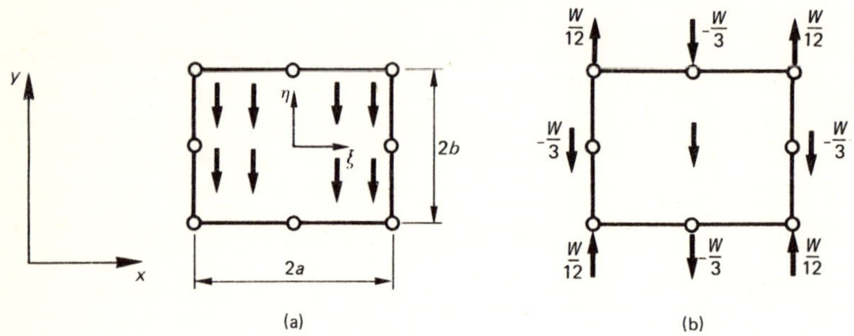

Fig.3.8 Gravity loading and equivalent nodal forces

$$P_{y5} = \frac{-W}{2}\int_{-1}^{1}\frac{1}{4}(1 + \xi\xi_i)(1 + \eta\eta_i)(\xi\xi_i + \eta\eta_i - 1)\,d\xi \tag{3.42a}$$

$$P_{y6} = \frac{-W}{2}\int_{-1}^{1}\frac{1}{2}(1 + \eta\eta_i)(1 - \xi^2)\,d\xi \tag{3.42b}$$

$$P_{y7} = \frac{-W}{2}\int_{-1}^{1}\frac{1}{4}(1 + \xi\xi_i)(1 + \eta\eta_i)(\xi\xi_i + \eta\eta_i - 1)\,d\xi \tag{3.42c}$$

In the above equation $\eta = 1$ for all the nodes along the top edge, and therefore, for node 5 $\xi_i = 1$, $\eta_i = 1$, for node 6 $\xi_i = 0$, $\eta_i = 1$, and for node 7 $\xi_i = -1$, $\eta_i = 1$. Substituting for ξ_i, η_i and η in Eqn.(3.42) gives:

$$\left.\begin{aligned}
P_{y5} &= \frac{-W}{2}\int_{-1}^{1}\frac{1}{4}(1 + \xi)(2)(\xi + 1 - 1)\,d\xi \\
P_{y6} &= \frac{-W}{2}\int_{-1}^{1}\frac{1}{2}(2)(1 - \xi^2)\,d\xi \\
P_{y7} &= \frac{-W}{2}\int_{-1}^{1}\frac{1}{2}(1 - \xi)(2)(- \xi + 1 - 1)\,d\xi
\end{aligned}\right\} \tag{3.43}$$

Evaluating these integrals gives:

$$P_{y5} = -W/6$$

$$P_{y6} = -2W/3$$

$$P_{y7} = -W/6$$

Hence a uniformly distributed load along an edge is not distributed in the intuitive ratio of $\frac{1}{4} : \frac{1}{2} : \frac{1}{4}$, but in the ratio $\frac{1}{6} : \frac{2}{3} : \frac{1}{6}$ (Fig.3.7(b)).

If the same procedure is used to determine the equivalent nodal force for gravity load (downwards negative) on the rectangular element (Fig. 3.8(a)), then by virtue of Eqn.(3.41),

$$\{P\} = \int_{-1}^{1}\int_{-1}^{1} m[N]^T \begin{Bmatrix} 0 \\ -g \end{Bmatrix} \det[J]\, d\xi d\eta \qquad (3.44)$$

This implies that all the equivalent nodal forces act in the y direction, and therefore Eqn.(3.44) can be expanded into

$$\begin{Bmatrix} P_{y1} \\ P_{y2} \\ \cdot \\ \cdot \\ \cdot \\ P_{y8} \end{Bmatrix} = -\int_{-1}^{1}\int_{-1}^{1} mg \begin{Bmatrix} N_1 \\ N_2 \\ \cdot \\ \cdot \\ \cdot \\ N_8 \end{Bmatrix} \det[J]\, d\xi d\eta$$

For a corner node i the integral becomes

$$P_{yi} = -\int_{-1}^{1}\int_{-1}^{1} mg\, \frac{1}{4}(1 + \xi\xi_i)(1 + \eta\eta_i)(\xi\xi_i + \eta\eta_i - 1)\det[J]\, d\xi d\eta$$

where the values of ξ_i and η_i are determined by the location of the corner node considered.

For a rectangular element, the Jacobian matrix is very simple, and

$$\det[J] = \begin{bmatrix} a & 0 \\ 0 & b \end{bmatrix} = ab$$

Hence

$$P_{yi} = \frac{-mgab}{4}\left[\int_{-1}^{1}\int_{-1}^{1}(1 + \xi\xi_i)(1 + \eta\eta_i)\xi\xi_i\, d\xi d\eta \right.$$

$$\left. + \int_{-1}^{1}\int_{-1}^{1}(1 + \xi\xi_i)(1 + \eta\eta_i)\eta\eta_i\, d\xi d\eta - \int_{-1}^{1}\int_{-1}^{1}(1 + \xi\xi_i)(1 + \eta\eta_i)\, d\xi d\eta\right]$$

Writing $mg(2a)(2b) = W$, which is the total weight of the element,

$$P_{yi} = \frac{-W}{16}\left[I_1 + I_2 + I_3\right]$$

where

$$I_1 = \int_{-1}^{1}\int_{-1}^{1}(1 + \xi\xi_i)(1 + \eta\eta_i)\xi\xi_i d\xi d\eta$$

$$= \int_{-1}^{1}(1 + \xi\xi_i)\xi\xi_i d\xi\left[\eta + \frac{\eta^2\eta_i}{2}\right]_{-1}^{1}$$

$$= 2\xi_i\left[\left(\frac{\xi^2}{2} + \frac{\xi^3}{3}\xi_i\right)\right]_{-1}^{1}$$

$$= \frac{4}{3}\xi_i^2$$

$$= \frac{4}{3}$$

since $\xi_i = \pm 1$ for all corner nodes.

$$I_2 = \int_{-1}^{1}\int_{-1}^{1}(1 + \xi\xi_i)(1 + \eta\eta_i)\eta\eta_i d\xi d\eta$$

$$= \frac{4}{3}\eta_i^2$$

$$= \frac{4}{3}$$

since $\eta_i = \pm 1$ for all corner nodes.

Finally,

$$I_3 = -\int_{-1}^{1}\int_{-1}^{1}(1 + \xi\xi_i)(1 + \eta\eta_i)d\xi d\eta$$

$$= -\int_{-1}^{1}(1 + \xi\xi_i)d\xi\int_{-1}^{1}(1 + \eta\eta_i)d\eta$$

$$= -\left[\xi + \frac{\xi^2\xi_i}{2}\right]_{-1}^{1} * \left[\eta + \frac{\eta^2\eta_i}{2}\right]_{-1}^{1}$$

$$= 4.$$

Therefore, the equivalent nodal forces for gravity loading at a corner node i can be given as

$$P_{yi} = \frac{-W}{16}\left[I_1 + I_2 + I_3\right]$$

$$= \frac{-W}{16}\left[\frac{4}{3} + \frac{4}{3} - 4\right]$$

$$= \frac{+W}{12}$$

For a midside node i with $\xi_i = 0$, the integral is

$$P_{yi} = -\int_{-1}^{1}\int_{-1}^{1} m\frac{1}{2}(1 - \xi^2)(1 + \eta\eta_i)\det\left[J\right]d\xi d\eta$$

$$= \frac{-W}{8}\int_{-1}^{1}(1 + \eta\eta_i)d\eta\int_{-1}^{1}(1 - \xi^2)d\xi$$

$$= \frac{-W}{8}\left[\eta + \frac{\eta^2}{2}\eta_i\right]_{-1}^{1} \times \left[\xi - \frac{\xi^3}{3}\right]_{-1}^{1}$$

$$= \frac{-W}{8}\left[2\right]\left[1 - \frac{1}{3} + 1 - \frac{1}{3}\right]$$

$$= \frac{-W}{3}$$

For a midside node i with $\eta_i = 0$, the integral is

$$P_{yi} = -\int_{-1}^{1}\int_{-1}^{1} m\frac{1}{2}(1 - \eta^2)(1 + \xi\xi_i)\det\left[J\right]d\xi d\eta = \frac{-W}{3}$$

Hence for gravity loading the equivalent nodal forces are as shown in Fig.3.8(b), and are thus very different from the values of $-W/12$ and $-W/6$, which would have been assigned intuitively to the corner and midside nodes respectively. The latter type of assigned loading will in fact yield very much poorer results in a finite element analysis using quadratic isoparametric elements.

References

1. O.C. Zienkiewicz, B.M. Irons, J. Ergatoudis, S. Ahmad, and F.C. Scott. Isoparametric and associated element family for two- and three-dimensional analysis. In *Finite Element Methods in Stress Analysis* (I. Holland and K. Bell, eds). New York, Tapir, 1969.
2. Z. Kopal. *Numerical Analysis,* 2nd ed. London, Chapman & Hall, 1961.
3. O.C. Zienkiewicz. *Finite Element Method in Engineering Science*. New York, McGraw-Hill Book Co., 1971.
4. K.J. Bathe and E.L. Wilson. *Numerical Methods in Finite Element Analysis*. Englewood Cliffs, N.J., Prentice-Hall, 1976.
5. K.A. Gupta and B. Mohraz. A method of computing numerically integrated stiffness matrices. *International Journal for Numerical Methods in Engineering,* vol.5, no.1, 1972, pp.83-89.

4 The Front Solver

4.1 Introduction

In the simple program presented in Chapter 2 a full matrix solution and an in-core band solver were used for solving the set of simultaneous equations which was formed by assembling the stiffness matrices of all the elements, and equating the internal forces corresponding to the displacements to the external loads. Such an approach naturally would severely restrict the size of the problem that can be tackled, even on the biggest computer available.

To overcome this limitation it is necessary either to store the structural stiffness matrix and load matrix on disc files and transfer to core memory when needed, or to carry out at any one time a simultaneous assembly and reduction on part of the structural stiffness matrix, in which case no intermediate storage is needed. Two suitable subroutines which will be described here are the out-of-core band solver and the front solver.

The in-core band solver was described in Section 2.9 and the mathematical details will not be repeated here. For an out-of-core band solver the diagonal coefficients are no longer shifted to the left as shown in Fig.2.10. Therefore instead of storing the whole matrix in a rectangular array, only a part of the equations are stored in a one-dimensional array, as shown in Fig.4.1(b). Furthermore, to use the first equation for the reductions only the coefficients of that equation need to be fully assembled and, furthermore, only the area to the left of the broken line are involved in the process. This implies that only the elements connected to node 1 have to be assembled before the equation (or equations) for node 1 is used for the reduction and so on for other nodes. Therefore the storage required at all times is $\frac{1}{2}n^2$, in which n is the half bandwidth of the stiffness matrix.

The out-of-core solver described above can be used to solve problems involving a large number of degrees of freedom, but it is still inefficient in core storage and computation in that all the zero coefficients inside the band are stored. For most problems the stiffness matrices are very sparse and even inside the band the zero coefficients very often far outnumber the non-zero coefficients.

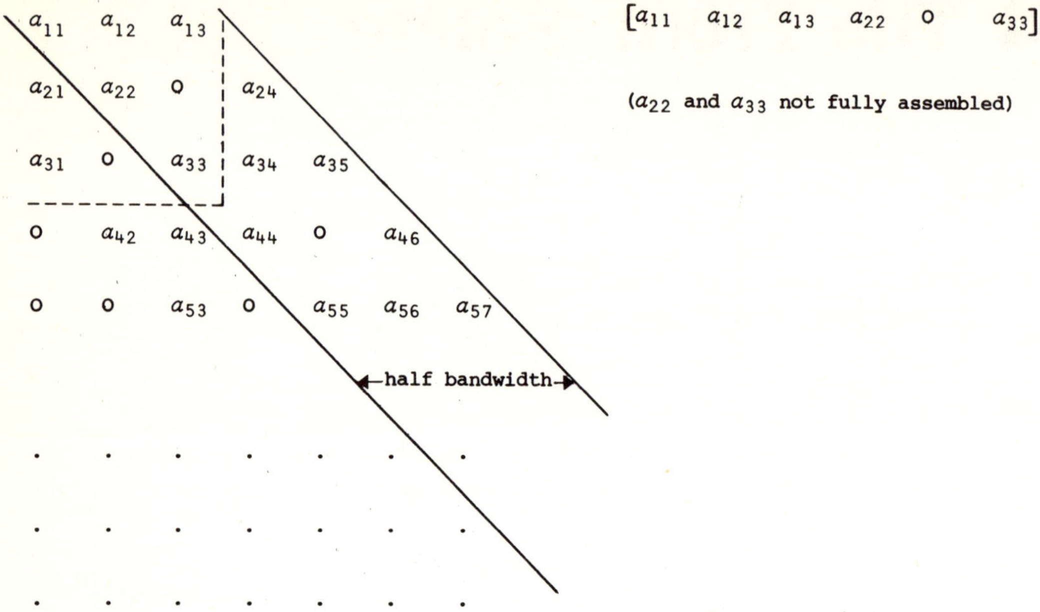

(a) Original band matrix

(b) Storage of coefficients in a vector

Fig.4.1 Out-of-core band solver storage scheme

A more advanced problem which takes advantage of this property is the front solver which will be discussed fully in this chapter.

The front solver is never less efficient than the band solver and for the majority of cases is very much more efficient, especially for problems in which elements with midside nodes are used. The front solver does not store as many zero coefficients inside the band, since unlike the band solver in which the nodal numbers determine the storage locations, it allocates storage in the order in which the nodes are presented for assembly. Thus for the front solver node numbering is irrelevant and it is the element numbering that is all important. This is of great significance when a re-analysis has to be made with local refinement of the finite element mesh.

4.2 The front solver [1]

The operations of a front solver can be split up into three logical parts:

(i) prefront;
(ii) reduction and pre-constraints;
(iii) backsubstitution and post-constraints.

The pre-constraints and post-constraints impose specified displacements on the nodal variables and because these subroutines are self-contained within themselves, they will be omitted for the time being in the small problem used here to demonstrate the operation of the front solver.

Figure 4.2 represents a solid structure, modelled by three elements and constrained against moving both horizontally and vertically at nodes 4 and 2 and loaded by a single vertical load W at node 11 as shown.

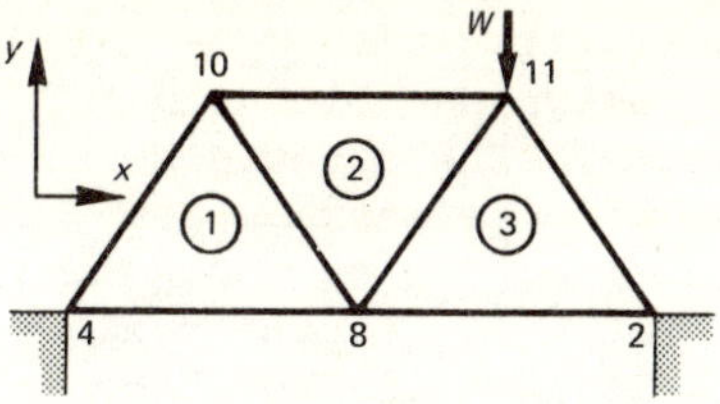

Fig.4.2 An example problem

The element definitions for the structure are:

Element number	Node i	Node j	Node m
1	4	8	10
2	8	11	10
3	2	11	8

Some of the nodal numbers have been missed out purposely to demonstrate the superiority of the front solver over the band solver which cannot cope with such a situation.

4.2.1 PREFRONT

This subroutine determines the nodal destination vector and the maximum front width, which is the equivalent to the half bandwidth for the band solver. The nodal destination vector contains the locations allocated to each node and it is from this vector that the element destination vectors are eventually determined.

The first step is to pin-point the position in which a node number appears for the last time in the element definition list by inserting a negative sign in front of the node number in the appropriate element. This is required because these nodal displacements can be eliminated at this stage as shown below. Thus the element definition vectors for this simple problem would finally look as follows:

$$[-4 \quad 8 \quad 10]$$

$$[8 \quad 11 \quad -10]$$

$$[-2 \quad -11 \quad -8]$$

It is now possible to determine the nodal and element destination vectors. At the start the displacement vector is completely empty and unallocated, and therefore the element destination vector of the first element is simply $[1, \ 2, \ 3]$ and this means that the first node, node 4, is allocated to the first unallocated location in the displacement vector, i.e. location 1, and node 8 and node 10 the next unallocated locations, i.e. location 2 and location 3 respectively. The fourth, eighth and tenth coefficients of the nodal destination vector are therefore set to 1, 2 and 3 respectively.

The displacement vector at this stage contains $[\delta_4 \quad \delta_8 \quad \delta_{10}]^T$.

The negative sign in front of node 4 in the definition vector means that all equations relating to δ_4 have now been fully assembled and hence δ_4 can be eliminated from the equation system. The equations used to effect this elimination are then removed to disc storage and the displacement vector after this reduction contains $[0 \quad \delta_8 \quad \delta_{10}]^T$.

Element 2 with element definition vector [8, 11, -10] is the next element presented for assembly. Node 8 and node 10 have already been allocated and since node 11 has not appeared previously it is allocated to the first unallocated location which in this instance is location 1. The eleventh coefficient of the nodal destination vector is therefore set to 1. Hence after assembly of element 2 the displacement vector contains $[\delta_{11} \quad \delta_8 \quad \delta_{10}]^T$ and from the nodal destination vector the element destination vector can be determined as [2, 1, 3].

The equations relating to node 10 have now been fully assembled and after reduction and removal to disc storage the displacement vector contains $[\delta_{11} \quad \delta_8 \quad 0]^T$.

Finally, the third element with element definition vector [-2, -11, -8] is presented for assembly. Node 2 is allocated to location 3 and the second coefficient of the nodal destination vector is set to 3. The displacement vector then contains $[\delta_{11} \quad \delta_8 \quad \delta_2]^T$ and the element destination vector is [3, 1, 2].

The equations are used in turn in the reduction process since element 3 is the last element. The sequence is indicated by the element definition vector from left to right and after reduction of the equations corresponding to δ_2 and then δ_{11} the remaining equations will only have coefficients relating to δ_8. This δ_8 can be determined and used as a starting point for the backsubstitution.

It should be mentioned that, as an example, δ_8 refers to all the components of the displacement of node 8, hence for the present problem it will include δ_{x8} and δ_{y8}. Similarly the stiffness coefficients shown as k_{ij} in the following section stand in the present problem for 2 × 2 submatrices.

In conclusion, after the pre-front subroutine the nodal destination vector contains

$$[0 \quad 3 \quad 0 \quad 1 \quad 0 \quad 0 \quad 0 \quad 2 \quad 0 \quad 3 \quad 1]$$

The zero coefficients indicate node numbers not used in the problem.

Subsequently, from the nodal destination vector subroutine, STIFN sets up the element destination vectors as

$$[1 \quad 2 \quad 3]$$

$$[2 \quad 1 \quad 3]$$

$$[3 \quad 1 \quad 2]$$

The maximum nodal front width is found to be 3, which is the highest number appearing in the nodal destination vector. The actual front width of the problem, defining the size of matrix required is obviously the maximum nodal front width multiplied by the number of variables (displacements) per node.

Subroutine PREFNT

```
C     INSERT COMMON BLOCK

      DO 1 I=1,MAXNW
      NW(I)=0
    1 CONTINUE
      DO 10 NEL=1,MAXNEL
      DO 5 I=1,NNODZ
      NIC=LDEF(NEL,I)
      LDEST(NIC)=NEL                                            (1)
    5 CONTINUE
   10 CONTINUE
      DO 20 NIC=1,MAXNOD
      NEL=LDEST(NIC)
      IF(NEL.EQ.0) GO TO 20
      DO 15 I=1,NNODZ
      IF(LDEF(NEL,I).NE.NIC) GO TO 15
      LDEF(NEL,I)=-NIC                                          (2)
      LDEST(NIC)=0
      GO TO 20
   15 CONTINUE
   20 CONTINUE
      DO 100 NEL=1,MAXNEL                                       (3)
      DO 50 I=1,NNODZ
      NIC=IABS(LDEF(NEL,I))
      IF(LDEST(NIC).NE.0) GO TO 50                              (4)
      DO 30 J=1,MAXNW
      IF(NW(J).NE.0) GO TO 30
      LDEST(NIC)=J
      NW(J)=NIC                                                 (5)
      IF(MAXFW.LT.J) MAXFW=J
      GO TO 50
   30 CONTINUE
      WRITE(6,1000)
      STOP
   50 CONTINUE
      DO 70 I=1,NNODZ
      NIC=LDEF(NEL,I)
      IF(NIC.GT.0) GO TO 70                                     (6)
      N1=LDEST(-NIC)
      NW(N1)=0
   70 CONTINUE
  100 CONTINUE
      MAXFW=MAXFW*NVABZ                                         (7)
      WRITE(6,1001) MAXFW
      RETURN
 1000 FORMAT(62H1MAXIMUM FRONT WIDTH DURING PREFNT EXCEEDS LENGTH OF NW
     1VECTOR)
 1001 FORMAT(/////,5X,19HMAXIMUM FRONT WIDTH,2X,I3)
      END
```

(1) Determine the element in which each node appears last.
(2) Set last appearance of node in the element definition list to a negative number.
(3) Loop 100 sets up the nodal destination vector.
(4) Checking for prior allocation of a node.
(5) Determining the first unallocated column and updating the maximum nodal front width.
(6) The last appearance of a node releases the column for reallocation.
(7) Maximum front width.

4.2.2 ASSEMBLY AND REDUCTION

The reduction procedure in a front solver is virtually the same as that in a band solver, except that the equation being used to reduce the other equations within the front area may occur anywhere in that area and is very often not the top equation.

To continue with the example problem, we have now the element destination vectors determined and the maximum front width calculated as equal to 3. The first element stiffness matrix is now calculated according to Eqn. (3.28) and is of the form

$$\begin{bmatrix} k_{ii}^1 & k_{ij}^1 & k_{im}^1 \\ & k_{jj}^1 & k_{jm}^1 \\ \text{symmetric} & & k_{mm}^1 \end{bmatrix} \begin{Bmatrix} \delta_4 \\ \delta_8 \\ \delta_{10} \end{Bmatrix} \begin{Bmatrix} p_4^1 \\ p_8^1 \\ p_{10}^1 \end{Bmatrix} \qquad (4.1)$$

where for the present two-dimensional problem all of the ks are 2×2 submatrices.

Because there are no edge loads or body forces we have $p_4^1 = p_8^1 = p_{10}^1 = 0$. However for the sake of generality p_4^1, p_8^1 and p_{10}^1 will be retained. Note that the equations are actually stored in one-dimensional arrays: one for the stiffness matrix and one for the load vector.

$(k_{ii}^1 \quad k_{ij}^1 \quad k_{im}^1 \quad k_{jj}^1 \quad k_{jm}^1 \quad k_{mm}^1)$ Structural stiffness vector;

$(p_4^1 \quad p_8^1 \quad p_{10}^1)$ Load vector.

The order of storing the actual coefficients of the stiffness matrix is most easily seen by referring to the intermediate output from the program in Chapter 5.

Using the element destination vector $[1, \quad 2, \quad 3]$ the element stiffness is assembled (see Section 2.6) into the structural stiffness matrix which appears as shown in Eqn. (4.1) at this stage.

From the element definition vector $[-4, \quad 8, \quad 10]$, it is seen that node 4 has appeared for the last time. The set of equations associated with node 4 can now be used to eliminate δ_4 from the equation system. Following the Gaussian elimination procedure described in Section 2.9 and transferring the coefficients (k_{ii}, k_{ij}, k_{im}, p_4^1) to disc storage we have

$$\begin{bmatrix} 0 & 0 & 0 \\ & a_{jj}^* & a_{jm}^* \\ \text{symmetric} & & a_{mm}^* \end{bmatrix} \begin{Bmatrix} 0 \\ \delta_8 \\ \delta_{10} \end{Bmatrix} = \begin{Bmatrix} 0 \\ p_8^{1*} \\ p_{10}^{1*} \end{Bmatrix} \qquad (4.2)$$

where a_{jj}^* was substituted for the reduced coefficient k_{jj}^{1*}, etc., to preserve similarity with the change to b_{jm}, c_{jm} in subsequent steps (*see* Eqns. (4.5) and (4.7b)).

Although the transferred coefficients have been written as k_{ii} etc. for neatness of presentation, because the program considers each equation separately and not the set of equations relating to δ_4 then although the equation relating to δ_{4x} will be transferred unaltered the equation relating to δ_{4y} will have been reduced to eliminate δ_{4x}, hence the transferred equation (k_{ii}, k_{ij}, k_{im}, p_4^1) implies that the pivotal submatrix k_{ii} has been triangularised and the other coefficients reduced accordingly.

The stiffness matrix of the second element is now calculated.

$$\begin{array}{c} \\ 2 \\ 1 \\ 3 \end{array}\begin{array}{c} \begin{array}{ccc} 2 & 1 & 3 \end{array} \\ \begin{bmatrix} k_{ii}^2 & k_{ij}^2 & k_{im}^2 \\ & k_{jj}^2 & k_{jm}^2 \\ \text{symmetric} & & k_{mm}^2 \end{bmatrix} \end{array} \begin{Bmatrix} \delta_8 \\ \delta_{11} \\ \delta_{10} \end{Bmatrix} = \begin{Bmatrix} p_8^2 \\ p_{11}^2 \\ p_{10}^2 \end{Bmatrix} \tag{4.3}$$

This time there is one non-zero load coefficient since $p_{11}^2 = \begin{vmatrix} 0 \\ W \end{vmatrix}$. The assembly follows on the usual way and the structural stiffness is

$$\begin{bmatrix} k_{jj}^2 & k_{ij}^2 & k_{jm}^2 \\ & a_{jj}^* + k_{ii}^2 & a_{jm}^* + k_{im}^2 \\ \text{symmetric} & & a_{mm}^* + k_{mm}^2 \end{bmatrix} \begin{Bmatrix} \delta_{11} \\ \delta_8 \\ \delta_{10} \end{Bmatrix} = \begin{Bmatrix} p_{11}^2 \\ p_8^{1*} + p_8^2 \\ p_{10}^{1*} + p_{10}^2 \end{Bmatrix} \tag{4.4a}$$

or

$$\begin{bmatrix} b_{ii} & b_{ij} & b_{im} \\ & b_{jj} & b_{jm} \\ \text{symmetric} & & b_{mm} \end{bmatrix} \begin{Bmatrix} \delta_{11} \\ \delta_8 \\ \delta_{10} \end{Bmatrix} = \begin{Bmatrix} p_{11}^2 \\ p_8^{1*} + p_8^2 \\ p_{10}^{1*} + p_{10}^2 \end{Bmatrix} \tag{4.4b}$$

Note that Eqn.(4.4b) has the same form as Eqn.(4.1) and that in general k_{ij}^2 in Eqn.(4.4a), being usually of submatrix form, should be $[k_{ij}^2]^T$, but the superscript T has been omitted here for ease of presentation. The element definition vector [8, 11, -10] shows that node 10 can be used for reduction and consequently after removing for storage the appropriate coefficients

$$\begin{bmatrix} b_{ii}^* & b_{ij}^* & 0 \\ & b_{jj}^* & 0 \\ & & 0 \end{bmatrix} \begin{Bmatrix} \delta_{11} \\ \delta_8 \\ 0 \end{Bmatrix} = \begin{Bmatrix} p_{11}^{2*} \\ (p_8^{1*} + p_8^2)^* \\ 0 \end{Bmatrix} \tag{4.5}$$

The coefficients which have been removed for disc storage are

$(b_{im}, b_{jm}, b_{mm}, (p_{10}^{1*} + p_{10}^2))$.

The third element is now processed in the same way and the element stiffness matrix is

$$\begin{matrix} & 3 & 1 & 2 \\ 3 \\ 1 \\ 2 \end{matrix}\begin{bmatrix} k^3_{ii} & k^3_{ij} & k^3_{im} \\ & k^3_{jj} & k^3_{jm} \\ \text{sym} & & k^3_{mm} \end{bmatrix} \begin{Bmatrix} \delta_2 \\ \delta_{11} \\ \delta_8 \end{Bmatrix} = \begin{Bmatrix} p^3_2 \\ p^3_{11} \\ p^3_8 \end{Bmatrix} \tag{4.6}$$

All the load coefficients are zero, including p^3_{11}, since the load W at node 11 was assembled previously with element 2.

The element stiffness matrix is assembled into the structural stiffness matrix in accordance with the element destination vector and the structural stiffness matrix is

$$\begin{bmatrix} b^*_{ii} + k^3_{jj} & b^*_{ij} + k^3_{jm} & k^3_{ij} \\ & b^*_{jj} + k^3_{mm} & k^3_{im} \\ \text{symmetric} & & k^3_{ii} \end{bmatrix} \begin{Bmatrix} \delta_{11} \\ \delta_8 \\ \delta_2 \end{Bmatrix} = \begin{Bmatrix} p^{2*}_{11} + p^3_{11} \\ (p^{1*}_8 + p^2_8)^* + p^3_8 \\ p^3_2 \end{Bmatrix} \tag{4.7a}$$

or

$$\begin{bmatrix} c_{ii} & c_{ij} & c_{im} \\ & c_{jj} & c_{jm} \\ & & c_{mm} \end{bmatrix} \begin{Bmatrix} \delta_{11} \\ \delta_8 \\ \delta_2 \end{Bmatrix} = \begin{Bmatrix} p^*_{11} \\ p^*_8 \\ p^*_2 \end{Bmatrix} \tag{4.7b}$$

The element definition vector is $[-2, \ -11, \ -8]$ and therefore δ_2 is eliminated first, δ_{11} next, finally ending with equation(s) containing δ_8 only.

Thus the first step yields

$$\begin{bmatrix} c^*_{ii} & c^*_{ij} & 0 \\ & c^*_{jj} & 0 \\ & & 0 \end{bmatrix} \begin{Bmatrix} \delta_{11} \\ \delta_8 \\ 0 \end{Bmatrix} = \begin{Bmatrix} p^{**}_{11} \\ p^{**}_8 \\ 0 \end{Bmatrix} \tag{4.8}$$

after (c_{im}, c_{jm}, c_{mm}, p^*_2) were removed to disc storage.

The second step yields

$$\begin{bmatrix} 0 & 0 & 0 \\ & c^{**}_{jj} & 0 \\ & & 0 \end{bmatrix} \begin{Bmatrix} 0 \\ \delta_8 \\ 0 \end{Bmatrix} = \begin{Bmatrix} 0 \\ p^{***}_8 \\ 0 \end{Bmatrix} \tag{4.9}$$

and the coefficients removed to disc storage were (c^*_{ii}, c^*_{ij}, 0, p^{**}_{11}).

The final step involves the transfer of the remaining equation(s) to disc storage ($0, c_{jj}^{**}, 0, p_8^{***}$). This then locates all the reduced equations in one position, the disc storage, to enable the use of a simple and repetitive backsubstitution procedure.

In the example problem only three sets of equations were dealt with at each step, each 'set' consisting of two equations corresponding to the two components, δ_x and δ_y, for each displacement.

For a bigger structural stiffness matrix, say 6 × 6 with the fourth equation being fully assembled, the reduced form would be

$$\begin{bmatrix} k_{11} & k_{12} & k_{13} & 0 & k_{15} & k_{16} \\ & k_{22} & k_{23} & 0 & k_{25} & k_{26} \\ & & k_{33} & 0 & k_{35} & k_{36} \\ & & & 0 & 0 & 0 \\ & & & & k_{55} & k_{56} \\ \text{symmetric} & & & & & k_{66} \end{bmatrix} \tag{4.10}$$

and the coefficients removed to core storage would be (k_{14}, k_{24}, k_{34}, k_{44}, k_{45}, k_{46}, p_4).

Subroutine ASMBLE

```
C     INSERT COMMON BLOCK

      N1=0
      DO 24 INOD=1,NNODZ
      IDES=NELDES(INOD)
      I1=(IDES-1)*NVABZ ---------------------------------- (1)
      I2=(INOD-1)*NVABZ ---------------------------------- (2)
      DO 3 I=1,NVABZ
      SRS(I1+I)=SRS(I1+I)+RS(I2+I) ----------------------- (3)
    3 CONTINUE
      DO 23 JNOD=INOD,NNODZ
      JDES=NELDES(JNOD)
      I3=(JDES-1)*NVABZ ---------------------------------- (4)
      I4=(JNOD-1)*NVABZ ---------------------------------- (5)
      DO 22 I=1,NVABZ
      ISS=I1+I
      NW(ISS)=1 ------------------------------------------ (6)
      IS=I2+I
      DO 20 J=1,NVABZ
      JSS=I3+J
      IF(IDES.GT.JDES) GO TO 19 -------------------------- (7)
      JS=I4+J
      IF(IS.GT.JS) GO TO 20
      N1=N1+1 -------------------------------------------- (8)
      LOC=JW(ISS)+JSS-ISS -------------------------------- (9)
      SS(LOC)=SS(LOC)+S(N1) ------------------------------ (10)
      GO TO 20
   19 N1=N1+1 -------------------------------------------- (8)
      LOC=JW(JSS)+ISS-JSS -------------------------------- (9)
      SS(LOC)=SS(LOC)+S(N1) ------------------------------ (11)
```

```
   20 CONTINUE
   22 CONTINUE
   23 CONTINUE
   24 CONTINUE
      RETURN
      END
```

(1) Vertical (I) destination in the structural stiffness matrix.

(2) Vertical (I) location of the next element coefficients to be assembled.

(3) Assembling the local vector.

(4),(5) As for (1) and (2), except horizontal (J) components.

(6) Setting corresponding location in heading vector to non-zero.

(7) If destination in structural stiffness is in the lower triangle then the transformed location is used.

(8) N1 is the location in the element stiffness vector of the next coefficient to be assembled.

(9) LOC is the location in the structural stiffness vector into which the element stiffness coefficient is to be assembled.

(10) Assembly.

(11) Assembly in the transformed location.

Subroutine FRONT

```
C     INSERT COMMON BLOCK


      REWIND 2
      REWIND 4
      DO 1 I=1,MAXFW
      NW(I)=0
    1 CONTINUE
      JW(1)=1
      DO 2 I=2,MAXFW                                  (1)
      JW(I)=JW(I-1)+MAXFW+2-I
    2 CONTINUE
      MZM=0
      NZN=0
      DO 42 NEL=1,MAXNEL                              (2)
      READ(4) S,RS,NELDEF,NELDES                      (3)
      CALL ASMBLE
      NSNW=1
      NFNW=MAXEW
      DO 40 NOD=1,NNODZ
      NIC=-NELDEF(NOD)                                (4)
      IF(NIC.LE.0) GO TO 40
      LIN=NELDES(NOD)                                 (5)
      CALL PRECON
      DO 38 NODV=1,NVABZ
    9 IF(NW(NSNW).NE.0) GO TO 10
      NSNW=NSNW+1
      GO TO 9
   10 IF(NW(NFNW).NE.0) GO TO 11                      (6)
      NFNW=NFNW-1
      GO TO 10
   11 LIV=(LIN-1)*NVABZ+NODV
      IF((NZN-NSNW+NFNW+1).LT.MAXRED) GO TO 12
      WRITE(2)  MZM,NZN,REQ,LRED                      (7)
      MZM=0
      NZN=0
   12 N1=JW(LIV)
      PIVOT=SS(N1)
```

```
      IF(PIVOT.LE.0.0) WRITE(6,1050) NEL,NIC,NODV,PIVOT
      N5=NZN
      DO 26 I=NSNW,LIV
      NZN=NZN+1
      N3=JW(I)+LIV-I
      REQ(NZN)=SS(N3)
      SS(N3)=0.0
   26 CONTINUE
      N7=NZN
      REQ(NZN)=0.0                                        (9)
      IF(LIV.EQ.NFNW) GO TO 29
      N3=JW(LIV)+1                                                  (8)
      N4=JW(LIV)+NFNW-LIV
      DO 28 I=N3,N4
      NZN=NZN+1
      REQ(NZN)=SS(I)
      SS(I)=0.0
   28 CONTINUE
   29 NZN=NZN+1
      REQ(NZN)=SRS(LIV)
      SRS(LIV)=0.0
      LREQ(MZM+1)=NSNW
      LREQ(MZM+2)=NFNW                                              (10)
      LREQ(MZM+3)=LIV
      MZM=MZM+3
      NW(LIV)=0
      DO 36 J=NSNW,NFNW
      N5=N5+1
      IF(NW(J).EQ.0) GO TO 36
      CFACT=REQ(N5)/PIVOT
      N6=N5-1
      LOC=JW(J)-1
      DO 32 I=J,NFNW
      N6=N6+1                                                       (11)
      LOC=LOC+1
      SS(LOC)=SS(LOC)-CFACT*REQ(N6)
   32 CONTINUE
      SRS(J)=SRS(J)-CFACT*REQ(NZN)
   36 CONTINUE
      REQ(N7)=PIVOT                                                 (12)
   38 CONTINUE
   40 CONTINUE
   42 CONTINUE
      RETURN
 1050 FORMAT(//,5X,22HZERO OR NEGATIVE PIVOT,/,05X,10HELEMENT NO,I5,/,
     1            5X,7HNODE NO,I8,/,5X,5HNVABZ,I10,/,5X,7HPIVOT =,E10.3)
      END
```

(1) Determine the locations of the pivots in the structural stiffness vector.
(2) Loop 42 is the element loop.
(3) For each element in turn read the stiffness, load vector, definition vector and destination vector.
(4) Check for the last appearance of each of the element nodes.
(5) Location of the pivotal equations.
(6) Reset the start and finish flags for the heading vector.
(7) If the temporary storage buffer for reduced equations is full, the buffer is transferred to permanent storage.
(8) The next equation to be used for reduction is stored in the temporary storage buffer.
(9) Set the pivotal coefficient in the stored equation to zero.
(10) Store the start and finish flags and the position of the pivot.
(11) Reduce the stiffness and load vectors.
(12) Replace the value of the pivot in the reduced equation.

4.2.3 BACKSUBSTITUTION

The backsubstitution is a more complicated procedure in the front solver than in the band solver because of the variable location of the pivotal coefficients in the stored equations. The non-sequential nature of the displacement vector likewise complicates the procedure. However, if a reverse procedure to reduction is used to determine the location of the pivotal coefficient in each of the stored equations and also the associated nodal number, then a single repetitive system can be set up to determine the displacements.

The element destination vectors are used to indicate the position of the pivots and the element definition vectors to determine the displacement associated with a particular pivot.

The data which is stored after the completion of reduction is listed in Table 4.1.

TABLE 4.1 *Data stored for use in backsubstitution*

Element definition vector	-4	8	10
	8	11	-10
	-2	-11	-8
Element destination vector	1	2	3
	2	1	3
	3	1	2
Reduced equations	$(k^1_{ii}, k^1_{ij}, k^1_{im}, p^1_4)$		
	$(b_{im}, b_{jm}, b_{mm}, (p^{1*}_{10}+p^2_{10}))$		
	$(c_{im}, c_{jm}, c_{mm}, p^*_2)$		
	$(c^*_{ii}, c^*_{ij}, 0, p^{**}_{11})$		
	$(0, c^*_{jj}, 0, p^{***}_8)$		

The backsubstitution procedure must be made automatic and repetitive so that an identical process is carried out to calculate the value of each nodal displacement. This programming philosophy saves a great amount of complicated FORTRAN coding and hence execution time. The method implemented in the program which is listed at the end of this section uses a working vector equal in length to the front width. For the two-dimensional problem this is equal to the nodal front width multiplied by 2, i.e. 3 × 2 = 6 in the present case.

Initially the vector contains only zeros but later on this changes to the vector of displacements which are being continuously updated as the backsubstitution progresses.

The procedure commences by searching the element definition vectors for negative node numbers starting with the last element and working from the right to left of each definition vector. This operation determines the reverse order to which the nodal equations were eliminated. The first negative node number detected is that of node 8, and from Table 4.1 it is seen that the corresponding coefficient in the destination vector is 2, indicating that the second coefficient, c^{**}_{jj}, in the stored equation is the pivot.

As indicated earlier in both the reduction and backsubstitution stages of the program each equation of the set of equations related to a node is considered separately. Considering the last equation $c^{**}_{jj}\delta_8 = p^{***}_8$ in its expanded form it can be represented as

$$\begin{bmatrix} {}_{11}c^{**}_{jj} & {}_{12}c^{**}_{jj} \\ 0 & ({}_{22}c^{**}_{jj})^* \end{bmatrix} \begin{bmatrix} \delta_{8x} \\ \delta_{8y} \end{bmatrix} = \begin{bmatrix} p^{***}_{8x} \\ (p^{***}_{8y})^* \end{bmatrix}$$

The stored equation relating to δ_{8y} is therefore

$$[0, \; 0, \; 0, \; ({}_{22}c^{**}_{jj})^*, \; 0, \; 0, \; (p^{***}_{8y})^*]$$

To make the process automatic a local variable PIVOT (see computer program) is set equal to the value of the pivotal coefficient which in turn is set equal to zero. Therefore at this instant

$$\text{PIVOT} = ({}_{22}c^{**}_{jj})^*$$

and the stored equation becomes $(0, 0, 0, 0, 0, 0, (p^{***}_{8y})^*)$. The reason for setting the pivotal coefficient to zero will become clear later.

The standard equation for calculating a displacement can be established as

$$\text{Nodal displacement} = \frac{\text{Load - (Stored coefficients)(Running vector of displacements)}}{\text{PIVOT}}$$

The above equation is applicable to all nodes, and this includes the last node since

$$\delta_{8y} = \frac{\left((p^{***}_{8y})^* - [0, 0, 0, 0, 0, 0][0, 0, 0, 0, 0, 0]^T\right)}{({}_{22}c^{**}_{jj})^*}$$

is the result that would have been obtained if the last stored equation had been used directly.

The value of δ_{8y} is now stored in the appropriate location in the running vector of displacements as determined by the destination vector and the running vector changes from $(0, 0, 0, 0, 0, 0)$ to $(0, 0, 0, \delta_{8y}, 0, 0)$.

The same approach is used to determine δ_{8x}.

PIVOT = ${}_{11}c^{**}_{jj}$

and the stored equation is modified into $(0, 0, 0, {}_{12}c^{**}_{jj}, 0, 0, p^{***}_{8x})$.
Hence δ_{8x} is determined by the standard equation,

$$\delta_{8x} = \frac{p^{***}_{8x} - [0, 0, 0, {}_{12}c^{**}_{jj}, 0, 0][0, 0, 0, \delta_{8y}, 0, 0]^T}{{}_{11}c^{**}_{jj}}$$

and the running vector of displacements is updated to $(0, 0, \delta_{8x}, \delta_{8y}, 0, 0)$.

To avoid continuous repetition the operations to calculate the x and y components of the displacement at a node will be written in terms of the submatrices. The above two operations to calculate δ_{8x} and δ_{8y} could be written as

Stored Pivot = c^{**}_{jj} and

$$\delta_8 = [c^{**}_{jj}]^{-1}(p^{***}_8 - [0, 0, 0][0, 0, 0]^T)$$

and the value of δ_8 stored in the running vector changes the running vector to $(0, \delta_8, 0)$.

The search continues through the definition vectors for the next negative node which is node 11 with a destination of 1. Using the same process as for node 8

Stored pivot = c^{**}_{ii}

and the stored equation is modified into $(0, c^{*}_{ij}, 0, p^{**}_{11})$.
Hence δ_{11} is determined by the standard equation

$$\delta_{11} = [c^{*}_{ii}]^{-1}(p^{**}_{11} - [0, c^{*}_{ij}, 0][0, \delta_8, 0]^T)$$

and the running vector of displacements is updated to $(\delta_{11}, \delta_8, 0)$.

Node 2 is the next negative node and has a destination of 3. Following the same procedure we have

Stored Pivot = c_{mm}

Modified equation = $(c_{im}, c_{jm}, 0, p^{*}_2)$

$$\delta_2 = [c_{mm}]^{-1}(p^{*}_2 - [c_{im}, c_{jm}, 0][\delta_{11}, \delta_8, 0]^T)$$

and the running vector of displacements is now $(\delta_{11}, \delta_8, \delta_2)$.

The next node with a negative node number is node 10 in element 2 with a destination of 3, giving

Stored Pivot = b_{mm}

Modified equation = $(b_{im}, b_{jm}, 0, (p^{1*}_{10} + p^{2}_{10}))$

$$\delta_{10} = [b_{mm}]^{-1}\left((p^{1*}_{10} + p^{2}_{10}) - [b_{im}, b_{jm}, 0][\delta_{11}, \delta_8, \delta_2]^T\right)$$

The need to set the pivotal coefficient to zero is now seen clearly. Had this not been done the vector multiplication in the above equation would have produced a term containing δ_2 which is not found in the original equation (*see* Eqn.(4.4b)).

The running vector now becomes (δ_{11}, δ_8, δ_{10}) and the last node is node 4 from element 1 with a destination of 1. The final step of the backsubstitution procedure ends with

Stored Pivot = k^1_{ii}

Modified equation = $(0,\ k^1_{ij},\ k^1_{im},\ p^1_4)$

$$\delta_4 = [k^1_{ii}]^{-1}\left(p^1_4 - [0,\ k^1_{ij},\ k^1_{im}][\delta_{11},\ \delta_8,\ \delta_{10}]^T\right)$$

and the running vector becomes (δ_4 δ_8 δ_{10}).

Subroutine BAKSUB

```
C     INSERT COMMON BLOCK


      DO 300 NEL=1,MAXNEL
      MEL=MAXNEL+1-NEL
      NNN=1
      DO 299 I=1,NNODZ
      N1=NNODZ+1-I
      NIC=-LDEF(MEL,N1)
      IF(NIC.LE.0) GO TO 299 ---------------- (1)
      DO 230 NODV=1,NVABZ
      IF(NZN.NE.0) GO TO 219       --+
      BACKSPACE 2                    |  (2)
      READ(2) MZM,NZN,REQ,LREQ       |
      BACKSPACE 2                  --+
  219 MZM=MZM-3                    --+
      NSNW=LREQ(MZM+1)               |  (3)
      NFNW=LREQ(MZM+2)               |
      LIV=LREQ(MZM+3)              --+
      GASH=REQ(NZN) ------------------------ (4)
      PIVOT=REQ(NZN-NFNW-1+LIV)
      REQ(NZN-NFNW-1+LIV)=0.0 -------------- (5)
      N2=NZN-NFNW-2+NSNW
      NZN=N2                       --+
      DO 220 J=NSNW,NFNW             |
      N2=N2+1                        |  (6)
      GASH=GASH-SRS(J)*REQ(N2)       |
  220 CONTINUE                       |
      SRS(LIV)=GASH/PIVOT          --+
      N3=NVABZ+1-NODV
      DISPL(NIC,N3)=GASH/PIVOT ------------- (7)
  230 CONTINUE
      CALL POSTCN(NNN,MEL)
  299 CONTINUE
  300 CONTINUE
      IF(MAXTNS.NE.0) CALL BTRANS
      RETURN
      END
```

(1) Determine the nodal order of back substitution.
(2) Refill the temporary storage buffer if empty.
(3) Location of the pivot, start and finish flags.
(4) Load coefficient.
(5) Set the pivot to zero.
(6) Calculate the displacement.
(7) Store the calculated displacement.

4.2.4 PRE-CONSTRAINTS AND POST-CONSTRAINTS

The technique for imposing specified displacements for the simple program in Chapter 2 will also be used in the front solver, and it was seen that in the simple program all the nodes were 'aware' of a specified displacement, say δ_n, because the load terms for all the equations were modified simultaneously. However, for a front solver only a small number of equations are in core at any one time and equation n is only fully assembled immediately prior to being used for reduction, which means that it is not possible to specify δ_n beforehand. Hence although all the nodes in core at that time and all subsequent nodes brought into core will be 'aware' of the specified δ_n, nodes eliminated prior to that time have no such 'awareness' since their load terms have not been modified.

The operations for imposing specified displacements is therefore split into two parts. The pre-constraint subroutine PRECON is called during the reduction process immediately before a node is eliminated. If node n has a specified displacement then the current load vector is modified and the pivot is increased by a big spring stiffness, in accordance with Eqn.(2.39).

The post-constraint subroutine POSTCN is called immediately after the computation of a nodal displacement during the backsubstitution process. When it comes across node n with specified displacement δ_n it will firstly calculate the reaction (see Method 4, Section 2.7) and secondly replace the small calculated displacement by δ_n in the running displacement vector. In this way the previously eliminated nodes are also made 'aware' of the specified displacement since δ_n will actually be used in calculating the displacements of these nodes.

Subroutine PRECON

```
C     INSERT COMMON BLOCK


      IF(MAXDIS.EQ.0) GO TO 175                         (1)
      DO 170 I=1,MAXDIS
      IF(NIC.NE.NDISP(I,3)) GO TO 170
      DO 160 J=1,NVABZ
      IF(NDISP(I,J).EQ.0) GO TO 160                     (2)
      DSP=DISP(I,J)
      N1=(LIN-1)*NVABZ+J
      DO 140 K=NSNW,N1
      N2=JW(K)
      SRS(K)=SRS(K)-DSP*SS(N2+N1-K)
  140 CONTINUE                                          (3)
      IF(N1.EQ.NFNW) GO TO 159
```

```
      N3=N1+1
      DO 150 K=N3,NFNW
      SRS(K)=SRS(K)-DSP*SS(N2-N1+K)
  150 CONTINUE
  159 SS(N2)=1.0E+50+SS(N2) ------------------------------- (4)
  160 CONTINUE
  170 CONTINUE
  175 IF(MAXNDZ.EQ.0) RETURN ------------------------------ (5)
      DO 190 I=1,MAXNDZ
      IF(NIC.NE.NDISPZ(I,3)) GO TO 190
      DO 180 J=1,NVABZ
      IF(NDISPZ(I,J).EQ.0) GO TO 180 ---------------------- (6)
      N1=(LIN-1)*NVABZ+J
      N2=JW(N1)
      SS(N2)=SS(N2)+1.0E+50 ------------------------------- (7)
  180 CONTINUE
      RETURN
  190 CONTINUE
      RETURN
      END
```

(1) Check if any of the nodes have specified non-zero displacements.
(2) Determine which variables have been specified.
(3) Subtrace from the load vector the specified displacement multiplied by the associated column of the stiffness matrix.
(4) Add the big spring stiffness to the pivotal coefficient.
(5) Check if any of the nodes have specified earthed displacements.
(6) Determine which variables have been specified.
(7) Add the big spring stiffness to the pivotal coefficient.

Subroutine POSTCN(NNN, MEL)

```
C     INSERT COMMON BLOCK

      DIMENSION RACTN(2)
      DO 232 I=1,NVABZ
      RACTN(I)=0.0
  232 CONTINUE
      IF(MAXDIS.EQ.0) GO TO 251 --------------------------- (1)
      DO 250 I=1,MAXDIS
      IF(NIC.NE.NDISP(I,3)) GO TO 250
      DO 240 J=1,NVABZ
      IF(NDISP(I,J).EQ.0) GO TO 240 ----------------------- (2)
      DSP=DISP(I,J)
      N1=LIV+J-1
      RACTN(J)=-SRS(N1)*1.0E+50 --------------------------- (3)
      SRS(N1)=DSP
      DISPL(NIC,J)=DSP ------------------------------------ (4)
  240 CONTINUE
      IF(NNN.EQ.1) WRITE(6,1005) MEL
      NNN=0
      WRITE(6,1200) NIC,RACTN
  250 CONTINUE
  251 IF(MAXNDZ.EQ.0) RETURN ------------------------------ (5)
```

```
      DO 270 I=1,MAXNDZ
      IF(NIC.NE.NDISPZ(I,3)) GO TO 270
      DO 260 J=1,NVABZ
      IF(NDISPZ(I,J).EQ.0) GO TO 260 ------------------------(6)
      N1=LIV+J-1
      RACTN(J)=-SRS(N1)*1.0E+50 -----------------------------(7)
      SRS(N1)=0.0
      DISPL(NIC,J)=0.0 --------------------------------------(8)
  260 CONTINUE
      IF(NNN.EQ.1) WRITE(6,1005) MEL
      NNN=0
      WRITE(6,1200) NIC,RACTN
  270 CONTINUE
      RETURN
 1005 FORMAT(///,5X,18HREACTIONS ELEMENT ,I3,//,
     1       7X,*NODE*,7X,*1ST COMP*,7X,*2ND COMP*)
 1200 FORMAT(/,6X,I5,3(5X,E10.3))
      END
```

(1) Check if any node has any non-zero displacements.
(2) Determine which variables have been specified.
(3) Calculate the reaction.
(4) Replace the small calculated displacement by the specified displacement.
(5) Check if any node has specified earthed displacements.
(6) Determine which variables have been specified.
(7) Calculate the reaction.
(8) Replace the small calculated displacement by zero.

4.3 Transformed constraints

It is very often possible to reduce the size of the structure to be modelled by using locally oriented displacements. For example the simple problem in Chapter 2 is axisymmetric, and instead of analysing a quarter of the disc it is really only necessary to use a small segment for a complete analysis, as shown in Fig.4.3.

From Fig.4.3(b) it is seen that some of the nodes are now defined with

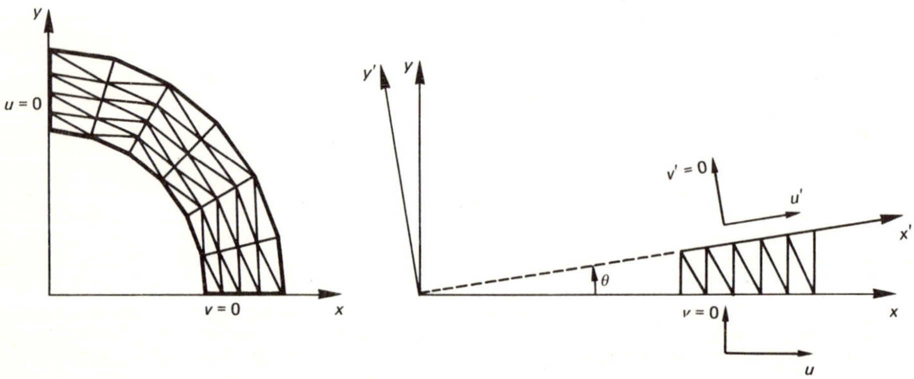

(a) Mesh layout for standard constraints (b) Mesh layout for transformed constraints

Fig.4.3 Two different mesh layouts for an axisymmetric problem

respect to their local axes, and Eqn.(1.45) which was originally given in global axes must undergo some sort of transformation process. The transformation between two sets of coordinate systems is well known [2] and is given by

$$\{\delta_i\} = \begin{Bmatrix} u_i \\ v_i \end{Bmatrix} = [L]\begin{Bmatrix} u_i' \\ v_i' \end{Bmatrix} = [L]\{\delta_i'\}$$

$$\{F_i'\} = \begin{Bmatrix} F_{xi}' \\ F_{yi}' \end{Bmatrix} = [L]^T\begin{Bmatrix} F_{xi} \\ F_{yi} \end{Bmatrix} = [L]^T\{F_i\} \qquad (4.11)$$

and

$$[L] = \begin{bmatrix} \cos\theta & -\sin\theta \\ \sin\theta & \cos\theta \end{bmatrix}$$

In the above equations the global variables are non-primed and the local variables are primed, with θ taken as positive as shown by the arrow in Fig.4.3(b).

It is now assumed that the displacements of node i are locally oriented, then Eqn.(1.45) should be rewritten as

$$\begin{Bmatrix} F_i \\ F_{i-1} \\ F_i' \\ F_{i+1} \\ F_{11} \end{Bmatrix} = \begin{bmatrix} I & & & & \\ & I & & & \\ & & L^T & & \\ & & & I & \\ & & & & I \end{bmatrix}\begin{Bmatrix} F_1 \\ F_{i-1} \\ F_i \\ F_{i+1} \\ F_{12} \end{Bmatrix}$$

$$= \begin{bmatrix} I & & & & \\ & I & & & \\ & & L^T & & \\ & & & I & \\ & & & & I \end{bmatrix}[K]\begin{Bmatrix} \delta_1 \\ \delta_{i-1} \\ \delta_i \\ \delta_{i+1} \\ \delta_n \end{Bmatrix}$$

$$= \begin{bmatrix} I & & & & \\ & I & & & \\ & & L^T & & \\ & & & I & \\ & & & & I \end{bmatrix}[K]\begin{bmatrix} I & & & & \\ & I & & & \\ & & L & & \\ & & & I & \\ & & & & I \end{bmatrix}\begin{Bmatrix} \delta_1 \\ \delta_{i-1} \\ \delta_i' \\ \delta_{i+1} \\ \delta_n \end{Bmatrix}$$

$$
= \begin{bmatrix}
K_{11} & \cdots & K_{1,i-1} & K_{1,i}L & K_{1,i+1} & \cdots & K_{1,n} \\
\vdots & & \vdots & \vdots & \vdots & & \vdots \\
K_{i-1,1} & \cdots & K_{i-1,i-1} & K_{i-1,i}L & K_{i-1,i+1} & \cdots & K_{i-1,n} \\
L^T K_{i,1} & \cdots & L^T K_{i,i-1} & L^T K_{i,i}L & L^T K_{i,i+1} & \cdots & L^T K_{i,n} \\
K_{i+1,1} & \cdots & K_{i+1,i-1} & K_{i+1,i}L & K_{i+1,i+1} & \cdots & K_{i+1,n} \\
\vdots & & \vdots & \vdots & \vdots & & \vdots \\
K_{n,1} & \cdots & K_{n,i-1} & K_{n,i}L & K_{n,i+1} & \cdots & K_{n}
\end{bmatrix}
\begin{Bmatrix}
\delta_1 \\ \vdots \\ \delta_{i-1} \\ \delta_i' \\ \delta_{i+1} \\ \vdots \\ \delta_n
\end{Bmatrix}
\tag{4.12}
$$

Hence whenever a variable is transformed into a local set of axes, all that is required in carrying out the transformation is to post-multiply the appropriate column by $[L]$ and then pre-multiply the appropriate row by $[L]^T$. This pre- and post-multiplication is carried out by subroutine TRANSF. After the solution of the equations the local displacements should be reorientated into the global system before element stresses are computed from the displacements. The reorientation of local displacements is carried out by the subroutine BTRANS.

Subroutine TRANSF

```
C     INSERT COMMON BLOCK


      DO 500 I=1,MAXTNS
      NIC=NTRANS(I,1)
      DO 400 J=1,NNODZ
      IF(NELDEF(J).NE.NIC) GO TO 400                                  (1)
      NTR=NTRANS(I,2)                                                 (2)
      NLOC=0
      IF(J.EQ.1) GO TO 150
      N1=J-1
      DO 100 I1=1,N1                                                  (3)
      LOC=NLOC+3+(J-I1-1)*4                                           (4)
      DO 50 I2=1,4
      CW(I2)=S(I2+LOC)
   50 CONTINUE
      S(LOC+1)=CW(1)*TRANS(NTR,1)+CW(2)*TRANS(NTR,3)
      S(LOC+2)=CW(1)*TRANS(NTR,2)+CW(2)*TRANS(NTR,4)                  (5)
      S(LOC+3)=CW(3)*TRANS(NTR,1)+CW(4)*TRANS(NTR,3)
      S(LOC+4)=CW(3)*TRANS(NTR,2)+CW(4)*TRANS(NTR,4)
      NLOC=NLOC+3+(8-I1)*4
  100 CONTINUE
  150 DO 160 I2=1,3
      CW(I2)=S(I2+NLOC)
  160 CONTINUE
      A2=CW(1)*TRANS(NTR,2)+CW(2)*TRANS(NTR,4)
      A5=CW(2)*TRANS(NTR,2)+CW(3)*TRANS(NTR,4)
      S(NLOC+2)=TRANS(NTR,1)*A2+TRANS(NTR,3)*A5                       (6)
      S(NLOC+3)=TRANS(NTR,2)*A2+TRANS(NTR,4)*A5
      S(NLOC+1)=TRANS(NTR,1)*(CW(1)*TRANS(NTR,1)+CW(2)*TRANS(NTR,3))+
     2          TRANS(NTR,3)*(CW(2)*TRANS(NTR,1)+CW(3)*TRANS(NTR,3))
```

```
      NLOC=NLOC+3
      N2=(J-1)*NVABZ
      A1=RS(N2+1)
      A2=RS(N2+2)
      RS(N2+1)=TRANS(NTR,1)*A1+TRANS(NTR,3)*A2        (7)
      RS(N2+2)=TRANS(NTR,2)*A1+TRANS(NTR,4)*A2
      IF(J.EQ.NNODZ) GO TO 500
      N3=J+1
      DO 300 I3=N3,NNODZ                              (8)
      DO 250 I4=1,4
      CW(I4)=S(NLOC+I4)
  250 CONTINUE
      S(NLOC+1)=TRANS(NTR,1)*CW(1)+TRANS(NTR,3)*CW(3)
      S(NLOC+2)=TRANS(NTR,1)*CW(2)+TRANS(NTR,3)*CW(4)
      S(NLOC+3)=TRANS(NTR,2)*CW(1)+TRANS(NTR,4)*CW(3) (9)
      S(NLOC+4)=TRANS(NTR,2)*CW(2)+TRANS(NTR,4)*CW(4)
      NLOC=NLOC+4
  300 CONTINUE
      GO TO 500
  400 CONTINUE
  500 CONTINUE
      RETURN
      END
```

(1) Check if any of the element nodes have locally defined axes.
(2) NTR is the number of the transformation matrix.
(3) Loop 100 transforms the column.
(4) Location is SS( ) vector of each 2 × 2 submatrix in turn.
(5) Transformation of each 2 × 2 matrix.
(6) Transformation of the upper triangle of the 2 × 2 pivotal matrix.
(7) Transformation of the load vector.
(8) Loop 300 transforms the row.
(9) Transformation of each 2 × 2 matrix along the row.

Subroutine BTRANS

```
C     INSERT COMMON BLOCK

      DO 10 I=1,MAXTNS                                (1)
      NIC=NTRANS(I,1)
      NTR=NTRANS(I,2)                                 (2)
      A1=DISPL(NIC,1)
      A2=DISPL(NIC,2)
      DISPL(NIC,1)=A1*TRANS(NTR,1)+A2*TRANS(NTR,2)    (3)
      DISPL(NIC,2)=A1*TRANS(NTR,3)+A2*TRANS(NTR,4)
   10 CONTINUE
      RETURN
      END
```

(1) Loop 10 considers each of the transformed nodes in turn.
(2) NTR is the number of the transformation matrix.
(3) Transformation of the displacements back into the global system.

4.4 Improvements to the efficiency of execution

4.4.1 HEADING VECTOR

During various stages in the reduction process the front area occasionally contains blocks of zero coefficients. It is possible to increase the efficiency of execution by using an indicator called the heading vector to identify the locations of these zero blocks so that they can be completely ignored for computational purposes. It is also possible via the heading vector to store a variable length equation that does not contain any zero coefficients which occur either at the beginning or the end of the equation, so that data transfer between core and disc can be reduced.

The heading vector contains as many locations as the maximum front width of the problem. If a column, and hence the corresponding row, contains only zero coefficients then the associated location in the heading vector is set to zero. Likewise non-zero coefficients in the heading vector indicate non-zero coefficients in the associated column and row.

Consider the frontal area of a structural stiffness matrix, as shown in Fig.4.4 in which for argument's sake only the crosses represent non-zero numbers

$$[0\ 0\ 0\ 1\ 1\ 1\ 1\ 1\ 1\ 0\ 0\ 0] \quad \text{Heading vector}$$

$$\begin{bmatrix}
0&0&0&0&0&0&0&0&0&0&0&0\\
&0&0&0&0&0&0&0&0&0&0&0\\
&&0&0&0&0&0&0&0&0&0&0\\
&&&\times&\times&\times&\times&\times&\times&0&0&0\\
&&&&\times&\times&\times&\times&\times&0&0&0\\
&&&&&\times&\times&\times&\times&0&0&0\\
&&&&&&\times&\times&\times&0&0&0\\
&&&&&&&\times&\times&0&0&0\\
&&&&&&&&\times&0&0&0\\
&&&&&&&&&0&0&0\\
&&&&&&&&&&0&0\\
&&&&&&&&&&&0
\end{bmatrix}
\begin{Bmatrix} -\\-\\-\\ \delta_4\\ \delta_5\\ \delta_6\\ \delta_7\\ \delta_8\\ \delta_9\\ -\\-\\- \end{Bmatrix}
=
\begin{Bmatrix} -\\-\\-\\ F_4\\ F_5\\ F_6\\ F_7\\ F_8\\ F_9\\ -\\-\\- \end{Bmatrix}$$

Fig.4.4 Use of the heading vector in a front solver

If the next variable to be eliminated is to be δ_6 then the usual method would be to store the coefficient relating to node 6

$$(0 \quad 0 \quad 0 \quad k_{46} \quad k_{56} \quad k_{66} \quad k_{67} \quad k_{68} \quad k_{69} \quad 0 \quad 0 \quad 0 \quad F_6)$$

This is wasteful of transfer time as there are six zero coefficients in the equation. If the heading vector is utilized and, indicating the first and last column and row positions with non-zero coefficient are kept updated after each elimination, then only part of the row and column between the start and finish flags need be considered. For the above matrix the start flag NSNW=4 and the finish flag NFNW=9 effectively contract the front area to that enclosed by the broken line. The coefficients for node 6 that are stored are then

$(k_{46} \quad k_{56} \quad k_{66} \quad k_{67} \quad k_{68} \quad k_{69} \quad F_6)$ NSNW NFNW.

The two flags also have to be retained at the end of each equation to indicate how many coefficients have been deleted.

Blocks of zeros occurring in the middle of non-zero coefficients must be stored since the above technique requires storage of all coefficients between the first and last non-zero coefficients of an equation. But by using the heading vector rows containing only zero coefficients can be ignored thus reducing the number of coefficients to be operated upon during the reduction process (see discussion below concerning the elimination of δ_6, δ_4 and δ_3, using the normal front).

The efficiency of such a scheme can be seen by comparing the number of non-zero coefficients in the matrix that would be operated upon when using the heading vector compared to the total number of coefficients in the front area.

The technique of using the heading vector has been implemented into the solution subroutines of the listed program.

4.4.2 LONGEVITY FRONT [3]

The front solver in its present stage of development still occasionally allows columns containing only zero coefficients to occur in the front area among columns containing non-zero coefficients. However, the frequency of such occurrence can be reduced by rearranging the storage of the nodes in the front area according to their 'life span'. Nodes which exist in the front area over the greatest number of elements, the longest lived nodes, are assigned storage locations before nodes with shorter lives.

The easiest means of showing the effect of this longevity is to perform a simple operation count on a two element problem solved firstly by a normal front solver and then by a longevity front.

Consider the two element problem in Fig.4.5. It is assumed that both types of solver have the heading vector facility implemented.

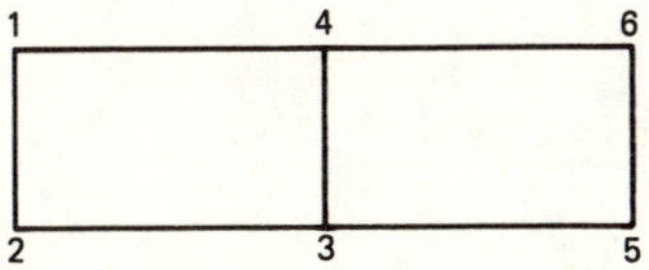

Fig.4.5 Example problem for longevity front

The element definitions are for the normal front

Element 1 [4 -1 -2 3]

2 [-5 -6 -4 -3]

and the element destinations are

Element 1 [1 2 3 4]

2 [2 3 1 4]

The operations required to reduce the upper triangle of the stiffness matrix and the load vector produced by this problem using the normal front are shown below.

First, element 1 is assembled and reduced

$$\begin{bmatrix} \times & \times & \times & \times \\ & \times & \times & \times \\ & & \times & \times \\ & & & \times \end{bmatrix} \begin{Bmatrix} \delta_4 \\ \delta_1 \\ \delta_2 \\ \delta_3 \end{Bmatrix} = \begin{Bmatrix} F_4 \\ F_1 \\ F_2 \\ F_3 \end{Bmatrix}$$

The reduction is carried out in rows so the heading vector will only check on rows not on columns.

(i) δ_1 is eliminated (negative sign for node 1),
5 sets of coefficients written to store,
14 sets of coefficients modified (including load terms).

(ii) δ_2 is eliminated (negative sign for node 2),
5 sets of coefficients written to store,
10 sets of coefficients modified.

Second, element 2 is assembled,

$$\begin{bmatrix} \times & \times & \times & \times \\ & \times & \times & \times \\ & & \times & \times \\ & & & \times \end{bmatrix} \begin{Bmatrix} \delta_4 \\ \delta_5 \\ \delta_6 \\ \delta_3 \end{Bmatrix} = \begin{Bmatrix} F_4 \\ F_5 \\ F_6 \\ F_3 \end{Bmatrix}$$

(i) δ_5 is eliminated,
5 sets of coefficients written to store,
14 sets of coefficients modified.

(ii) δ_6 is eliminated,
5 sets of coefficients written to store,
10 sets of coefficients modified (row 2 contains only zeros after reduction in (i) and is ignored completely through the use of the heading vector).

(iii) δ_4 is eliminated,
5 sets of coefficients written to store,
7 sets of coefficients modified (rows two and three ignored).

(iv) δ_3 is eliminated,
2 sets of coefficients written to store,
2 sets of coefficients modified (rows one, two and three ignored).

Thus we find that the reduction process for the normal front solver requires 84 operations in which 27 sets of coefficients are stored and 57 sets of coefficients modified.

Repeating the operation using the longevity front solver results in a lower operations count as is detailed below. It should be realized that when two nodes have the same element life, the node that occurs farthest towards the right of the definition vector in which both appear *last* is the node with the longer life.

The element definitions are unchanged:

Element 1 [4 -1 -2 3]

2 [-5 -6 -4 -3]

The element destination vectors for the longevity front are different from the normal front and are given as

Element 1 [3 1 2 4]

2 [1 2 3 4]

It is seen that nodes 3 and 4 have the longest life among the six nodes.

First, element 1 is assembled and reduced

$$\begin{bmatrix} \times & \times & \times & \times \\ & \times & \times & \times \\ & & \times & \times \\ & & & \times \end{bmatrix} \begin{Bmatrix} \delta_1 \\ \delta_2 \\ \delta_4 \\ \delta_3 \end{Bmatrix} = \begin{Bmatrix} F_1 \\ F_2 \\ F_4 \\ F_3 \end{Bmatrix}$$

(i) δ_1 is eliminated,
5 sets of coefficients written to store,
14 sets of coeffients modified.

Because the first row and column now contain only zero coefficients the front area contracts.

(ii) δ_2 is eliminated,
4 sets of coefficients written to store,
9 sets of coefficients modified.

Second, element 2 is assembled and reduced.

$$\begin{bmatrix} \times & \times & \times & \times \\ & \times & \times & \times \\ & & \times & \times \\ & & & \times \end{bmatrix} \begin{Bmatrix} \delta_5 \\ \delta_6 \\ \delta_4 \\ \delta_3 \end{Bmatrix} = \begin{Bmatrix} F_5 \\ F_6 \\ F_4 \\ F_3 \end{Bmatrix}$$

(i) δ_5 is eliminated,
5 sets of coefficients written to store,
14 sets of coefficients modified and front area contracts.

(ii) δ_6 is eliminated,
4 sets of coefficients written to store,
9 sets of coefficients modified and front area contracts further.

(iii) δ_4 is eliminated,
3 sets of coefficients written to store,
5 sets of coefficients modified and front area contracts even further.

(iv) δ_3 is eliminated,
2 sets of coefficients written to store,
2 sets of coefficients modified.

For the front solver using the longevity technique only 76 operations are required and so even on a small problem there is a saving of approximately 10% on a simple operations count. The backsubstitution would likewise be more efficient because of the shorter length of the equations. However, the saving in time during the backsubstitution stage is very small compared with the time saved during the reduction.

A word of caution is probably appropriate at this stage. To implement longevity considerations in the front solver is to proceed along the same path that occurred in the band solver when complete nodal renumbering schemes were introduced for minimizing the bandwidth and it is not impossible to spend more execution time in setting up the improved destination vectors than that which can be saved through a reduction in the number of operations. The authors' experience suggests that little advantage is gained using a longevity front solver for two-dimensional problems but significant savings can be made in three-dimensional analysis particularly where front widths in excess of 200 occur.

References

1. B. M. Irons. A frontal solution program for finite element analysis. *International Journal for Numerical Methods in Engineering*, vol.2, no.1, 1970, pp.5-32.
2. A. K. Gupta and B. M. Mohray. A method for computing numerically integrated stiffness matrices. *International Journal for Numerical Methods in Engineering*, vol.5, no.1, 1972, pp.83-89.
3. M. F. Yeo. A more efficient front solution: allocating assembly locations by longevity considerations. *International Journal for Numerical Methods in Engineering*, vol.7, no.4, 1973, pp.570-573.

5 Computer Program Listing and Computations Using Isoparametric Elements

5.1 Computer program listing

Most of the subroutines associated with the eight-node isoparametric element program have been listed in Chapters 3 and 4. The remaining subroutines, which will be described in this chapter, are the master driving program, BLOCK DATA, subroutine INTCRD to interpolate for the nodal coordinates of midside nodes on straight edges and subroutine INDAT which reads the data.

5.1.1 THE MASTER DRIVING PROGRAM

The master program or control program is normally a very simple routine whose main function is to call the various subroutines so that the sequence of operations is performed in the correct order. The secondary function of this routine is to initialize the variables, matrices and vectors to either zero or some predetermined non-zero value.

Program ISOP

```
      PROGRAM  ISOP(INPUT,OUTPUT,TAPE5=INPUT,TAPE6=OUTPUT,TAPE2,TAPE3,
     1                    TAPE4)

      COMMON/MFY1/INT1(4),INT2(4),INT3(4),XX(8),YY(8)
      COMMON/MFY2/WTFUN(3),VECTLC(3)
      COMMON RS(16),AW(4),S(136),SHP(8),P(8),CW(4),X(8),Y(8)
      COMMON DISP(100,2),TRANS(10,4),PRES(100),REACT(100,2)
      COMMON NELDES(8),NELDEF(8),LDEF(10,8)
      COMMON DX(8),DY(8),U(8),V(8),SIGMA(3),D(5),YM(5),PR(5),WT(5)
      COMMON SRS(150),SS(4000),CORD(100,2),MAT( 10),REQ(2000),LREQ(300)   (1)
      COMMON NDISP(100,3),DISPL(100,2),NW(150),JW(150),NDISPZ(100,3)
      COMMON LDEST(100),NTRANS(100,2),NPRES(100),NREACT(100)
      COMMON NVABZ,MAXDIS,NIC,LIV,NNODZ,NSNW,NFNW,MAXNDZ,MAXNEL,NEL,NZN
      COMMON MZM,MAXFW,MAXREQ,MAXNW,MAXNOD,MAXMAT,MAXPRS,MAXRCT
      COMMON YMOD,PRAT,WEIGHT,WX,WY,XL,YL,PIVOT,DETJ,WIRL,GRAVX,GRAVY
      COMMON GRAVTY,NPUT,NRULE,LIN,NSTOP,MAXSS,MAXTNS
```

```
      NSTOP=0
      WIRL=0.0
      GRAVX=0.0
      GRAVY=0.0
      MAXPRS=0
      MAXRCT=0
      MAXTNS=0
      MAXFW=0
      NRULE=3
      MAXMAT=1
      NPUT=0
      MAXNEL=0
      MAXNOD=0
      NVABZ=2                                                    (2)
      NNODZ=8
      MAXSS=4000
      MAXREQ=2000
      MAXDIS=0
      MAXNDZ=0
      DO 2 I=1,100
      DO 1 J=1,2
      CORD(I,J)=0.0
    1 CONTINUE
    2 CONTINUE
      DO 5 I=1,MAXSS
      SS(I)=0.0
    5 CONTINUE
      DO 10 I=1,150
      SRS(I)=0.0
   10 CONTINUE
      MAXNW=150
      CALL INDAT                                                 (3)
      CALL PREFNT                                                (4)
      CALL STIFN                                                 (5)
      CALL FRONT                                                 (6)
      CALL BAKSUB                                                (7)
      CALL STRESS                                                (8)
      STOP
      END
```

(1) Common block which should be inserted in all subroutines.
(2) Initialization of variables, matrices and vectors.
(3) INDAT reads the data for the problem.
(4) PREFNT determines the nodal destinations and the maximum front width.
(5) STIFN calculates the element stiffnesses and load vectors.
(6) FRONT assembles and reduces the structural stiffness matrix and load vector.
(7) BAKSUB calculates the nodal displacements from the reduced equations set up by FRONT.
(8) STRESS determines the element stresses.

5.1.2 BLOCK DATA

Subroutine BLOCK DATA is used, in this case, to initialize vectors efficiently. To carry out this initialization in the master driving program would have necessitated considerably more FORTRAN statements. There is no call to this subroutine as it is called automatically by the computer executive system before execution commences.

Subroutine BLOCK DATA

```
C     INSERT COMMON BLOCK


      DATA XX/1.0,1.0,1.0,0.0,-1.0,-1.0,-1.0,0.0/
      DATA YY/-1.0,0.0,1.0,1.0,1.0,0.0,-1.0,-1.0/
      DATA INT1/1,3,5,7/
      DATA INT2/2,4,6,8/
      DATA INT3/3,5,7,1/
      DATA WTFUN/0.55555555556,0.88888888889,0.55555555556/
      DATA VECTLC/-0.77459666924,0.0,0.77459666924/
      END
```

(1) XX( ) and YY( ) contain the local coordinates of each of the eight nodes.
(2) INT2( ) contains the midside node number between the two corner nodes INT1( ) and INT3( ).
(3) WTFUN( ) is the weighting values for the Gauss points, the coordinates of which are defined by VECTLC( ).

5.1.3 NODAL COORDINATE INTERPOLATION SUBROUTINE

Subroutine INTCRD checks that the nodal coordinate of each of the midside nodes in an element have been specified. Coordinates not specified are calculated by a linear interpolation from the coordinates of the adjacent corner nodes. The coordinates are taken to be unspecified if both the x and y coordinates of the node are zero.

This subroutine is called by subroutine STIFN.

Subroutine INTCRD

```
C     INSERT COMMON BLOCK


      DO 10 I=1,4
      N2=INT2(I)
      IF(X(N2).NE.0.0.OR.Y(N2).NE.0.0) GO TO 10
      N1=INT1(I)
      N3=INT3(I)
      X(N2)=(X(N1)+X(N3))*0.5
      Y(N2)=(Y(N1)+Y(N3))*0.5
      NIC=NELDEF(N2)
      CORD(NIC,1)=X(N2)
      CORD(NIC,2)=Y(N2)
   10 CONTINUE
      RETURN
      END
```

(1) N2 is a midside node.
(2) Checking if the coordinates of N2 are specified.
(3) N1 and N3 are the corner node numbers adjacent to N2.
(4) Interpolating for the coordinates of N2.
(5) NIC is the actual node number of N2.
(6) Storing the coordinates of NIC.

5.1.4 INPUT SUBROUTINE

An input subroutine must be able to serve the following two functions: (i) to enable the user to prepare the data as easily as possible and (ii) to present the requisite information to the computer to enable it to solve the problem. For a finite element program, the data can be subdivided into five types, each of which has already been discussed in detail in previous chapters. These five groups of data, which include element definitions, nodal coordinates, nodal fixity, material properties and applied loads, are read by subroutine INDAT in the isoparametric element program. In INDAT, standard format has been used instead of free format since the latter is machine dependent. However, apart from this limitation, INDAT has been made quite general and is capable of accepting the various types and subtypes of data in any order provided each type or subtype is complete in itself. Inside each subtype the data may be presented in any order and cards do not have to be counted beforehand. Furthermore, it is not necessary to specify the maximum number of elements, highest node number, number of loaded nodes, etc., since the input subroutine works all these out for itself.

This method of input is achieved by allocating the first four columns on every data card to two variables ICODE and NCODE specified in 2I2 format. Each set of data is then preceded by a card containing only the first four columns punched specifying values for ICODE and NCODE. No other information is allowed on this card. The variable ICODE indicates the type of data to follow, e.g. nodal coordinates, element definitions, etc., and the variable NCODE indicates the particular subtype of the data, e.g. Cartesian or polar coordinates. Each card in the subsequent data then has columns 1 to 4 unpunched. To change the type of data it is only necessary to insert another card containing a non-zero ICODE. The rest of this card is read by the program in the same FORMAT as that used to read the current type of data. However, other than the values of ICODE and NCODE, the rest of the data on this card is ignored. To avoid field errors in the input it is therefore necessary to leave blank columns 5 to 80 on this card.

To indicate that the data input is complete it is only necessary to punch 99 in the first two columns of a card which is then inserted at the end of all the data.

A detailed input manual for the illustrative program, defining the values of ICODE and NCODE for each type of data, is given below.

ICODE	NCODE	
01	01	User headings READ ( ) HEAD FORMAT (5X, 9A8) The user may insert as many cards as required with the alphanumeric text to identify the particular analysis.
02	01	Nodal stress averaging READ ( ) NPUT FORMAT(5X, I5) To obtain nodal stress averaging the variable NPUT is set to a non-zero integer value. Default is NPUT = 0 giving no nodal stress averaging.

03 01 Element material properties
READ ( ) NMAT, YMOD, PRAT, WEIGHT
FORMAT (5X, I5, 3E10.3)
NMAT = material number
YMOD = Young's modulus
PRAT = Poisson's ratio
WEIGHT = Weight per unit volume } for the material number defined by NMAT

03 02 Material disposition
READ ( ) NMAT, (NW(I),I=1,14)
FORMAT (5X, I5, 14I5)
NMAT = material number
NW( ) = the numbers of up to 14 elements made from the material defined by NMAT. The default value for elements omitted from this section is material number 1.

04 01 Cartesian coordinates.
READ ( ) NIC, COX, COY
FORMAT (5X, I5, 2E10.3)
NIC = node number
COX = X Cartesian coordinate
COY = Y Cartesian coordinate.

04 02 Polar coordinates
READ ( ) NIC, RAD, ANG, COX, COY
FORMAT (5X, I5, 4E10.3)
NIC = node number
RAD = radius
ANG = angle in degrees } for polar coordinates
COX =
COY = } Cartesian coordinates of origin of polar coordinate system. *See* Fig.5.1.

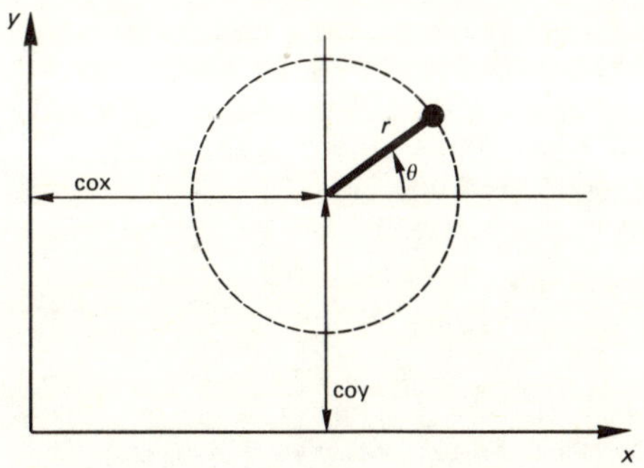

Fig.5.1 Polar coordinates and centre of polar system

05 01 Element definitions
READ ( ) (NW(I),I=1,8)
FORMAT (10X, 8I5)
NW( ) = eight node numbers defining an element. The first element defined in the data list is element number 1. The second element number 2, etc.

06 01 Specified nodal displacements.

```
READ (   ) NFIXX, NFIXY, NIC, DISX, DISY
FORMAT (5X, 2I1, 3X, I5, 5X, 2F10.5)
```

NFIXX = } Nodal constraint code
NFIXY = } *See* Section 2.4.5.
DISX = specified displacement in x direction
DISY = specified displacement in y direction
DISX or DISY may be left undefined when the associated constraint code is zero.

06 02 Earthed nodes.

```
READ (   ) NFIXX, NFIXY, (NW(I),I=1,14)
FORMAT (5X, 2I1, 3X, 14I5)
```

NFIXX = } Nodal constraint code
NFIXY = } *See* Section 2.4.5.
NW( ) = a list of up to 14 node numbers that have zero specified displacements and a constraint code defined by NFIXX, NFIXY.

07 01 Nodal point loads

```
READ (   ) NIC, REX, REY.
FORMAT (5X, I5, 2E10.3)
```

NIC = node number
REX = x component of nodal point load applied at node NIC
REY = y component

08 01 Edge pressure

```
READ (   ) NIC, PRE
FORMAT (5X, I5, E10.3)
```

NIC = node number
PRE = edge pressure at node NIC

09 01 Body forces

```
READ (   ) WIRL, GRAVX, GRAVY
FORMAT (5X, 3F10.5)
```

WIRL = angular rotation of structure about global origins in radians per second
GRAVX = } components of gravity in the x and y
GRAVY = } directions.

10 01 Transformation matrices

```
READ (   ) (AW(I),I=1,4)
FORMAT (10X,4F10.3)
```

AW( ) = direction cosine matrix (see example problem)
The first transformation matrix in the data is matrix No.1. The second is No.2, etc. The matrix is presented in the form

$$|L| = \begin{bmatrix} 1,1 & 1,2 \\ 2,1 & 2,2 \end{bmatrix} \rightarrow \text{AW( )}$$

$$= [L(1,1),\ L(1,2),\ L(2,1),\ L(2,2)]$$

10 02 Nodes with locally defined axes (*see* Section 4.3)

READ ( ) NTR, (NW(I),I=1,14)

FORMAT (5X,I5, 14I5)

NTR = transformation matrix number

NW( ) = a list of up to 14 node numbers whose displacements are to be defined relative to the local coordinate system defined by transformation matrix number NTR.

99 Data complete

Subroutine INDAT

```
C     INSERT COMMON BLOCK
      DIMENSION HEAD(9)
    1 READ(5,1002) ICODE,NCODE
    2 WRITE(6,1083) ICODE,NCODE                                    (1)
      IF(ICODE.EQ.99) RETURN                                       (2)
      GO TO(100,150,200,250,300,350,400,450,500,550),ICODE         (3)
  100 WRITE(6,2001)
  101 READ(5,1002) ICODE,NCODE,HEAD
      IF(ICODE.NE.0) GO TO 2
      WRITE(6,1003) HEAD                                           (4)
      GO TO 101
  150 WRITE(6,2002)
      READ(5,1004) NPUT
      WRITE(6,1005) NPUT                                           (5)
      GO TO 1
  200 GO TO(201,210),NCODE
  201 WRITE(6,2003)
  202 READ(5,1006) ICODE,NCODE,NMAT,YMOD,PRAT,WEIGHT
      IF(ICODE.NE.0) GO TO 2
      WRITE(6,1007) NMAT,YMOD,PRAT,WEIGHT
      YM(NMAT)=YMOD                                                (6)
      PR(NMAT)=PRAT
      WT(NMAT)=WEIGHT
      GO TO 202
  210 WRITE(6,2004)
  211 READ(5,1014) ICODE,NCODE,NMAT,(NW(I),I=1,14)
      IF(ICODE.NE.0) GO TO 2
      WRITE(6,1015) NMAT,(NW(I),I=1,14)
      IF(MAXMAT.LT.NMAT) MAXMAT=NMAT
      DO 220 I=1,14
      NEL=NW(I)                                                    (7)
      IF(NEL.EQ.0) GO TO 220
      MAT(NEL)=NMAT
  220 CONTINUE
      GO TO 211
  250 GO TO(251,260),NCODE
  251 WRITE(6,2005)
  252 READ(5,1006) ICODE,NCODE,NIC,COX,COY
      IF(ICODE.NE.0) GO TO 2
      WRITE(6,1007) NIC,COX,COY                                    (8)
      CORD(NIC,1)=COX
      CORD(NIC,2)=COY
      GO TO 252
  260 WRITE(6,2006)
  261 READ(5,1006) ICODE,NCODE,NIC,RAD,ANG,COX,COY
      IF(ICODE.NE.0) GO TO 2
      WRITE(6,1007) NIC,RAD,ANG,COX,COY
      A1=0.0174532925*ANG                                          (9)
      CORD(NIC,1)=RAD*COS(A1)+COX
      CORD(NIC,2)=RAD*SIN(A1)+COY
      GO TO 261
```

```
300 NEL=0
    WRITE(6,2007)
301 NEL=NEL+1
    READ(5,1018) ICODE,NCODE,(NW(I),I=1,8)
    IF(ICODE.NE.0) GO TO 303
    DO 302 I=1,8                                                    (10)
    LDEF(NEL,I)=NW(I)
302 CONTINUE
    WRITE(6,1015) NEL,(LDEF(NEL,I),I=1,NNODZ)
    MAXNEL=NEL
    GO TO 301
303 DO 305 NEL=1,MAXNEL
    IF(MAT(NEL).EQ.0) MAT(NEL)=1
    DO 304 LNOD=1,NNODZ
    IF(MAXNOD.LT.LDEF(NEL,LNOD)) MAXNOD=LDEF(NEL,LNOD)              (11)
304 CONTINUE
305 CONTINUE
    GO TO 2
350 NCOUNT=0
    GO TO(351,360),NCODE
351 WRITE(6,2008)
352 NCOUNT=NCOUNT+1
    READ(5,1026) ICODE,NCODE,NFIXX,NFIXY,NIC,DIS1,DIS2
    IF(ICODE.NE.0) GO TO 2
    NDISP(NCOUNT,1)=NFIXX
    NDISP(NCOUNT,2)=NFIXY
    NDISP(NCOUNT,3)=NIC                                             (12)
    DISP(NCOUNT,1)=DIS1
    DISP(NCOUNT,2)=DIS2
    WRITE(6,1027) (NDISP(NCOUNT,I),I=1,3),(DISP(NCOUNT,I),I=1,2)
    MAXDIS=NCOUNT
    GO TO 352
360 WRITE(6,2009)
361 READ(5,1008) ICODE,NCODE,NFIXX,NFIXY,(NW(I),I=1,14)
    IF(ICODE.NE.0) GO TO 2
    WRITE(6,1009) NFIXX,NFIXY,(NW(I),I=1,14)
    DO 365 I=1,14
    IF(NW(I).EQ.0) GO TO 365
    NCOUNT=NCOUNT+1                                                 (13)
    NDISPZ(NCOUNT,3)=NW(I)
    NDISPZ(NCOUNT,1)=NFIXX
    NDISPZ(NCOUNT,2)=NFIXY
    MAXNDZ=NCOUNT
365 CONTINUE
    GO TO 361
400 NCOUNT=0
    WRITE(6,2010)
401 NCOUNT=NCOUNT+1
    READ(5,1006) ICODE,NCODE,NIC,REX,REY
    IF(ICODE.NE.0) GO TO 2
    NREACT(NCOUNT)=NIC                                              (14)
    REACT(NCOUNT,1)=REX
    REACT(NCOUNT,2)=REY
    WRITE(6,1007) NREACT(NCOUNT),(REACT(NCOUNT,I),I=1,2)
    MAXRCT=NCOUNT
    GO TO 401
450 NCOUNT=0
    WRITE(6,2011)
451 NCOUNT=NCOUNT+1
    READ(5,1006) ICODE,NCODE,NIC,PRE
    IF(ICODE.NE.0) GO TO 2
    NPRES(NCOUNT)=NIC                                               (15)
    PRES(NCOUNT)=PRE
    WRITE(6,1007) NPRES(NCOUNT),PRES(NCOUNT)
    MAXPRS=NCOUNT
    GO TO 451
```

```
  500 WRITE(6,2012)
      READ(5,1012) WIRL,GRAVX,GRAVY
      WRITE(6,1013) WIRL,GRAVX,GRAVY                  (16)
      GO TO 1
  550 NCOUNT=0
      GO TO(551,560),NCODE
  551 WRITE(6,2014)
  552 NCOUNT=NCOUNT+1
      READ(5,1016) ICODE,NCODE,(AW(I),I=1,4)
      IF(ICODE.NE.0) GO TO 2
      WRITE(6,1017) (AW(I),I=1,4)                     (17)
      DO 555 I=1,4
      A1=AW(I)*0.0174532925
      TRANS(NCOUNT,I)=COS(A1)
  555 CONTINUE
      GO TO 552
  560 WRITE(6,2015)
  561 READ(5,1014) ICODE,NCODE,NTR,(NW(I),I=1,14)
      IF(ICODE.NE.0) GO TO 2
      WRITE(6,1015) NTR,(NW(I),I=1,14)
      DO 570 I=1,14
      IF(NW(I).EQ.0) GO TO 570
      NCOUNT=NCOUNT+1                                 (18)
      NTRANS(NCOUNT,1)=NW(I)
      NTRANS(NCOUNT,2)=NTR
      MAXTNS=NCOUNT
  570 CONTINUE
      GO TO 561
 1002 FORMAT(2I2,1X,9A8)
 1003 FORMAT(6X,9A8)
 1004 FORMAT(5X,6I5)
 1005 FORMAT(6X,6I5)
 1006 FORMAT(2I2,1X,I5,7E10.3)
 1007 FORMAT(6X,I5,7E10.3)
 1008 FORMAT(2I2,1X,2I1,3X,14I5)
 1009 FORMAT(6X,2I1,3X,14I5)
 1012 FORMAT(5X,4F10.5)
 1013 FORMAT(6X,4F10.5)
 1014 FORMAT(2I2,1X,15I5)
 1015 FORMAT(6X,15I5)
 1016 FORMAT(2I2,6X,4F10.3)
 1017 FORMAT(11X,4F10.3)
 1018 FORMAT(2I2,6X,14I5)
 1026 FORMAT(2I2,1X,2I1,3X,I5,5X,3F10.5)
 1027 FORMAT(6X,2I1,3X,I5,5X,3F10.5)
 1083 FORMAT(1X,2I2)
```

(1) Print out the value of ICODE and NCODE associated with the set of data to follow.

(2) If ICODE = 99 then all the data has been read.

(3) The value of ICODE determines the type of data to follow and hence control is transferred to that part of the subroutine capable of reading that data.

(4) Section to read user alphanumeric headings.

(5) NPUT is the flag to indicate request for average nodal stresses.

(6) Material properties.

(7) Material disposition.

(8) Nodal Cartesian coordinates.

(9) Nodal polar coordinates read in and transformed to Cartesian coordinates before storage.

(10) Element definitions and determination of number of elements.

(11) Determination of highest node number and setting default value for undefined element material disposition.

(12) Specified non-zero displacements.
(13) Earthed nodes.
(14) Nodal point loads.
(15) Nodal edge pressures.
(16) Angular velocity and components of gravity.
(17) Transformation matrices.
(18) Nodes with locally defined axes.

5.2 Example problem using isoparametric elements

To enable a comparison to be made between the elementary program described in Chapter 2 and the isoparametric element program, the same problem of a thick cylinder subjected to internal pressure will be analysed. Also to provide a direct comparison between the two programs the same quarter of a cylinder will be analysed using the two mesh divisions shown in Fig.5.2(a), (b). To illustrate the use of the facility in the isoparametric program allowing nodes to be given locally defined axes, two additional mesh divisions will also be analysed (Fig. 5.2(c), (d)) which represent a 6° segment of the cylinder. The 6° segment was arbitrarily chosen.

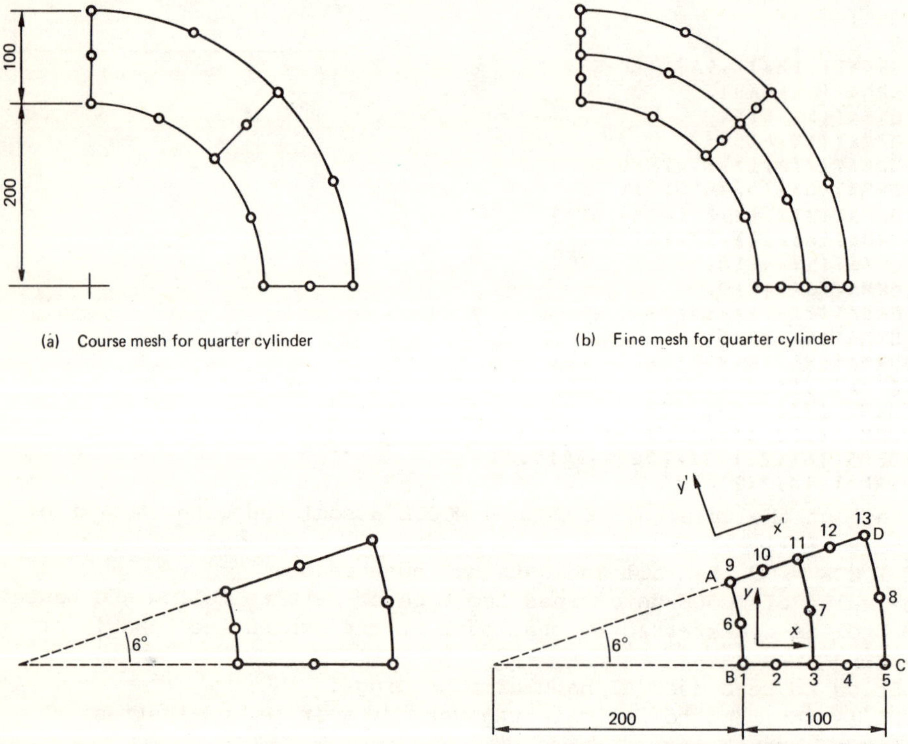

(a) Course mesh for quarter cylinder

(b) Fine mesh for quarter cylinder

(c) Course mesh for 6° segment

(d) Fine mesh for 6° segment

Fig.5.2 Isoparametric mesh divisions for a thick cylinder

5.3 Input data for the 6° segment

To minimize the possibility of errors in data generation the first task is to draw the structure to scale and then subdivide it into elements. Once this has been done it is possible to locate the nodes and to give each a number as shown in Fig.5.2(d).

The nodal coordinates can be specified in either Cartesian or polar coordinates. For this mesh it is obviously easier to use polar coordinates with the origin at the axis of tne cylinder. For example, for node 9 the data card would contain the following information

NODE	RAD	ANG	COX	COY
9	200.0	6	0.0	0.0

COX and COY are the Cartesian coordinates of the centre of the polar system (*see* Fig.5.1).

If any of the midside nodal coordinates are omitted from the data list then subroutine INTCRD automatically carries out a linear interpolation for these coordinates from the coordinates of the adjacent corner nodes. Hence the coordinates of nodes 2, 4, 10 and 12 which are situated on straight edges can be omitted from the input, and only midside nodes situated on curved edges need to have their coordinates specified.

The element definitions for the two elements are:

Element 1	3	4	5	8	13	12	11	7
Element 2	1	2	3	7	11	10	9	6

The material properties associated with these elements are:

Young's modulus	0.1×10^6 MPa
Poisson's ratio	0.25

Since there is only one material this can be material No.1. This is the default number for element material properties and therefore the material disposition does not have to be specified.

The material properties are input as

NMAT	YMOD	PRAT	WEIGHT
1	0.100E06	0.250	0.0

The problem does not involve body forces so the weight per unit volume can be omitted and is zero by default.

All specified displacements for this problem are zero displacements. However, before all the fixities can be introduced the displacements of all the nodes along AD in Fig.5.2(d) must be locally orientated, so that the axisymmetric situation can be modelled by having roller supports along AD. To enforce $V = 0$ for nodes on BC and $V' = 0$ for nodes on AD a constraint code of 01 is required. This data can be input on one card as

Constraint code	Nodes associated with the constraint code
01	1 2 3 4 5 9 10 11 12 13

The only applied loading is an edge pressure of 100 MPa on face AB, i.e. on nodes 1, 6 and 9. This is input as

```
1    0.100 E03
6    0.100 E03
9    0.100 E03
```

It only remains to define the local coordinate system for AD to complete the data. For this purpose, it is desirable to draw the two coordinate systems superimposed and the definition of the DC matrix, i.e.

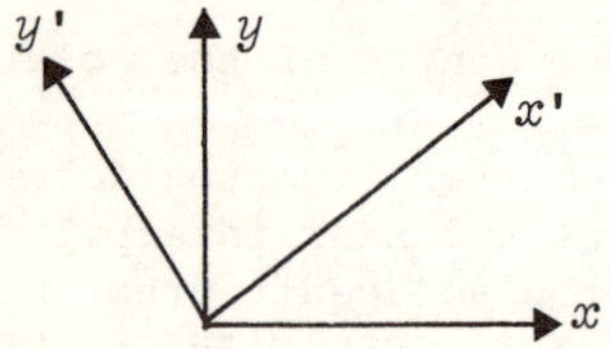

$$[L] = \begin{bmatrix} XX' & XY' \\ YX' & YY' \end{bmatrix}$$

The angles corresponding to YX' for example are measured in an anti-clockwise direction from the first axis Y to the second axis X', i.e.

$$[L] = \begin{bmatrix} \cos 6^\circ & \cos 96^\circ \\ \cos 276^\circ & \cos 6^\circ \end{bmatrix} = \begin{bmatrix} \cos 6^\circ & -\sin 6^\circ \\ \sin 6^\circ & \cos 6^\circ \end{bmatrix}$$

This DC matrix is input as

XX'	XY'	YX'	YY'
6.0	96.0	276.0	6.0

The nodes affected by this transformation matrix are

9 10 11 12 13

Since only one DC matrix has been defined in the input it is DC matrix no.1. The transformation data is therefore input as

DC matrix no.	Nodes associated with this DC matrix				
1	9	10	11	12	13

The end of data is indicated by punching 99 in the first two columns of a card and inserting at the end of the data.

Program data echo

```
1 1       USER HEADINGS
    TWO ELEMENT 6-DEG WEDGE
3 1       MATERIAL PROPERTIES
        1  .100E+06  .250E+00  .100E+03
```

```
 4 2      POLAR CO-ORDINATES
        1  .200E+03 0.          0.          0.
        6  .200E+03  .300E+01 0.          0.
        9  .200E+03  .600E+01 0.          0.
        3  .250E+03 0.          0.          0.
        7  .250E+03  .300E+01 0.          0.
       11  .250E+03  .600E+01 0.          0.
        5  .300E+03 0.          0.          0.
        8  .300E+03  .300E+01 0.          0.
       13  .300E+03  .600E+01 0.          0.
 5 1      ELEMENT DEFINITIONS
        1    3    4    5    8   13   12   11    7
        2    1    2    3    7   11   10    9    6
 6 2      EARTHED NODES
     01        1    2    3    4    5    9   10   11   12   13    0    0    0    0
 8 1      EDGE PRESSURES
        1  .100E+03
        6  .100E+03
        9  .100E+03
10 1      TRANSFORMATION MATRICES
              6.000     96.000    276.000      6.000
10 2      TRANSFORMED NODES
        1    9   10   11   12   13    0    0    0    0    0    0    0    0    0
99 0
```

Once the input has been accepted the first operation carried out automatically by the program is to determine the nodal destination vector and the front width. The front width for this problem is 16.

The information required by the front solver is then set up by subroutine STIFN, i.e. the element stiffness, load vectors, definition vector and destination vectors. The element stiffness is stored in the S( ) vector. For element 1 the S( ) vector written in the form

```
      WRITE (6,5001) S
 5001 FORMAT (40 (/, 5(2X, E10.3)))
```

contains the following coefficients

```
  .755E+05    .297E+05    .127E+06   -.476E+05   -.192E+05
 -.103E+05   -.718E+04    .323E+05    .351E+03   -.187E+04
  .421E+05   -.336E+05   -.470E+04   -.470E+04   -.836E+05
  .333E+05    .843E+04    .175E+05    .530E+05   -.330E+05
 -.584E+04   -.117E+05   -.228E+05    .346E+05   -.624E+04
  .624E+04    .628E+05   -.617E+05   -.619E+04   -.151E+05
 -.172E+06    .146E+06   -.144E+04    .140E+06   -.525E+05
  .104E+05    .193E+05   -.882E+04   -.653E+03   -.323E+05
 -.323E+05   -.276E+04   -.306E+05   -.331E+04   -.886E+04
 -.221E+05    .176E+05    .318E+04   -.318E+04   -.781E+05
 -.330E+05    .117E+05    .584E+04   -.228E+05    .122E+04
  .269E+05    .269E+05    .220E+04    .763E+05   -.330E+05
  .114E+06   -.546E+05    .143E+05    .231E+05   -.155E+06
  .315E+05   -.373E+04    .373E+04    .607E+05   -.306E+05
  .886E+04    .331E+04   -.221E+05    .333E+05   -.175E+05
 -.843E+04    .530E+05   -.369E+05    .101E+05    .101E+05
 -.838E+05    .154E+06   -.900E+04    .325E+06   -.528E+05
 -.684E+04   -.199E+05   -.156E+06   -.402E+04    .324E+05
  .320E+05    .627E+03   -.339E+05    .134E+05    .116E+04
 -.827E+05    .296E+05   -.644E+04   -.644E+04    .152E+06
  .763E+05    .330E+05    .114E+06   -.525E+05   -.193E+05
 -.104E+05   -.882E+04    .323E+05    .187E+04   -.351E+03
  .421E+05   -.356E+05   -.139E+05   -.125E+04   -.844E+05
  .146E+06    .144E+04    .140E+06   -.476E+05    .103E+05
  .192E+05   -.718E+04    .402E+04   -.266E+05   -.270E+05
 -.627E+03    .755E+05   -.297E+05    .127E+06   -.620E+05
 -.287E+03    .330E+05   -.169E+06    .161E+06   -.104E+05
  .358E+06
```

This represents the upper triangle of the element stiffness shown in Fig.5.3.

75500.	29700.	-47600.	-19200.	32300.	351.	-33600.	-4700.	33300.	8430.	-33000.	-5840.	34600.	-6240.	-61700.	-6190.
	127000.	-10300.	-7180.	-1870.	42100.	-4700.	-83600.	17500.	53000.	-11700.	-22800.	6240.	62800.	-15100.	-172000.
		146000.	-1440.	-52500.	10400.	-653.	-32300.	-30600.	-3310.	17600.	3180.	-33000.	11700.	1220.	26900.
			140000.	19300.	-8820.	-32300.	-2760.	-8860.	-22100.	-3180.	-78100.	5840.	-22800.	26900.	2200.
				76300.	-33000.	-54600.	14300.	31500.	-3730.	-30600.	8860.	33300.	-17500.	-36900.	10100.
					114000.	23100.	-155000.	3730.	60700.	3310.	-22100.	-8430.	53000.	10100.	-83800.
						154000.	-9000.	-52800.	-6840.	-4020.	32400.	-33900.	13400.	29600.	-6440.
							325000.	-19900.	-156000.	32000.	627.	1160.	-82700.	-6440.	152000.
								76300.	33000.	-52500.	-19300.	32300.	1870.	-35600.	-13900.
									114000.	-10400.	-8820.	-351.	42100.	-1250.	-84400.
										146000.	1440.	-47600.	10300.	4020.	-26600.
											140000.	19200.	-7180.	-27000.	-627.
												75500.	-29700.	-62000.	-287.
													127000.	33000.	-169000.
														161000.	-10400.
															358000.

Fig.5.3 Upper triangle of the stiffness matrix of element No.1

```
 -.476E+05   -.103E+05    .146E+06   -.144E+04   -.525E+05    .104E+05   -.653E+03   -.323E+05   -.306E+05   -.331E+04
  .176E+05    .318E+04   -.330E+05    .117E+05    .122E+04    .269E+05   C.
NSNW =    1  NFNW =   16  PIVOT POSITION =    3

 -.197E+05   -.728E+04   0.           .100E+51    .188E+05   -.871E+04   -.323E+05   -.308E+04   -.916E+04   -.222E+05
 -.301E+04   -.781E+05    .552E+04   -.227E+05    .269E+05    .246E+04   C.
NSNW =    1  NFNW =   16  PIVOT POSITION =    4

  .153E+05   -.556E+04   0.          0.           .575E+05   -.293E+05   -.548E+05    .270E+04    .205E+05   -.492E+04
 -.243E+05    .100E+05    .215E+05   -.133E+05   -.364E+05    .197E+05   0.
NSNW =    1  NFNW =   16  PIVOT POSITION =    5

  .115E+05    .400E+05   0.          0.          0.           .100E+51   -.471E+04   -.151E+06    .164E+05    .584E+05
 -.103E+05   -.173E+05    .482E+04    .454E+05   -.856E+04   -.757E+05   0.
NSNW =    1  NFNW =   16  PIVOT POSITION =    6

 -.193E+05   -.100E+05   0.          0.          0.          0.           .101E+06   -.657E+04   -.334E+05   -.115E+05
 -.271E+05    .419E+05   -.136E+05    .784E+03   -.515E+04    .125E+05   0.
NSNW =    1  NFNW =   16  PIVOT POSITION =    7

 -.171E+05   -.863E+05   0.          0.          0.          0.          0.           .317E+06   -.298E+05   -.158E+06
  .353E+05    .358E+04   -.799E+04   -.794E+05   -.479E+04    .158E+06   0.
NSNW =    1  NFNW =   16  PIVOT POSITION =    8

  .993E+04    .593E+04   0.          0.          0.          0.          0.          0.           .488E+05    .155E+05
 -.457E+05   -.805E+04    .125E+05    .188E+04   -.245E+05    .363E+04   0.
NSNW =    1  NFNW =   16  PIVOT POSITION =    9

 -.520E+04    .641E+04   0.          0.          0.          0.          0.          0.          0.           .100E+51
  .169E+05    .121E+04   -.876E+04    .124E+04    .455E+03   -.341E+04   0.
NSNW =    1  NFNW =   16  PIVOT POSITION =   10

 -.148E+05   -.351E+03   0.          0.          0.          0.          0.          0.          0.          0.
  .801E+05    .855E+04   -.256E+05    .141E+05   -.353E+05   -.324E+05   0.
NSNW =    1  NFNW =   16  PIVOT POSITION =   11

  .391E+04   -.155E+05   0.          0.          0.          0.          0.          0.          0.          0.
 0.           .100E+51    .267E+05   -.574E+04   -.188E+05   -.752E+04   0.
NSNW =    1  NFNW =   16  PIVOT POSITION =   12
```

Fig.5.4 Stored equations after assembly of element No.1

```
  .318E+05    .280E+02    .791E+05    .291E+05   -.364E+05   -.188E+05   -.293E+05   -.627E+04    .377E+05   -.709E+04
 -.778E+05   -.522E+04    .344E+05    .789E+04   -.398E+05   -.411E+04    .349E+03
NSNW =    1  NFNW =   16  PIVOT POSITION =    3

 -.139E+05    .486E+05   0.           .100E+51    .347E+04    .686E+04   -.113E+04   -.220E+05   -.678E+04    .785E+05
  .145E+05   -.211E+06    .543E+04    .596E+05    .105E+05   -.100E+06   -.129E+03
NSNW =    1  NFNW =   16  PIVOT POSITION =    4

 -.271E+05    .108E+05   0.          0.           .118E+06   -.112E+05   -.140E+05    .259E+04   -.119E+05    .865E+04
 -.345E+05    .239E+05   -.107E+05    .415E+03   -.189E+05   -.347E+05    .161E+03
NSNW =    1  NFNW =   16  PIVOT POSITION =    5

  .247E+05   -.180E+04   0.          0.          0.           .100E+51   -.138E+05   -.105E+06    .141E+05   -.252E+05
  .451E+04    .314E+04   -.120E+04   -.206E+05   -.441E+05   -.708E+04    .984E+02
NSNW =    1  NFNW =   16  PIVOT POSITION =    6

  .106E+06   -.111E+05   0.          0.          0.          0.          -.179E+05    .115E+05    .165E+05   -.132E+05
 -.197E+05    .183E+05    .255E+05   -.834E+04   -.111E+06    .103E+05   -.104E+03
NSNW =    1  NFNW =   16  PIVOT POSITION =    1

  .100E+51   0.          0.          0.          0.           .262E+04   -.216E+05   -.508E+04    .603E+05    .118E+05
 -.102E+06    .890E+04    .111E+06   -.340E+04   -.308E+06   -.257E+02
NSNW =    2  NFNW =   16  PIVOT POSITION =    2

 -.398E+05    .422E+05   -.568E+04   -.133E+04   -.159E+05    .129E+05   -.775E+05    .215E+05    .144E+06   -.157E+05
  .931E+02
NSNW =    7  NFNW =   16  PIVOT POSITION =   15
```

```
  .244E+05    .456E+04   -.385E+04   -.974E+05   -.215E+05    .197E+06   -.416E+05   -.308E+06   0.          .629E+06
  .858E+02
NSNW =    7  NFNW =   16  PIVOT POSITION =   16

 -.458E+05    .361E+04    .704E+04    .209E+04   -.163E+05    .489E+04    .579E+05    .341E+04   -.565E+02
NSNW =    7  NFNW =   14  PIVOT POSITION =   13

  .114E+05   -.561E+04   -.389E+04    .602E+03   -.807E+03   -.619E+04   0.           .100E+51   -.122E+02
NSNW =    7  NFNW =   14  PIVOT POSITION =   14

  .713E+05    .168E+05   -.169E+05    .111E+05   -.486E+05   -.222E+05    .109E+03
NSNW =    7  NFNW =   12  PIVOT POSITION =    7

  .100E+51    .255E+05   -.103E+04   -.123E+05   -.384E+04   -.146E+02
NSNW =    8  NFNW =   12  PIVOT POSITION =    8

  .522E+05   -.210E+05   -.516E+05   -.506E+04    .252E+03
NSNW =    9  NFNW =   12  PIVOT POSITION =    9

  .100E+51    .133E+05   -.178E+06    .107E+03
NSNW =   10  NFNW =   12  PIVOT POSITION =   10

  .464E+04   -.186E+05    .209E+04
NSNW =   11  NFNW =   12  PIVOT POSITION =   11

  .284E+06    .847E+04
NSNW =   12  NFNW =   12  PIVOT POSITION =   12
```

Fig.5.5 Stored equations after assembly of element No.2

After element 1 has been assembled the variables relating to nodes that do not appear in element 2 can be used for reduction. The coefficients of the reduced equations, including the control parameters, stored by the front solver are shown in Fig.5.4. The last coefficient in each equation is the load term. After assembly of element 2 the reduced equations stored are shown in Fig.5.5.

The effect of the heading vector on the equation length can be seen from the reduced equations after assembly of the second element.

The constant length of the reduced equations, before assembly of the second element, would not have occurred if a longevity prefront had been implemented. The nodes causing the non-zero coefficients at the end of the equation would then have been allocated storage towards the left-hand side of the stiffness matrix which would enable the equations to contract in length.

The stored reduced equations are then used for backsubstitution to calculate the displacements. During this operation the reactions at fixed nodes are calculated. However as the backsubstitution proceeds the displacements and forces at transformed nodes are in the locally orientated directions and hence *the reactions at these nodes will also be in a locally orientated direction.*

Reactions printed out by the program are

REACTIONS ELEMENT 2

NODE	1ST COMP	2ND COMP
9	0.	.215E+04
10	0.	.744E+04
11	0.	.323E+04
3	0.	-.323E+04
2	0.	-.744E+04
1	0.	-.215E+04

REACTIONS ELEMENT 1

NODE	1ST COMP	2ND COMP
12	0.	.585E+04
13	0.	.133E+04
5	0.	-.133E+04
4	0.	-.585E+04

After completion of the backsubstitution the locally orientated displacements are reorientated into the global direction and the global displacements of all nodes are printed out.

NODAL DISPLACEMENTS

NODE	X-COMP	Y-COMP
1	.5699737	0.0000000
2	.5349848	0.0000000
3	.5099799	0.0000000
4	.4922556	0.0000000
5	.4799816	0.0000000
6	.5691878	.0298299
7	.5092776	.0266901
8	.4793219	.0251202
9	.5668513	.0595785
10	.5320541	.0559211
11	.5071862	.0533074
12	.4895590	.0514547
13	.4773522	.0501717

The final operation in the program is to use the global displacements to calculate the element stresses which are printed out.

STRESSES ELEMENT 1

NODE	SIGMA X-X	SIGMA Y-Y	TAU X-Y
3	-32.9	195.8	-.0
4	-16.3	174.9	.0
5	1.9	160.5	-.0
8	2.3	160.0	-8.3
13	3.7	158.7	-16.5
12	-14.2	172.8	-19.9
11	-30.4	193.3	-23.7
7	-32.2	195.2	-11.9

STRESSES ELEMENT 2

NODE	SIGMA X-X	SIGMA Y-Y	TAU X-Y
1	-94.6	261.3	-.0
2	-64.6	221.6	.0
3	-31.0	196.2	-.0
7	-30.4	195.6	-11.9
11	-28.5	193.8	-23.6
10	-61.5	218.5	-29.8
9	-90.7	257.5	-37.0
6	-93.6	260.4	-18.6

The results obtained from the four meshes used to represent the thick cylinder subjected to 100 MPa internal pressure are given in Table 5.1. It can be seen that each of the meshes give good agreement with the theoretical tangential stress. However there must be at least two elements in the radial direction, meshes (b) and (d), before reasonable agreement is obtained with the theoretical radial stress. Nevertheless comparing the results obtained using the constant strain triangle with those obtained using the isoparametric element illustrates the efficiency of the second element compared with the first.

TABLE 5.1 *Comparison of the stresses obtained for a thick cylinder under internal pressure using different meshes*

Radius	Theoretical stress	Mesh (Fig.5.2) (a)	(b)	(c)	(d)
Radial stress (MPa)					
200	-100.0	-86.7	-96.6	-83.4	-96.4
225	- 62.2		-64.4		-64.5
250	- 35.2	-41.4	-31.3	-41.4	-31.8
275	- 15.2		-16.1		-16.2
300	0	13.4	3.1	10.2	1.9
Tangential stress (MPa)					
200	260.0	262.8	260.9	263.9	261.3
225	222.2		222.0		221.6
250	195.2	193.9	196.2	193.5	195.9
275	175.2		175.1		174.9
300	160.0	163.1	160.8	162.4	160.4

6 Automatic Mesh Generation

6.1 Introduction

The example problem used in this book is a very simple one and little difficulty has been encountered in the data preparation. However, for many practical problems, thousands of elements and nodes are involved and the task of preparing the data becomes extremely lengthy and tedious. Moreover, during the preparation of thousands of data cards some human error may be introduced and remain undetected in spite of the checks which are usually made. The presence of such errors will inevitably bring about incorrect results and if detected at that stage it would mean another run on the computer after correcting the data, which is tiresome, but not disastrous. However, if the errors stay undetected and such incorrect results are used for making decisions and judgements, there may be serious repercussions. It is therefore important to eliminate such data error, and this can be achieved to a large extent by automatic mesh generation, in which nodal numbers and their coordinates, together with element numbers and their definitions, are prepared automatically by the computer, using as input data the minimal amount of information necessary to describe the geometry of the domain and the desired fineness of the mesh divisions.

For non-linear problems, which sometimes involve the creation of new mesh divisions based on the results of the previous analysis, automatic mesh generation is essential, especially for problems involving many steps or iterations.

In the last few years, considerable effort has been expended in developing mesh generation routines in order to eliminate the drudgery of working out the data and at times punching out the data cards, and to minimize data errors. An early survey of a number of routines was given by Buell and Bush [1].

6.2 Mesh generation

6.2.1 TRIANGULAR ELEMENTS

An elementary routine for generating a triangular mesh is presented here. The input data consists of information concerning the total number of generating lines (NY), a weighting factor (CON), and for each generating line the number of intervals (NX(I)) required, together with the coordinates of the two end points (XF(I), YF(I), XL(I), YL(I)). The output consists of firstly the x and y coordinates of each point with its corresponding number, and secondly the element number with its element definition. Both nodal numbers and element numbers are arranged in sequential order.

To optimize the band width the generating lines should always traverse the shorter direction of the domain.

The number of divisions in adjacent generating lines can be equal or differ by one, so that the mesh can vary according to the specific requirement of the problem. On Fig.6.1, there are two divisions in line A, but the neighbouring line B is allowed to have one, two or three divisions.

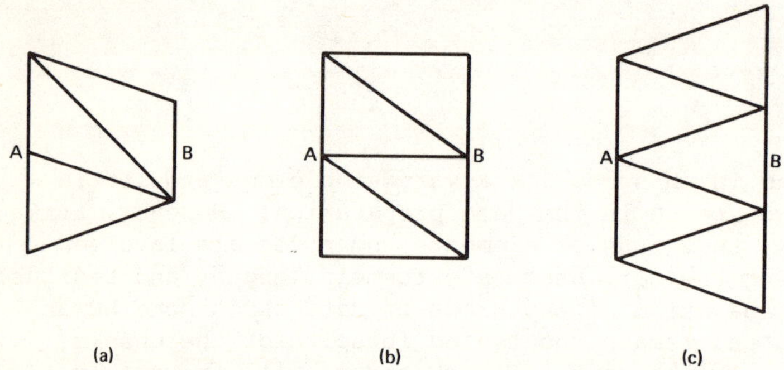

Fig.6.1 Possible triangular element divisions

Another refinement incorporated into the program is the weighting factor, CON. Depending on whether the weighting factor is < 1, = 1 or > 1, the intervals along a generating line will become progressively shorter, stay equal, or become progressively longer.

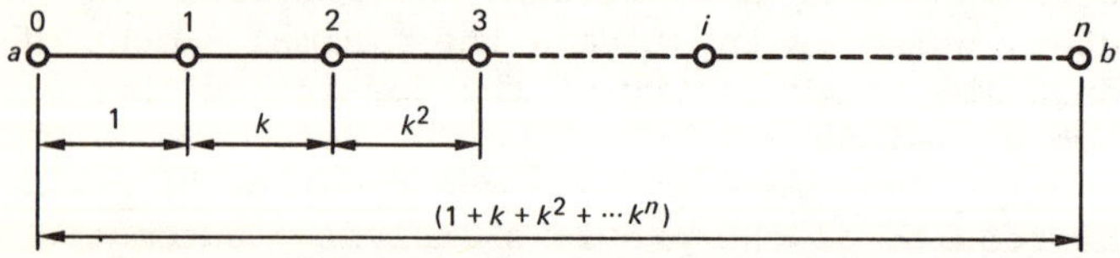

Fig.6.2 Effect of weighting factor upon nodal intervals

From Fig.6.2 it can be seen that the coordinates of a point i along a generating line ab can be computed by

$$x_i = (x_b - x_a)\frac{\sum_{j=1}^{i} k^{j-1}}{\sum_{j=1}^{n} k^{j-1}} + x_a$$

or, summing the geometric progression,

$$x_i = (x_b - x_a)\frac{k^i - 1}{k^n - 1} + x_a$$

only if $k \neq 1$

and

$$y_i = (y_b - y_a)\frac{\sum_{j=1}^{i} k^{j-1}}{\sum_{j=1}^{n} k^{j-1}} + y_a = (y_b - y_a)\frac{k^i - 1}{k^n - 1} + y_a$$

The node numbers will be labelled correctly as we go from one generating line to the next, by always increasing the number by one every time we come across a point.

The element definitions are worked out by taking the mth and the $(m + 1)$th generating lines, starting from $m = 1$, and forming n quadrilaterals along the jth line. Referring to Fig.6.1(a), in which the $(m + 1)$th line has one less division than the mth line, the last (nth) quadrilateral degenerates into a triangle, while, for Fig.6.1(b), the mth and the $(m + 1)$th lines have the same number of intervals and each of the n quadrilaterals is simply split into two triangles. For Fig. 6.1(c), however, the $(m + 1)$th line has now one more division than the jth line, and one extra triangle must be added to the n quadrilaterals which have been established.

A number of features can be added to the subroutine without a great deal of difficulty. The shape of the triangles can be improved by always using the shorter diagonal of a quadrilateral when splitting it into two triangles. For layered materials with intermediate boundaries it is only necessary to incorporate intermediate points into a generating line at the interfaces so as to ensure that no element will cross over a boundary. For such a case a generating line may take up the form of a polygon with the nodes at the two ends and at the interfaces.

6.2.2 COMPUTER SUBROUTINE

The complete subroutine listing together with the output for a sample problem (Fig.6.3) are given below.

```
      PROGRAM TGEN(INPUT,OUTPUT,TAPE5=INPUT,TAPE6=OUTPUT)
      DIMENSION NX(20),YL(20),XF(20),YF(20),NOD(4),SUM1(15),XL(20)
      READ(5,1000) NY,CON                                          (1)
      WRITE(6,1001) NY,CON
      DO 100 I=1,NY
      READ(5,1000) NX(I),XF(I),YF(I),XL(I),YL(I)                   (2)
      WRITE(6,1000) NX(I),XF(I),YF(I),XL(I),YL(I)
  100 CONTINUE
      WRITE(6,1431)
      N=0
```

```
      DO 350 I=1,NY
      NXI=NX(I)+1
      SUM1(1)=0.0
      SUM1(2)=1.0
      SUM=1.0
      IF(NXI-2) 190,291,190
  190 DO 250 K=3,NXI
      SUM1(K)=SUM1(K-1)*CON
      SUM=SUM+SUM1(K)
  250 CONTINUE
  291 CONTINUE                                                    (3)
      X=XF(I)
      Y=YF(I)
      DO 300 J=1,NXI
      N=N+1
      X=(XL(I)-XF(I))*SUM1(J)/SUM+X
      Y=(YL(I)-YF(I))*SUM1(J)/SUM+Y
      WRITE(6,1430) X,Y,N
  300 CONTINUE
  350 CONTINUE
      N=0
      NSUM=0
      NYI=NY-1
      WRITE(6,1432)
      DO 600 I=1,NYI
      NXI=NX(I)
      DO 500 J=1,NXI
      IF(J-NXI) 379,371,379
  371 IF(NX(I+1)-NX(I)) 380,379,401
  379 NOD(1)=J+NSUM
      NOD(2)=NOD(1)+1
      NOD(3)=NOD(2)+NXI+1                                         (4)
      NOD(4)=NOD(3)-1
      GO TO 412
  380 NOD(1)=NOD(2)
      NOD(2)=NOD(1)+1
      NOD(4)=0                                                    (5)
      GO TO 412
  401 NOD(1)=J+NSUM
      NOD(2)=NOD(1)+1
      NOD(3)=NOD(2)+NXI+1
      NOD(4)=NOD(3)-1
      N=N+1
      WRITE(6,1470) N,NOD(1),NOD(2),NOD(3)                        (6)
      N=N+1
      WRITE(6,1470) N,NOD(1),NOD(3),NOD(4)
      NOD(1)=NOD(2)
      NOD(2)=NOD(3)+1
      NOD(4)=0
  412 N=N+1
      WRITE(6,1470) N,NOD(1),NOD(2),NOD(3)
      IF(NOD(4))433,434,433                                       (7)
  433 N=N+1
      WRITE(6,1470) N,NOD(1),NOD(3),NOD(4)
  434 CONTINUE
  500 CONTINUE
      NSUM=NSUM+NXI+1
  600 CONTINUE
      STOP
 1000 FORMAT(I5,5X,6F10.5)
 1001 FORMAT(*1 CONTROL DATA*,/,I5,5X,6F10.5)
 1431 FORMAT(*1  X CO-ORDINATE   Y CO-ORDINATE   NODAL POINT*)
 1430 FORMAT(2F16.8,I4)
 1432 FORMAT(*1  ELEMENT NO    NODAL POINTS OF THE ELEMENT*)
 1470 FORMAT(I8,5X,3I8)
      END
```

CONTROL DATA

5	1.20000			
5	0.00000	4.00000	0.00000	12.00000
5	1.00000	4.00000	2.30000	12.00000
5	2.00000	4.00000	5.00000	12.00000
6	3.00000	4.00000	8.00000	12.00000
7	4.00000	4.00000	12.00000	12.00000

X CO-ORDINATE	Y CO-ORDINATE	NODAL POINT
0.00000000	4.00000000	1
0.00000000	5.07503763	2
0.00000000	6.36508278	3
0.00000000	7.91313696	4
0.00000000	9.77080198	5
0.00000000	12.00000000	6
1.00000000	4.00000000	7
1.17469361	5.07503763	8
1.38432595	6.36508278	9
1.63588476	7.91313696	10
1.93775532	9.77080198	11
2.30000000	12.00000000	12
2.00000000	4.00000000	13
2.40313911	5.07503763	14
2.88690604	6.36508278	15
3.46742636	7.91313696	16
4.16405074	9.77080198	17
5.00000000	12.00000000	18
3.00000000	4.00000000	19
3.50352873	4.80564597	20
4.10776320	5.77242113	21
4.83284457	6.93255132	22
5.70294222	8.32470755	23
6.74705939	9.99529503	24
8.00000000	12.00000000	25
4.00000000	4.00000000	26
4.61939141	4.61939141	27
5.36266110	5.36266110	28
6.25458474	6.25458474	29
7.32489309	7.32489309	30
8.60926312	8.60926312	31
10.15050716	10.15050716	32
12.00000000	12.00000000	33

ELEMENT NO	NODAL POINTS OF THE ELEMENT		
1	1	2	8
2	1	8	7
3	2	3	9
4	2	9	8
5	3	4	10
6	3	10	9
7	4	5	11
8	4	11	10
9	5	6	12
10	5	12	11
11	7	8	14
12	7	14	13
13	8	9	15
14	8	15	14
15	9	10	16
16	9	16	15
17	10	11	17
18	10	17	16
19	11	12	18
20	11	18	17

21	13	14	20
22	13	20	19
23	14	15	21
24	14	21	20
25	15	16	22
26	15	22	21
27	16	17	23
28	16	23	22
29	17	18	24
30	17	24	23
31	18	25	24
32	19	20	27
33	19	27	26
34	20	21	28
35	20	28	27
36	21	22	29
37	21	29	28
38	22	23	30
39	22	30	29
40	23	24	31
41	23	31	30
42	24	25	32
43	24	32	31
44	25	33	32

(1) Read/write number of lines and weighting factor.
(2) Read/write number of intervals, coordinates of starting and ending points of generating lines.
(3) Calculate node numbers and their corresponding x, y coordinates.
(4) Mesh variation in Fig.6.1(b).
(5) Mesh variation in Fig.6.1(a).
(6) Mesh variation in Fig.6.1(c).
(7) Skip a triangle if fourth node of quadrilateral is zero.

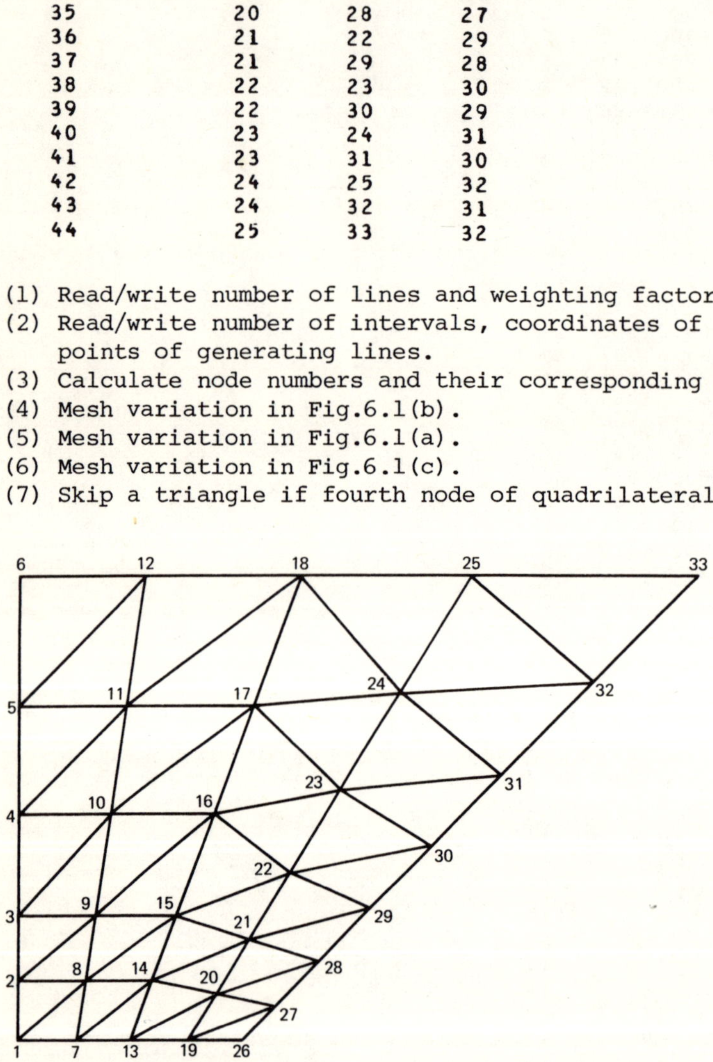

Fig.6.3 Example problem

6.2.3 ISOPARAMETRIC ELEMENTS

In this section a simple but effective mesh generation scheme using the isoparametric concept and suitable for band or front solvers will be described first. Later on a slightly more elaborate scheme suitable for front solver only will be presented to deal with mesh generation for domains with complex shapes and for multiply-connected domains.

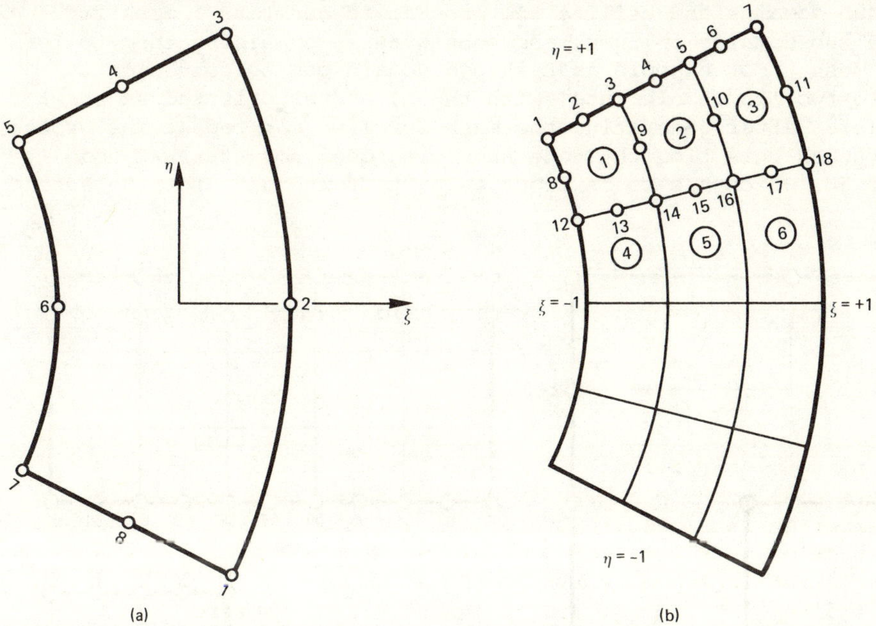

Fig.6.4 Curved bridge deck and example mesh

Using the curved bridge deck in Fig.6.4(a) as an example it is seen that the whole domain can be regarded as one big isoparametric element [2] and the shape is adequately defined by the eight master nodes. (Note that the circular edge is now replaced by a parabola.) A suitable mesh is then obtained by using lines of constant ξ and η to divide the 'large element' into n elements in the ξ direction and n elements in the η directions. The coordinates of any point j are given as

$$x_j = \sum_{i=1}^{8} N_i(\xi_j, \eta_j) x_i$$

$$y_j = \sum_{i=1}^{8} N_i(\xi_j, \eta_j) y_i$$

(*see* Eqns.(3.6b) and (3.6c)).

In the illustrative subroutine, the node numbering starts from $\xi = -1$, $\eta = +1$ and proceeds along the line of constant η until $\xi = +1$ is reached. For the odd numbered generating lines there are m element divisions and therefore $2m + 1$ nodes because of the presence of both corner and midside nodes. The numbering then recommences at $\xi = -1$ on the next line of constant η and the process is repeated, although this time an even numbered generating line is involved and there are only

m + 1 nodes because only midside nodes are present. All node numbers are arrayed in ascending order as indicated in Fig.6.4(b). There is of course no reason why the first node has to be node number 1, and in fact the subroutine allows the user to define any number as the first node number and then all nodes will follow sequentially. This facility is necessary for handling domains with complex shapes.

With this simple subroutine it is possible to divide a domain into a number of sub-domains and utilize the program to generate a mesh for each of the sub-domains in turn, and subsequently combining them to form a complete mesh. For example an L-shaped domain can be conveniently divided into two 'large elements' with the ξ, η axes oriented as shown in Fig.6.5(a). After generating the mesh for I we can repeat the process for II, although this time the node numbering does not start at node 1, but at node 58, which has to be input as part of the data for the second run.

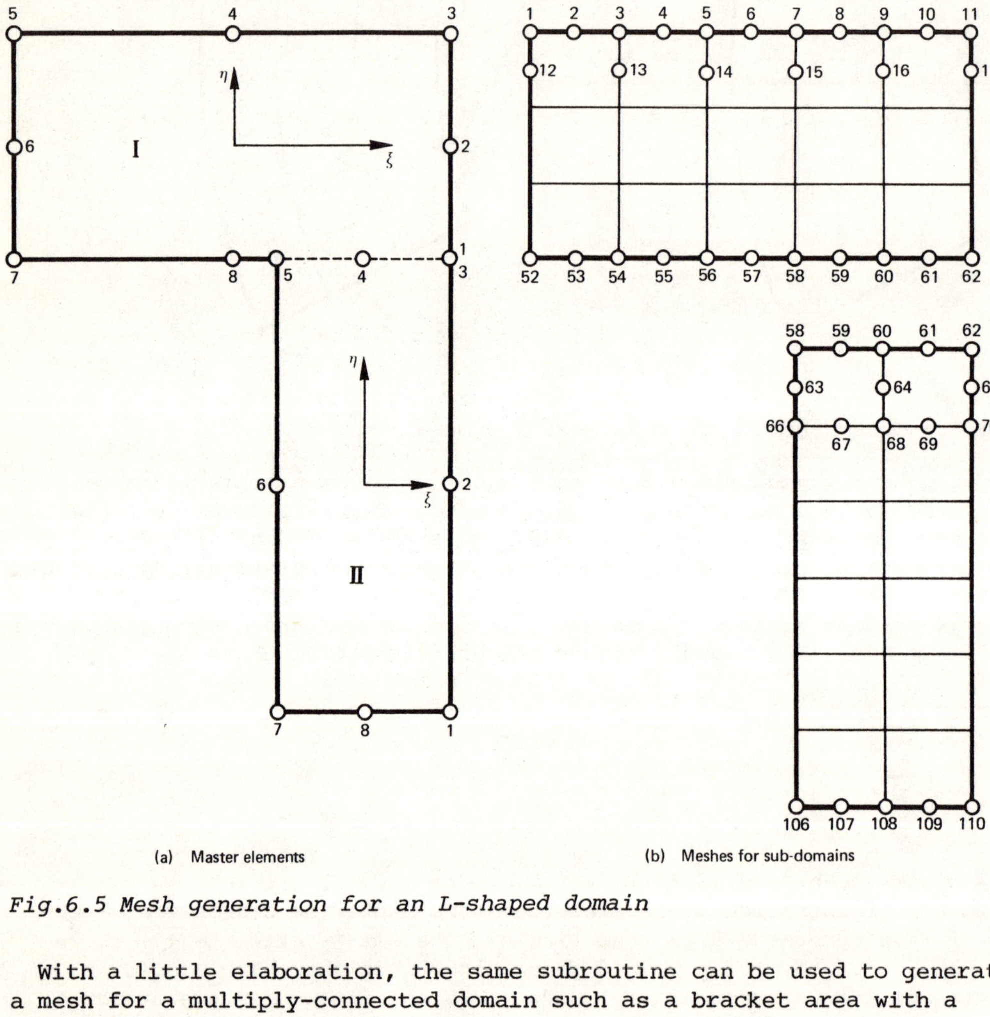

Fig.6.5 Mesh generation for an L-shaped domain

With a little elaboration, the same subroutine can be used to generate a mesh for a multiply-connected domain such as a bracket area with a hole [3] shown in Fig.6.6. As can be seen, the bracket area has been divided into seven sub-domains.

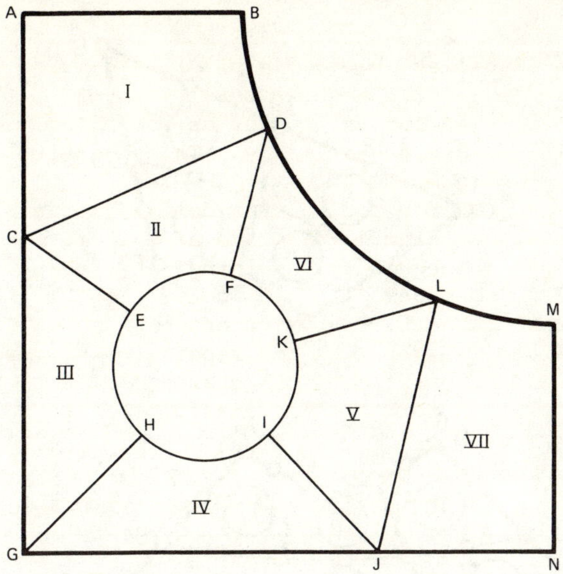

Fig.6.6 Bracket area with a hole

The subroutine is now used repeatedly to generate meshes for the seven sub-domains. The nodal numbers of the first generating line of II corresponding to the nodal numbers of the last generating line of I, and therefore no modification is necessary at the interface between I and II. For III, the first generating line does not correspond to the last generating line of II and therefore the first node number must be sequential to the last node number of II, and so on for all the other sub-domains.

In order to connect the meshes of all the sub-domains together, it is necessary to comply with the following conditions:

Along CE	17 = 38,	22 = 39,	25 = 40,	30 = 41,	33 = 42
Along GH	62 = 89,	63 = 85,	64 - 78,	65 - 74,	66 - 67
Along IJ	73 = 112,	77 = 113,	84 = 114,	88 = 115,	95 = 116
Along KL	96 = 141,	97 = 142,	98 = 143,	99 = 144,	100 = 145
Along FD	37 = 117,	32 = 118,	29 = 119,	24 = 120,	21 = 121
Along LJ	100 = 146,	103 = 151,	108 = 154,	111 = 159,	95 = 162

Note that points J and L are common to three sub-domains and therefore there are altogether three nodal numbers for each of these two points.

It is a simple matter to write several Fortran statements which will instruct the computer to replace the node numbers on the right side of the equations by the corresponding node numbers on the left side of the equations. In this way the meshes for the seven sub-domains (Fig.6.7) are connected together to form one single mesh for the multiply-connected domain.

Using the technique described above it is obvious that quite a few node numbers have been eliminated altogether. This means that a mesh generated this way is unsuitable for a band solution scheme, but is on

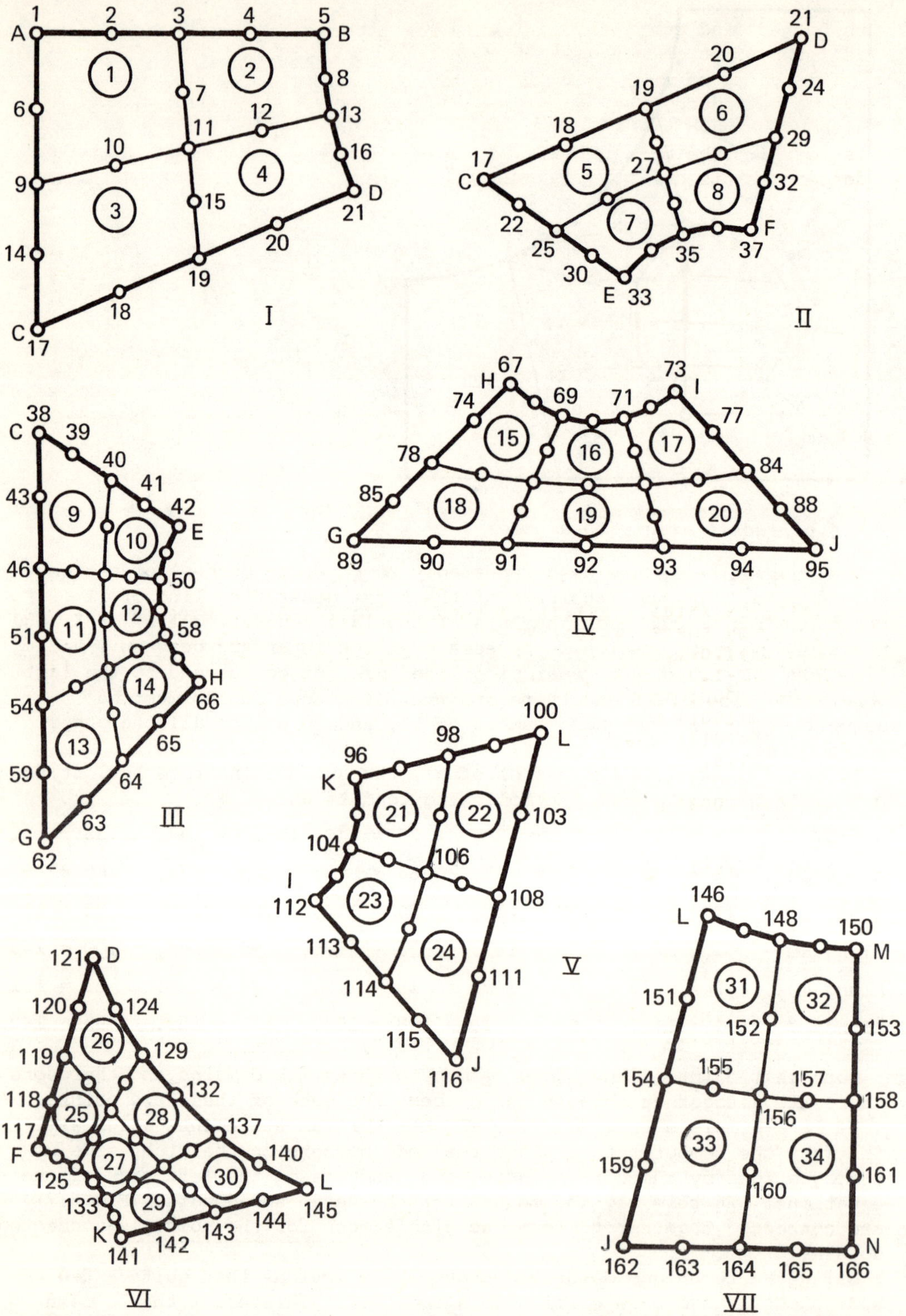

Fig.6.7 Sub-domains and their meshes

the other hand completely suitable for a front solution scheme, which is independent of node numbering.

If grading of mesh is required for the analysis, weighting factors can be introduced to create unequal divisions inside a 'large element', although a much simpler way, which is satisfactory for most problems, is to shift the origin of the ξ, η axes towards the corner in which a denser mesh is desired, as shown in Fig.6.8.

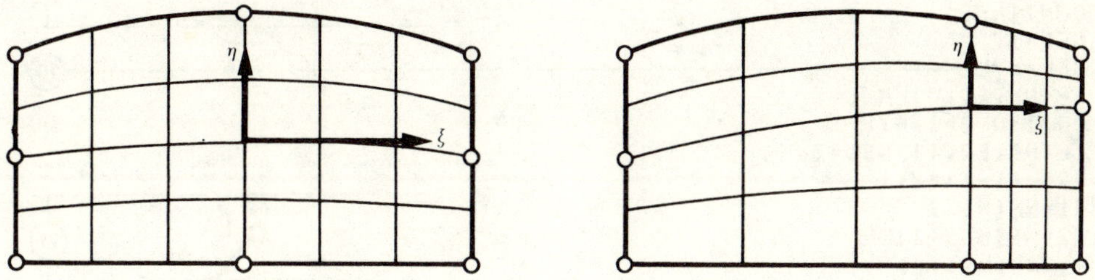

(a) Normal mesh (b) Graded mesh

Fig.6.8 Graded mesh within a large element

Program GEN

```
      DIMENSION XM(8),YM(8),X(1000),Y(1000),LDEF(200,8),NS(200)
      DIMENSION SHP(8),XX(8),YY(8)
      DATA XX/1.0,1.0,1.0,0.0,-1.0,-1.0,-1.0,0.0/                  (1)
      DATA YY/-1.0,0.0,1.0,1.0,1.0,0.0,-1.0,-1.0/
      READ(5,1000) NXL,NYL,NN,ND                                   (2)
      WRITE(6,1001) NXL,NYL,NN,ND
      WRITE(6,1013)
      DO 5 I=1,8
      READ(5,1002) XM(I),YM(I)                                     (3)
      WRITE(6,1003) I,XM(I),YM(I)
    5 CONTINUE
      II=0
      NSF1=2*NYL+1
      DO 50 I=1,NSF1
      YL=1.0-2.0*(I-1)/(2*NYL)                                     (4)
      II=II+1
      NS(I)=NN                                                     (5)
      IF(II.EQ.3) II=1
      NSF2=2*NXL+1
      DO 45 J=1,NSF2,II
      XL=-1.0+2.0*(J-1)/(2*NXL)                                    (6)
      DO 25 IJ=1,8
      GO TO(10,15,10,20,10,15,10,20),IJ
   10 SHP(IJ)=0.25*(1.0+XL*XX(IJ))*(1.0+YL*YY(IJ))*
     1                   (XL*XX(IJ)+YL*YY(IJ)-1.0)
      GO TO 25                                                     (7)
   15 SHP(IJ)=0.5*(1.0+XL*XX(IJ))*(1.0-YL*YL)
      GO TO 25
   20 SHP(IJ)=0.5*(1.0+YL*YY(IJ))*(1.0-XL*XL)
   25 CONTINUE
      XXX=0.0
      YYY=0.0
      DO 30 K=1,8
      XXX=XXX+SHP(K)*XM(K)                                         (8)
      YYY=YYY+SHP(K)*YM(K)
   30 CONTINUE
```

```
      X(NN)=XXX
      Y(NN)=YYY
      NN=NN+1
   45 CONTINUE
   50 CONTINUE
      MAXNOD=NN-1
      NSS=NS(1)
      WRITE(6,1014)
      DO 60 NIC=NSS,MAXNOD
      WRITE(6,1020) NIC,X(NIC),Y(NIC)                                  (9)
   60 CONTINUE
      NEL=1
      MEL=NYL                                                          (10)
      IF(ND.EQ.0) MEL=1
      DO 150 I=1,NYL
      IF(ND.EQ.1) NEL=I
      NSS=(I-1)*2+1
      N1=NS(NSS)
      N2=NS(NSS+1)                                                     (11)
      N3=NS(NSS+2)
      DO 140 J=1,NXL
      LDEF(NEL,1)=N3  +(J-1)*2
      LDEF(NEL,2)=N3+1+(J-1)*2
      LDEF(NEL,3)=N3+2+(J-1)*2
      LDEF(NEL,4)=N2+1+J-1
      LDEF(NEL,5)=N1+2+(J-1)*2                                         (12)
      LDEF(NEL,6)=N1+1+(J-1)*2
      LDEF(NEL,7)=N1  +(J-1)*2
      LDEF(NEL,8)=N2  +J-1
      NEL=NEL+MEL
  140 CONTINUE
  150 CONTINUE
      MAXNEL=NXL*NYL
      WRITE(6,1015)
      DO 170 NEL=1,MAXNEL
      WRITE(6,1010) NEL,(LDEF(NEL,I),I=1,8)                            (13)
  170 CONTINUE
      STOP
 1000 FORMAT(4I10)
 1001 FORMAT(1H1,4X,*NUMBER OF DIVISIONS IN LOCAL X DIRECTION =*,I3,
     1  //,5X,*NUMBER OF DIVISIONS IN LOCAL Y DIRECTION =*,I3,
     2  //,5X,*FIRST NODE NUMBER =*,I3,
     3  //,5X,*DIRECTION PARAMETER =*,I3)
 1002 FORMAT(2F10.3)
 1003 FORMAT(1X,I5,4F10.3)
 1010 FORMAT(10X,9I5)
 1013 FORMAT(//////,5X,*MASTER NODAL CO-ORDINATES*,//,
     1  2X,*NODE*,9X,1HX,9X,1HY)
 1014 FORMAT(//////,5X,*NODAL CO-ORDINATES*,//,
     1   7X,*NODE*,9X,1HX,9X,1HY)
 1015 FORMAT(//////,5X,*ELEMENT DEFINITIONS*)
 1020 FORMAT(6X,I5,2E10.3)
      END
```

Program output for Area II of Fig. 6. 5.

```
NUMBER OF DIVISIONS IN LOCAL X DIRECTION =  2

NUMBER OF DIVISIONS IN LOCAL Y DIRECTION =  6

FIRST NODE NUMBER = 58

DIRECTION PARAMETER =  0
```

MASTER NODAL CO-ORDINATES

NODE	X	Y
1	30.000	20.000
2	20.000	20.000
3	10.000	20.000
4	10.000	16.000
5	10.000	12.000
6	20.000	12.000
7	30.000	12.000
8	30.000	16.000

NODAL CO-ORDINATES

NODE	X	Y
58	.100E+02	.120E+02
59	.100E+02	.140E+02
60	.100E+02	.160E+02
61	.100E+02	.180E+02
62	.100E+02	.200E+02
63	.117E+02	.120E+02
64	.117E+02	.160E+02
65	.117E+02	.200E+02
66	.133E+02	.120E+02
67	.133E+02	.140E+02
68	.133E+02	.160E+02
69	.133E+02	.180E+02
70	.133E+02	.200E+02
71	.150E+02	.120E+02
72	.150E+02	.160E+02
73	.150E+02	.200E+02
74	.167E+02	.120E+02
75	.167E+02	.140E+02
76	.167E+02	.160E+02
77	.167E+02	.180E+02
78	.167E+02	.200E+02
79	.183E+02	.120E+02
80	.183E+02	.160E+02
81	.183E+02	.200E+02
82	.200E+02	.120E+02
83	.200E+02	.140E+02
84	.200E+02	.160E+02
85	.200E+02	.180E+02
86	.200E+02	.200E+02
87	.217E+02	.120E+02
88	.217E+02	.160E+02
89	.217E+02	.200E+02
90	.233E+02	.120E+02
91	.233E+02	.140E+02
92	.233E+02	.160E+02
93	.233E+02	.180E+02
94	.233E+02	.200E+02
95	.250E+02	.120E+02
96	.250E+02	.160E+02
97	.250E+02	.200E+02
98	.267E+02	.120E+02
99	.267E+02	.140E+02
100	.267E+02	.160E+02
101	.267E+02	.180E+02

```
102   .267E+02   .200E+02
103   .283E+02   .120E+02
104   .283E+02   .160E+02
105   .283E+02   .200E+02
106   .300E+02   .120E+02
107   .300E+02   .140E+02
108   .300E+02   .160E+02
109   .300E+02   .180E+02
110   .300E+02   .200E+02


ELEMENT DEFINITIONS
          1    66    67    68    64    60    59    58    63
          2    68    69    70    65    62    61    60    64
          3    74    75    76    72    68    67    66    71
          4    76    77    78    73    70    69    68    72
          5    82    83    84    80    76    75    74    79
          6    84    85    86    81    78    77    76    80
          7    90    91    92    88    84    83    82    87
          8    92    93    94    89    86    85    84    88
          9    98    99   100    96    92    91    90    95
         10   100   101   102    97    94    93    92    96
         11   106   107   108   104   100    99    98   103
         12   108   109   110   105   102   101   100   104
```

(1) ξ, η coordinates of the master nodes.

(2) Read/write control data.
NXL = number of elements in ξ direction
NYL = number of elements in η direction
NN = first node number
ND = direction parameter. If ND = 1 then elements are numbered sequentially along lines of constant ξ. If ND = 0 the sequential numbering is along lines of constant η.

(3) Read/write x, y coordinates of master nodes.

(4) η coordinates of each generated node.

(5) NS( ) stores a list of the node numbers on ξ = - 1.

(6) ξ coordinates of each generated node.

(7) Calculate the shape functions for the ξ, η coordinates of each generated node.

(8) Calculate the x, y coordinates of each node from

$$\begin{Bmatrix} X \\ Y \end{Bmatrix} = \sum_{i=1}^{8} N_i \begin{Bmatrix} X_{mi} \\ Y_{mi} \end{Bmatrix}$$

(9) Write out the nodal coordinates.

(10) MEL is the increment in the element numbers in ξ direction.

(11) Node numbers along edge ξ = - 1 for use in generating the first element definition of a row of elements in the ξ direction.

(12) Node numbers of elements along a row of elements in the ξ direction using the sequential numbering of the nodes.

(13) Write out the element definitions.

References

1. W.R. Buell and B.A. Bush. Mesh generation - a Survey. *Journal of Engineering for Industry, Transactions of the American Society of Mechanical Engineers, Series B*, vol.95, no.1, Feb 1973, pp.332-338.
2. O.C. Zienkiewicz and D.V. Phillips. An automatic mesh generation scheme for plane and curved surfaces by isoparametric coordinates, *International Journal for Numerical Methods in Engineering*, vol.3, 1971, pp.519-528.
3. J. Suhara and J. Fukuda. Automatic mesh generation for finite element analysis. In *Advances in Computational Methods in Structural Mechanics and Design* (J.T. Oden, R.W. Clough and Y. Yamamoto, eds). UAH Press, Alabama, 1972, pp.607-624.

Index

भूख तथा अन्य कहानियाँ

(कहानी-संग्रह)

भूख तथा अन्य कहानियाँ

सुभाष शर्मा

भारतीय ज्ञानपीठ

लोकोदय ग्रन्थमाला : ग्रन्थांक 1029

ग्रन्थमाला सम्पादक

रवीन्द्र कालिया

सह-सम्पादक

गुलाबचन्द्र जैन

ISBN 978-81-263-1855-1

प्रकाशक :
भारतीय ज्ञानपीठ
18, इन्स्टीट्यूशनल एरिया, लोदी रोड
नयी दिल्ली-110 003

मुद्रक : विकास कम्प्यूटर ऐंड प्रिंटर्स, दिल्ली-110032

आवरण-चित्र : राधेश्याम अग्रवाल

आवरण-सज्जा : ज्ञानपीठ कला प्रभाग

दूसरा संस्करण : 2011
मूल्य : 200 रुपये

BHOOKH TATHA ANYA KAHANIYAN
(Short stories)
by Subhash Sharma

Published by
Bharatiya Jnanpith
18, Institutional Area, Lodi Road
New Delhi-110 003
e-mail : jnanpith@satyam.net.in, sales@jnanpith.net
website : www.jnanpith.net

Second Edition : 2011
Price : Rs. 200

समाज के उन लोगों को
जो हाशिए पर हैं
मगर इन कहानियों में
अपना स्थान बना चुके हैं

अनुक्रम

भूख

प्रदेश स्तर के सभी अख़बारों के मुखपृष्ठ पर पहली ख़बर छपी थी—'सरकार बेनकाब।' सभी समाचारपत्रों ने अपने संवाददाताओं का हवाला देते हुए दिल दहला देने वाली यह ख़बर प्रमुखता से छापी थी। गाँव के मुखिया से लेकर सूबे में विपक्ष के नेता तक ने इस घटना की कड़ी निन्दा की थी। एक संवाददाता की हैसियत से मुझे उस गाँव में जाने का मौका मिला। दरअसल उत्तर भारत का एक गाँव है जीवनपुर, जिसमें कुल तीन टोले हैं—बाभन टोला, यादव टोला और मुसहर टोला। गाँव के भीतर-बाहर वाले यहाँ के निवासियों की निजी पहचान न करके जातिगत पहचान करते हैं। ये टोले भारत की सनातन परम्परा को निभाते आ रहे हैं। विभिन्न जातियों में बँटे इस गाँव में सत्यमूर्ति जी का अकेला घर है, जिनकी पहचान किसी जाति के नाम पर नहीं होती क्योंकि वह उन्नीस सौ इकहत्तर में बन रहे बांगलादेश से आकर यहाँ बसे थे। मगर इस गाँव वाले इन टोलों को दूसरे नाम से भी पहचानते हैं। बाभन टोला और यादव टोला को मुसहर टोला के जयकरन माँझी वगैरह क्रमशः 'बड़े मालिक' और 'छोटे मालिक' टोला कहते है जबकि मुसहर टोला को बाभन टोला और यादव टोला वाले 'मजूर टोला' कहते हैं।

जब सूबे की हवाओं में भी जाति का असर हो तो गाँव का भूगोल भी जाति पर आधारित होना असमंजस में नहीं डालता। सबसे पश्चिम में बाभन टोला, बीच में यादव टोला तथा सबसे पूरब में मुसहर टोला स्थित है। सोचता रहता है जयकरन कि पहले अक्सर पछुआ बयार की धूल-धक्कड़ से पूरब वाले बाशिन्दे तबाह रहते थे, मगर जब से लाल झंडा पार्टी ने इस इलाके में डेरा डाला है, तब से पुरवैया बयार लगातार जोर से बह रही है, जिससे बाभन टोले के लोगों को काफी कमर दर्द हो गया है। यादव टोला दोनों के बीचोंबीच बिल्कुल सुरक्षित है जैसे भूकम्परोधी मकान सुरक्षित होते हैं। लाल झंडा पार्टी उन्हें अपना दुश्मन मानकर लक्षित नहीं करती। उन्हें 'पेटी बुर्जुआ' मानती है मगर बाभन टोले को

'कुलक' मानती है। 'कुलक' ही हमारा वर्ग-शत्रु है। ऐसा उसके गाँव के नेता कामरेड अजय कार्यकर्त्ताओं को समझाते हैं। लाल झंडा पार्टी ने मुसहरों को गोलबन्द कर लिया है। जयकरन देखता है कि पहले से कई चीज़ें बदल गयी हैं—अब मुसहर टोले के नौजवान बाभन टोले और यादव टोले के बड़े-बुजुर्गों को प्रणाम-पाती नहीं करते; उनके आने पर चारपाई से नहीं उठते; उनके गाली-गलौज का जवाब उसी भाषा में, उसी मुद्रा में और उसी तीव्रता के साथ देते हैं। मुसहर टोले के नौजवानों ने एक बार मजदूरी बढ़ाने के सवाल पर काम रोको अभियान लागू कर दिया था। उनका नारा था—'इस साल के भादों में, गोरी कलइया कादो में।' उस जीवनपुर सहित करीब एक दर्जन गाँवों में कोई मजदूर दूसरों के खेतों पर काम करने नहीं गया। यादव टोला के लोगों ने अपनी खेती पूरी तरह खुद कर ली मगर बाभन टोले के नौजवानों ने खेती का काम खुद नहीं किया। बाहरी मजदूरों के लाने पर पूरी तरह से रोक लगा दी गयी थी, सो उस साल खेती नहीं हुई बाभन टोले में। मगर इस नाकेबन्दी से मुसहर टोले के जयकरन जैसे लोगों को भी नुकसान हुआ। उन्हें काम नहीं मिला, सो मजूरी नहीं मिली और इसलिए उनके पास बाजार से कुछ भी खरीदने के लिए पैसे नहीं थे। चारों ओर बाजार फूल की तरह खिले थे मगर उन्हें तोड़ना गरीबों के बूते की बात नहीं थी। घर में पिछले साल की कमाई से थोड़ा भी अनाज नहीं बचा था। उनकी पुरानी जीवन-शैली रोज कुआँ खोदकर रोज पानी पीने की थी। दूसरी ओर बाभन टोले और यादव टोले के लोगों के पास अपनी ज़मीन थी, पिछले साल की उपज के बचे अनाज थे और खेती के अलावा दूसरी कमाई भी कई परिवारों में मनीआर्डर से आ जाती थी। लेकिन जयकरन जैसे लोगों को किसी मनीआर्डर की आशा नहीं थी। लाल झंडा पार्टी ने एक अभियान चलाकर चन्दा इकट्ठा करके मुसहरों की कुछ मदद ज़रूर की थी, मगर वह नाकाफी थी। उसने काम देने के लिए ब्लॉक पर प्रदर्शन भी किया था, मगर वही ढाक के तीन पात वाली मसल हुई।

फिर पड़ गया सूखा। जीवनपुर में सूखे की यह तीसरी सालगिरह है। जयकरन माँझी जैसे गरीबों के पास खेत भले नहीं है, खेत के काम उन्हें कमोबेश ज़रूर मिलते रहे हैं। फिर बाजार दर पर पूरी मजूरी न सही, आधी-तिहाई तो मिल ही जाती रही है। खेती-बाड़ी के मालिक भले ही गरम मिजाज के हैं, मगर वे काम तो जयकरन माँझी वगैरह से ही कराते रहे हैं। लेकिन जब तीन साल से बारिश ही नहीं हो रही है और बरसाती नदी में चिड़िया के दूध भर भी पानी नहीं है, तो खेती हो तो कैसे हो? खेतों में बड़ी-बड़ी दरारें पड़ गयी हैं मानो इन्सानों और जानवरों की कब्रें थोक में तैयार की गयी हों। जयकरन सोचता है कि एक दिन वह इन्हीं दरारों में समा जाएगा। देखते-देखते आसपास

के कई गाछ सूख गये। उन सूखे गाछों में सूखे के मारे लोग अपनी नियति देख रहे हैं। वे भी सूख जाएँगे एक दिन खड़े-खड़े, बैठे-बैठे या लेटे-लेटे। फिर लेटे भी तो नहीं रहा जाता। इतनी गर्मी में बेचैनी ने जयकरन की नींद हराम कर दी। नींद भी गरीबों के साथ भेदभाव करती है, ऐसा उसे लगा। जानवरों को घास-फूस भी खाने को नहीं मिल रही। लोग पेड़ों की पत्तियाँ तोड़कर उन्हें खिला रहे हैं। कुओं का तल पाताल में चला गया है। पीने के पानी की भी किल्लत है, नहाने-धोने की कौन कहे! सोचता रहता है जयकरन माँझी कि यह कोई पंजाब, हरियाणा और पश्चिमी उत्तर प्रदेश जैसा खुशहाल इलाका तो है नहीं कि नहर या ट्यूबवेल के ज़रिये सुनिश्चित सिंचाई की सुविधा गाँवों में हो। यह तो मौत की घाटी बना जीवनपुर है जहाँ सिर्फ़ छोटी-मोटी खेती होती है बारिश के सहारे। सो रात-दिन करवटें बदलता रहता है जयकरन माँझी मगर उसकी किस्मत करवट बदलने का नाम ही नहीं ले रही। उसे ऐसा लगता है कि उसकी किस्मत किसी गड्ढे में औंधी गिरी पड़ी है और कुछ टटोल रही है। तीन साल से तो प्रकृति ने ही प्रतिबन्ध लगा दिया था खेती पर। धान के खेत में धूल की होली। बगल गाँव के पुरोहित ने घोषित कर दिया है, "कलियुग है यह। इस धरती पर इतना पाप हो गया है कि इन्द्र देवता कुपित हो गये हैं।" मगर इससे हटकर जयकरन माँझी याद करते हैं गोविन्द भाई की बात, "जलवायु बदल रही है क्योंकि वनों का नाश हो गया है। धरती पर जहरीली गैस की मात्रा बहुत बढ़ गयी है।" कार्बन या ऐसा ही कुछ बढ़ने की बात वह कह रहा था, याद करता है जयकरन माँझी।

सूखे की मार झेल रहे मुसहर टोले के कई जवान दिल्ली, पंजाब, हरियाणा, मुम्बई, वगैरह कमाने के लिए चले गये। उनमें से कुछ को बेहतर मजूरी वाला काम मिला और कुछ को अपना पेट पालने भर काम। मगर जयकरन सहित कुछ परिवारों से सिर्फ़ बूढ़े, बच्चे या महिलाएँ थीं, जो बाहर काम करने नहीं जा सकती थीं। सो उन्होंने बकरी और सूअर पालना शुरू किया। उनके बच्चे बकरी का दूध पीकर गुजारा करते। जयकरन माँझी टोले के सबसे बुजुर्ग आदमी हैं। पाँच वर्ष पहले उनका इकलौता बेटा एक बस दुर्घटना में मर गया था जिसकी याद में वह तड़पते रहते हैं। सो घर में उनकी विधवा पतोहू और उसके तीन छोटे-छोटे बच्चे हैं। कमाने वाला कोई नहीं है। पहले विधवा पतोहू भी काम करती थी मगर एक साल से वह गठिया बीमारी से बुरी तरह पीड़ित है। उसको अपने पैरों से चलकर दिशा-मैदान जाना भी मुश्किल हो गया है। जयकरन माँझी ने घर के खाने-खर्च के लिए बारी-बारी से अपने पाँचों सूअरों को बेच दिया। उससे खाने-पीने का जुगाड़ कुछ दिनों तक हुआ। मगर उस बिक्री के पैसे भी एक दिन खत्म हो

गये। अब करे तो क्या करे जयकरन माँझी? आख़िर वह एक दिन कई बागों में घूम-घूमकर आम की गुठलियाँ खोजकर लाया। उन्हें फोड़कर और पीसकर उसने पानी और नमक में उबालकर लपसी तैयार की। पूरे परिवार ने उसे तीन दिनों तक थोड़ा-थोड़ा करके खाया। इससे उनके पेट की आग कुछ बुझ गयी।

एक दिन बकरी ने भी दूध देना बन्द कर दिया। वह गाभिन हो गयी थी। सम्पत्ति के नाम पर मात्र वह बकरी और तीन छोटे-छोटे मेमने थे। जयकरन माँझी की पोती-पोते बकरी के मेमनों से खेलते रहते थे मगर जयकरन ने एक दिन गौर से देखा कि खाली पेट खेलने पर उनका खेल देर तक नहीं चल पाता। जयकरन माँझी ने बाभन टोले और यादव टोले के कुछ किसानों से पहले ही जो अनाज उधार लिया था, उसे चाहकर भी लौटा नहीं पाया। वह सोचने लगा अब किससे और किस मुँह से उधार माँगू। कई दिनों से सत्यमूर्ति जी भी दिखायी नहीं दे रहे हैं। शायद वह अपने बेटे के पास इलाज कराने के लिए मुम्बई चले गये हैं क्योंकि उन्हें कैंसर हो गया है। वह होते तो मुझे देखते ही मेरी पीड़ा समझ लेते। दुनिया में कोई ऐसा भी हो सकता है जो दूर से ही किसी के मन की बात समझ ले। बच्चों की मौत तो पिछले महीने ही हो गयी होती जब कई दिनों की भूख से बच्चे तड़फड़ा रहे थे। उन्होंने दो सौ रुपये दिये थे। पहले मूस के शिकार से दो-एक शाम गुजारा हो जाता था। पर अब दिनभर खेत कोड़ने पर भी एक मूस पकड़ में नहीं आता। खेत में लगातार कोई फसल हो ही नहीं रही है तो मूस क्यों मिले? शायद यह इलाका छोड़कर वे अन्न की खोज में दूसरी जगह भाग गये हैं। मगर हम तो ऐसा भी नहीं कर सकते। अपाहिज पतोहू और अबोध पोती-पोते को छोड़कर हम जाएँ तो जाएँ कहाँ? मैं जिस गगरी में हाथ डालता हूँ वहीं छूँछी मिलती है। सो अब बकरी और उसके तीनों मेमनों को बेचने का फैसला मैंने कर लिया है।

जब अगले दिन एक खरीददार चिकवा बकरी और मेमने लेने के लिए जयकरन माँझी के द्वार पर पहुँचा और सौदा पट जाने के बाद रस्सी में बाँधकर उन्हें ले जाने लगा, तो उसकी पोती-पोते जोर-जोर से रोने लगे। उन सभी ने मेमनों को अपनी-अपनी गोद में उठा लिया। वे उन्हें चूमने लगे, हाथ से सहलाने लगे और उनकी नकल करके 'ऊँ म्याँ म्याँ' बोलने लगे। जयकरन माँझी बच्चों को समझाने लगा, "तुम सबके लिए खूब बढ़िया-सा खस्सी लाएँगे, बड़ी-बड़ी सींगों वाला, चौड़े माथे वाला और घुँघराले बालों वाला। ये तो छोटे-छोटे हैं, किसी से लड़ भी नहीं सकते।" मगर बच्चों की दुनिया बिल्कुल अलग थी। जयकरन की झूठी दिलासा बच्चों को अपने पक्ष में नहीं कर सकी। उसकी सबसे छोटी पोती ने कहा, "मुझे तो छोटा बच्चा ही प्यारा लगता है। कितना मीठा बोलता

है, कितना उछलता है, कितना सीधा-सादा है। मुझे बड़ा-बड़ा खस्सी नहीं चाहिए। मैं इसे किसी को नहीं दूँगी।''

यह सुनकर जयकरन को बहुत गुस्सा आया। उसने अपनी पोती-पोते को झिड़का और दो-दो चटकने लगा दिये। मगर थोड़ी देर में उसकी आँखों में भी आँसू आ गये। उसने अपना घर देखा। उसका खपड़पोस मकान विधवा विलाप-सा कर रहा था। सिवाय सूराखों के, बाँस-बल्लियों से गिरते घुन के और दीवारों से गिरती मिट्टी की परतों के, और कुछ नहीं था। मगर तब भी वे सब उसे घर ही कहने को मजबूर थे। घर के ढाँचे से निराश होकर वह फिर अपनी पोती-पोते के सिर सहलाने लगा। फिर भी बच्चों ने उन मेमनों को अपनी गोद से जाने नहीं दिया। सो लाचार होकर जयकरन ने खरीददार चिकवा से मेमनों के लिए हुए रुपये उसे लौट दिये। सिर्फ़ बकरी ही बिकी दो सौ रुपये में। घर का माहौल गमगीन हो गया। जयकरन हताशा और गुस्से में काँप रहा था। उसे लगा कि उसका घर जोर-जोर से हिल रहा था। मानो भुँइडोल आ गया हो। सारे बच्चे रोने लगे। बच्चों की रुलाई ने उनकी माँ को भी रुला दिया। कुछ देर तक वह अपने को रोके रही। फिर वह रोते-रोते कहने लगी, ''हे भगवान! तूने हम लोगों को जनम क्यों दिया? जब हमें मुँह और पेट दिया है, तो हमारे खाने का जुगाड़ क्यों नहीं करता? हे दाताराम! जब तू हमें खाने को नहीं देता, तो हमारा प्राण क्यों नहीं हर लेता?''

मगर भगवान ने तत्काल न तो खाने का ज़ुगाड़ किया और न यमराज को भेजकर उनकी ज़िन्दगी का खाता ही बन्द किया। सरकार और समाज ने माना कि यही नियति है।

जयकरन ने ओसारे से पतोहू की ये डरावनी बातें सुन लीं। वह अन्दर से काँप गया मौत का नाम सुनकर। उसके भीतर एक हचका-सा लगा कि जीवनपुर अब मरनपुर होने वाला है। सो वह भागकर बाभन टोले में गया। दिन भर वह दर्जनों दरवाजों को खटखटाता रहा। एक ने कहा, ''क्या हुई तुम लोगों की नाकेबन्दी? हम लोगों को सताओगे, तो ऊपर वाला तुम्हें भी सताएगा।''

''मालिक! वह तो लाल झंडा पार्टी का सामूहिक फैसला था। हम लोग तो बराबर काम करते रहे हैं, भले हमें कम मजूरी मिली।''

''अच्छा, तो अब पूरी मजूरी लो। रस्सी जल गयी, मगर ऐंठन नहीं गयी।'' दो टूक जवाब मिला। फिर पूरे गाँव का उसने चक्कर लगाया। वह दया की भीख नहीं, बल्कि उधार चाहता था। आख़िर में जब आशा का सूरज अस्त होने वाला था, तभी सालभर में ड्योढ़ा चुकाने की शर्त पर महाजनी का काम करने वाले एक बाभन से वह एक बोरी में बीस किलो मकई का दाना उधार पा गया और

फिर उसे बच्चों की मदद से उठा लाया। घर में आये दाने देखकर बच्चों की आँखों की रोशनी दुगुनी हो गयी। दानों को देखकर मौत का भूत दबे पाँव भाग गया। छिपते सूरज की लालिमा में जयकरन को भोर की किरन-सी दिखायी पड़ी। तवे पर बालू में जयकरन ने मकई के दानों को भूँजा। बड़े-बड़े फूले हुए लावे उसकी आँखों के धुँधलेपन को अनोखी चमक में बदल दिये। जयकरन, उसकी पतोहू और बच्चों ने भरपेट लावा चबाया। भूख और नमक के साथ ने उसका स्वाद चौगुना कर दिया था। पेट में दाने जाने के बाद जयकरन को फिर ज़िन्दगी नयी नवेली दुल्हन-सी लौटती हुई दिखी। उसने सोचा कि धीरे-धीरे किसी तरह अब बेड़ा पार लग जाएगा और उसके परिवार का कोई भी सदस्य भूखा नहीं मरेगा। मगर बगल के पड़ोसी और चचेरे भाई सीताराम माँझी को ताजा भूँजे हुए लावे की गन्ध ने भीतर से बेचैन कर दिया। सीताराम ने एक बार सोचा कि वह जयकरन से मकई के दाने उधार माँग ले मगर उसे शंका हुई कि जयकरन का परिवार खुद कई दिनों से भूखा रहा है, तो एक भूखा परिवार दूसरे भूखे परिवार को कोई दुर्लभ चीज कैसे देगा? सीताराम के घर का चूल्हा एक हफ्ते से नहीं जला था। वह देर तक सोचता रहा। उसने अपने को जोर से डाँटा—धत्! चौथेपन में चोरी? भूख से रात में उसे नींद नहीं आयी। पेट की आग होली की तरह धधकने लगी। भूख और ज़मीर की लड़ाई साँप-नेवले की तरह देर तक चलती रही। फिर भूख ने उसे चोरी करने के लिए आख़िर में मजबूर कर दिया। वह आधी रात को चुपके से अपने बिस्तर से उठा और अँधेरे में जयकरन के आँगन मे कूद पड़ा। वह काम खास मुश्किल नहीं था क्योंकि उसके आँगन की एक तरफ की दीवार कम ऊँची थी और धीरे-धीरे वह गलकर और नीची हो गयी थी। कई दिनों की फाँकाकशी के बाद पेट भरने के कारण जयकरन का समूचा परिवार घोड़े बेचकर सो रहा था। सीताराम को जयकरन के घर के भीतर का पूरा हुलिया पहले से पता था। वह आँगन से धीरे-धीरे दरवाजे में चला गया। जहाँ एक बोरी में मकई के दाने रखे थे। एक बार उसने पूरी बोरी उठा ले जाने का मन बनाया लेकिन अचानक फिल्म की रील की तरह उसके सामने बड़े भाई जयकरन का चेहरा कौंध गया। फिर सीताराम ने सोचा कि पूरी बोरी ले जाने से उस पर शंका बढ़ सकती है और खोजबीन करने पर उसकी बदनामी और दुर्दशा भी हो जाएगी। फिर लेने के देने पड़ेंगे। वह समाज में मुँह दिखाने लायक भी नहीं रहेगा। इसके अलावा उसने बड़े भाई जयकरन के बारे में भी सोचा। सो उसने उसमें से आधा ही अपने गमछे में बाँधा और बाकी दानों को बोरी में ही छोड़ दिया। फिर वह आहिस्ता-आहिस्ता आँगन की दीवार फाँदकर बाहर निकल आया।

अगले दिन जयकरन सुबह उठकर दिशा-मैदान करने के लिए पूरब के तालाब पर चला गया। उसका मन बिलकुल हरा-भरा था। उसे लगा कि पतझड़ के बाद वसन्त ज़रूर आता है और उसकी ज़िन्दगी में भी आएगा। वह वहाँ मिले लोगों से खूब हँस-हँसकर बतिया रहा था। एक ज़माना गुजरने के बाद उसे जवानों से हँसी-मजाक सूझ रहे थे। वहाँ से लौटने के बाद जयकरन अपने घर के भीतर घुसा तो उसकी निगाह बरबस बोरी में रखे मकई के दानों की ओर चली गयी। मकई के दाने बोरी में कुलबुला रहे थे मानो कह रहे हों कि हमारे भाइयों को कौन ले गया? कुछ खोजबीन करो उनकी। जयकरन ने उस बोरी को घूरा तो उसका माथा ठनका। बढ़ती उम्र और अनुभव के कारण उसे ताड़ते देर नहीं लगी कि किसी ने उस बोरी में से कुछ दाने ज़रूर निकाले हैं। वह एक-एक करके अपने पोती-पोतों से गहरी पूछताछ करने लगा। पहले उसने नरमी दिखायी, मगर कोई फल नहीं निकला। फिर वह रात के तवे-सा गरम हो गया। सभी बच्चों के कान उमेठने के बाद उसने उठक-बैठक करवायी। मगर तब भी कोई फल नहीं निकला। फिर उसने उन सबको मुर्गा-मुर्गी बनवाया। फिर भी कोई फल नहीं निकला। वह हैरान हो गया। तभी उसके मन में यह विचार आया कि हो सकता है कि उसकी पतोहू ने कुछ दाने बच्चों से छुपाकर कहीं सुरक्षित रख दिये हों। सो वह उससे भी इस बाबत पूछताछ करने लगा। वह रो-रोकर कहने लगी, "भगवान कसम। मैंने जो एक भी दाने को हाथ लगाया हो। कौन ऐसी कठोर माँ होगी जो अपने ही बच्चों के पेट के दाने बेचेगी? पता नहीं क्यों, भूखे भात में खुदा का साझा हो रहा है?"

"तो क्या हुआ बाकी दानों का? क्या उन्हें धरती लील गयी?" जयकरन गुस्से में काँप रहा था।

"मेरे बच्चे मर जाएँ यदि मैंने एक भी दाना चुराया हो।" यह कहकर उसकी पतोहू फूट-फूटकर रोने लगी।

जयकरन की समझ में कुछ नहीं आ रहा था। वह अपने पर झुंझलाने लगा। "काश! मैं पाँच-पाँच किलो दाने कई दिनों में घर लाता। न ज़्यादा दाने होते, न चोरी की घटना होती।" माथे पर हाथ रखे हुए वहाँ से उठकर वह घर से बाहर निकला। उसने अचानक देखा कि सीताराम के घर का एक बच्चा छोटू मकई के लावे फाँकते हुए उसी के घर की ओर बढ़ा आ रहा है। उसके पीछे सीताराम बेतहाशा दौड़ रहा था। सीताराम के मन को यह शंका कँपा गयी थी कि यदि वह बच्चा जयकरन के आँगन में गया तो उसकी चोरी पकड़ ली जाएगी। अचानक बच्चे के सामने जयकरन को देखकर वह सकपका गया। जयकरन सीताराम के चेहरे की हवाइयाँ पढ़ने लगा। जयकरन ने पूछा, "आख़िर क्या लेकर भाग रहा

है छोटू?'' मगर जवाब देने की बजाय वह सिर झुकाकर अपने घर में घुस गया। जयकरन ने छोटू के नजदीक जाकर उससे कहा, ''जाओ, लावे चबाकर मेरे घर आना।'' उसने सोचा सीताराम ने चोरी करके बड़ी गलती की है मगर छोटू को देखते ही पतोहू सारा माजरा समझ जाएगी और आकाश-पाताल एक कर देगी। फिर जयकरन सोचने लगा कि भूखा आदमी कौन-सा पाप नहीं करता? आख़िर सीताराम के बच्चों की भी तो ज़िन्दगी बचानी है, आख़िर सबको जिन्दगी की जंग जीतनी है मौत से। फिर सीताराम का परिवार भी तो अपना ही खून है।

मगर दो हफ्ते में मकई के दाने भी चुक गये। अब कोई उपाय उसे नहीं सूझ रहा था। उसने रात में एक सपना देखा कि उसकी लाश को चील-कौवे, कुत्ते-सियार खा रहे हैं। वह उन्हें भगा रहा है, मगर उनकी संख्या बढ़ती जा रही है। फिर उसके परिवार के सभी लोगों को चील-कौवे, कुत्ते-सियार खाने लगे...। वह नहीं-नहीं कहता हुआ उठ बैठा। एक बार मेमनों को बेचने की बात उसने फिर सोची मगर बच्चों की बेपनाह चाहत के आगे उसने वह विचार त्याग दिया। सूखे के कारण मेमनों के चारे की भी कमी हो गयी थी। इस बार अकाल, जलकाल और तिनकाल एक साथ लोगों की छाती पर मूँग दल रहे थे। कई दिनों तक मेमनों के लिए चारे का कोई जुगाड़ नहीं हो सका। सो सबसे छोटे मेमने ने भूख से दम तोड़ दिया। बच्चे-बड़े-बूढ़े सभी रोने लगे। मगर कोई फ़ायदा नहीं था। जयकरन उस मेमने को फेंकने जाने लगा, तभी तीनों बच्चों ने उसके हाथ-पाँव पकड़कर रोते हुए कहा, ''बाबा! इसे मत फेंको। इसे जमीन में गाड़ दो। एक दिन गेहूँ की तरह यह भी जम जाएगा। तब ढेर सारे मेमने हो जाएँगे।''

यह सुनकर जयकरन ने दुख में भी फिच्च से हँस दिया। इस बाबत वह बच्चों को कुछ भी समझाता मगर बच्चे मानने को तैयार नहीं थे। सो हार मानकर बच्चों की ज़िद उसने पूरी कर दी। उसके बाद बच्चों ने खुद एक गड्ढा खोदा और उस मरे मेमने को उसमें आहिस्ता-आहिस्ता लिटा दिया। बच्चों ने उसके ऊपर फूल चढ़ा दिये। उसके बाद मिट्टी से आहिस्ते से उसे ढँक दिया। फिर उस मिट्टी को उन सभी ने अपने आँसुओं से सींच दिया जिससे नमी पाकर अंकुर जल्दी निकले, ऐसा बच्चों का विश्वास था।

अगले दिन जयकरन ब्लॉक गया। वहाँ उसने काम पाने के लिए ग्रामसेवक और बी.डी.ओ. साहब से भी भेंट की। बी.डी.ओ. ने कहा, ''यदि आपको काम करना हो तो आवेदन देने पर पन्द्रह दिनों में काम मिल जाएगा। मिट्टी की सड़क बनानी है मगर इस बुढ़ापे में आप जैसे लँगड़े से कोई काम कैसे होगा? सारी मजूरी बेकार हो जाएगी। अन्नपूर्णा योजना में आपका लाल कार्ड बनवा देते हैं।'' बी.डी.ओ. ने ग्रामसेवक को लाल कार्ड बनाने की कार्रवाई करने का निर्देश दिया।

ग्रामसेवक, मुखिया और बी.डी.ओ. के बीच दौड़ते-दौड़ते जयकरन थक गया मगर आशा की पहली डोर के सहारे वह जीता रहा। इसी बीच एक दिन वह ब्लॉक में था तो बी.डी.ओ. ने जयकरन की समस्या बाल विकास परियोजना अधिकारी से बतायी। उन्होंने रास्ता निकाला, "जीवनपुर में आँगनबाड़ी केन्द्र स्वीकृत हुआ है। वहाँ के मुखिया का एक चेला उसे चला रहा है। यदि जयकरन अपने मुसहर टोले के लोगों से दस्तख़त करा दे कि छह महीने से बच्चों को खिचड़ी-हलवा वगैरह मिला है, तो उसे बीस किलो चावल मिल जाएगा।" जयकरन कुछ देर चुप रहा। मगर जब बी.डी.ओ. ने इशारा किया, तो उसने धीरे से हामी भर ली। उसने सोचा कि कोई-न-कोई तो ऐसा करेगा ही मजबूरी में। बाद में वहीं ब्लॉक पर जयकरन ने हाथ और पैर की सभी उँगलियों से कई ठप्पे लगा दिये। मजमून बाद में तैयार कर लिया परियोजना अधिकारी ने। जयरकन को थोड़ी देर में दस किलो चावल मिल गया और बाकी बाद में देने का आश्वासन मिला। वह किसी तरह चावल लेकर घर लौटा। घर पर पतोहू ने चावल देखा, तो वह खुशी से उछल पड़ी। मगर बच्चे उसे दिखायी नहीं दिये। थोड़ी देर में उसकी पोती-पोते वहाँ आ गये और बोले, "बाबा! अभी-अभी हम लोग तालाब के पास से आ रहे हैं। अभी तक मेमना नहीं उगा है। हम लोगों ने ऊपर से एक लोटा पानी वहाँ डाल दिया है जिससे जल्दी अंकुर निकले।" यह सुनते ही जयकरन हँसने लगा और बोला, "अब वह मेमना कभी नहीं उगेगा।" यह सुनकर बच्चे निराश हो गये। फिर जयकरन ने चावल को धोकर उबाला और सभी को परोस दिया। भात और नमक से पेट भर गया सबका। सभी ने कई बार डकारें लीं। बच्चों ने दोनों मेमनों को बचा-खुचा चावल खिलाया और पानी पिलाया। मगर अब चारे की समस्या बहुत गम्भीर हो गयी। सूखे और उबलती गर्मी से सारी घासें जल गयीं। मेमने कमजोर हो गये और एक दिन बाकी दोनों मेमने भी मर गये। उन्हें भी बच्चों ने मिट्टी में गाड़ दिया और उसके ऊपर एक लोटा पानी डाल दिया।

अगले दिन जयकरन फिर ब्लॉक गया और लाल कार्ड के बारे में वह दरयाफ़्त करने लगा। ग्रामसेवक ने उसे समझाते हुए कहा, "तुम्हारा लाल कार्ड चम्मन साव की दुकान पर है। वहीं जाओ और अपना अनाज ले लो।"

जयकरन सिर पर पाँव रखकर भागा। चम्मन साव मोटी चमड़ी के आदमी थे। उन्होंने जयकरन को देखते ही कहा, "मैंने तुम्हें बुलाने के लिए तुम्हारे गाँव के तीन लोगों से सन्देश भिजवाया था मगर अब तुम आये जब राशन खतम हो गया है। खैर...कोई बात नहीं। दो हफ्ते बाद आकर राशन ले जाओ।"

"मगर तब तक हम लोग क्या खायें?" जयकरन के सब्र का बाँध टूट

गया।

"तो हम कहाँ से पैदा करें चावल-गेहूँ? मेरा गोदाम खुला है। अपनी आँखों से तुम खुद देख लो।" यह कहते हुए चम्मन साव ने गोदाम का किवाड़ खोल दिया। और सचमुच वहाँ अनाज नहीं था। फिर जयकरन ने अपना लाल कार्ड चम्मन साव से माँगा। वह ज़िन्दगी की उस आस को देखना चाहता था, छूना चाहता था और सीने से लगाना चाहता था। मगर चम्मन साव कच्चा खिलाड़ी नहीं था। उसने तुरुप का इक्का चल दिया, "फूस के घर में लाल कार्ड कहाँ रखोगे? फिर कोई उसे चुरा लेगा, तो अनाज की सुविधा से हाथ धो लोगो। यहाँ सबका कार्ड सुरक्षित रखा है। सप्लाई इंस्पेक्टर साहब जब भी जाँच में आते हैं, तो लाल कार्ड भी चेक करते हैं। तब तुमको खोजने हम कहाँ-कहाँ जाएँगे? फिर इस बुढ़ापे में बुलाने पर तुमको आने में तकलीफ होगी कि नहीं?"

अब जयकरन के पास कोई जवाब नहीं था। वह पन्द्रह दिनों के बाद आने पर अनाज पाने की आशा को पगुराते हुए वहाँ से चल दिया।

कुछ दिनों के बाद पहले वाला चावल खत्म हो गया। भूख से ज़्यादा खाया जाता है या चावल ही कम था, इसे जयकरन नहीं समझ सका। न तो परियोजना अधिकारी से बकाया चावल मिला और न चम्मन साव की दुकान पर अन्नपूर्णा योजना का गेहूँ-चावल आया। सो जयकरन निःसहाय हो गया। उस रात जोर का तूफान आया। हवा, धूल, बादलों की गर्जना और बिजली की चमक से अमावस की रात और भयावह हो गयी। घर में खाने को कुछ भी नहीं था। जयकरन ने पूरी ताकत लगाकर सीताराम-सीताराम कहकर कई बार पुकारा, मगर उस अंधड़ में कुछ सुनायी नहीं दिया। जयकरन का छप्पर हवा में उड़ गया। वह बचाओ-बचाओ कहता रहा, मगर कोई नहीं आया। शायद किसी ने उसकी आवाज़ नहीं सुनी अथवा सभी को उस तूफान में पहले अपने को बचाने की पड़ी थी। बच्चों ने उसे एक लोटा पानी दिया। मगर पानी पीते ही वह उल्टी करने लगा। उसने खाने के लिए कुछ माँगा। मगर माटी, पानी के अलावा कुछ भी नहीं था। उसे खून की उल्टियाँ होने लगीं और आख़िर में उसने दम तोड़ दिया। उसकी पतोहू और बच्चे रो-रोकर बेहाल हो गये। अगले दिन सुबह सीताराम ने गाँववालों की मदद से किसी तरह उसका क्रिया-कर्म किया।

एक-दो पड़ोसियों ने थोड़ा भात बच्चों को दे दिया। वे नमक-भात खाये और सो गये। अलगे दिन बच्चे भूख से बिलबिलाने लगे। माँ की आँखें भूख के बढ़ते आकार-प्रकार को देखकर डर गयीं। अब बाभन टोला और यादव टोला से कोई सूद पर भी अनाज या रुपये देने को तैयार नहीं था। पहले वाला महाजन तगादा करने आ गया था मगर मौत देखकर वह चुप रह गया। मगर एक दिन

उसने जयकरन के घर-वार वाली जमीन पर अपने बकाये के नाम पर कब्जा कर लिया। सो उससे माँगने का अब सवाल ही नहीं था। मुसहर टोला कुछ देने लायक नहीं था। वे पहले अपना पेट किसी तरह भर लेते। सो रात में बड़ा लड़का अपने बिस्तर से धीरे से उठा। उसने समूचे उजड़े घर में खोजा मगर एक दाना भी नहीं मिला। उसने अपने हाथ में एक भगोना, एक खुरपी तथा एक छुरी ली। उसने सोचा कि यदि कोई उसे देख लेगा तो पकड़ लेगा। सो अपने मुँह पर कपड़ा बाँधकर वह चुपके से उस जगह पर गया जहाँ हाल में दोनों मेमनों को गाड़ा गया था। उसने कब्र पर से खुरपी से धीरे-धीरे मिट्टी निकाली। तभी उसका हाथ काँपने लगा। वह रुककर दम भरने लगा। फिर कुछ देर बाद उसने गड्ढे से मेमनों को निकाला। वे सूख रहे थे। उसने अपने आँसुओं से उनका स्वागत किया। उसे लगा कि मेमने ऊँ म्याँ म्याँ कर रहे और घर चलने को कह रहे हैं। फिर उसने मेमनों को अपने भगोने में धीरे से रखा और कपड़े से ढँक दिया। इस दौरान वह हाँफने लगा मानो वह किसी बड़े अखाड़े में मौत से कुश्ती लड़ रहा हो। उसे लगा कि वह मौत से जूझ रहा है लेकिन किसी भी कीमत पर मौत को हराकर ही दम लेगा। फिर वह धीरे-से वहाँ से घर की ओर चल दिया।

थोड़ी देर में अपने घर पहुँचकर उसने चूल्हा जलाया। उसने मेमनों को धोकर काटा और चूल्हे पर मेमनों का मांस पकाना शुरू किया। आग की गर्मी और मांस की गन्ध पूरे घर में जोर से फैल गयी। उसके छोटे भाई-बहन चूल्हे के आसपास बैठ गये। उनकी माँ भी थोड़ी देर में आ गयी। एक बार उसने सोचा कि वह मांस खाना ठीक नहीं मगर कोई उपाय न देखकर वह चुप ही रही। उधर बच्चों की सूनी आँखों में उस रात एक अजीब चमक आ गयी। जब मांस पककर तैयार हो गया, तो बड़ा भाई छोटे भाई-बहन को प्रेम से खिलाने लगा। खाते-खाते सबसे छोटे भाई की आँखें मुँदने लगीं। बड़े भाई और बहन ने उसे उठाकर लिटा दिया। सोचा पेट भरने पर बहुत दिनों के बाद उसे गहरी नींद आयी है। सभी ने छककर मांस खाया। उनकी माँ को तभी सुनी-सुनायी बात याद आयी। एक बार विश्वामित्र नामक मुनि भूख से व्याकुल थे। वह एक चांडाल के घर पहुँचे और बाहर पड़े डाँगर का मांस खाने लगे। उसे लगा कि जब ऋषि-मुनि जीने के लिए ऐसा कर सकते हैं, तो हम लोग कोई अपराध नहीं कर रहे हैं। फिर तीनों डकारें लेकर अँगड़ाई लेते हुए सोने चले गये। रातभर वे सभी घोड़े बेचकर सोते रहे। जब सूरज अपने बिस्तर से उठा, तब अन्य दोनों बच्चों ने पेट में दर्द की शिकायत की। थोड़ी देर में उन दोनों को उल्टी होने लगी। फिर उन्हें कै-दस्त होने लगी। घर में दवा के लिए पैसे नहीं थे। घरेलू इलाज के लिए टोले में जाना चाहती थी वह मगर उसे भी कै-दस्त होने लगी। सीताराम के घरवालों

ने तीनों को पुदीना पिलाया मगर वह नहीं पचा। थोड़ी देर में पूरे टोले के लोग जुट गये। चारों ओर बदबू का आलम था। सीताराम के परिवारवालों ने बहू और बच्चों की सफाई की। वह बार-बार कहती रही, ''मुझे बचा लो।'' मगर सीताराम मदद करने लायक नहीं था। उसने कुछ घरेलू इलाज किया। उसके अनुरोध करने पर जनसशक्तीकरण प्रोजेक्ट चलाने वाले गोविन्द भाई ने कहा, ''हमारी संस्था सिर्फ़ एडवोकेसी करती है, जनता को शिक्षा और स्वास्थ्य के प्रति जागरूक करती है। इस प्रोजेक्ट में दवा आदि का प्रावधान नहीं है।'' सो सीताराम सिर्फ़ हाथ मलता रहा। उधर पुलिस की बढ़ती दबिश पर लाल झंडा पार्टी के नेता-कार्यकर्त्ता एक रणनीति के तहत गाँव छोड़कर जंगल में चले गये थे सो उनसे भी कोई मदद नहीं मिल सकी। और शाम घिरते-घिरते एक-एक कर परिवार के चारों सदस्यों को मौत ने अपने आगोश में ले लिया। इससे पूरा गाँव सकते में आ गया। मुसहर टोले के बाकी लोग अपनी ऐसी ही नियति सोचने लगे। टोले के एक आदमी ने थाने को ख़बर देने में ही अपने कर्तव्य की इतिश्री समझ ली। स्थानीय थाने ने तत्काल एक चौकीदार को वहाँ तैनात कर दिया। फिर थाने में डी.एम. और एस.पी. को वायरलेस से इस घटना की सूचना दे दी। मगर वहाँ भीड़ बढ़ती गयी। जो लोग पहले तटस्थ थे, वे भी सरकार के खिलाफ नारे लगाने लगे सो थाना प्रभारी और अंचल अधिकारी को नाजुक स्थिति देखते हुए घटनास्थल पर जाना पड़ा। उन पर गाँववालों ने सवालों की बौछार लगा दी और लाश उठाने से मना कर दिया। वे दोनों बिल्कुल असहाय थे। वे लोगों को समझा रहे थे। दारोगा कह रहा था, ''होनी कौन टाल सकता है? इस क्षणभंगुर दुनिया में कोई अमर नहीं है।'' मगर युवा वर्ग यह सब मानने को बिलकुल तैयार नहीं था। वैसे कुछ बूढ़े लोग अपना भविष्य अन्धकारमय देखकर सिर हिलाकर हामी-सी भरने लगे। कुछ लोग एक-एक लाख मुआवजे की माँग करने लगे। अंचल अधिकारी ने जवाब के रूप में एक सवाल दागा, ''मगर मुआवजा लेने को बचा ही कौन है?'' इस पर चारों ओर सन्नाटा छा गया।

अगले दिन सभी अख़बारों के पहले पन्ने पर मुख्य ख़बर छपी, ''जीवनपुर में भूख से पाँच मौतें, सरकार बेनकाब।'' यह ख़बर पढ़कर शासन-प्रशासन के कान खड़े हो गये। मगर थेथर को शर्म नहीं आती। वे घोंटकर पी गये शर्म को भाँग की तरह। जिला प्रशासन ने एक बार फिर सिद्ध कर दिया कि शासन लोहे से चलता है और गाँव से चारों लाशें मँगवाकर पुलिस सुरक्षा में डॉक्टरों की एक टीम पोस्टमार्टम जाँच के लिए तैनात कर दी। डॉक्टरों ने आनन-फानन में पोस्टमार्टम किया और जाँच रिपोर्ट सौंप दी। उसमें साफ-साफ दिखा था, ''ये मौतें खाने में ज़हर और इंफेक्शन से हुईं।'' मगर उसमें इंफेक्शन के कारणों का कोई जिक्र नहीं था। शासन-प्रशासन ने भी गहराई

में जाकर उन कारणों की तहकीकात नहीं की। उधर सरकारी गोदामों में रखे हुए अनाजों को सड़ने से बचाने में पूरा महकमा लगा रहा। संविधान उकड़ू बैठा था मगर तरह-तरह के सवाल गाँव के लोग बार-बार कर रहे थे। सूबे की 'न्यायप्रिय' सरकार के मौखिक आदेश का तहेदिल से पालन करते हुए उसी शाम को जिलाधिकारी ने सभी अख़बारों को जो खंडन भेजा, वह अगले दिन छपा गया, "डॉक्टरों की पोस्टमार्टम रिपोर्ट के अनुसार जीवनपुर में हुई पाँचों मौतों का कारण खाने में ज़हर और इंफेक्शन है। इस प्रकार ये मौतें भूख से कतई नहीं हुई हैं। प्रशासन आम नागरिकों के सर्वांगीण कल्याण के लिए कटिबद्ध है। राज्य सरकार के निर्देशानुसार किसी भी व्यक्ति को भूख से नहीं मरने दिया जाएगा। सरकार का साफ निर्देश है कि यदि किसी पंचायत में भूख से कोई मौत होगी, तो उसके लिए उस ग्राम पंचायत के मुखिया को पूर्णत: ज़िम्मेदार माना जाएगा। अभी सरकारी गोदामों में लाखों टन अनाज उपलब्ध है। शासन के राशन के आगे भूख कैसे टिक सकेगी? भूख से मरने की ख़बर पूरी तरह भ्रामक एवं निराधार है।"

मगर यह खंडन पढ़-सुनकर ठठाकर हँस रहा है जीवनपुर का बूढ़ा और विशाल पीपल का गाछ—कभी डॉक्टर पर, कभी जिला प्रशासन पर, कभी सूबे की सरकार पर और कभी 'जनसशक्तीकरण' प्रोजेक्ट के स्वयंसेवी गोविन्द भाई पर। बूढ़े पीपल को लग रहा है कि जिलाधिकारी के खंडन का एक-एक शब्द इन पाँचों के खून से लिखा गया था। देखते-ही-देखते पीपल की सारी पत्तियाँ झड़ गयीं और कल का हरा-भरा गाछ अब मात्र एक ठूँठ बनकर रह गया मानो वह उन सबका श्राद्ध करने के लिए अपना भी तर्पण कर रहा हो। मगर सामूहिक मौतों के बाद से जीवनपुर में विभिन्न दलों के नेताओं का ताँता रोज लगा रहता है मानो उनकी चिर-प्रतीक्षित मनौती पूरी हुई हो और वे गिद्धों की भूमिका खुद निभा रहे हों।

ख़ारिज़

भूल यही हुई कि उनके साथ मैं भी चला गया था, कन्या देखने। यह न हुआ होता तो आज मेरी यह स्थिति न हुई होती। हुआ यह कि मेरे भतीजे की शादी तय होने वाली थी। इसके लिए हमें विराटनगर जाना था। कन्या पक्ष को वर यानी मेरा भतीजा पसन्द था। मेरे ज़माने की बात और थी, तब शादियों के तौर-तरीके अलग थे। तब बचपन में शादी होती थी और वर-वधू देखने की सारी ज़िम्मेदारी बुजुर्गों पर होती थी। शादी तब सचमुच गुड्डे-गुड्डी का खेल होता था। मसलन, मेरी शादी मेरे दादा जी ने पक्की की थी और आप यदि उम्र जानना चाहेंगे तो मुझे पेरशानी होगी, उस वक्त मेरी उम्र 13 साल की थी। ठीक वही उम्र जो शादी के समय हमारे राष्ट्रपिता की थी। होश आने पर मुझे कई बार यह बतलाया गया कि शादी तय करने आये मेहमान के सामने जाने में मैं किस तरह लजाता था। तब वर देखने का ही रिवाज था। आजकल की कन्या देखने का तब रिवाज नहीं था। लेकिन अब तो भतीजा सयाना है—यही कोई तीस साल का। तीस साल ज़िन्दगी की एक-तिहाई उम्र होती है। ज़िन्दगी अब तक जाने कितने मोड़ ले चुकी होती है। सपने में बहुत सारी लड़कियाँ, बहुत सारी तारिकाएँ आ चुकी होती हैं। मेरा भतीजा निर्द्वन्द्व रूप से खुद शादी की बात में शरीक होता है। शादी करने आये भावी ससुर-साले के सामने जाने और साक्षात्कार देने से न कतराता है, न पीछे हटता है। इतना ही नहीं, घरवालों के अलावा वह भी शादी के पहले अपनी प्रस्तावित पत्नी से मिलना चाहता है, बतलाना चाहता है और साझा सपना देखना चाहता है।

उसी पृष्ठभूमि में अपने भतीजे, भाई साहब और भाभी जी के साथ हमने विराटनगर के लिए रविवार को प्रस्थान किया। मेरे नगर से वहाँ के लिए एक रेलगाड़ी सप्ताह में तीन दिन जाती थी, जिसका नाम था गंगा-ताप्ती एक्सप्रेस। यह गाड़ी हाल में शुरू हुई थी। नगर के लोग इस गाड़ी को शुरू करने के कारण रेल मन्त्री से काफी खुश थे। इसकी पहली वजह थी कि हमारे सूबे से मजदूरों

की बिन भर्ती वाली फौज़ नियमित रूप से वहाँ जाती रहती थी, जो एक तेजी से विकसित हो रहा औद्योगिक शहर था। वहाँ दर्जनों कपड़ा मिलें थीं, चमड़ा और हीरा घिसाई के भी कारखाने थे। जब विवाह प्रस्तावक, जिसे हम बरतुहार कहते हैं, ने बताया था कि उस नगर से हर साल करीब पचीस हजार करोड़ रुपये मूल्य का हीरा निर्यात किया जाता है, तो मेरी भाभी को लगा था कि विवाह में ज़रूर हीरे का भी कुछ जोड़-तोड़ होने वाला है।

वह सपना देखने लगी थी कि उन्हें शादी में भले कोई और जवाहरात न मिले लेकिन बहू को उसका बाप कुछ न कुछ हीरा ज़रूर देगा। खुश मैं भी हुआ था। लेकिन हमारी खुशी राष्ट्रीय थी। मैं सोच रहा था कि भारत जैसे अल्प विकसित देश में कम से कम ऐसा कोई शहर तो है, जो विकसित देशों को क़ीमती चीज़ निर्यात करता है। 'यूँ गम मिटाने को गालिब ख्याल अच्छा है' के अन्दाज में मेरी पत्नी कहा करती थी कि अपने यहाँ से भी लीची और सत्तू का निर्यात विदेशों में होता है। लेकिन भाभी जी इसे काटते हुए कहतीं, ''कहाँ हीरा, कहाँ सत्तू। कहाँ राजा भोज, कहाँ गंगू तेली।'' देवरानी-जेठानी में हीरा और सत्तू को लेकर कई बार 'तू-तू मैं-मैं' हो चुकी थी। दरअसल वह खाये-अघाये घर की बतकही होती थी समय काटने का साधन। बात का न कोई ओर होता, न छोर। फिर बात चलती रहती। अब सत्तू और हीरा को एक साथ बात में रखे का कोई मतलब था क्या? लेकिन इसे लेकर भी घंटों तक गप होती और कभी-कभी तो मेरे भाईसाहब भी, जिन्हें हम सब तजुर्बेकार समझते थे, इस बात में शामिल हो जाते थे। एक बार इसी द्वन्द्वयुद्ध में कूदकर उन्होंने सत्तू की महिमा पर एक लाबा भाषण ही दे दिया था।

ओह, हम तो भूल ही गये कि हम यात्रा पर थे। खैर...30-32 घंटे की यात्रा के बाद गन्तव्य वाला स्टेशन शीघ्र आने वाला था। भाभीजी ने सख़्त हिदायत दी थी कि दूसरे दर्जे के डिब्बे से फौरन उतर कर वातानुकूलित डिब्बे के पास हमलोग खड़े हो जाएँगे जिससे कन्यापक्ष वाले समझें कि हमने वातानुकूलित दर्जे में सफर किया है। दरअसल कन्यापक्ष वालों ने हमें सेकेंड ए.सी. का किराया दिया था आने-जाने के लिए । हम लोग उसमें भी डंडी मारना चाहते थे। तय यह हुआ कि साधारण दूसरे दर्जे से यात्रा कर प्लेटफार्म पर थोड़ी होशियारी कर ली जाए। इतने में बहुत सारे पैसे बच जाने थे। हालाँकि व्यक्तिगत रूप से मैंने विरोध किया था कि यह अनैतिक है। हम चाहे जिस श्रेणी में चलें, नाटक न करें, तो अच्छा रहेगा। साधारण दूसरी श्रेणी में चलकर ही कुछ पैसे बचा लेते हैं, तो उससे कन्या के लिए उपहार ले लें। यह किसी को बुरा नहीं लगेगा। इस पर लम्बा वाद-विवाद चला, तब जाकर कहीं इस बात पर हम सहमत हुए

कि बचे पैसे से उपहार ले लेने में कोई बुराई नहीं है।

इस लम्बे सफर के आख़िर तक हम सबकी हालत खस्ता थी। हम सब थकान से भरे थे। सबसे ज़्यादा खामियाजा भाभीजी को भुगतना पड़ा था क्योंकि धूल, धूप, गन्दगी और भीड़ की वजह से उन्हें काफी कष्ट हो रहा था। गर्म चाय पीने के सिवाय हमारे पास कोई दवा न थी और न स्टेशन पर मिल सकती थी। हमने डिस्प्रिन-विस्प्रिन की टिकिया तलाशी थी, लेकिन कहीं कुछ न मिला। भाभी जी अपने सिर-दर्द का सारा गुस्सा रेल प्रशासन और मन्त्री पर उतार रही थीं। कैसा है रेल प्रशासन कि गाड़ी में दवा की कोई व्यवस्था नहीं करता? मैंने चुटकी ली, ''क्या मूँगफली की तरह डिस्प्रिन, क्रोसीन, एनासिन, कालपोल की गोलियाँ भी खोंमचे वाले बेचने लगें?'' इस पर मेरी ओर उन्होंने आँखें तरेरीं दुर्वासा की तरह और भयवश मैंने नज़र फेर ली। फिर वे भाईसाहब पर बरस पड़ीं, ''इसीलिए मैं आपसे कहती थी कि मेरी तबियत ठीक नहीं रहती, मुझे मत ले चलो। लेकिन आप माने नहीं और बहू दिखाने में मुझे बीमार कर दिये।'' भाईसाहब ने उसका कोई जवाब उन्हें नहीं दिया, किन्तु जब वे शौचालय गयीं तो मुझसे कहने लगे, ''देखा तुमने इनका पिचाल? यदि बहू में कोई कमी होती, तो ये मुझे ही कोसतीं। फिर बहू कोई सामान तो है नहीं कि अदला-बदली हो जाये। क्या जमाना है जब मुझे कोई तकलीफ होती है तो उसकी जड़ ये मुझे मानती हैं और जब परिवार का कोई दूसरा तकलीफ पाता है, तो उसके लिए भी मुझे ज़िम्मेदार ठहराया जाता है। हे भगवान, सबका भला कर।'' तब तक भाभी भी वापस आ गयीं और भाईसाहब की सिट्टी-पिट्टी गुल हो गयी।

जैसे ही गाड़ी स्टेशन पर पहुँची, हम लोग प्लेटफार्म पर झटपट उतर गये। फिर ऊपर के आदेश पर दौड़ते हुए वातानुकूलित डिब्बे के सामने खड़े हो गये। कुछ सेकेंडों में ही कन्या के पिता और चाचा आ गये। उन्होंने भाईसाहब के चरण-कमल छुए और भाभीजी की ओर भी झुके। मैंने भी उन दोनों के चरण-कमलों का सादर स्पर्श किया। फिर कुली के साथ हम लोग स्टेशन से बाहर आ गये।

हमें एक खूबसूरत गेस्ट हाउस में ठहराने की व्यवस्था थी। दो कमरे उन्होंने बुक कराये थे दो दिनों के लिए। अपनी आदत से मजबूर भाभी ने पहले उनका किराया पता किया। उसका किराया चार हजार रुपये रोज़ाना था। वे वातानुकूलित कमरे थे। कमरे की सफाई और व्यवस्था देखकर हमारी आधी थकान मिट गयी। लेकिन हमने गौर किया, भाभीजी की त्यौरियाँ कुछ तन गयी थीं। वह कुछ भुनभुनाये जा रही थीं। दरअसल उन्हें कुछ भी अच्छा नहीं लग रहा था। कन्यापक्ष वालों ने हमें तैयार होने के लिए छोड़ दिया। उनके विदा होते ही भाभीजी शुरू हुईं, ''घर का संस्कार देख लिया। चार अतिथियों के लिए घर में जगह नहीं

है।'' हमने विरोध किया, ''यह सब तो हमारी सुविधा के लिए किया गया है।'' मगर भाभीजी तनी थीं, ''सब समझती हूँ। यह सब हम पर रोब जमाने का तरीका है। मैं झाँसे में नहीं आने वाली।'' तब तक हमारे लिए चाय आ चुकी थी। चाय लेकर हम सबने थोड़ी राहत की साँस ली फिर एक-एक कर तैयार हुए। स्वादिष्ट नाश्ता हमारा इन्तज़ार कर रहा था। हम नाश्ता कर पूरी तरह तैयार हो गये।

उसके बाद हम लोग कन्या के घर गये। वहाँ फिर सूखे मेवे, ताजे फल, जूस, मिठाई आदि से हमारा स्वागत किया गया। मैंने मन ही मन ईश्वर को धन्यवाद दिया। उनकी आवभगत से हम भावविभोर थे। जब वहाँ कन्यापक्ष का कोई नहीं था, तो भाभी जी फिर फुसफुसाईं, ''इतना खर्च कर रहे हैं, दाल में काला तो नहीं है?''

''हम उनके नये मेहमान हैं, स्वागत तो होता ही है।'' भाईसाहब ने कहा।

''आप तो मोटे शरीर और मोटी बुद्धि के आदमी हैं। आप बारीक चीज़ें नहीं समझ सकते, क्यों देवर जी?

''जब कन्या हमारे सामने आ जाएगी, तो सारी बातें साफ हो जाएँगी।'' मैंने कहा। तभी कन्या को उसकी माँ और चाची लेकर आयीं। मैं एक बार फिर चौंक गया। महाकाव्यों में वर्णित विवाह की झाँकियाँ मेरी स्मृति में कौंध गयीं। शकुन्तला को लेकर उसकी सखियाँ दुष्यन्त के पास आयी हैं या फिर सहेलियों के साथ सीता अपने बाग में विचरण कर रही है। सहेलियों की जगह माँ और चाची मुझे जँच नहीं रही थीं। लेकिन कन्या की छवि ऐसी थी कि हमारी बोलती बन्द हो गयी। वह लड़की, एक मूरत-सी थी। हम बस देखते जा रहे थे।

'क्या रूप निखर आया है', 'रेशम-सी त्वचा', 'चौदहवीं का चाँद', वगैरह-वगैरह, मैं अपने मन में सोचने लगा। उसे व्यक्त करना इसलिए ठीक नहीं था कि वह मेरी नहीं, मेरे भतीजे की होने वाली बहू थी। लेकिन भाभीजी यहाँ भी सन्तुष्ट नहीं थीं। ऐसी मूरतों में ही फरेब होता है। उन्होंने उसे कुछ दूर तक चलवाया—लंगड़ी नहीं थी। वह उसके कदम, गति, लय, अन्दाज़ आदि परख रही थीं। लेकिन कोई दिखाई देने वाली कमी नहीं मिली। फिर वे उसे लेकर बाथरूम में चली गईं। वहाँ उन्होंने उसके बाह्य वस्त्रों को उतरवा कर देखा। उसके गले पर एक निशान था। वह उसके बारे में तहकीकात करने लगीं। वह चोट का निशान था, मगर उन्हें शंका हुई कि कन्या की बाईं आँख दाईं आँख से कुछ छोटी है और बोलते समय वह अपना मुँह तिरछा कर लेती है। मगर आईना के सामने ले जाकर तीन-चार बार देखने-सुनने के बाद उनका भ्रम टूटा।

खैर...उसके बाद मेरे भतीजे का नम्बर आया। भतीजा पढ़ा-लिखा था मगर गाँववालों की समझ में कलियुगी था। उसने लड़की से उसकी पढ़ाई के सारे

कागज़ात माँगे। हाई स्कूल से लेकर एम.ए. तक की मार्कशीट उसने देखी। सबमें वह प्रथम श्रेणी में उत्तीर्ण हुई थी किन्तु मेरा भतीजा इस पर भी मुँह बिचका रहा था। उसके कहने का आशय यह था कि जब तक कोई रोजगार न मिले, ये कागज़ की डिग्रियाँ बिल्कुल बेकार हैं। फिर आजकल तो नकल करके भी कई लोग अच्छे अंक ले लेते हैं। यह सुनकर वह लड़की अन्दर ही अन्दर बहुत दुखी और नाराज हुई। भतीजा सरकारी महकमे में लिपिक था लेकिन वह उस कन्या के सामने ऐसी शेखी बघार रहा था मानो वह प्रथम श्रेणी का अधिकारी हो। फिर वह उससे कुछ निजी बातें करने लगा। लड़की को वे बातें पसन्द नहीं थीं, वह भी बड़ों के सामने तो बिल्कुल नहीं। फिर भी उसने बर्दाश्त किया। फिर मेरी बारी आयी। मैंने तपाक से कहा, ''बहुत अच्छी लड़की है। ब्यूटी विथ ब्रेन।'' लेकिन भाभीजी सीधा जवाब नहीं देना चाहती थीं। सो उन्होंने कह दिया, 'ऐसा है कि मेरे देवर जी बहुत सीधे-सादे हैं। पेशे से अध्यापक हैं, इसलिए उन्हें सारे बच्चे-बच्चियाँ प्रिय लगते हैं। अपने भाई की तरह ये भी दूर तक नहीं सोचते।''

मेरी बात खत्म होने के बाद भाई साहब का नम्बर आया। उन्होंने सबसे अच्छा काम यह किया कि लड़की को वहाँ से जाने को कहा। फिर वे लड़की के बाप से बतियाने लगे, ''पंडितजी! यूँ तो निजी तौर पर दहेज़ के सख़्त ख़िलाफ़ रहा हूँ। मगर आज का ज़माना ऐसा है कि यदि आप अपने बेटे की शादी में कुछ नहीं लें, तो बेटी की शादी में कैसे देंगे? बेटी का हिस्सा उसके माँ-बाप की ज़ायदाद में होता है, जो मायके वाले बेटी को दान देते हैं। सो आपने भी तो कुछ संकल्प लिया होगा दान करने के लिए?''

कन्या का बाप चुप था। थोड़ी देर में उसने मेरे भाई साहब से ही खुलासा करने को कहा। भाई साहब अब पूरी तरह खुले, ''मैं अपनी बेटी की शादी करने के लिए वर ढूँढ़ने गया था। नेवी में एक नया रंगरूट पाँच लाख रुपये तिलक माँग रहा है। नेवी की नौकरी पन्द्रह साल की और मेरे बेटे की नौकरी तीस साल की। आप तो हिसाब-किताब जानते ही हैं...''

''लेकिन मेरी औकात इतनी नहीं है, भाईसाहब!'' कन्या के बाप ने विनम्रता-पूर्वक कहा, ''हमने अपनी बेटी को बेटे की तरह ही पढ़ाया है। हमने पढ़ाई पर खर्च करने में कभी कोताही नहीं की। दहेज़ के लिए बहुत ग़ौर नहीं किया। आप जो कह रहे हैं वह हमारी सीमा में कही नहीं है।''

''तो ये सेकेंड ए.सी. की टिकटें और ए.सी. कमरे आपकी सीमा में था जो आपने खर्च किया।''—हमारी भाभीजी ने तुरन्त जवाब जड़ दिया। उनके जवाब ने सबको एकबारगी चुप रहने के लिए मजबूर कर दिया। वातावरण में एक सन्नाटा

पसर कर बैठ गया।

कन्या की माँ मायूस थी। उसके अन्दर बहुत कुछ उमड़-घुमड़ रहा था। उसकी आँखों से आँसू गिरने लगे। फिर वह अपने को नहीं रोक सकी और भाभी के पैरों पर गिर कर सुबकने लगी। इस सुबकने ने वातावरण को कुछ भावुक बना दिया। भावना की झुरझुरी भाभीजी के दिल में भी उठी। कहा, ''क्या करूँ कुछ समझ में नहीं आता। इसमें कोई दो राय नहीं कि आपकी बेटी सुन्दर है। लेकिन खाली सुन्दर होना तो काफी नहीं होता न। यदि कन्या ने बी.एड. किया होता तो उसे किसी विद्यालय में अध्यापिका की नौकरी मिलने की उम्मीद बनती। यदि आप उसे बी.एड.करा दें, तो बात बन सकती है।''

यह सुनते ही कन्या की माँ की बची-खुची आशा निराशा में बदल गयी। वह सोचने लगी कि शादी के लिए लड़का खोजते-खोजते उसकी बेटी की उम्र यूँ ही सत्ताइस वर्ष हो चुकी है। फिर एक साल की बी.एड. की ट्रेनिंग कैसे हो सकेगी? ट्रेनिंग कॉलेज में पहले से ही कम सीटें हैं। फिर उसमें आरक्षण-फारक्षण और पैरवी भी। फिर सीटें बचती ही कहाँ हैं? प्राइवेट ट्रेनिंग कॉलेज में लाखों में डोनेशन लगता है। फिर बी.एड. करने के बाद नौकरी मिलने की गारंटी भी नहीं है। मगर पूर्व में भी कई शादियों के प्रयास विफल हो चुके थे। सो कन्या के बाप ने मजबूर होकर उसे बी.एड. कराने का वादा किया। यद्यपि मुझे यह सब अच्छा नहीं लग रहा था मगर भाभी जी के रुख़ को जानते हुए मैं कायर बन चुपचाप बैठा रहा।

अगले दिन हम लोगों को वापस लौटना था। वे लोग एक मोटरकार में हमें स्टेशन तक छोड़ने आये। गाड़ी से सामान निकालकर वे अन्दर प्लेटफार्म पर जाने को अग्रसर हुए किन्तु भाभीजी ने टोका, ''भाईसाहब! आप जाइए। हम लोग आराम से चले जाएँगे। गाड़ी आने में अभी कुछ विलम्ब है। हम लोग घर-परिवार की कुछ निजी बातें करेंगे।'' वे मेहमान वापस चले गये। फिर क्या था? भाभी जी ने मुझे और भाईसाहब को कोसना शुरू किया, ''दोनों भाई कितने बुद्धू हो। कुछ ख्याल नहीं है कि हम लोगों की टिकटें सेकेंड क्लास स्लीपर की हैं।'' जबकि मेहमान ने दोनों तरफ के लिए वातानुकूलित दर्ज़े का किराया अग्रिम के रूप में दे दिया था। आते समय तो किसी प्रकार से हमारी इज़्ज़त बच गयी थी मगर इस बार आप लोगों की बेवकूफी से सारी पोल खुल जाती और हमारी इतनी भद्द होती कि समाज में मुँह दिखाने लायक हम नहीं रह पाते।''

हम लोग भाभी जी का आक्रोशपूर्ण प्रवचन सुनते रहे। उनकी कुशाग्र बुद्धि के सामने हम सब नतमस्तक थे। हमें उनका लोहा मानना पड़ा। मुझे याद आया

कि यह भाभी जी की ही बुद्धि का कमाल है कि वे अब तक एक दर्जन से अधिक लड़कियाँ ख़ारिज़ कर चुकी हैं और उन सभी में उनका एक पैसा भी खर्च नहीं हुआ, बल्कि उनका निजी बैंक बैलेंस बढ़ा ही है। कम से कम विदाई में मिली साड़ियों से उनका ट्रंक तो भर ही गया।

थोड़ी देर में गाड़ी आ गयी और हम लोग अपने डिब्बे में ठुंस कर बैठे गये। गाड़ी के गार्ड ने सीटी दी और फिर गाड़ी धीरे-धीरे सरकने लगी। तभी हाँफता हुआ एक अधेड़ आदमी आया और बोला, ''पंडित जी! आप लोग तो यह उपहार भूल ही गये थे। मैं ए.सी. कोच में आप लोगों को काफी देर तक ढूँढ़ता रहा...'' यह कहते-कहते वह ओझल हो गया गाड़ी की गति तेज हो गयी थी और हम लोगों के दिलों की धड़कन भी। भाभी जी को काटो तो खून नहीं। मानो वे पोल खुल जाने के कारण बेहोश हो जाएँगी। उन्हें पहली बार धक्का लगा था। रास्ते भर हम बेचैन रहे। मगर इसके लिए भी भाभी जी ने मुझे और भाईसाहब को दोषी ठहराया।

हम लोग घर पहुँच गये थे लेकिन हमारे मध्यवर्गीय मन को जो शर्मिन्दगी उठानी पड़ी थी उससे उबरे नहीं थे। हज़ार लोगों से हज़ार बार हम यही विचार करते रहे कि कन्यापक्ष वालों को किस तरह सफाई दी जाये। ऐसा लगता था मानो पूरा कन्यापक्ष हमारे सामने खड़ा होकर स्पष्टीकरण माँग रहा हो—तो ये है आपका बड़प्पन? किराया चाहिए सेकेंड ए.सी. का और यात्रा साधारण दर्जे में। थू-थू। बेईमानी की भी हद होती है। अन्ततः एक युक्ति तय की गयी। एक पत्र लिखा जाए जिसमें यह स्पष्ट किया जाए कि ए.सी. श्रेणी का टिकट नहीं मिलने के कारण हम लोग द्वितीय श्रेणी के स्लीपर में यात्रा किये थे। फिर उसमें शादी का एक आश्वासन भी हो, मगर बी.एड. करने के बाद। इस पक्ष के लिए मेरे भतीजे का ही जोर ज़्यादा था। मुझे समझते देर नहीं लगी कि लड़की उसे भा गयी है। भाईसाहब उसकी बात नहीं टाल सके और अगले दिन एक चिट्ठी भेज दी गयी। पन्द्रह दिन बीत गये, परन्तु कोई जवाब नहीं आया। फिर भाभी जी ने दूसरी चिट्ठी स्मार-पत्र के रूप में भेजने को कहा। उसमें इतना संशोधन था बी.एड. शादी के बाद भी हो सकता है। हाँ, उसका खर्च कन्यापक्ष वाले वहन करें। हमने इसे अपनी उदारता की हद समझी। खुद को गौरवान्वित भी किया कि हम कन्यापक्ष को सताने वाले लोग नहीं हैं। वे और लोग होते होंगे। हम तो मानवीय लोग हैं। पीर पराई जानने वाले वैष्णव जन।

करीब डेढ़ माह बाद जवाब आया। साफ-साफ लिखा था, ''पंडित जी! प्रणाम। आप लोगों की हमने यथोचित सेवा की। भूल-चूक के लिए माफ़ करेंगे।

जहाँ तक शादी का सवाल है, मैं आपके कहे हुए 'संकल्प' की याद आपको दिलाना चाहता हूँ..." तभी भाभी जी ने बीच में ही टोका, "कितने लाख रुपये का संकल्प किया है?

मैंने फिर पत्र पढ़ना शुरू किया, "मेरी बेटी ने संकल्प किया है कि वह ऐसे लालची परिवार में शादी नहीं करेगी। उसने दूसरा संकल्प यह किया है कि वह अपने बूते पर कम्पीटीशन के आधार पर रोज़गार प्राप्त करेगी और तभी अपने लायक वर से ही शादी करेगी।"

यह दूसरा धक्का था। पहला धक्का तो भीतर से लगा था। हम खुद शर्मिन्दा थे कि ए.सी. का किराया लेकर साधारण स्लीपर में सवार थे। लेकिन यह तो बाहर का धक्का था। पूरे वजूद पर धक्का। हम ख़ारिज़ कर दिये गये थे। हमें लालची बतलाया गया था। पता नहीं लालची लिखने के पहले हमारे बारे में कितनी टिप्पणियाँ हुई होंगी। हमें किन-किन संज्ञाओं से नवाजा गया होगा। इस लगातार सोचे जा रहे थे और उसी अनुपात में छोटे भी होते जा रहे थे।

हमारी भाभी जी ने ही इस बार भी सन्नाटा तोड़ा। वे बुदबुदाईं, "निगोड़ी लड़की, सूरत इतनी सलोनी और भीतर इतना ज़हर। संकल्प करती है कलमुँही। ज़िन्दगी भर बाप के खूँटे पर न रही तो मेरा नाम कलावती नहीं। पढ़ाई का घमंड हो गया है। गुमान बोल रहा है उसका। रावण का गुमान तो रहा नहीं, वह है किस गली की ठिकरी।"

मेरा भतीजा मुँह लटकाए बैठा था। वह कुछ बोल तो नहीं रहा था, किन्तु उसके चेहरे पर उदासी की जो परतें बिछी थीं, उन्हें कोई भी पढ़ सकता था। ऐसा लगता था मानो उसके मन में भावनाओं की आँधी उठी हो। मेरे मन में भी भावनाएँ उमड़ रही थीं। कितनी खूबसूरत लड़की थी भला। दोष तो हमारा ही था। उसके संकल्प और साहस ने मेरी दृष्टि में उसके व्यक्तित्व को और सुन्दर बना दिया था। मैं सोच रहा था—काश! मेरा भतीजा भी कोई संकल्प लेता तो कितना अच्छा होता। फिलवक्त मुझे लड़की देखने जाने का अफ़सोस था। काश मैं न गया होता, तो मुझे शर्मिन्दगी नहीं झेलनी पड़ती।

अन्तराल

पन्द्रह अगस्त का दिन था। राजधानी दुल्हन की तरह सजी थी। जिला प्रशासन ने मुख्यमन्त्री के झण्डा मैदान में आने के पहले पूरी व्यवस्था कर ली थी। भारत के कोने-कोने में तिरंगा झंडा फहरा रहा था। कोई खादी कपड़े का, तो कोई काग़ज़ का झंडा बनाकर अपने घर, साइकिल, मोटरसाइकिल, मोटरगाड़ी, रिक्शा, ठेला, कार्यालय भवन, दुकान आदि जगहों पर फहरा रहा था। मुख्यमन्त्री ने जोशीला भाषण दिया, ''भाइयो और बहनो, हम देश की आजादी में कुर्बानी देने वाले शहीदों की स्मृति को प्रणाम करते हैं। हमारी सरकार ग़रीब-गुरबों के चौमुखी विकास के लिए कृतसंकल्प है। सभी दलितों को इन्दिरा आवास दिया जाएगा। महात्मा गाँधी और बाबा साहब अम्बेडकर के सपनों को पूरा किया जाएगा। ग़रीब बच्चों की शिक्षा का बन्दोबस्त किया जाएगा। हमारी सरकार ने अपराध पर काबू पा लिया है। हमने सामन्ती अपराधियों की नाक में नकेल कस दी है। बच्चे हमारे देश के भावी कर्णधार हैं। जयहिन्द!''

उस दिन सारे सरकारी गैर सरकारी विद्यालय खुले थे। शिक्षकों ने हिदायत दे दी थी कि जो विद्यार्थी विद्यालय नहीं आयेगा, उसके अंक काट लिये जाएँगे। सो किशन भी पूरी तैयारी के साथ स्वतन्त्रता दिवस समारोह में भाग लेने के लिए स्कूल गया। अन्त्याक्षरी में उसकी टीम अव्वल आयी और टीम की ओर से प्रधानाध्यापिका के हाथों से उसे ही पुरस्कार मिला। वह खुशी से झूम रहा था। फिर खेल-कूद का कार्यक्रम शुरू हुआ। उसमें उसकी क्रिकेट टीम दूसरे नम्बर पर रही। फिर उसमें भी पुरस्कार मिला। उसके बाद सांस्कृतिक कार्यक्रम सम्पन्न हुआ। 'सैंया भये कोतवाल' नामक नाटक में वह नेता की भूमिका निभा रहा था। उसने पहले काफी अभ्यास किया था। नेताओं की नकल उसने काफी बारीकी से उतारी थी जिसके कारण उसे सर्वोत्तम अभिनेता का पुरस्कार मिला। सभी बच्चों के साथ उसे भी मिठाई मिली। उत्सव का माहौल था। मगर उस दिन उसके माँ-बाप स्कूल नहीं गये थे, सो उनकी कमी उसे खल रही थी। स्कूल में सारा

कार्यक्रम ख़त्म हो जाने के बाद स्कूल की बस के आने का वह परिसर के बाहर इन्तज़ार कर रहा था। तभी एक नौजवान उसके पास गया और बोला, ''तुम्हारे पापा के दोस्त तुम्हें बुला रहे हैं। वे तुम्हारे काम से खुश होकर तुम्हें इनाम देना चाहते हैं।''

किशन जलसे के मूड में अभी भी था। उसने आगंतुक से कुछ पूछताछ नहीं की। शायद उसके पापा ने उन्हें भेजा हो लाने के लिए। सो वह उसके साथ हो लिया। थोड़ी दूर पर एक कार में एक आदमी सूट-बूट पहने बैठा था। किशन को देखते ही उसने कहा, ''आओ बेटे! मैं तुम्हारे पापा का दोस्त हूँ। मैं तुम्हें आज के शुभ अवसर पर यह अनमोल जैकेट उपहार में देता हूँ। मैं तुम्हें तुम्हारे घर पर छोड़ दूँगा।'' और किशन उसकी कार में खुशी-खुशी आराम से बैठ गया। उसके घर रोज़ाना पापा से मिलने तरह-तरह के लोग आते थे। थोड़ी दूर जाने के बाद कार में पीछे बैठे आदमी ने उसकी आँख और मुँह पर कपड़ा बाँध दिया। गाड़ी में पहले ही शीशे पर काली फिल्म चढ़ी थी। उसमें ए.सी. भी लगा हुआ था। किशन की समझ में नहीं आ रहा था कि उसके अच्छे काम के लिए यह दंड क्यों? उसके पापा के दोस्त ऐसा क्यों करते हैं? उसका माथा चकराने लगा। उसे लग रहा था कि सारी धरती उसके चारों ओर घूम रही है।

करीब एक घंटे के बाद वह कार एक सुनसान जंगल में रुकी। किशन के मुँह-आँख पर से पट्टी खोल दी गयी। उसने अपने को बन्दूक, पिस्तौल से लैस आधा दर्जन नौजवानों से घिरा पाया। उसकी सिट्टी-पिट्टी गुम हो गयी थी। अब न तो उसकी अन्त्याक्षरी काम आ रही थी, न क्रिकेट और न नाटक। सो वह फूट-फूटकर रोने लगा। तभी उसमें से एक मुछमुंडे नौजवान ने उसे घुड़कते हुए कहा, ''चुप हो जा, वरना मुझे अपनी एक गोली खराब करनी पड़ेगी।''

यह सुनते ही किशन के आधे आँसू आँखों में और आधे गालों पर ठहर से गये। शायद गोली के डर ने आँसुओं पर ज़ोर से विराम लगा दिया। फिर दूसरे ने पूछा, ''तेरा बाप क्या काम करता है, किशन?''

''मेरे पापा सेल्स टैक्स डिपार्टमेंट में डिप्टी कमिश्नर हैं।'' किशन ने बताया मगर उसे ताज्जुब हुआ कि मेरा नाम इन लोगों को कैसे मालूम है?

''तुम कहाँ रहते हो?''

''राम बाग की अपनी कोठी में।''

''अच्छा, अच्छा! उस तीन मंजिला कोठी में? उसका नाम शायद किशन कुंज है मगर लोग उसे 'किशन पैलेस' कहते हैं।'' एक ने कहा।

''जी!'' किशन को फिर आश्चर्य हुआ।

''और तुम पढ़ते कहाँ हो?''

"एम. पी. एस. कान्वेंट में—मेट्रो पब्लिक स्कूल में।"

"कितनी फीस लगती है उसमें?"

"दो हज़ार रुपये महीना।"

"और भर्ती के समय डोनेशन कितना दिए?"

"मुझे नहीं मालूम, सर!" किशन ने विनम्रता से कहा।

"तो मैं बताता हूँ। दो लाख रुपये डोनेशन।" एक ने ऐसे कहा मानो वह उस समय मौजूद था।

"और तुम्हारी बहन कहाँ पढ़ती है?"

"वह बंगलौर में एम.बी.बी.एस. कर रही है।"

"उसकी कितनी फीस लगती है?"

"दस हज़ार रुपये महीना।"

"और उसकी भर्ती के समय कितना डोनेशन दिया था तुम्हारे बाप ने?" एक ने अकड़ते हुए पूछा।

"मुझे कुछ नहीं मालूम, सर।"

"तो मैं बताता हूँ। बीस लाख रुपये डोनेशन दिया था तुम्हारे बाप ने। क्या तुमने कभी उनसे पूछा कि ये सारे रुपये कहाँ से आते हैं?"

"उनको सरकार से वेतन मिलता है, सर!" किशन ने कहा।

"नहीं, नहीं, नहीं। वेतन से वह तुम्हारे परिवार को दाल-भात तरकारी खिला सकता है, कुछ कपड़े पहना सकता है, सरकारी स्कूल में तुम लोगों को पढ़ा सकता है। मगर मुल्क के नम्बर वन स्कूल में नहीं पढ़ा सकता और न प्राइवेट मेडिकल कॉलेज में पढ़ा सकता है।" मुछमुंडे ने कहा।

"और फिर तेरी माँ, आयी मीन योर मम्मी, जो अशोक होटल में स्विमिंग पूल में नहाने जाती है, पार्टियाँ देती है, क्लब एटेंड करती है, बार-बार मार्केटिंग करती है, सो सारे खर्च कहाँ से आते हैं?" दूसरे ने इन शब्दों की गालियों से किशन को दागा। किशन बिल्कुल चुप था मगर उसके दिल के एक किनारे में बातें सवाल की शक्ल ले रही थीं। उसे याद आया कि रमेन्द्र अंकल भी सेल्स टैक्स डिपार्टमेंट में हैं, मगर उनका रहन-सहन, खाना-पीना, बिल्कुल सादा रहता है। बीस वर्षों की सरकारी नौकरी के बाद भी वह सरकारी क्वार्टर में रहते हैं। उन्होंने अपना कोई मकान नहीं बनाया है। उनके बच्चे आस-पास के सरकारी स्कूल में पढ़ते हैं। और आंटीजी तो पहनावे से एकदम देहाती भुच्चड़-सी लगती हैं। उनका घर अफसरों का-सा लगता ही नहीं। अभी भी उनके यहाँ चौदह इंच का ही एक रंगीन टी.वी. है जबकि मेरे यहाँ हर कमरे में चालीस इंच का फ्लैट रंगीन टी.वी. प्लाज्मा है। उनके किचेन और टॉयलेट में मार्बल की फर्श तक नहीं

है। उनके यहाँ कूलर तक नहीं है, जबकि मेरे यहाँ हर कमरे में ए.सी. है। उनके बच्चे साइकिल से स्कूल जाते हैं जबकि मैं बस या अपनी कार में आराम से जाता हूँ। वे टेम्पो पर सवार होकर रोज़ाना ऑफिस जाते हैं, जबकि मेरे पापा अपनी सेन्तरो कार में ऑफिस जाते हैं। मेरे घर के सारे कपड़े बैंड बॉक्स में धुलकर आते हैं और उनके यहाँ वाशिंग मशीन तक नहीं है। मेरे यहाँ बड़ा-सा नया फ्रिज है, मगर उनके यहाँ बाबा आदम के जमाने का छोटा-सा फ्रिज है।

तभी उस गिरोह के एक नौजवान ने दहाड़ा, ''क्यों चुप हो? क्या तुम्हें नहीं मालूम कि सरकारी नौकरों को महीने में एक ही दिन वेतन मिलता है मगर तुम्हारे पापा तीसों दिन वेतन वसूलते हैं व्यापारियों से। बिना जेब गरम किये वह फाइल को नहीं छूते। क्या तुम्हें मालूम है कि तुम्हारे पापा ने ईमानदार व्यापारियों को भी नहीं बख़्शा और बेईमान व्यापारियों के साथ मिलकर इस मुल्क के करोड़ों रुपयों का बन्दर-बाँट किया है? फिर ऐसे करोड़पति से कोई दस-बीस लाख वसूल ही ले तो क्या फ़र्क़ पड़ता है।'' किशन बिल्कुल बुत-सा बना रहा। उसके भीतर एक द्वन्द्व चल रहा था। उसे अपने भ्रष्ट बाप पर शर्म आ रही थी। उसे लगा कि काश उसे किसी ने पहले ही उनकी करतूतें बतायी होतीं। उसकी आँखों से आँसू टपकने लगे। मगर अपहरणकर्त्ताओं ने समझा कि शायद किशन अपहरण के कारण भयभीत है, कि वह माँ-बाप की याद कर रहा है कि उसे अपने जीवन पर ख़तरा नज़र आ रहा है, कि वह भागने की कोई तरकीब सोच रहा है, कि वह अपने स्कूल के सहपाठियों को याद कर रहा है, वगैरह-वगैरह। सो एक अपहरणकर्त्ता ने कहा, ''किशन! हम लोगों से कोई चालाकी नहीं करना। अगर ऐसी कोशिश की तो मारे जाओगे। अब मैं तुम्हारे पापा से फोन पर बात करा देता हूँ। तुम अपनी बात उनसे कह सकते हो मगर यह कतई नहीं बताना कि तुम अभी कहाँ हो।'' और नम्बर लगाकर मोबाइल उसने किशन को पकड़ा दिया,

''हेलो, पापा!''

''हाँ बेटे, तुम कैसे हो?'' उसके पापा की आवाज़ पूस की रात की तरह ठिठुर रही थी।

''अभी तक मैं ठीक हूँ। अंकल ने मुझे एक चॉकलेट अभी खाने को दी है।''

''मगर किशन! तुम उनकी दी हुई कोई चीज़ मत खाना। उसमें ज़हर हो सकता है।'' उसके पापा ने बुदबुदाते हुए कहा।

''ज़हर! मगर चॉकलेट तो पैक में है।'' किशन ने तर्क दिया।

''बेटे वे लोग बहुत बड़े अपराधी हैं। उनके झाँसे में मत आना।''

''लेकिन पापा अंकल कह रहे थे कि आप लोगों का पैसा...''

तभी एक नौजवान ने मोबाइल किशन से छीन लिया और कहने लगा, ''रस्तोगी साहब! अब मेरी बात ध्यान से सुनिए। आपको जवाब देने के लिए मिलेंगे सिर्फ़ दो मिनट। इसमें सोच-समझकर उत्तर देना, वरना बाद में पछताओगे। और हाँ पुलिस को ख़बर मत करना। वैसे ख़बर भी कर दोगे तो कोई ख़ास फ़र्क़ नहीं पड़ेगा। वो खचड़ा एस.पी. तो बदल गया न। अब तो अपना आदमी है ऊपर से नीचे तक। हाँ, सुनो कान खोलकर। पचास लाख रुपये लेकर कल शाम पाँच बजे तक मंगल तालाब के पूर्वी किनारे पर स्थित मन्दिर पर तुम अकेले पहुँच जाना। वैसे भी आपके लिए यह तो छोटी-मोटी रकम है। आप तो करोड़पति अधिकारी हैं। थोड़ा दान दे देंगे तो बहुत पुण्य मिलेगा और पाप धुल जाएगा। वरना क्या जवाब देंगे ऊपर जाने पर? मुझे पता है कि आप सेल्स टैक्स विभाग के मन्त्री के बहुत करीबी रहे हैं। उसका भी आपने कल्याण किया और अपना भी। अब यदि बच्चा चाहिए तो इस पर अमल करिए बच्चे को दफनाने के लिए कब्र खोदकर तैयार रखिए।''

''देखिए, मेरी आपसे हाथ जोड़कर विनती है कि मेरे बेटे को छोड़ दीजिए। वह मेरा इकलौता बेटा है। उसकी मम्मी उसके बिना पागल हो जाएगी...मेरा यकीन कीजिए...'' कहते हुए रस्तोगी रो पड़े और आगे नहीं बोल पाये।

''जब आपने इकलौते बेटे के भविष्य के लिए करोड़ों रुपये का जुगाड़ कर लिया है, तो दूसरों के बेटों-बेटियों के बारे में भी थोड़ा सोचिए। हम सब बेरोज़गार नौजवान हैं। आपकी मेम साहब तो बड़े-बड़े होटलों और दुकानों पर खाने-पीने और ख़रीदने के लिए बराबर जाती हैं, मगर हम लोगों के घर पर अक्सर फाँका होता है और हमारे परिवार वालों पर बाज़ार मुँह चिढ़ाता रहता है दाँत चियारकर।''

''लेकिन भगवान की कसम, मैंने किसी का दिल नहीं दुखाया।''—रस्तोगी साहब गिड़गिड़ाए।

''बिल्कुल सही। आपने सिर्फ़ लोगों की जेबें हल्की कीं, दिल कहाँ दुखाया?''

''मैंने कभी किसी से जोर-जबर्दस्ती नहीं की। यदि किसी ने थोड़ा बहुत उपहार दिया तो मैंने उसे स्वीकार कर लिया, बस।'' रस्तोगी घिघियाने लगे।

''उपहार ही सही। उसी में से दान-पुण्य कर दीजिए। एक करोड़ का आधा पचास लाख।''

''नहीं, भाई जी! मेरे पास उतने पैसे नहीं हैं। बच्चों की पढ़ाई में लाखों रुपये चले गये। सालों ने बहुत सारा डोनेशन ले लिया। अभी दस-बीस हजार दे सकता हूँ।'' किशन के पापा ने कहा। उधर से आवाज आयी, ''हम कोई भीख

माँग रहे हैं क्या? आपका समय खत्म हो गया।''

सारा वाकया सुन-समझकर मिसेज रस्तोगी रोने लगीं, ''मुझे मेरा बेटा चाहिए। मुझे अब दौलत नहीं चाहिए। मैं दौलत के पीछे अन्धी हो गयी थी। मगर बेटे के बिना यह सारी दौलत दो कौड़ी की है। मुझे मेरा बेटा दिलवा दो, भले यह महल बिक जाये, भले मेरे सारे गहने बेच दो, भले घर को, ज़मीन को बेच दो...''

''तुमने मुझे नाजायज़ कमाने के लिए बार-बार उकसाया। मेरी सिधाई का तुमने नाज़ायज फ़ायदा उठाया। इसीलिए तुमने मेरी माँ से झगड़ा किया। उनकी बातों को न तुमने माना और न मुझे मानने दिया।'' रस्तोगी साहब गुस्से से काँप रहे थे।

''मुझे मारो-पीटो, गालियाँ दो और सराप दो। मगर किसी भी कीमत पर किशन को घर वापस ले आओ।'' मिसेज रस्तोगी रस्तोगी साहब के पैर पकड़कर गिड़गिड़ा रही थीं।

''यह घर बेच दीजिए। सूर्या ज्वेलर्स, सोना ज्वेलर्स, रत्ना ज्वेलर्स, मैक्सा फार्मा, डिलक्स फार्मा वगैरह से उधार ले लीजिए। उनके हाथ घर के गहने बेच दीजिए। उन्हें मकान गिरवी रख दीजिए या बेच दीजिए। जाइये, फौरन कुछ कीजिए। हमारे पास समय बिल्कुल नहीं है।'' मिसेज रस्तोगी रोते-रोते बदहवास हो गयी थीं। रस्तोगी साहब कोई चारा न देख दौड़-दौड़ उन व्यापारियों के यहाँ गये। रत्ना ज्वेलर्स के मालिक भगवान जौहरी ने कहा, ''सर, अब आया ऊँट पहाड़ के नीचे। मेरा बिजनेस घाटे में चल रहा है। जैसा कर्म करोगे वैसा फल देगा भगवान। आपने मेरे साथ कभी कोई रियायत नहीं की। हर महीने आप मेरी खाल उधेड़ते रहे। सो मैं आपकी बुरी नज़र के कारण तबाह हूँ। मेरी देह के अलावा मेरे पास देने को कुछ नहीं है।''

रस्तोगी साहब सोचने लगे—भगवान जौहरी ने मेरे घाव पर नमक छिड़क दिया है। मुझे नंगा कर दिया उसने। मगर इस मौके पर मैं उसका कुछ भी नहीं बिगाड़ सकता, उसका एक रोआँ भी टेढ़ा नहीं कर सकता। फिर वह सूर्या ज्वेलर्स के मालिक रतन दास के वहाँ पहुँचे और रोते-रोते अपना दुखड़ा सुनाया। रतन दास ने पूरी सहानुभूति जतायी मगर जब पैसे देने की बात आयी, तो बोले, ''साहब जी! मेरी तो सारी पूँजी फँस गयी है। मुख्यमन्त्री की बेटी की शादी थी न, उसमें मेरी दुकान से बीस लाख के जेवर उनके परिवार वाले ले गये थे। पहले उनके अपनों ने शादी के बाद पैसा देने को कहा था। फिर छः महीने बाद पैसा देने को कहा था। फिर एक साल बाद देने को कहा मगर कुछ नहीं दिया। सो एक दिन मैं मुख्यमन्त्री जी से खुद मिला और अपनी दुर्दशा उन्हें सुनाई। वे बोले,

"बीस लाख जेवर का सेल्स टैक्स आपने दिया? उसका सेल्स टैक्स और इन्कम टैक्स कितना होगा। आपने कितना चुकाया है टैक्स? दोनों महकमों की एक-एक टीम बुलाकर जाँच करवा देते हैं, फिर आपका यह भुगतान हो जाएगा। यह सुनकर मुझे जूड़ी-बुखार आ गया और मैं दबे पाँव वहाँ से भाग आया। सो मेरी ज़िन्दगी भर की कमाई एकमुश्त चली गयी। अब किससे कहूँ? आज पहली बार आपसे कहने की हिम्मत की क्योंकि आपसे पुराना सम्बन्ध है। मगर भगवान के वास्ते आप इसे किसी से मत कहना वरना मेरा बचा-खुचा हाड़-मांस भी वे लोग नोच लेंगे।"

रस्तोगी साहब को लगा कि रतन दास सचमुच ज़माने की मार झेल चुका है। सो वह सोना ज्वेलर्स के मालिक मोहनका के पास गये। उनकी बात सुनकर मोहनका ने कहा, "पिछले महीने अतिक्रमण हटाने का अभियान चला था। नगर निगम की मकान तोड़ने वाली गाड़ी ने मेरी दुकान पर हाथ लगा दिया। मैंने बहुत विनती की मगर उसका इंचार्ज नहीं माना। मैंने पाँच लाख रुपये नगद और पाँच लाख के जेवर दिये, तब वह वहाँ से हटा और मुझे एक फर्जी स्टे आर्डर बनाकर दे गया। मेरी तो समूची लुटिया डूब गयी हुजूर! आप लोग कितने अच्छे हैं पच्चीस-पच्चीस हजार में ही हम लोगों का काम कर देते हैं।" रस्तोगी साहब को क्षण भर के लिए महसूस हुआ कि भले ही वह रिश्वत लेते हैं लेकिन एक हद में रहते हैं किसी को खंखोरते नहीं। मगर अभी किशन के अलावा और कुछ भी सोचने का मौका नहीं था। फिर जाएँ तो कहाँ जाएँ? उनका मन हुआ कि वह खूब रोएँ और अपने पापों का प्रायश्चित करें। यह सोचते-सोचते वे खाली हाथ घर लौटे। रात के बारह बज चुके थे। उन्हें निरुपाय पाकर उनकी पत्नी और हताश हो गयी। रसोई में बर्तनों ने मौन व्रत धारण कर लिया था और चूल्हा बिल्कुल उदास था।

रात भर वे तेजी से करवटें बदलते रहे। उन दोनों ने अपनी बेटी से किशन के खोने का जिक्र नहीं किया। रात अपने भयानक रूप में उन दोनों को निगलती जा रही थी। हाथ मलते और करवटें बदलते-बदलते रात भी थक कर चूर हो गयी। रात तो बीत गयी किसी तरह मगर असली सुबह नहीं हुई। रस्तोगी साहब मैक्स फार्मा के मालिक के पास गये। उसका मालिक सोहन शाह बोला, "सर! मैंने आपको बताया नहीं। मेरा मकान तो स्थानीय सांसद के आदमियों ने हथिया लिया है। जब मैंने सांसद महोदय से बातचीत की तो उन्होंने अपने कार्यकर्ताओं को फटकारा और पूरा मकान अपनी पत्नी के नाम मुझसे महज़ एक लाख रुपये में रजिस्ट्री लिखा लिया। वह मकान पच्चीस लाख से कम का नहीं था। उन्होंने एक लाख भी नहीं दिया और कह दिया— उसे पार्टी फंड में चन्दा समझ लीजिए।

सो इस राज में यहाँ से भागने के सिवाय कोई रास्ता नहीं है। अपराधी लोगों ने पहले राजनीति पर क़ब्ज़ा जमाया, अब व्यापार भी उनकी ज़द में है। पहले राजनीति का अपराधीकरण हुआ, अब व्यापार का भी। अभी हम लोग मजबूरन किराये के मकान में रह रहे हैं।''

रस्तोगी साहब का मुँह कद्दू जैसा लटक गया। वह वहाँ से निराशा की झोली कन्धे पर लटकाये वापस चले गये। उनके घर पर उनके साले दुख बँटाने आ गये थे। साले ने राय दी, ''मेरी मानिये, जीजाजी! एक बार आप मुख्यमन्त्री से खुद मिल लीजिए। आख़िर मुख्यमन्त्री तो सारे लोगों का रक्षक होता है। धरती पर वही ब्रह्मा, विष्णु और महेश होता है। यदि कहीं पत्ता भी खड़कता है तो उसे मालूम हो जाता है। वे एक बार जोर लगा देंगे, तो सारे अपराधी टेकुआ की तरह सीधे हो जाएँगे।''

''मगर समय बहुत कम है। फिरौती वाले गिरोह ने पुलिस को कुछ भी नहीं बताने को कहा है वरना किशन के साथ कुछ भी हो सकता है।'' रस्तोगी साहब अपनी बेबसी व्यक्त करने लगे।

''देखिए, जीजाजी, मुख्यमन्त्री को इतना कमअक्ल का मत समझिए। वे भले जोकर जैसा लगते हैं मगर वह मसखरापन एक रणनीति के रूप में इस्तेमाल करते हैं। वह ऐसी जुगत लगाएँगे कि साँप भी मर जाएगा और लाठी भी नहीं टूटेगी। चलिए, उनके पी.ए. से मेरी जान-पहचान है। हर महीने उन्हें रेड लेबल मैं देता हूँ।'' साले ने सहानुभूति दर्शाते हुए कहा।

रस्तोगी की समझ में बात आ गयी। वे अपने साले के साथ मुख्यमन्त्री के पी.ए. से बात करने के लिए उनके आवास पर दस बजते ही पहुँच गये। उन्हें करीब एक घंटे तक इन्तजार करना पड़ा। मुख्यमन्त्री जी को उनके पी.ए. ने किशन के अपहरण की घटना पहले ही ब्रीफ कर दी थी।

मुख्यमन्त्री बोले, ''सेल्स टैक्स के डिप्टी कमिश्नर होकर आप मुझसे कभी मिलने नहीं आये और मीडिया में आँय-बाँय-साँय बयान देने लगे। सरकारी सेवा शर्तों का आपको कुछ ख्याल है कि नहीं?''

''ग़लती हो गयी, सर! हम लोग अपहरण के बाद अपनी सुधि-बुधि भूल गये थे। मुझे माफ कर दीजिए।'' रस्तोगी साहब ने हाथ जोड़कर कहा। ''ठीक है, माफ कर दिया। डी.जी.पी. को फोन लगाओ, फोन ड्यूटी।'' जब फोन ड्यूटी ने चोंगा मुख्यमन्त्री जी को थमाया तो वह दहाड़ने लगे, ''आपको डी.जी.पी. बना दिया अपहरण कराने के लिए? किशन को शाम तक बरामद कीजिए, वरना आप सब पर गाज गिरेगी।''

''सर! फत्तन गोप का गैंग इसमें शामिल हुआ लगता है। कल से छापेमारी

हो रही है मगर अभी सफलता नहीं मिल पायी है।'' डी.जी.पी. ने कराहते हुए कहा।

मगर शाम तक किशन बरामद नहीं हो सका। बरसों की लगी जंग पाँच-छह घंटों में नहीं छूट सकती। खैनी ठोंक-ठोंककर खाने वाले, भारी भरकम तोंदवाले, चौराहे पर ट्रैफिक से वसूली करने वाले, अपराधियों से हफ्ता लेने वाले और निर्दोष नागरिकों को फँसाने वाले दारोगा-सिपाही फत्तन गोप के गिरोह से मुकाबला नहीं कर सके। हालाँकि उन सभी ने कई जगह छापामारी की। पूर्व मुख्यमन्त्री और विपक्ष के नेता मनोहर दास ने प्रेस को बयान दिया 'मेरा किशन कहाँ है? मुझे मेरा किशन चाहिए। सूबे की सरकार और पुलिस अपराधी गिरोहों से मिली है, इसलिए किशन को बरामद नहीं कर सकी। सिर्फ़ दिखाने के लिए छापेमारी की जा रही है। छापेमारी करने के पहले ही पुलिस का बयान अख़बारों और टी.वी. चैनलों पर आ जाता है जिससे अपराधी गिरोह सतर्क हो जाता है और अपना ठिकाना बदल देता है। हमारी माँग है कि जनता को सुरक्षा देने में विफल होने पर यह सरकार नैतिकता के आधार पर इस्तीफा दे दे।''

इस तरह राजनैतिक माहौल गरम हो गया। जब दो बड़े-बड़े हाथी लड़ते हैं, तो उसका खामियाज़ा मैदान की घास को भुगतना पड़ता है। बिल्कुल वही हुआ। दुबारा रस्तोगी साहब मुख्यमन्त्री से मिले, तो वह उबल पड़े, ''तुम मेरे दुश्मनों के इशारे पर नाच रहे हो। सिर्फ़ तुम्हारे ही बेटे का अपहरण तो नहीं हुआ। एक दर्जन बच्चे पिछले एक माह से लापता हैं। मगर किसी ने इतनी हाय-तौबा नहीं मचायी। आख़िर तुम पैसे वाले हो न? सुना है कि तुम्हारा किशन पैलेस करोड़ों रुपये का बना है। कहाँ से आये थे ये रुपये? नौकरी में आने से पहले तुम्हारी क्या औकात थी? तुमने सूबे का ख़ज़ाना खाली किया है और अपना घर भर लिया है।''

रस्तोगी साहब की दशा बिगड़ गयी। वह कुर्सी पर लुढ़क गये। पानी का छींटा उनके मुँह पर बार-बार मारा गया और पंखे से हवा की जाने लगी। वैसे ए.सी. पहले से चल रहा था। मगर रस्तोगी साहब की आँखें मुँदी थीं और धड़कन तेज हो गयी थी। सो मुख्यमन्त्री को लगा कि कहीं कुछ एड़-बेड़ हो गया, तो मीडिया और विपक्षी दल तिल का ताड़ बना देंगे। उन्होंने तुरन्त डॉक्टर को बुलाया। एक डॉक्टर उनके परिसर में सुबह-शाम ड्यूटी करता था और उनका रोज़ाना चेक-अप करता था। उसने ब्लड प्रेशर चेक किया, वह कुछ बढ़ गया था। सो उसने एक गोली उन्हें खिला दी। आधे घंटे में वह साामन्य हो गये।

अब मुख्यमन्त्री ने फोन पर डी.जी.पी. को फिर डाँटना शुरू किया, ''आपको डी.जी.पी. बनाना बेहूदगी हो गयी। एक बच्चे को बरामद नहीं कर सकते, तो

फिर वर्दी पहनने का क्या मतलब? अब आपको अपना बोरिया-बिस्तर बाँधना होगा। हम तो आपको अपना समझकर, कमज़ोर वर्ग का जानकर डी.जी.पी बनाये थे, मगर आपसे पुलिस महकमा नहीं सम्भला। जो आप नहीं कर सके, अब उसे मैं करके दिखाता हूँ। जहाँ न पहुँचे पुलिस, वहाँ पहुँचे सरकार। कौन कर सकता है मुकाबला सरकार से?''

उधर से कोई आवाज़ नहीं आयी। डी.जी.पी. की अक्षमता खामोशी में तब्दील हो गयी थी। मुख्यमन्त्री ने अपनी पार्टी के कई नेताओं और कार्यकर्त्ताओं को फोन पर सख़्त हिदायत दी कि हर हालत में सुबह तक किशन की बरामदगी होनी चाहिए।

दूसरी ओर सभी प्राइवेट स्कूलों ने किशन की बरामदगी के लिए अगले दिन बन्द की घोषणा की। सभी बच्चों, शिक्षकों और अभिभावकों ने किशन के सकुशल वापस लौटने के लिए सर्वधर्म प्रार्थना की। कई विद्यालयों और स्वयं सेवी संस्थाओं ने एक दिन का उपवास भी रखा। और सचमुच किशन सुबह एक मन्त्री के आवास से दो सौ मीटर की दूरी पर उस गिरोह द्वारा छोड़ा हुआ मिला। मुख्यमन्त्री ने प्रेस कांफ्रेंस की। किशन को कुछ भी बोलने से मना कर दिया गया। उन सभी ने आत्म-प्रशंसा के गीत गाये। उसके बाद रस्तोगी दम्पती ने भाव-विह्वल होकर मुख्यमन्त्री के चरणों में साष्टांग प्रणाम किया।

मुख्यमन्त्री ने दोनों की पीठों को थपथपाकर आशीर्वाद दिया। उनके पी.ए. ने रस्तोगी साहब को एक तरफ ले जाकर कहा, ''इस बार साहब ने किसी तरह छुड़ाया है। फत्तन गिरोह उनके बीच में पड़ने के कारण सिर्फ़ बीस लाख रुपये के सौदे पर राजी हो गया है। उसने बताया कि वह दूसरे गिरोह को पन्द्रह लाख रुपये दे चुका है। सो जल्दी से बीस लाख रुपये का जुगाड़ कर दीजिए वरना उस गिरोह को तो आप लोग जानते ही हैं कि कितना ख़तरनाक है। रस्तोगी साहब! सरकार ने आपके बेटे को बचा लिया है, अब आप सरकार के वादे की लाज बचा लीजिए।''

यह सुनते ही रस्तोगी साहब के चेहरे की खुशी कपूर-सी उड़ गयी। वह खामोश हो गये। उसका सिर चकराने लगा कि सरकार की घुसपैठ गिरोह में है या गिरोह की सरकार में। तभी उनकी पत्नी ने चलते समय हाथ जोड़कर कहा, ''सर! हम लोग आपका यह एहसान ज़िन्दगी भर नहीं भूलेंगे। जब तक साँसें रहेंगी आपका नाम जबान पर हर दम रहेगा।'' चहकते हुए वह किशन के सिर पर हाथ फेरने लगीं। तभी रस्तोगी साहब जोर से चीखे, ''चोप्प! अभी कहानी खत्म कहाँ हुई? यह तो सिर्फ़ अन्तराल है।'' और अचानक वह धम्म-से नीचे गिर पड़े। ऐसा लगा कि मानो अपनी जड़ से पेड़ उखड़ गया हो।

चलो, कहीं और चलें

जैसे ह्वांग हो नदी को 'चीन का शोक' माना जाता है, ठीक वैसे ही कोशी को बिहार का शोक। थोड़ी बारिस में भी बाढ़ आ जाती है कोशी नदी में। लोग-बाग कहते है कि कोशी की कोख छोटी है जिससे वह बड़ी सन्तान नहीं सम्भाल पाती। बाँध बनने से लोगों की मुसीबतें और बढ़ गयी हैं। पहले बरसात में पानी का बहाव आता था और कुछ दिनों तक तंग-ओ-तबाह करने के बाद वापस चला जाता था। मगर अब जो पानी बाढ़ से आता है, वह बाहर नहीं जा पाता। वह तबाही का आलम बन जाता है। जल-जमाव की समस्या विकराल रूप ले चुकी है जिसके कारण छः माह तक लोग घर से बाहर नहीं निकल पाते। उनकी हर साँस में बाढ़ की लीला थिरकती है और खुलेआम ललकारती है। वे नाव पर चढ़कर पानी में टट्टी करते हैं। और छिः छिः नहीं, सचमुच उसी बाढ़ का पानी आँख मूँदकर पी लेते हैं—चाहे हँसकर पियें, चाहे रोकर। इसी का नाम है ज़िन्दगी। बाढ़ अपने साथ नाना प्रकार की बीमारियाँ भी लाती है। बच्चे-बुतरू कै-दस्त से बड़ी तादाद में मर जाते हैं। बूढ़े लोग भी पानी की मार लम्बे समय तक नहीं झेल पाते और 'टें' हो जाते हैं। उन्हें जलाने का इन्तज़ाम भी नहीं हो पाता बारिश में। सो कोशी में ही उन्हें बहा दिया जाता है भगवान का नाम लेकर, वैसे ही जैसे डाँगर बहा दिया जाता है। एक ही फ़र्क़ रहता है दोनों में। आदमी की लाश को कपड़े के कफ़न में लपेट दिया जाता है, जानवर की खाल इतनी मोटी होती है कि उसे कफ़न की ज़रूरत नहीं पड़ती। मरने वालों की तादाद इतनी ज़्यादा होती है कि राहत के रूप में कफ़न भी बाँटा जाता है। कुछ लोग 'धराऊँ' कफ़न रख लेते हैं। ऐसा सोचता रहता है चन्नर। हर साल बरसात के दौरान वह यही सब सोचता है मगर उसके बाद भूल जाता है जैसे बच्चे अपनी चोट।

इस साल बाढ़ आयी, तो बल्लमपुर दियारा के चन्नर दुसाध को नहीं छोड़ी। पिछले चार-पाँच साल से वह चन्नर को बीमार करती थी, चलने-फिरने से लाचार करती थी मगर आख़िर में उसके बाल-बच्चों पर रहम करके उसकी जान बख़्श

देती थी। फिर सरकार की ओर से 'तीसरी फसल' काटने को मिलती। जिला प्रशासन के हाकिंम-हुक्काम से राहत के नाम पर गेहूँ, किरासन तेल, पोलिथिन, कपड़ा सत्तू-चूड़ा, आदि मिलता था। चुनाव नज़दीक होने के कारण विभिन्न पार्टियों के नेताओं ने, छुटभैये से बड़भैये तक, पिछले साल जमकर राहत सामग्री बाँटे थे। गाँव के कई लोगों ने एक आदमी के नाम पर तीन-चार बार सामान लिए। मगर यह साल चुनाव का समय नहीं था, सो कोई नेता दियारा में सपने में भी नहीं गया। जियो तो वाह-वाह, मरो तो वाह-वाह। चन्नर को टी.बी. हो गयी थी, मगर उसके पास लम्बे समय तक इलाज कराने के पैसे नहीं थे। वैसे लोग रोजाना वो इबारत देखते थे जो सरकार ने दीवारों पर लिखवा दी थी— टी.बी. लाइलाज नहीं है। बाढ़ आने से उसकी खाँसी और बढ़ गयी। खाँसी बढ़ने का मतलब ज़्यादा बलगम और ज़्यादा खून का निकलना था। सो वह खाँसते-खाँसते एक दिन सरग सिधार गया। मगर अपने पीछे छोड़ गया पाँच बच्चों और उनकी महतारी को। चन्नर की लुगाई झिनका रो-रोकर झाँखर हो गयी। अब किस बिल में घुस जाए या किस गगरी में हाथ डाले कि बच्चों को खाने-पीने को कुछ मिल जाये। मगर कुछ नहीं मिलता था। वह चन्नर की याद में परेशान हो उठी। चन्नर बीमारी के बावजूद कुछ-न-कुछ करता रहता था। जब हल जोतने और फावड़ा-कुदाल चलाने से वह लाचार हो गया था, तो वह घर बैठे-बैठे चारपाई बुनता, चटाई बुनता, खाँची-पलड़ी बनाता, नरिया-खपड़ा पाथता, रस्सी बटता, गाय को चारा खिलाता आदि। इससे वह खाने का कुछ-न-कुछ जुगाड़ ज़रूर कर लेता था। मगर अब कौन करेगा? झिनका के मन में ये बातें बार-बार कौंधतीं तो वह कलपने लगती। मगर बाद में वह यह भी सोचती कि उसका मर्द टी.बी. की गम्भीर बीमारी कई सालों तक झेलता रहा, सो उसे अब जाकर मुक्ति मिली। मगर तभी उसके सामने बिलबिलाते बच्चे नज़र आते, तो वह दहाड़ माकर रोने लगती। गाँव के दुसाध टोले, कायस्थ टोले, राजपूत टोले और ब्राह्मण टोले की महिलाएँ उसे बार-बार नाना प्रकार से सान्त्वना दे रही थीं। दुसाध टोले की महिलाएँ उसके आँसू पोंछ रही थीं, उसका हाथ-पैर पकड़ रही थीं, उसका कपड़ा ठीक कर रही थीं और समझा-बुझा भी रही थीं। दूसरी महिलाएँ बगल खड़ी होकर सिर्फ़ शब्दों से उसे समझा रही थीं।

थोड़ी देर में चन्नर की लाश की अरथी श्मशान घाट ले जायी गयी। झिनका उसके पीछे-पीछे दौड़ने लगी। कई महिलाओं ने उसे पकड़कर बिठाया। पड़ोस की महिलाओं के अलावा दूसरे टोलों से भी उसके बच्चों के खाने के लिए रोटी, दाल-भात, सब्जी वगैरह आ गयी। भूखे बच्चे खाने पर टूट पड़े । पड़ोस की एक महिला ने उन सभी का हाथ-मुँह धुलाकर खाना परोस दिया। वे बड़े चाव से खाने लगे। वे बच्चे शायद अपने पिता के मरने की बात भूल गये थे अथवा

मरने का अर्थ ही नहीं समझ पाये थे।

दूसरे दिन से भीड़ कम हो गयी। कुछ पड़ोसियों और रिश्तेदारों ने मिलकर चन्नर का क्रिया-कर्म सम्पन्न किया। घर में तो भूँजी भाँग भी नहीं थी। मगर ऐसे में भी बिरादरी को तो कच्चा-पक्का भोज खिलाना ही था, वरना मृत्यु जैसी अशुभ चीज का पाप कैसे मिटेगा? पुश्तैनी खेत तो था नहीं जिसे बेचकर जुगाड़ किया जाता। गहना-गुरिया भी नहीं था झिनका के पास। दोनों गायों को बेच दिया गया। फिर भी दो हजार रुपये का कर्ज हो गया। जो चन्नर जीते जी कर्ज़दार नहीं था, मरने पर कर्ज़ा का गट्ठर दे गया।

बिरादरी के भोज के बाद ज़िन्दगी अपने असली रूप में प्रकट होने लगी। झिनका को अपने तथा बच्चों के खाने-पीने, कपड़े-लत्ते, दवा-दारू का इन्तज़ाम करना था। चूँकि झिनका का कोई बेटा-बेटी पढ़ने नहीं जाता था, सो वैसे उपरफट्टू खर्च की चिन्ता नहीं थी।

एक महीने के बाद उस बल्लमपुर दियारा में नये बी.डी.ओ. साहब अपनी लाव-लश्कर लेकर आये और इस टोले की सबसे ज़्यादा उजड़ी झोपड़ी के सामने गाड़ी रोक दिये। उतरते ही उन्होंने आस-पास के लोगों को बुलाया, "देखो, भाइयो और बहनो! हम आप सबके बच्चों को पढ़ाना चाहते हैं।"

"क्या पढ़ाएँगे हुजूर? मास्टर ही स्कूल नहीं आते?" एक नौजवान जियावन ने तुरन्त टोका।

"देखिए, जो पहले हुआ, सो हुआ। उसे भूल जाइये। अब मास्टर लोग रोज़ाना पढ़ाने आएँगे। आप सभी अपने बच्चों को स्कूल भेजिए।" बी.डी.ओ. ने आश्वस्त करना चाहा।

मगर तभी एक औरत बोल पड़ी, "स्लेट-पेंसिल खरीदने में भी तो पैसा लगता है। हमारे दुसाध टोला के लोगों के पास उसका इन्तज़ाम होता, तो हमारे बच्चे भी स्कूल में पढ़ने पहले ही चले गये होते।"

तब बी.डी.ओ. साहब ने विस्तार से बताना शुरू किया, "देखिए, अब सरकार ने फैसला किया है कि चौदह वर्ष के सभी बच्चों को स्कूल में नाम लिखाना है। कॉपी-किताब, स्लेट-पेंसिल सब कुछ सरकार देगी। आप लोगों को कुछ भी खर्च नहीं करना है। बस, अपने बच्चों को नहला-धुलाकर स्कूल भेज देना है।"

"मगर जो बच्चे मजूरी करके अपने परिवार का खर्चा चलाते हैं, वो कैसे पढ़ने जाएँगे? वे पढ़ेंगे, तो पेट कैसे भरेगा उनके परिवार का?"—एक बुजुर्ग ने बात की डंडी मारी।

"और भूखे भजन न होई गोपाला। वे दिन भर स्कूल में रहेंगे बिना खाये-पिये? यह तो मुश्किल है।"—एक औरत ने पानी का छींटा मारकर पूरे प्रस्ताव

की रोशनी बुझा दी। तब बी.डी.ओ. साहब ने योजनाओं की पिटारी खोल दी, ''सभी बच्चों को दोपहर का भोजन मिलेगा—खिचड़ी या हलवा। सभी बच्चों को वज़ीफ़ा भी मिलेगा। सरकार ने इतनी व्यवस्था की है, तो आप लोग इसका फ़ायदा उठाइये वरना आपके बच्चे आगे नहीं बढ़ पाएँगे। जो पढ़ेगा, आगे बढ़ेगा। दुनिया तेजी से बदल रही है। जो ज़माने की दौड़ में पीछे हो गया, वो हमेशा के लिए अपने को कमज़ोर बना लेगा, गुलाम बना लेगा। बच्चों का बचपन वापस नहीं आएगा। तब पछताने के सिवाय कुछ भी हाथ नहीं लगेगा।''

इतना सुनते ही झिनका के शरीर में जोर से कम्पन हुआ। वह सोचने लगी कि वह पहले ही पिछड़ गयी है। उसके पास एक धुर भी ज़मीन नहीं है। अपना मकान नहीं है। उसके बच्चे अनपढ़ गँवार हैं। जब सरकार पूरा खर्च उठाने को तैयार है, तो मुँह मोड़ने की बात कहाँ? वह धीरे-धीरे बी.डी.ओ. साहब की ओर बढ़ी और बोल पड़ी, ''हम पढ़ाएँगे अपने सभी बच्चों को। उनको जानवर से इनसान बनाएँगे।''

इस पर बी. डी. ओ. साहब खुश होकर तालियाँ बजाने लगे। सो सारे लोग तालियाँ बजाने लगे। सभी ने अपने बच्चों को स्कूल भेजने की बात कही। बी. डी. ओ. के साथ गये किरानी बाबू ने दुसाध टोला के सभी लड़के-लड़कियों का नाम रजिस्टर में लिखा। कुल साठ बच्चे नाम लिखवाये। बी. डी. ओ. साहब ने झिनका को जवाबदेही दी कि वह अपने टोला के सभी बच्चों को स्कूल भिजवाये। यदि नियमित रूप से बच्चे स्कूल जाएँगे, तो झिनका को इनाम मिलेगा। उन्होंने उसको विधवा पेंशन स्वीकृत करने का वादा किया।

झिनका को लगा कि उसकी चोंच में ज़िन्दगी का तिनका पकड़ में आ रहा है। आदमी किसी सहारे की अपेक्षा उसकी आशा पर ज़्यादा जीता है। उसने एक पढ़े-लिखे लड़के की मदद से अगले दिन विधवा पेंशन का एक आवेदन ब्लॉक में भिजवा दिया। मगर बी. डी. ओ. ने कहा कि उस पर मुखिया का हस्ताक्षर और मुहर चाहिए कि ये सारी बातें सही हैं। झिनका अगले दिन मुखिया के यहाँ गयी। पता चला कि वह सिवान में धान की रोपनी करवा रहे हैं। वह दौड़कर वहाँ पहुँची और मुखिया के सामने अपनी समस्या पेश कर दी। मुखिया ने कहा, ''पहले मेरे खेत में धान की रोपनी करो। फिर मैं तुम्हारा सारा काम कर दूँगा। तुम्हें ब्लॉक भी नहीं जाना पड़ेगा।'' सो झिनका भी दर्जनभर स्त्री-पुरुष मजदूरों के साथ उस काम में लग गयी। मुखिया ने उसका आवेदन अपनी जेब में रख लिया। मुखिया उसे बार-बार निहारता रहा जब वह झुककर धान की जरई लगा रही थी। शाम को मुखिया ने सभी मज़दूरों को मजदूरी दे दी मगर झिनका को बाद में देने का वादा किया। झिनका को अपनी पेंशन बँधवाने की समस्या याद थी, सो वह दबाव नहीं दे सकी।

तीन दिनों के बाद झिनका फिर गयी मुखिया के दरबार में। वह अपनी दालान में कुछ छुटभैयों से घिरे थे। एक आदमी ने कहा, "अब मुखिया के यहाँ आ गयी हो, तो तुम्हारा कल्याण हो जाएगा।" थोड़ी देर तक इन्तज़ार करने के बाद उसने बात करने की हिम्मत जुटायी, तो मुखिया ने घुड़की दी, "पूरी दुनिया में तुम्हारा ही अकेले काम नहीं है। मेरे ऊपर पूरी ग्राम पंचायत का बोझ है। तुम आकर मेरे घर का झाड़ू-पोंछा, दुआर-गोबर कर दिया करो, फिर तुम्हारा काम टंच हो जाएगा।" वह खाली हाथ आयी थी, खाली हाथ लौट गयी। उसकी हिम्मत नहीं हुई कि वह धान की रोपनी की बकाया मजूरी माँगे। वह डर रही थी कि मजूरी माँगने से मुखिया और नाराज़ हो सकता है, तब तो पेंशन का मिलना सपना ही हो जाएगा। वह अपने दुख की डगर पकड़े धीरे-धीरे वापस चली गयी।

करीब एक माह बाद झिनका फिर मुखिया के दरबार में फरियाद करने गयी। मुखिया ने उस फरियादी को अगले दिन ब्लॉक चलने को कहा। भाड़े के पैसे उसके पास नहीं थे। सो अगले दिन वह तैयार होकर जियावन की साइकिल पर बैठकर ब्लॉक पहुँची। उसने कार्यालय से पता किया तो वहाँ न तो वह मुखिया था और न उसका आवेदन था। सो वह फिर बी. डी. ओ. से हाथ जोड़कर गिड़गिड़ाने लगी। बी. डी. ओ. ने मुखिया से उसके आवेदन पर हस्ताक्षर करवाने और पेंशन स्वीकृत करने का वादा किया। मगर फिर एक माह बीत गया। इस बीच उसके घर की स्थिति बद से बदतर हो गयी। उसने अब तक पाँच सौ रुपये का अनाज-पानी, नोन-तेल कर्ज पर लिया था। मगर भविष्य की बेदी पर वर्तमान का बलिदान करके वह अपने बच्चों को स्कूल भेजती रही। वहाँ बच्चों को दोपहर में खिचड़ी मिल जाती मगर सुबह-शाम के लिए कुछ जुगाड़ करना पड़ता था। वह कई घरों में सफाई करने, गोइंठा पाथने, पानी भरने, अनाज साफ करने आदि का काम करने लगी।

जिस अध्यापक के जिम्मे खिचड़ी बनाने की जवाबदेही थी, उसका तबादला हो गया तो अचानक खिचड़ी बनानी बन्द हो गयी। गरीब बच्चे फिर स्कूल से भागने लगे। झिनका ने बी. डी. ओ. साहब को सूचना दी। वह भी सर्व शिक्षा अभियान के प्रति समर्पित थे। सो उन्होंने अस्थायी व्यवस्था के तहत झिनका को ही खिचड़ी बनाने का जिम्मा दे दिया और इसके लिए उसे महीने में एक सौ रुपये देने को कहा। इस बात की ख़बर जैसे मुखिया को मिली, वह कुपित हो गया। उसने कहा, "एक दुसाध औरत की बनायी हुई खिचड़ी हमारे बच्चे कैसे खाएँगे?" सो उसने अड़ंगा लगा दिया कि जिसके बच्चे स्कूल में पढ़ते हैं, उसके खिचड़ी बनाने पर गड़बड़ी होगी क्योंकि वह अपने बच्चों को ज़्यादा खिचड़ी परोस देगी और बच्चों के जरिये सामानों की चोरी भी कर लेगी। सो मुखिया ने झिनका को यहाँ भी धोबिया पाट पटकनी दी।

अब झिनका बेहद परेशान हो उठी। मगर दुसाध टोले के ही जियावन नामक नौजवान ने उसे साइकिल पर बिठाकर ब्लॉक पहुँचाया तथा बी. डी. ओ. से मिलकर उसकी पेंशन स्वीकृत करने की मिन्नत करने लगा। जियावन ने उपमुखिया से हस्ताक्षर करा दिया था। सो बी. डी. ओ. ने कुछ ही दिनों में अपने उच्चाधिकारियों से मिलकर उसकी पेंशन स्वीकृत करवा दी। इससे झिनका के चेहरे पर खुशी के मेघ झमाझम बरसने लगे। उसे लगा कि उसकी ज़िन्दगी सही करवट ले रही है। जब मुखिया को उसकी जानकारी मिली तो उसने प्रचार कर दिया कि झिनका जियावन की रखैल बन गयी है।

उधर नौजवान जियावन को इस सफलता से इतनी ज़्यादा खुशी हुई थी कि उसने कुछ नौजवानों को बटोरकर, सलाह-मशविरा करके 'ग्राम विकास समिति' का गठन किया और दुसाध टोले के युवकों और युवतियों को सदस्य बनाया। यद्यपि झिनका पढ़ी-लिखी नहीं थी, मगर उसे भी उस समिति में रखा गया। इस समिति ने पूरे गाँव का सर्वे करना शुरू किया कि किसके पास क्या-क्या चल-अचल सम्पत्ति है, गाँव में पंचायत ने क्या-क्या योजनाएँ बनवायी हैं, उनकी गुणवत्ता क्या है, उनमें कितना खर्च हुआ है, वगैरह-वगैरह। करीब एक माह में ये सारे काम पूरे हो गये, तो जियावन और झिनका ने पूरी सूची बनाकर बी. डी. ओ. साहब को सौंप दी कि जिन-जिन योजनाओं में पैसे की लूट हुई है, उनकी जाँच कराकर दोषी मुखिया, ठेकेदार आदि पर कानूनी कार्रवाई की जाये। झिनका जियावन की दी हुई अकल पर अमल करने में नहीं चूकती थी।

मुखिया उस सारी घटना से बहुत चिन्तित हुआ। उसे लगा कि यदि जाँच हुई, तो वह जेल भी जा सकता है और लूट के पैसे की भरपाई भी करनी होगी। सो उसने दुसाध टोले के वैसे लोगों को बुलाया जो दारू-ताड़ी-भाँग आदि के शौकीन थे। उन सभी को अपने यहाँ उसने जमकर खिलाया-पिलाया और अन्त में झिनका-जियावन के खिलाफ एक आवेदन भी लिखवाया। मौजूद पंचों ने अपना-अपना हस्ताक्षर किया। उसमें लिखा था, ''पूरी पंचायत यह फैसला लेती है कि जियावन और झिनका के अवैध सम्बन्धों के कारण उनका सामाजिक बहिष्कार किया जाय। झिनका वेश्यावृत्ति करती है जिसके कारण पूरे गाँव का माहौल दूषित हो गया है।''

अगले दिन जियावन अपने तीन साथियों के साथ झिनका के यहाँ अगली कार्रवाई की योजना बना रहा था। उसे लगा कि बी. डी. ओ. साहब भी शायद मुखिया से मिल गये हैं, इसलिए कोई कार्रवाई नहीं कर रहे हैं। सो जिलाधिकारी के पास आवेदन देना चाहिए। यह योजना बनाते और नयी रणनीति तय करते-करते रात के दस बज गये। मुखिया ने उन युवा साथियों में से एक को पैसा देकर और दारू पिलाकर फोड़ लिया था। वह नशे में बहक रहा था और झिनका को

अनाप-शनाप कह रहा था जियावन ने उसे डाँटा-फटकारा, तो वह जियावन पर टूट पड़ा। उधर मुखिया ने कुछ लोगों के साथ झिनका के घर पर धावा बोल दिया। वहाँ रात में तीन-चार पुरुषों के साथ झिनका को पाकर गाँववालों ने बेहद नाराजगी जतायी और कुछ लोगों ने झिनका पर हाथ-पैर भी चला दिये। उनका पारा सातवें आसमान पर था।

फिर क्या था? पूरा गाँव मधुमक्खी की तरह भनभनाने लगा। लगभग सभी स्त्री-पुरुषों ने झिनका को कुलटा, छिनार, रंडी, पतुरिया, हरजाई आदि कहा। झिनका सोचने लगी— किसी बिगड़े मर्द पर कभी ऐसा आरोप नहीं लगता। मर्द ही तो सृष्टि चलाता है। बीज कभी अपवित्र नहीं माना जाता, सो मर्द अपवित्र नहीं हो सकता। वक्त की नज़ाकत समझकर और अपनी माँ के इशारे पर जियावन अचानक वहाँ से फूट लिया। सबकी राय थी कि राँड़ होने पर झिनका ने नया साँड़ खोज लिया है, कि उसके नखरे तीसरे आसमान पर रहते हैं, कि वह बी. डी. ओ.-सी. ओ. के यहाँ आँखें लड़ाने जाती है, कि वह 'भतरासी' है—भतार चाहिए रोज़-रोज़, नया-नया। सो गाँव वालों ने उसे गाँव से निकाल देने का फैसला किया। दुसाध टोला, राजपूत टोला, ब्राह्मण टोला और कायस्थ टोला किसी बात पर शायद ही कभी एकमत होते थे, मगर गाँव की इज़्ज़त-आबरू के नाम पर सब एक हो गये थे। गाँव की नाक का सवाल सबसे अहम होता है। महिलाएँ पुरुषों से भी चार कदम आगे थीं। वे झिनका के इतिहास का पोथन्ना सरेआम खोल रही थीं—कोई चौपाई कहती थी, तो कोई दोहा। कोई मुहावरा कहती थी, तो कोई पहेली बुझाती थी। महिलाओं, विशेषकर दुसाधिनों को, यह फिक्र थी कि महिला के नाम पर झिनका एक कलंक है जिसने अपनी वासना को नियन्त्रित नहीं किया। उसने पेट की आग बुझाने के लिए उसके नीचे का रास्ता अपनाया। अब कोई बिरादरी वाला यहाँ की लड़कियों से विवाह नहीं करेगा। सो उसके कुकर्म से अगली पीढ़ी का सत्यानाश हो जाएगा। सो दुसाध टोला की एक महिला फुलेना ने प्रस्ताव पेश किया, ''पंचो! झिनका ने जो कुछ किया है वह माफी जोग नहीं है। उसने पूरे गाँव-समाज की आबरू माटी में मिला दी है। हमारी बेटियों का हाथ कोई नहीं थामेगा। मैं भी राँड़-रेवा हूँ मगर अपनी देह अपने काबू में रखती हूँ। अपनी लगाम अपने पास रहनी चाहिए। सो झिनका को इस गाँव से निकाल दिया जाय। मगर उसके पहले उसका मूड़ मुड़ाकर उसे सरेआम नंगा घुमाया जाए पूरे गाँव-जवार में।''

इसे विशाल जन-समूह ने करतल ध्वनि से पारित कर दिया। मुखिया विलायती यादव ने ऐलान किया, ''जब सभी लोगों की यह राय है, तो मेरी भी यही राय है। दुष्ट के साथ दुष्टता बरतनी चाहिए—शठे शाठ्यम् समाचरेत्।'' झिनका सिहर उठी। उसकी आँखों से झरने की तरह लोर टपकने लगे। वह जानती

थी कि फुलेना का चरित्र कैसा है, कि वह खुद अपना देह बेचती रही है सालों से, कि वह मुखिया विलायती यादव की भी खास पसन्द रही है, कि विधवा होने के डेढ़ साल बाद उसके अवैध लड़का पैदा हुआ था जो मुखिया के जानवरों को चराता है। मगर वह कैसे कहे? किसे कहे, है कोई सुनने वाला? है कोई साथ देने वाला? उसने चारों ओर निहारा। नीचे धरती से लेकर ऊपर आकाश तक कोई नहीं दिखा उसे। जियावन के साथी भी नहीं।

वह चिन्ता के अग्निकुंड में जल रही थी कि फुलेना ने उसकी पीठ पर एक लात मारा। देखते-देखते उस पर सैकड़ों लात-घूँसे पड़े। वह कराहती रही—कोई सुनने वाला और रोकने वाला नहीं था। तभी उसने चीख कर कहा, "मुखिया ने मेरी मजूरी नहीं दी। जब मैं माँगती हूँ तो वह आनाकानी करता है। मेरी पेंशन भी इसने नहीं बँधवायी। मुझ पर इसकी निगाह पहले से बुरी है..."

मगर लोगों ने इसे बहाना माना—सोलहे आने झूठ। एक रत्ती भी सच नहीं। लोग हीं हीं करके हँसते रहे और तर्भ। मुखिया के एक खास आदमी बिन्देसर ने झिनका का झोंटा पकड़कर घसीट लिया। वहाँ मौजूद नाई ने उस्तरे से मूड़ का बाल छील दिया। वह रोती-बिलखती रही। तभी मुखिया ने भारत की महान प्राचीन परम्परा याद करते हुए दु:शासन की भूमिका निभाते हुए झिनका को निर्वस्त्र कर दिया। वहाँ मौजूद स्त्री-पुरुष-बच्चे तालियाँ बजा-बजाकर, फिकरे कस-कस कर मज़ा लेने लगे। वह अपने हाथों से अपने अंगों को ढकने का प्रयास कर रही थी। तब दो जवानों ने उसके हाथ दोनों तरफ से पकड़ लिए और उसे खड़ी होकर चलने को बोले। वह रोती हुई बैठ जाती, मगर वे उसे घसीटकर ले जाते। कोई आगे से, तो कोई पीछे से उसकी देह पर हाथ या डंडे से खोदकर हो-हो करता। यह प्रक्रिया दिन भर चलती रही। अन्त में वह गिर पड़ी और बेहोश हो गयी।

शाम में टी. वी. के एक स्थानीय चैनल ने इस घटना का रस ले-लेकर प्रसारण किया, तब पुलिस की गहरी नींद टूटी। थाने के पुलिस दारोगा के गाँव में आठ बजे रात को पहुँचने के समय सारे लोग झिनका को पूरब के पोखर पर अधमरी छोड़कर जा चुके थे। वह बुत की तरह पड़ी थी। उसके बच्चे बिलख रहे थे। उनके आँसुओं से पोखर भरता जा रहा था। दारोगा ने कलम चलाई और अज्ञात लोगों के नाम मामला दर्ज किया। मगर उसे झिंनका नहीं मिली। उसके बच्चों ने कहा, "हम्मर माई के भितरिया मार मर लो है।" दो दिन बाद राष्ट्रीय महिला आयोग की टीम ने घटना-स्थल का औचक निरीक्षण किया। बल्लमपुर दियारा के सभी पुरुषों ने पूर्व नियोजित तरीके से झिनका के बारे में नयी-नयी कहानियाँ गढ़कर उसे सुनायीं। जब टीम ने मुखिया को बुलवाया, तो पता चला कि वह अपनी ससुराल गया है एक शादी में। दारोगा ने जाँच टीम को बताया

कि मुखिया जी शरीफ़ आदमी हैं, कि वह गरीबों की मदद करते हैं, कि वह महिलाओं के विकास के कार्यक्रम करते हैं, कि वह ईमानदारी की मूर्ति हैं, कि उनकी ज़िन्दगी एक खुली किताब है। तभी अचानक जियावन वहाँ आ पहुँचा। उसने जाँच टीम के सामने सब कुछ सही-सही बयान कर दिया। मगर जाँच टीम ने विश्वास नहीं किया बल्कि उसने झिनका का बयान दर्ज़ करना चाहा। सो जियावन उस टीम को लेकर जिला मुख्यालय के लिए रवाना हो गया। वहाँ एक झोपड़ी से झिनका निकली। उसने चीखते हुए कहा, "मुखिया ने मुझे पेंशन नहीं दिलवाई। पेंशन के नाम पर उसने मुझसे बेगार करवाया। हम लोगों ने उसके खिलाफ जाँच की अर्जी दी, तो दबाने के लिए उसने मेरे साथ मुँह काला करने का प्रयास किया। जब मैंने गाँव-समाज को यह घटना बता देने की धमकी दी, तो उसने अपने आदमियों से मिलकर मेरी यह दुर्दशा कर दी है।"

राष्ट्रीय महिला आयोग की जाँच टीम ने जब दारोगा को बाघ की तरह घूरा, तो उसने अपनी आँखें नीचे कर लीं। फिर उसने कहा, "गाँव वालों ने मुझे बताया था कि इस औरत का चाल-चलन ठीक नहीं है।" आयोग की जाँच टीम ने पूछा, "और तुम्हारा?" दारोगा को यह सुनते ही गस आ गया। मगर उस जाँच टीम को फुर्सत नहीं थी कि इसकी पूरी तहकीकात करे। उसके सामने समूचा देश और उसकी बड़ी-बड़ी राष्ट्रीय समस्याएँ थीं। उसकी निगाह में यह स्थानीय समस्या थी। सो उसने सूबे के पुलिस महानिदेशक को जाँच कर कार्रवाई करने का आदेश दे दिया। पुलिस महानिदेशक अभी तक उसकी छान-बीन कर रहे हैं। उन्होंने एक सौ से अधिक महिलाओं और पुरुषों से बयान अपने अधीनस्थों के ज़रिये ले लिया है। जियावन को छोड़कर सारे लोगों ने झिनका के चाल-चलन को व्यभिचारी बताया है। मीडिया के द्वारा सवाल दागे जाने पर पुलिस महानिदेशक ने सिर्फ़ इतना ही कहा, "गाँव के लगभग सभी लोगों ने झिनका के खिलाफ़ गवाही दी है। उसकी तरफ सिर्फ़ एक आदमी है, और उस पर भी गाँव वालों ने झिनका के साथ अवैध सम्बन्ध का आरोप लगाया है। और लोकतन्त्र में बहुमत ही सब कुछ होता है। फिर भी जाँच जारी है।" अख़बारवालों ने जियावन से खोद-खोदकर पूछा, तो वह बमक उठा, "सच का साथ देने वाला कोई नहीं है। आज भी भारत में सिर्फ़ संविधान में लोकतन्त्र है, सरज़मीन पर नहीं। कानून बड़े लोगों के लिए बना है। वह उनकी ही हिफ़ाजत करता है, गरीबों की नहीं। चलो भाभी, हम लोग कहीं और चलें जहाँ न मुखिया हो न उसकी रखैल पुलिस हो। यह गाँव हमारे रहने लायक नहीं रहा।" झिनका धीरे-धीरे आगे बढ़ी, उसके पीछे-पीछे जियावन बढ़ा और उसके पीछे झिनका के बच्चे बढ़ने लगे। जमीन काँप रही थी उनके चलते समय। और रो रहे थे आस-पास के गाछ।

ठीहे पर आदमी

उस दिन विद्यालय में विवेकानन्द जयन्ती मनायी जा रही थी। विद्यालय परिसर काफी साफ-सुथरा था। फूल-पौधों की क्यारियाँ सज-धजकर तैयार थीं मानो आने वाली बारात का इन्तजार कर रही हों। हम विद्यार्थीगण खुशी से उछल-कूद रहे थे। पहली बेला में संगीत-नाटक आदि का कार्यक्रम धूमधाम से चला। कई छात्रों ने विवेकानन्द की जीवनी पर प्रकाश डाला। विवेकानन्द से प्रेरणा लेते हुए छात्रों ने राष्ट्र के विकास और एकता की प्रतिज्ञा की। इसके बाद विद्यालय के प्रधानाध्यापक श्रीमन्त जी ने विस्तार से विवेकानन्द की प्रासंगिकता की चर्चा की। सारे विद्यार्थी भाव-विभोर होकर सुन रहे थे। अधिकांश शिक्षकों ने भी प्रधानाध्यापक के भाषण को सराहा। मगर सूबेदार सिंह और उनके गुट के चार-पाँच शिक्षकों का मत नकारात्मक था। वे चाँद में भी दाग खोज रहे थे और चाँदनी को ही कोस रहे थे।

खैर...संगीत-नाटक आदि के बाद खेलों का दौर शुरू हुआ। दौड़, लम्बी कूद, ऊँची कूद, वॉलीबाल, फुटबाल, बास्केटबाल वगैरह-वगैरह। छात्रों के खेलने के बाद छात्रों और शिक्षकों के बीच वॉलीबाल खेला गया और छात्र खेल में जीत गये। श्रीमन्तजी ने लड़कों की पीठ थपथपाई और शाबाशी में कहा, ''मुझे खुशी है कि हमारे विद्यार्थीगण पढ़ने के साथ-साथ खेलने में भी आगे बढ़ रहे हैं। शारीरिक और दिमागी काम में सन्तुलन होना चाहिए। पर खेल को खेल की भावना से खेलना चाहिए। मुझे और भी खुशी है कि हमारे विद्यार्थीगण खेल में हम शिक्षकों से बेहतर हैं। सन्तानों और शिष्यों से हारने का सुख अद्‌भुत होता है...।'' यह कहते-कहते उनकी आँखों में आँसू आ गये और सभी ने तालियाँ बजाईं। अगले दिन वे हमारी कक्षा में अँग्रेजी कवि जॉन डन की कविता 'मृत्यु' पढ़ा रहे थे। वैसे दूसरे विद्यालयों के प्रधानाध्यापक कक्षाओं में नहीं पढ़ाते थे। उस कविता को उन्होंने हिन्दी साहित्य और समाज से जोड़ा। फिर गीता से जोड़ते हुए हमें बताया कि जिस तरह से पुराने कपड़ों को त्यागकर नये कपड़े पहने जाते

हैं, ठीक उसी प्रकार आत्मा एक शरीर को त्यागकर दूसरा शरीर धारण करती है। इसलिए कवि के द्वारा मृत्यु को ललकारने को उन्होंने सही सिद्ध किया। हम लोगों की शंकाओं और सवालों का उन्होंने स्वागत करते हुए जवाब दिया। कक्षा के बाद हम लोगों के बीच चर्चा होने लगी कि यदि ऐसे सवाल पिछले प्रधानाध्यापक से पूछे जाते, तो वे आग-बबूला हो जाते और पूछने वाले छात्र को बेंच पर खड़ा करके अपमानित करने में खुश होते। लेकिन श्रीमन्त जी ने अनुशासन के नाम पर आतंक नहीं फैलाया। छड़ी की जगह अपनी वाणी से ही विद्यार्थियों को सचेत कर देते थे। फिर वह एक जनवादी कवि भी थे जिससे कविताओं का मर्म खोलने में उन्हें महारत हासिल हो गयी थी। हम लोग खुशकिस्मत थे कि हमें अच्छे कवि, अच्छे इन्सान और अच्छे प्रधानाध्यापक एक साथ मिले थे।

जब कोई अभिभावक उनसे मिलने जाता, तो वह उसकी समस्या पूरे मनोयोग से सुनते और उसके समाधान का कोई-न-कोई उपाय ढूँढ़ते। अभिभावकों से उनकी ही बोली-बानी में बतियाने पर उन्हें इत्मीनान होता। वे स्वयं साधारण कुर्ता-धोती पहनते जैसे अधिकांश किसान पहनते थे। अगले दिन जब एक अभिभावक ने उनसे पूछा, ''आप इतना समय इन लोगों के लिए क्यों गँवाते हैं?''

उन्होंने कहा, ''मैं आप लोगों का समय लेता हूँ कुछ नयी चीज़ें सीखने के लिए। किसान और कवि ब्रह्मा की तरह स्रष्टा हैं बल्कि किसान तो ब्रह्मा की तरह स्रष्टा, विष्णु की तरह पालनहार और शंकर की तरह दुष्टों का संहारक है।''

और तभी एक नवयुवक बिना कुछ पूछे-पछोरे धड़ल्ले से उनके कमरे में घुस गया। वहाँ बैठा अभिभावक चकित रह गया। उस युवक के हाव-भाव से अभिभावक को अनिष्ट की आशंका हुई। तभी नवयुवक ने कागज का एक टुकड़ा प्रिंसिपल की ओर बढ़ाते हुए कहा, ''मास्टर साहब! इसे देख लीजिए।''

उन्होंने उसे पढ़े बिना उसे देख लेने की हुंकारी भर दी। वह चला गया। अभिभावक ने अपनी जिज्ञासा व्यक्त कर दी, ''इस लड़के का हाव-भाव ठीक नहीं। उसकी भौंहें तनी थीं। उनसे क्रोध की चिंगारी फूटती-सी लग रही थी।''

''अरे नहीं! मैंने इसे पढ़ाया है। हालाँकि यह लाल तीव्र बुद्धि का छात्र नहीं था, पर मेरे जैसे शिक्षक से इसकी क्या अदावत? ज़माना बदल रहा है, सो उसका असर नये खून पर ज़्यादा है। यह तो हमेशा होता आया है। हम लोग भी छात्र-जीवन में थोड़ा गरम मिज़ाज के थे। कभी-कभी हम लोग नासमझी के कारण अपने शिक्षकों से भी उलझ जाते थे,'' आश्वस्त होकर श्रीमन्त जी ने कहा।

मैं भूगोल शिक्षक के आदेश पर कार्यालय से भारत का नक्शा लाने के लिए वहाँ गया था। विद्यालय का चपरासी उसे कोने-अँतरे में खोज रहा था। मैं चुपचाप

प्रधानाध्यापक की बातों को सुन रहा था। करीब दस मिनट तक खोजने के बाद चपरासी ने मुझे नक्शा पकड़ाते हुए कहा, ''यह नक्शा जर्जर हो गया है। बहुतों ने इसके साथ छेड़छाड़ की है। सम्भालकर ले जाना वरना टुकड़े-टुकड़े हो जाएगा।''

श्रीमन्त जी ने वह काग़ज़ उप-प्रधानाध्यापक को सौंप दिया। उन्होंने उसे बड़े बाबू को और अन्ततः बड़े बाबू ने छोटे बाबू को दे दिया। छोटे बाबू के पास वे सारे काम थे जो बड़े बाबू से लेकर प्रधानाध्यापक तक को फाइल में पेश किये जाते थे, मगर इसके अलावा उन्हें फुटकर काम भी करने पड़ते थे। मसलन, शिक्षकों को खड़िया देना, हाजिरी बही हर कक्षा में भिजवाना, शिक्षण-सामग्री शिक्षकों को देना, विद्यार्थियों की छोटी-मोटी समस्याएँ सुलझाना, वगैरह-वगैरह। इसी वजह से वह काग़ज़ उसके पास महीने भर पड़ा रहा। सो एक दिन लाल प्रधानाध्यापक श्रीमन्त जी के कमरे में गरजा, तो बड़े बाबू और छोटे बाबू के कान खड़े हो गये। वे सभी काँपने लगे जब उसने दहाड़ा, ''आप लोगों ने मेरे काग़ज़ को रद्दी की टोकरी में फेंकने की हिम्मत कैसे की? क्या आप लोगों को यह मालूम नहीं था कि यह लाल बद्री विशाल कुंअर चन्द्र सेन प्रताप बहादुर सिंह का काग़ज़ है? मैं पूछता हूँ कि आप लोगों को इस हाईस्कूल में रहने का मन है या...'' क्रोध ने उसकी वाणी पर ब्रेक लगा दिया।

''ऐसा मत कहो लाल! यह विद्यालय सबका है। तुम्हारा, इनका और मेरा यानी हमारा विद्यालय है। तुम जानते हो कि यह नयी भर्ती करने का समय है। अभी काफ़ी समय इसी एक काम में जाया होता है। मैं इन सभी को सख़्त हिदायत देता हूँ कि जल्द तुम्हारी फाइल बढ़ा दें'', श्रीमन्तजी ने अपने पुत्रवत् शिष्य को समझाते हुए कहा।

लाल ने अपना सिर नीचे झुका लिया। थोड़ी देर बाद फिर बोला, ''गुरुजी! मैं आपको कुछ नहीं कह रहा हूँ मगर इन कामचोरों को सचेत कर रहा हूँ। पन्द्रह दिनों बाद मैं फिर यहाँ आऊँगा और इस काम के लिए वह मेरा आख़िरी आना होगा। नमस्ते।'' और वह तेजी से बाहर निकल गया।

श्रीमन्त जी को लगा कि वह परोक्ष रूप से उन्हें भी कामचोर कह रहा है। उन्हें भी धमकी और चेतावनी दे रहा है जैसे मुम्बई के अंडरवर्ल्ड वाले। उन्हें भीतर से कष्ट हुआ। फिर उन्होंने सोचा आख़िर ऐसा कौन-सा काम है जिसके लिए वह हाथ धोकर पीछे पड़ा है। सो तुरन्त सबने उस काग़ज़ की छानबीन शुरू कर दी। लाल ने स्पष्ट लिखा था कि पूर्व में उस विद्यालय में प्रबन्धक के एक आदमी को चतुर्थवर्गीय पद पर नियुक्त किया गया था, सो उनके भी एक आदमी को नियुक्त किया जाये। लम्बे अरसे से प्रबन्धन समिति में आपसी मनमुटाव

चल रहा था। यानी मूल प्रबन्धक के गुजर जाने के बाद उनके बेटों में खींचतान चल रही थी। लाल की दहशत से छोटे बाबू को कै होने लगी और बड़े बाबू लोटे में पानी लेकर मैदान की ओर भागे। शिक्षकगण अपनी-अपनी कक्षा में पढ़ा रहे थे। थोड़ी देर में संस्कृत के गुरुजी ने प्रधानाध्यापक के कमरे में हाजिरी बनाने के लिए प्रवेश किया, "प्रणाम! महोदय, आज मेरा हलवाहा काम पर नहीं आया। मैंने पड़ोसी के बैल मँगनी लिये थे। आषाढ़ माह खरीफ की फसल बोने का उपयुक्त मौसम होता है। तेरह कातिक तीन आषाढ़। घर पर समय का सदुपयोग करने के कारण विद्यालय आने में विलम्ब हुआ। इसके लिए क्षमाप्रार्थी हूँ।"

मगर श्रीमन्त जी ने कुछ नहीं कहा। संस्कृत गुरुजी को न टोकने पर आश्चर्य हुआ। उन्हें याद आया कि पिछली बार प्रधानाध्यापक के टोकने पर उन्होंने कह दिया था, "महोदय, उम्र की इस सन्ध्या बेला में मैं समय पर क्या आ पाऊँगा, जब मेरे परलोक जाने का समय आ गया है? अब मुझे पढ़ाने की कम, चिता के लिए चार बोझ लकड़ी की चिन्ता ज़्यादा है।" और यह सुनकर प्रधानाध्यापक का गुस्सा काफूर हो गया था। मगर आज प्रधानाध्यापक खुद उदास लग रहे थे, इसलिए गुरुजी ने प्रसंग आगे नहीं बढ़ाया। चुपचाप हाजिरी बही में हस्ताक्षर करके वह अपनी कक्षा के लिए चल दिए।

मगर उधर सूबेदार सिंह और उनकी मंडली के लोग खुश थे। सूबेदार सिंह चाहते थे कि प्रधानाध्यापक बन जाएँ, मगर वरीयता में वह श्रीमन्त जी के बाद थे। लाल गुट भी चाहता था कि सूबेदार सिंह प्रधानाध्यापक बनें जिससे जो कुछ वे चाहें, हो जायें। सूबेदार सिंह स्थानीय थे और श्रीमन्त जी दूरदराज के थे। सो सूबेदार सिंह गाँववालों के कान भी भरते थे जिससे गाँव में गन्दी राजनीति फैल गयी थी। श्रीमन्त जी के बेदाग आचरण के बावजूद कुछ लोगों के मन में शंका पैदा करने से सूबेदार सिंह सफल हो गये थे।

उस दिन फाइल प्रधानाध्यापक तक पहुँच गयी। श्रीमन्तजी ने वह आवेदन कई बार पढ़ा और एक-एक अक्षर चुभलाते रहे। मगर वे कुछ भी निर्णय लेने में समर्थ नहीं थे। एक तरफ उनका सिद्धान्त और नैतिकता थी, दूसरी ओर एक बिगड़े शिष्य का जबर्दस्त दबाव। उनका सिर चक्कर काटने लगा जैसे कीचड़ में फँसी गाड़ी का पहिया चक्कर काटता है। शाम को वे अपने घर चले गये। पत्नी ने यह सब सुना, तो अपशकुन की कल्पना करने लगीं। उन्होंने सलाह दी कि वह लम्बी छुट्टी पर चले जाएँ। मगर उन्होंने यह प्रस्ताव स्वीकार नहीं किया क्योंकि जुलाई माह भर्ती का समय होता है और प्रधानाध्यापक के न रहने पर ऊपर तक शिकायत होगी और तिल का ताड़ बना दिया जाएगा। उन्हें सेवानिवृत्त होने में तीन वर्ष का समय बाकी था। वे स्वैच्छिक सेवानिवृत्ति नहीं ले सकते

थे क्योंकि घर की आर्थिक स्थिति सुदृढ़ नहीं थी और छोटी बेटी की शादी करनी थी जो लग्गी की तरह बढ़कर सयानी हो गयी थी। पत्नी अक्सर उन्हें याद दिलाती थीं कि बेटी के हाथ पीले करने हैं। मगर दहेज-दानव की विकराल महामारी के कारण कहीं जुगाड़ नहीं बैठ रहा था। बेटे को एम. एस-सी. करने के बावजूद चौथी श्रेणी की भी नौकरी हासिल नहीं हुई थी। बाप-दादा ने जो खेत-बधार बनाये थे, उनमें पैदावार बढ़ने के बजाय घट रही थी क्योंकि सिंचाई का एकमात्र स्रोत सरकारी ट्यूबवेल अक्सर खराब रहता था और उस पर सरकारी ऑपरेटर, स्थानीय 'थोकदार' और शराबी-कबाबी लड़कों का प्रभुत्व था। सारी पारिवारिक बातें सोचते रहने के कारण रात में उन्हें नींद नहीं आयी। उन्होंने सुबह दिशा-फराकत के बाद दातून करके स्नान-ध्यान करके मुँह में दो दाने खर छुड़ाने के लिए डाल लिए और साइकिल से स्कूल के लिए रवाना हो गये। यद्यपि वे प्रधानाध्यापक थे मगर मोटरसाइकिल या स्कूटर खरीदने की व्यवस्था नहीं कर पाये थे।

कार्यालय में पहुँचने पर जब उन्होंने फाइल की ओर निहारा, तो उन्हें लगा कि फाइल उन्हें अपने जबड़े में भर लेना चाहती है। उन्होंने अपने दोनों पैर ज़मीन पर पटके, हाथ झटक कर सीधा किये और खंखार कर अपना अस्तित्व जताया। फिर फाइल खोलकर पढ़ने लगे। छोटे बाबू, बड़े बाबू और उपप्रधानाध्यापक ने कुछ भी लिखने की हिम्मत नहीं की थी। सबने सिर्फ़ एक ही शब्द लिखा था, "आदेशार्थ।" श्रीमन्त जी ने तीन बार वह आवेदन पढ़ा। फिर तीनों की टिप्पणियाँ पढ़ीं और तब काली स्याही से लिखा, "आवेदन अस्वीकृत। यह विद्यालय नियम से चलता है और नियम से ही चलता रहेगा। किसी की धमकी से नियम नहीं तोड़ा जाएगा।"

जैसे ही फ़ाइल कार्यालय में पहुँची, सारा स्टाफ भौंचक रह गया। मिनटों में यह बात पूरे विद्यालय में आग की तरह फैल गयी। ख़बर का धुआँ लाल तक पहुँच गया। वह हवा से बातें करता हुआ अपनी मोटरसाइकिल लेकर स्कूल पहुँच गया। प्रधानाध्यापक के कमरे में वह घुसा। उसके एक हाथ में गंड़ासा और दूसरे में पिस्तौल थी। दोनों हथियार प्रधानाध्यापक की ओर लहराते हुए उसने कहा, "तो आज आपने मुझे चुनौती दे ही दी। मैंने पिछली बार आपको एक मौका दिया था अपना गुरु होने के नाते। मगर आपकी बूढ़ी खोपड़ी ढीली हो गयी लगती है। बोलो, बजरंग बली की जय!" और पिस्तौल की तीन गोलियों से लाल ने अपने पुराने गुरु को भून दिया मानो गुरु-ऋण चुकाने का यह नया सूत्र था।

तुरन्त हाहाकार मच गया। आख़िर तूफान आ गया था और सबसे मजबूत

माने जाने वाला बरगद उखड़कर टुकड़े-टुकड़े हो गया था, जैसे हम स्कूली बच्चे रफ पन्नों को चिन्दी-चिन्दी कर फेंक देते हैं और वे टुकड़े हवा में गायब हो जाते हैं। चारों ओर हहोर हो गया। शिक्षक-विद्यार्थी अपनी-अपनी जान बचाकर भागने लगे। मैं भी सिर पर पाँव रखकर भागा। सिर्फ़ एक शिक्षक—हिन्दी वाले, हिम्मत जुटाकर आगे बढ़े, मगर तब तक श्रीमन्त जी का काम तमाम हो चुका था। उनके समझाने-बुझाने के पहले ही श्रीमन्त जी लाश में तब्दील हो गये थे। लाल की आँखों से अंगारे निकल रहे थे और वह उनका खून सचमुच चूसना चाह रहा था। वह लाश को घिसियाते हुए सड़क पर ले गया और जोर से माँ की गाली देते हुए गंड़ासे से उनका सिर धड़ से अलग कर दिया। फिर खोपड़ी को एक डंडे में घुसेड़कर नचाने लगा जैसे मेले में खिलौने बेचने वाले डुगडुगिया खिलौना नचाते हैं। लगता था कोई आनन्दमार्गी शव-साधना कर रहा है या शंकर तांडव-नृत्य कर रहे हैं। पर सोचने-बिसूरने की किसे फुर्सत थी। दूर-दूर सिर्फ़ आँखें-ही-आँखें। ये पत्थर की आँखें थीं जिनमें रोशनी नहीं थी। ये तांडव-नृत्य के पास की भीड़ थी। ये इन्सान नहीं रह गये थे।

अब तक हम लोगों ने टी.वी. पर ही ऐसे दृश्य देखे थे। हम सबकी चुप्पी पर श्रीमन्त जी की लाश उछल-उछल कर हमें धिक्कार रही थी कि तीस सालों तक की गयी उनकी निःस्वार्थ सेवा के लिए उनके सहयोगी, स्टाफ और शिष्य तनिक भी काम नहीं आये।

मगर यह हादसा था। सिर्फ़ विद्यालय तक सीमित नहीं रह सकता था। देश में कानून मौजूद था और कानून के रखवाले इस्त्री की हुई वर्दी में तैनात थे। सो हिन्दी शिक्षक के बयान पर थाने ने पहली रिपोर्ट दर्ज की। अख़बारों के पहले पृष्ठ पर इस नृशंस हत्या की ख़बर छपी। हल्ला हुआ कि लाल जेल की चक्की पीसते-पीसते बूढ़ा हो जाएगा। कानून के हाथ आदमी के हाथ से ज़्यादा लम्बे होते हैं। दो दिनों में ही लाल गिरफ़्तार हो गया। लोगों का कानून में विश्वास दुबारा हुआ। पुलिस अधिकारी ने जाँच-पड़ताल शुरू की। मगर हिन्दी शिक्षक के अलावा कोई शिक्षक गवाही देने के लिए तैयार नहीं हुआ। स्टाफ पर लाल का खौफ पहले से ही था। हम विद्यार्थीगण ने अपने अभिभावकों के दबाव के कारण गवाही नहीं दी। महीने भर पुलिस तफतीस करती रही। जिला शिक्षक संघ ने विरोध जुलूस निकाला और पर्चे बाँटे। जिलाधिकारी और पुलिस कप्तान को ज्ञापन सौंपे गये। पुलिस कप्तान ने स्पष्ट तौर पर कह दिया, "आख़िर वहाँ के शिक्षक भी तो आपके संघ के सदस्य हैं। जब वे आँखों देखी घटना में बयान नहीं दे रहे हैं, तो हम सुबूत कहाँ से जुटाएँ? अँग्रेजी राज तो है नहीं। अब न्यायपालिका ठोस सुबूत खोजती है, वरना पुलिस केस को झूठा करार देकर

मुलजिम को बरी कर दिया जाता है। हाँ, मैं इतना ज़रूर कर दूँगा कि वहाँ पुलिस के जवान तैनात रहेंगे जिससे भविष्य में ऐसी घटना की पुनरावृत्ति न हो।''

और शिक्षक संघ के प्रतिनिधियों ने इस पर मुँह सी लिए। बड़ी-बड़ी सभाओं में लम्बे-चौड़े भाषण देने वालों के पास कोई उपाय नहीं था। कुछ महीनों के बाद पुलिस ने रिपोर्ट सौंप दी, ''मामला सत्य मगर सूत्रहीन।'' न्यायालय ने भी अपनी प्रगति सूची में एक और मामले का निष्पादन दर्ज करने के लिए इस मामले को बन्द कर दिया और लाल जेल से छूट गया। लोगों का कानून पर से पुनः विश्वास उठ गया।

तब से गाँव-जवार में लाल घूम रहा है। वह छुट्टा साँड़ बन गया है। उसने काफी प्रगति की है। अब उसकी एक अच्छी-खासी टीम है, जिसमें अचूक निशानेबाज, हिप्पीबाज, दगाबाज, छुरेबाज, बलात्कारी और छलात्कारी शामिल हैं। गाँव-जवार के लोग उस हिन्दी शिक्षक को कोस रहे हैं, जिसने आख़िरी घड़ी में किसी दबाव के कारण अपनी गवाही बदल दी थी। वे बाकी शिक्षकों, स्टाफ और हम विद्यार्थियों को भी कोस रहे हैं जिन्होंने गवाही देने की हिम्मत नहीं जुटाई। वे लोग दबे मुँह कहते फिरते हैं कि यदि वे स्कूल में उस समय मौजूद होते, तो लाल के खिलाफ़ खुलकर सामने आते और पुलिस और न्यायालय में गवाही देते। यदि वे पुलिस महकमे में होते, तो लाल को किसी-न-किसी तरह जेल में ठूँसे रहते। यदि वे न्यायाधीश होते, तो लाल को किसी सूरत में बरी नहीं करते। ये बातें रोज सुबह-दोपहर-शाम सुनी जाती हैं। जिला पुलिस के सौजन्य से विद्यालय परिसर के बाहर पुलिस चौकी खुल गयी है और वर्दियाँ कुर्सी तोड़ रही हैं, विद्यालय भी चल रहा है। शुरू में कोई भी शिक्षक डर से प्रभारी प्रधानाध्यापक बनने को तैयार नहीं हुआ। सूबेदार सिंह कुछ दिनों के लिए छुट्टी पर चले गये। मगर धीरे-धीरे पद और प्रतिष्ठा के लालच ने उनका यह डर खत्म कर दिया। विद्यार्थी, शिक्षक और स्टाफ रोज़ाना सुबह स्कूल आते हैं। दिन में पढ़ाई-लिखाई के नाम पर कुछ-कुछ करते-धरते हैं और शाम चार बजे फुर्र हो जाते हैं। विद्यालय में स्थापित श्रीमन्त जी की खुली मूर्ति पर कौवे बैठकर काँव-काँव करते हैं मगर उससे किसी को कोई फ़र्क़ नहीं पड़ता। आख़िर विद्यालय की गाड़ी चल रही है न? भले फूल-पौधे मुरझाये हों। भले स्कूल सहमा हो। मगर हम विद्यार्थियों को श्रीमन्त जी की कमी खलती है। सूबेदार सिंह उनके पासंग के बराबर भी नहीं हैं। वे कभी कक्षा में पढ़ाने नहीं आते। वे स्कूल कभी-कभी आते हैं, अक्सर जिला मुख्यालय में सैर करते हैं। वहीं से बड़े बाबू को सन्देश भेजते रहते हैं और वहीं फाइलें मँगाकर निष्पादित करते रहते हैं।

गाँव-ज्वार के लोग तरह-तरह की बातें करते हैं। आख़िर उनके पास बचा

ही क्या है सिवाय बातें करने के। कुछ का कहना है कि यदि श्रीमन्त जी अकलियत के होते, तो प्रतिपक्ष काफी सक्रिय होता। दूसरों का कहना है कि यदि वह दलित होते तो सरकार कुछ-न-कुछ ज़रूर करती। तीसरी श्रेणी के लोगों का कहना है कि यदि वे अमीर होते तो मीडिया कुछ करता। मगर श्रीमन्त जी न सरकार के ढाँचे में फिट हुए, न प्रतिपक्ष के साँचे में और न मीडिया के खाँचे में। इसीलिए बाकी लोग अपनी राय देने के पहले सामने वाले को घूरकर पहचानने की कोशिश करते हैं, फिर अगल-बगल झाँकते हैं और आख़िर में खँखार-खँखार कर यह कहते से लगते हैं—वह जानते थे कि आजकल सच कहना गुनाह है, फिर भी उन्होंने गुनाह किया। वह जानते थे कि बेईमानी का विरोध करना मौत को न्यौता देना है, फिर भी उन्होंने विरोध किया। विद्यालय की धरती पर गिरे खून के कतरे अभी भी राहगीरों से कहते-से लगते हैं कि स्वाभिमानी श्रीमन्त जी मरे नहीं हैं। उनके निशान विद्यालय के कण-कण में रचे-बसे हैं।

अभयदान

कुछ साल पहले ग्राम प्रधान चुने गये थे हीरा सिंह। वही मुबारकपुर वाले। आसानी से जीत गये थे हीरा सिंह। न किसी पार्टी का झंझट, न चुनाव आयोग की जाँच-पड़ताल। फिर मुबाकरपुर में दबदबा था। उनकी जाति का। और चुनाव होता ही है संख्या बल पर। सो धनबल, बाहुबल और संख्याबल के आधार पर अच्छे-खासे बहुमत से जीत गये थे हीरा सिंह। फिर केन्द्र सरकार और राज्य सरकार की बहुत सारी योजनाएँ आयीं। सबका पैसा मिला ग्राम पंचायत को। हीरा सिंह की पढ़ाई काम आयी। वे पुराने ज़माने के इंटर पास हैं। इसलिए बीच-बीच में अँग्रेजी शब्दों का तड़का लगाने से नहीं चूकते। उनकी बोली से युवा वर्ग काफ़ी प्रभावित रहता है। इसलिए जब वे बोलने लगते हैं तो लोग सुनने के लिए जुट जाते हैं। सुनते कम, मनोरंजन ज़्यादा पाते हैं लोग। उनके साथ जो पंचायत सचिव तैनात था, वह शुरू में नियम कानून बघारता था मगर हीरा सिंह ने ऐसी घुड़की लगायी कि हमेशा के लिए सचिव ठीक हो गया। उसे वे अपने घर में एक कोठरी रहने के लिए दे दिये थे। उसे सरकारी काम-काज करने में ज़्यादा कष्ट न हो, इस लिहाज से वे उसका सारा काग़ज़ात खुद तैयार कर देते थे। उसे सिर्फ़ दस्तख़त करनी पड़ती थी। और दस्तख़त करने के नाम पर उसे वे हर माह दो हजार रुपये बख़्शीश के तौर पर देते थे। पंचायत सचिव उनसे इसलिए डरता था कि जवाहर रोजगार योजना, स्वर्ण जयन्ती रोजगार योजना वगैरह की जो राशि खर्च की जा रही थी वह सरज़मीन पर दिखायी नहीं पड़ती थी। मगर ग्राम प्रधान को इसकी फिक्र नहीं थी। एक दिन पंचायत सचिव अपनी पत्नी को बताने लगा, ''प्रधान जी योजना का अधिक-से-अधिक तीस प्रतिशत काम पर खर्च करते हैं, दो प्रतिशत अपने चेलों-चमचों पर खर्च करते हैं। दस प्रतिशत ब्लॉक के अधिकारियों के बीच नगद या उपहार के रूप में खर्च करते हैं। उनके हाथ की सफाई का बड़े-बड़े जादूगर भी लोहा मानते हैं। समूचे गाँव-जवार, आरी-पवस्त में यह बात फैल गयी है कि ग्राम समाजी तालाब के जीर्णोद्धार के नाम पर उन्होंने तालाब का तीन-

चौथाई हिस्सा यानी तीन बीघा जमीन अपने बाग में मिला ली और बाकी एक-चौथाई में खूब गहरी खाई खोद दी जिससे दूसरे किनारे के ग्रामवासी उसे न पाट सकें। वह 'सरकारी सम्पत्ति, अपनी सम्पत्ति' के हिमायती हैं। फिर वहाँ सात कमरोंवाला एक आलीशान मकान बनवा लिया। दरअसल ग्राम प्रधान का पुश्तैनी मकान गाँव के भीतर एक सँकरी गली के बाद स्थित है। सो अब सारी मजलिस नये मकान पर जुटती है।'' उसकी पत्नी ने आशंका भाँपकर तबादला कराने का सुझाव दिया। इस पर पंचायत सचिव ने निराश होकर कहा, ''इतना आसान नहीं है। उसके लिए भी पौवा चाहिए।'' उसने अपनी पत्नी को बताया कि हीरा सिंह ने अपने कार्यों को सरकारी विकास योजनाओं तक ही सीमित नहीं रखा है। राजस्व मामलों में भी उसकी अच्छी खासी पकड़ है। उसने ग्राम समाज की करीब पचास बीघा ज़मीन पट्टे पर दे दी है—जिसने ज़्यादा पैसा दिया उसे ज़्यादा और मौके की ज़मीन मिली। जिसने कम पैसा दिया, उसे थोड़ी दूरी पर ऊबड़-खाबड़ ज़मीन मिली। इतने से भी हीरा सिंह की सोहरत पूरी नहीं हुई। सो उसने गाँव-जवार में झगड़ा-झंझट होने पर सुलह कराने के लिए पैसा और कीमती चीज़ें लेनी शुरू कर दीं। इस हाथ पैसे दे, उस हाथ न्याय ले। मगर इससे भी उसे सन्तुष्टि नहीं मिली, तो उसने पुलिस थाने की दलाली करनी शुरू कर दी। यदि सुलह सपाटे की कोशिश से कोई मामला नहीं सुलटता, तो वह जिस पार्टी से ज़्यादा पैसा मिलता उसे लेकर थाना पहुँच जाता और थाने में रुपये दिलवाकर केस दर्ज कराता और उससे खुद भी पैसा ऐंठ लेता। थाने से बख़्शीश अलग से मिलती। थाने की दलाली का एक फ़ायदा और था। गाँव वाले इस चीज़ से वाकिफ़ थे कि हीरा सिंह प्रधान की पहुँच थाने तक है, इसलिए सरकारी योजनाओं में सुरसा के मुँह की तरह फैले भ्रष्टाचार के खिलाफ़ कोई मुँह खोलने की हिम्मत नहीं करता था। जनता में व्याप्त डर की पुनः पुष्टि हो जाती जब हीरा सिंह छट्ठी, बरही, तेरही, तिलक, ब्रह्मभोज आदि अवसरों पर थाना और ब्लॉक के अधिकारियों को सादर बुलाकर आदर-सत्कार करता।

पिछली बार हीरा सिंह के यहाँ आयोजित पार्टी में थाना और ब्लॉक के अलावा वन विभाग का रेंजर भी शामिल हुआ था। पार्टी का लुत्फ उठाते हुए थानेदार ने हीरा सिंह से कहा, ''प्रधान जी! मेरे घर में फर्नीचर टूट गये हैं। कुछ उपाय हो सकता है क्या?''

''साहब! फर्नीचर कौन सोने की मुर्गी है?''

''तो भई, कब तक आशा करूँ?''

''साहब! इसके लिए रेंजर साहब की मेहरबानी चाहिए।'' प्रधान ने रेंजर साहब की ओर देखते हुए कहा।

फिर थानेदार और रेंजर ने गुफ़्तगू की और सारी योजना तैयार हो गयी। पास के जंगल से नीम-शीशम-सागौन आदि की लकड़ियाँ तेजी से कटने लगीं। थानेदार और रेंजर की कृपा से प्रधान ने चौराहे पर लकड़ी का एक बड़ा स्टॉल खोल दिया और कुछ कारीगरों को लगाकर फर्नीचर बनवाने लगे।

एक साल भी नहीं बीता था कि गठबन्धन की सरकार की खिचड़ी पकते-पकते जल गयी और विधान सभा का चुनाव घोषित हो गया। ग्राम प्रधान के नाते उसकी जान-पहचान अधिकांश दलों के स्थानीय नेताओं से थी मगर किसी पार्टी से गहरा लगाव न था। उसमें विचारधारा या सिद्धान्त का कोई आग्रह कभी न था। जैसे दक्षिणपंथी दल, वैसे वामपंथी और वैसे ही बीच पंथी। यानी गाय भी गाभिन, बैल भी गाभिन।

पिछली रात उन्होंने एक सपना देखा कि वे भी चुनाव के मैदान में हैं। मगर उन्हें यह याद नहीं कि वे किसी पार्टी के उम्मीदवार थे अथवा स्वतन्त्र उम्मीदवार थे। उन्होंने इस सपने की चर्चा अपने घर में की। उनकी माँ ने कहा, "ग्राम प्रधान से पेट नहीं भरता क्या?" परन्तु उनकी पत्नी ने ढाँढ़स बँधाया, "मकर संक्रान्ति हमारा शुभ त्यौहार है और इस त्यौहार की रात में देखा गया सपना भी शुभ ही होगा।" मगर उसके अस्सी वर्षीय पिता भी चुनाव के पक्ष में नहीं थे। वे चुनावी राजनीति को फालतू कुतिया मानते थे जिसका सुहाग नहीं होता। प्रधान के बड़े बेटे से नहीं रहा गया और वह बोल पड़ा, "आपका वश चलता, तो पिताजी ग्राम प्रधानी का भी चुनाव नहीं लड़ते। अर्जुन सिंह, नटवर सिंह, मुलायम सिंह जैसे दिग्गजों को भी चुनाव की वैतरणी में डुबकी लगानी पड़ती है। दुनिया कितनी तेज रफ़्तार से आगे बढ़ रही है। मुसई पिताजी के साथ स्कूल में पढ़ता था मगर उसने एक पार्टी की पूँछ पकड़ ली और पिछले विधान सभा चुनाव में शोषित दल से चुनाव जीत गया। आजकल क्षेत्र में उसकी सोहरत है...।"
"पिताजी, शंकर ठीक ही कह रहा है। हम सुबह टट्टी खुले मैदान में जाते हैं और आड़ पाकर बैठ जाते हैं। फिर कुछ देर बाद जैसे-जैसे टट्टी करते हैं, आगे बढ़ते जाते हैं। कोई पीछे घूम कर नहीं देखता। यदि देखो तो गन्दगी से कै हो जाये।" हीरा सिंह ने समझाते हुए कहा।

यह सुनते ही उसके माँ-बाप ने बगल में किच्च से थूक दिया मानो सचमुच वे मैदान में हों। फिर हीरा सिंह ने अपने चेलों से सलाह ली। सबने पूर्व की भाँति हाँ-में-हाँ मिला दिया। मगर हीरा सिंह पार्टी से टिकट पाने में आश्वस्त नहीं था। सो वह स्थानीय बी.डी.ओ. के पास पहुँचे। बी.डी.ओ. से उसका गुड़-चींटा का रिश्ता था। बी.डी.ओ. ने कहा, "जनता की सेवा के लिए सदन में जाना चाहिए। जब ज़िन्दगी ही एक जुआ है तो क्यों न इसे जम कर खेला जाये?"

आगे बात बढ़ी, तो इस बिन्दु पर अटक गयी कि किस पार्टी से चुनाव लड़ा जाये। बी.डी.ओ. ने साफ कह दिया कि पार्टी के आधार, पार्टी फंड, प्रचार वगैरह के आगे किसी नेता की नैया स्वतन्त्र उम्मीदवार के रूप में चुनावी नदी में पार नहीं लग सकती। बी.डी.ओ. साहब के एक रिश्तेदार लम्बे अरसे से राजधानी में रहकर राजनीति कर रहे थे मगर बी.डी.ओ. साहब को अभी याद नहीं आ रहा था कि वे अभी किस पार्टी में हैं। पिछली बार की मुलाकात के समय उन्होंने जनता पार्टी का नाम लिया था मगर दो सालों में जनता पार्टी छः पार्टियों की माँ बन चुकी थी। सो अगले दिन राजधानी जाने का कार्यक्रम बन गया।

जब बी.डी.ओ. साहब अपने रिश्तेदार नेताजी के यहाँ पहुँचे तो वे बहुत सारे कार्यकर्ताओं से घिरे थे। घर के बाहर लगे बैनर और झंडे से मालूम हुआ कि वे अभी भारतीय न्याय पार्टी में हैं। बी.डी.ओ. ने नाश्ता-पानी करने के बाद हीरा सिंह की समस्या उनके सामने रख दी। नेताजी ने छूटते ही कहा, ''देखिए, वे ठाकुर हैं, हमारी पार्टी ब्राह्मण और ठाकुर को टिकट नहीं दे सकती। हमारी लड़ाई ऊँची जातियों से है। हम पिछड़ी जातियों को गोलबन्द कर रहे हैं।''

''ये पार्टी को चन्दा देना चाहते हैं।''

''क्या फ़र्क़ पड़ता है? हमारे यहाँ पिछड़े वर्गों में बहुत सारे कुर्मी-यादव उम्मीदवार हैं जो बीस-पच्चीस लाख रुपये देकर टिकट ले रहे हैं। सो पैसा टिकट देने की कोई क्राइटेरिया नहीं है। हमारे यहाँ जाति ही टिकट बाँटने का आधार है और ये तो सवर्ण हैं। हमारी लड़ाई तो सवर्णों से ही है। बहुत दिन तक राज किया इन लोगों ने सामन्ती ताकत के बल पर। अब लोहिया के सपनों को साकार करना है। जनतन्त्र को सही अर्थ देना है।'' नेताजी के दोटूक कथन ने उनका सारा गुड़ गोबर कर दिया।

वे दोनों वहाँ से निराश होकर लौटे और सोचने लगे कि पता नहीं, सुबह-सुबह किसका मुँह देखकर चले थे। तभी रास्ते में बी.डी.ओ. को एक नयी बात सूझी। उनका एक सहपाठी आजकल शोषित दल में है, क्यों न उसकी टोह ली जाये?

सो वे मेंहदी राय, प्रदेश महासचिव शोषित दल के पास पहुँचे। फिर बिना किसी लाग-लपेट के मेंहदी राय ने पूछा, ''टिकट चाहिए क्या?''

''हाँ-हाँ''—बी.डी.ओ. और हीरा सिंह को लगा कि बिल्ली के भाग्य से सिकहर टूट रहा है।

''मगर टिकट मुफ़्त में नहीं मिलता''—नेताजी ने चुटकी ली।

''मालूम है कि पार्टी के तमाम खर्चे हैं। सो हीरा सिंह जी अपनी शक्ति

भर चन्दा देंगे।''—बी.डी.ओ. ने कहा।

''ठीक है, शाम को सात बजे आइये। हाई कमान के फैसले का समय वही है।''

सो वे दोनों बाहर चले गये। जिस जलपान गृह पर वे पहुँचे, वहाँ दुखीराम दिख गये। बी.डी.ओ. और हीरा सिंह ने उनसे नमस्कार किया और बताया कि वे दोनों एक शादी में शरीक होने राजधानी आये हैं। फिर दुखीराम ने कहा, ''बी.डी.ओ. साहब! इस चुनाव में मेरी मदद कीजिएगा।''

''ज़रूर, ज़रूर!''—बी.डी.ओ. ने कहा।

जब दुखीराम वहाँ से चले गये तो बी. डी. ओ. ने कहा, ''प्रधान जी! यह दुखिया बहुत चालू है। उसने मेरी पोस्टिंग के लिए ग्रामीण विकास मन्त्री को खत लिखने के लिए मुझसे एक लाख रुपये ले लिया था। और मजबूरी में मुझे देना पड़ा क्योंकि ग्रामीण विकास विभाग ने यह नियम लागू कर दिया है कि जिस ब्लॉक में किसी बी.डी.ओ. को पदस्थापित करना है, वहाँ के स्थानीय माननीय विधायक की सहमति और अनुशंसा पत्र ज़रूरी है। इस माननीय विधायक ने अमानवीय सुविधा-शुल्क के बिना अपना पैड नहीं खोला। मैंने इसे कई बार चन्दा दिया है। इस चुनाव में इसे चन्दा तो क्या, फूटी कौड़ी भी नहीं दूँगा। भगवान ने चाहा तो इसका पत्ता कटवाकर आपको दिलाऊँगा।''

वे दोनों पौने सात बजे ही शोषित दल के दफ़्तर में पहुँच गये। करीब साढ़े तीन घंटे इन्तज़ार के बाद हीरा सिंह का नम्बर आया। वे अकेले जाने में सकपका रहे थे, मगर अकेले ही जाने का हुक्म था। अन्दर सिंहासन पर आसीन मूर्ति की ओर हाथ जोड़कर हीरा सिंह ने कहा, ''जी नमस्ते।''

''तुझे बोलने भी नहीं आता क्या? फिर यह शहर तहजीब के लिए जाना जाता है।''

''जी गलती हो गयी।''

''भविष्य में हुज़ूर, सरकार या मैडम कहना, समझे? मुझको अनपढ़-गंवार समझते हो?''

''जी, समझ गया मैडम।''

''उसके बाद पाँव छूकर, ज़मीन पर बिछी दरी पर कायदे से बैठ जाना। पान या खैनी खाकर मत आना, समझे?''

इतना सुनते ही हीरा सिंह ने मैडम के चरणों पर माथा टेक दिया जैसा वह महावीर मन्दिर में प्राय: करते थे। फिर दरी पर सिकुड़कर उकड़ूँ बैठ गये।

''बोल, तेरा क्या काम है?''

''जी मैडम, आपकी मेहरबांनी हो जाती इस गरीब पर...''

"जब तू भिखारी है तो चुनाव कैसे लड़ेगा? चुनाव में तो पैसा पानी की तरह बहता है। फिर और बहुत कुछ चाहिए..."

"जो हो सकेगा, मैं दूँगा मैडम!"

"देख मेरे पास ज़्यादा टाइम नहीं है। मोहनपुर विधानसभा क्षेत्र से पहले से दुखीराम मेरी पार्टी का विधायक है। तू जानता है कि वह शोषित वर्ग का है और तू ठाकुर है, यानी शोषक वर्ग का है। ठाकुर यानी पृथ्वीराज चौहान के वंशज! यदि वे शोषितों को साथ लेते, तो मुहम्मद गौरी से नहीं हारते। लेकिन ज़माने ने अब पलटा खाया है। आज उनका वंशज मेरे पास गिड़गिड़ा रहा है। तिलक, तराजू और तलवार हमारे पुश्तैनी दुश्मन हैं। अपने शोषित दल में किसी शोषक को शामिल करके मैं पार्टी की छवि धूमिल होने का जोखिम उठाऊँगी। सो इसकी कीमत तुझे चुकानी होगी। अगर तुझे मंजूर हो तो ठीक, वरना अपना रास्ता नाप। मेरे यहाँ टिकटार्थियों की लाइन लगी है। जल्दी फैसला कर। अभी तू बिल्कुल नया है। धीरे-धीरे सब सीख जाएगा।"

"मैडम, पाँच लाख दे दूँगा।"

"देख, यह कोई बनिया की दुकान नहीं है। यहाँ सौदेबाजी नहीं होती। आलाकमान का हुक्म आख़िरी होता है। तू जानता है कि दूसरी पार्टियों का रेट और महँगा है। मोहनपुर क्षेत्र में चालीस फीसद वोट शोषित वर्ग के हैं, चालीस फीसद ठाकुरों के हैं और बाक़ी बीस फीसद मिले-जुले हैं। यदि तुम दूसरी पार्टी से टिकट लेने में सफल भी हो जाओगे, तो हमारे चालीस फीसद वोटर तुम्हें जमकर टक्कर देंगे। पिछली बार हमारी पार्टी का दुखीराम वहाँ से जीता था क्योंकि सवर्णों के वोट बँट गये थे। इस बार जटिल समीकरण है। सो तू पन्द्रह लाख दे-दे।"

"मैडम! अभी तो पाँच लाख ही हैं..."

"देख, ये पाँच लाख मुझे देता जा। कल शाम तक बाकी दस लाख लेकर आना और टिकट ले जाना। और कहीं चूँ-चपड़ मत करना। पार्टी को देवता मानना, समझे?"

हीरा सिंह ने विनम्रता की चादर ओढ़कर वैसा ही करने का वचन दिया और पत्र-पुष्प चढ़ाकर माथा टेककर चला गया। दूसरे दिन हीरा सिंह के पहुँचने के पहले दुखीराम ने आलाकमान के चरणों में माथा टेका।

"बोल, फिर क्यों आया है"—आलाकमान ने पूछा।

"मैडम! इस चुनाव में पार्टी का टिकट मिल जाता, तो बहुत मेहरबानी होती।"

"दुखीराम! तुमको पार्टी ने एक बार मौका दिया। तुम नहीं, पार्टी जीती थी पिछले चुनाव में। पिछले पाँच साल तुम सिर्फ़ पैसा कमाने में लगे रहे, पार्टी

की तूने तनिक भी परवाह नहीं की। दूसरी बात यह कि शोषित दल सिर्फ़ शोषितों के बल पर सत्ता में नहीं आ सकता। जन जागरण के लिए सवर्णों और पिछड़ों का विरोध करना हमारी रणनीति थी। मगर आज हम कुछ कदम आगे बढ़ गये हैं। तूने इतना कमा लिया है कि तीन पीढ़ियाँ ऐश करेंगी। अब तू दूसरों को भी मौका दे। आख़िर जनतन्त्र है न? जा, अपने इलाके में कुछ समाजसेवा कर। अगली बार तुझे मौका मिल सकता है।''

यह सुनकर दुखीराम फूट-फूटकर रोने लगे। उनकी काली कमाई आँसू बनकर निकलने लगी। मगर मैडम के सामने तर्क-वितर्क करने की उनकी हिम्मत नहीं थी। सो वे दरी पर से उठकर दण्डवत् प्रणाम करते हुए कमरे से बाहर चले गये।

इस बीच हीरा सिंह और बी.डी.ओ. ने दस लाख रुपये की व्यवस्था कर ली थी। किसी को ज़मीन का पट्टा देने का वादा, तो किसी को ठेका देने का। पंचायत कोष का पैसा तो उसका अपना पैसा था ही। वे लोग रात में दस बजे आलाकमान के पास पहुँचे। हीरा सिंह ने रुपयों से भरा ब्रीफकेस मैडम को देते हुए दण्डवत् प्रणाम किया। मैडम ने नोटों को देखा तक नहीं। हीरा सिंह को लगा कि काश वह दो लाख कम देते। ब्लैक मनी गिनी थोड़ी जाती है। मैडम ने पूरी ईमानदारी से दूसरे हाथ से उन्हें टिकट थमा दिया। हीरा सिंह धन्य हो उठे। मानो उनका दूसरा जन्म होने वाला हो।

राजधानी से मुबारकपुर की यात्रा बातों-ही-बातों में ऐसे बीत गयी, मानो सपना हो। घर पहुँचते ही हीरा सिंह ने अपने माँ-बाप का चरण-स्पर्श किया। घर के बाकी सभी लोगों ने बारी-बारी से हीरा सिंह का चरण-स्पर्श किया। उसके बड़े बेटे ने चौराहे से फौरन फूलों का गजरा लाकर पहनाया और पूरे गाँव के सभी टोलों में उन्हें घुमाया।

उम्मीदवारी का पर्चा दाखिल करने के बाद प्रचार शुरू हो गया। इलाके के ठेकेदार, दुकानदार, व्यापारी, समाजसेवी सभी ने उसे चन्दा दिया। चुनाव आयोग के द्वारा निर्धारित खर्च की सीमा के बाहर पच्चीस लाख रुपये खर्च हुए। थानेदार ने अपने पाले-पोसे गुंडों का इस पावन-कर्म में जमकर उपयोग किया। शोषित वर्ग के छुटभैया नेताओं की हीरा सिंह ने शराब, अनाज, कपड़े और नगद रुपये से भरपूर सेवा की। बी.डी.ओ. की सलाह के मुताबिक उसने दुखीराम को भी खुश किया जिससे वह वोट न कटवाये। आख़िर वोटकटवा मुड़कटवा से कम ख़तरनाक नहीं होता। उसे प्रचार सभाओं में भी वह ले गया। जो दूसरी पार्टी के ठाकुर कार्यकर्त्ता थे, वे अपने आदमी को शोषित दल में पाकर एकजुट हो गये और पार्टी राजनीति के मतभेदों को भुलाकर हीरासिंह के लिए जी-जान से

काम करने लगे। कार्यकर्त्ताओं के खाने-पीने का अव्वल इन्तज़ाम हीरासिंह का बड़ा लड़का बखूबी कर रहा था और इस मौके पर भरपूर ट्रेनिंग लेकर भविष्य की सीढ़ियाँ चढ़ रहा था।

और हीरा सिंह अन्ततः चुनाव जीत गया। चारों ओर जुलूस निकाला गया। ढेर सारी मालाओं से गरुवाते हीरासिंह ने सभी लोगों का आभार जताया। बी.डी.ओ., रेंजर, थानेदार वगैरह को घर पर बुलाकर उसने आदर-सत्कार किया।

इस चुनाव में शोषित दल को बहुमत मिल गया। सो बी.डी.ओ. ने विधायक जी को चुपके से सलाह दी, ''मेरी मानिए मैडम से मिलिए।''

''किसलिए बी.डी.ओ. साहब?'' हीरासिंह ने अब बी.डी.ओ. साहब कहना शुरू किया क्योंकि पहले की तरह वे प्रधान नहीं थे कि साहब या सर कहते। ऊँट ने करवट बदल ली थी।

''ऐसा है, जब खुदा देता है तो छप्पर फाड़कर देता है। मैं तो ज्योतिष में विश्वास करता हूँ। यह ज्योति देने वाला विज्ञान है। अभी आपकी राशि का लग्नेश सूर्य है और उच्च स्थान पर है। कोई भी सोचा-विचारा काम हो जाएगा। सो एक बार आप मन्त्री पद के लिए प्रयास करें। कर्म प्रधान विश्व करि राखा।''

''इतना आसान नहीं है मन्त्री पद। करीब दो-तिहाई पार्टी विधायक मुझसे वरीय हैं। मैं तो एकदम नौसिखिया हूँ प्रदेश स्तरीय राजनीति के लिए। फिर मन्त्री पद के लिए मैं बहुत काबिल भी नहीं''—हीरासिंह ने शंका जतायी।

''अधिकांश लोग पहली बार मन्त्री बनते हैं। मन्त्री बनने के लिए काबिलियत नहीं आँकी जाती। फिर शोषित दल पहली बार सरकार बना रहा है। इस दल में ठाकुर बहुत कम हैं। जनता को यह दिखाने के लिए कि सत्ता में सभी जातियों, वर्गों-धर्मों की हिस्सेदारी है, आलाकमान आपको कैबिनेट नहीं तो राज्यमन्त्री का पद दे सकता है। इससे दूसरे दलों के ठाकुर नेताओं का कद भी छोटा हो सकता है। राजनीति में जोड़-घटाव, गुणा-भाग कम्प्यूटर से भी तेज गति से हो सकता है। गणित का असली इस्तेमाल विज्ञान में नहीं, राजनीति में होता है। चलिए, कल राजधानी चलें। मन्त्री पद नहीं मिलेगा, तो आपको कोई नुकसान भी तो नहीं होगा। भिखारी को कोई भीख नहीं देगा तो उसकी 'तुमड़ी' तो नहीं फोड़ देगा।''

अगले दिन वे दोनों शोषित दल के दफ़्तर पहुँच गये। चारों ओर विजय मनाने का उत्साह था। कोई फूलों से, कोई मिठाइयों से, तो कोई शब्दों से स्वागत कर रहा था। जिन लोगों ने चुनाव में चन्दा नहीं दिया था या कम दिया था, वे अब अपना कोटा पूरा कर रहे थे। आलाकमान से मिलना मुश्किल था। जनता की भीड़, कार्यकर्त्ताओं की भीड़, जीते हुए विधायकों की भीड़ और दूसरे दलों को छोड़कर आने वाले नेताओं की भीड़। बहुत कोशिश करने पर रात में साढ़े

बारह बजे हीरा सिंह को आलाकमान से मिलने का मौका मिला। भीतर पहुँचते ही वे मैडम के चरणों पर गिर पड़े और रोने लगे।

"तू तो जीत गया हीरा सिंह! फिर क्यों रो रहा है?"

"मैडम आपकी मेहरबानी से। ये खुशी के आँसू हैं?"

"तो बोल, दुबारा क्यों आया है?"

"मैडम, इस बार भी कृपा कर देंती, तो मैं सरकार की और आपकी सदा सेवा करता रहता।"

"तू जानता है, मन्त्री पद का भार बहुत ज़्यादा होता है। तू पहली बार प्रदेश की राजनीति में चुनकर आया है। यह कोई ग्राम प्रधानी नहीं कि गाँव वालों को धकिया कर काम कर लें। यहाँ विपक्षी दलों के धुरन्धर नेता हैं, मीडिया की हज़ारों आँखें हैं, बहुत सारी पढ़ी-लिखी जनता है। सो मन्त्री का सपना मत देख।"

"मैडम, एक बार मौका देतीं, तो अच्छा होता। हमारे गरीब इलाके का प्रतिनिधित्व आजादी के पाँच दशकों के बाद भी सरकार में कभी नहीं हुआ।"

"सो तो मैं समझती हूँ। तेरे आने से सरकार में ठाकुरों का भी प्रतिनिधित्व हो जाएगा। मगर तू जानता है, पार्टी ने चुनाव में सारा पैसा खर्च कर दिया है। हमारा खजाना ही खाली हो गया है। अभी तू नया है। धीरे-धीरे समझ जाएगा।"

"मैडम मैं कुछ चन्दा भी दे दूँगा" हीरा सिंह ने ताव में कहा।

"तो ठीक है, पन्द्रह लाख ले आ। तेरा काम हो जाएगा।"

"मैडम अभी तो नहीं है। कल तक जुगाड़ करूँगा।"

"ठीक है पन्द्रह लाख देने पर ही तुझको शपथ ग्रहण समारोह में बुलाया जाएगा। और सुन, तुझे न्यास का राज्य मन्त्री बनाऊँगी...."

"लेकिन मैडम, उसमें तो कुछ कमाने की गुंजाइश भी नहीं रहेगी। जिस तालाब में पानी न हो, उसमें मछली क्या मारेंगे? दूसरा विभाग मिल जाता, तो कुछ ज़्यादा भी देने का जुगाड़ करता।"

"देख, हीरा सिंह! ज़्यादा होशियार और महत्त्वाकांक्षी न बन। तेरे जैसे पचासों लोग मन्त्री बनने के लिए सड़कों पर कदमताल कर रहे हैं। तू क्यों भूलता है कि मन्त्री की लाल बत्ती लगी गाड़ी के कारण तुम्हारा स्टेटस बढ़ेगा और तुम्हारा लकड़ी का धन्धा लोकल से ग्लोबल हो जाएगा। तू अब पड़ोसी देश से कीमती लकड़ियाँ बेरोकटोक लाना। तुझे वन विभाग और पुलिसवाले कुछ कहने की हिम्मत नहीं करेंगे। फिर यह भी सोचो कि तुम्हारा काम-काज ठीक रहा तो अगले मन्त्रिमंडल विस्तार में तुम कैबिनेट मन्त्री बन जाओगे। इतना ही नहीं, अगले चुनाव में टिकट मिलना असान हो जाएगा और यदि तुम अगला चुनाव हार गये, तो भी तुम भूतपूर्व मन्त्री कहलाओगे। और यदि जीत गये तो मन्त्री पद के प्रबल

दावेदार हो जाओगे। फिर मत भूलो कि लोग बालू से तेल निकाल लेते हैं। तुम्हें अकबर-बीरबल का किस्सा याद है न? बीरबल ने समुद्र की लहरें गिनने में भी पैसा कमाने का गुर दिखा दिया था। तुम भी न्यासों के विभिन्न दावेदारों को लड़ाकर अपना उल्लू सीधा कर सकते हो। अभी तू नया है, धीरे-धीरे समझ जाएगा।''

और हीरा सिंह की समझ में बहुत सारी बातें आ गयीं। उन्होंने दण्डवत् प्रणाम किया और बाहर निकल गये। इसके बाद बी.डी.ओ., रेंजर, थानेदार, ठेकेदारों, जुआरिओं और व्यापारियों से पूरी मदद पाकर अगले दिन वे पार्टी मुख्यालय में सूटकेस सहित हाज़िर हुए। मैडम उसे देखकर मुस्करायीं। मानो मुरझाई कली पानी मिलने से खिल गयी हो। मैडम ने पूरी ईमानदारी बरती—एक हाथ से सूटकेस ले लिया और दूसरे हाथ से राज्य मन्त्री का शपथ लेने का पत्र उसे दे दिया।

शपथ-ग्रहण समारोह में हीरा सिंह को देखने न केवल उनका परिवार आया, बल्कि बी.डी.ओ., थानेदार, रेंजर और उनके चेले-चपाटे भी आये। गजब का जलसा था। पूरा अम्बेडकर मैदान दुल्हन की तरह सजा था। धत् दुल्हन भी झख मारे ऐसे सिंगार पर। लगा कि सत्ता-परिवर्तन हो रहा है। आख़िर राजभवन के बन्द पिंजड़े से निकलकर खुले आसमान के नीचे सैकड़ों माइक, हजारों कुर्सियों और दर्जनों सामियानों में सैकड़ों तरह के फूल, दर्जनों किस्मों की इत्र अपनी सुगन्ध समूची राजधानी में बिखेर रही थी।

अगले दिन से माननीय राज्यमन्त्री हीरा सिंह जी का कारोबार शुरू हो गया। लाल बत्ती वाली गाड़ी, दो सशस्त्र पुलिस जवान सुरक्षा हेतु। बी.डी.ओ. साहब उनके निजी सचिव बन गये। उनके पीछे चेलों की गाड़ियों का कारवाँ। धीरे-धीरे लकड़ी का धन्धा गोरखधन्धा बन गया। महोगनी, शीशम, सागौन के पेड़ सड़कों से गायब हो गये और ठाकुर टिम्बर कम्पनी में रोजाना हजारों फर्नीचर बिकने लगे। पड़ोसी देश से सीधा सम्पर्क काफ़ी आसान हो गया। राज्यमन्त्री के आदमियों ने पुलिस और वन विभाग के अधिकारियों और कर्मचारियों को खुश कर दिया। सब कुछ ठीक चल रहा था। राजा राज कर रहा था और प्रजा सुख-मन्त्री हीरा सिंह और उनके आदमियों की निगाह में।

मगर अपशकुन यह हुआ कि सीमावर्ती वन डिवीजन में एक तेज तर्रार युवा डी.एफ.ओ. की पोस्टिंग हो गयी। राज्यमन्त्री हीरा सिंह के आदमियों ने उसे भी खरीदना चाहा। मगर उसने खुले बाजार में अपना सौदा नहीं किया। मन्त्री को बुरा लगा कि यह कैसी अँधेर नगरी है कि सरकार अपने को रोज़ाना बेच रही है मगर उसका एक अदना नौकर बिकने से इनकार कर रहा है। वह ज़िद्दी और पागल-सा था। उसने एक रात छापेमारी के लिए पूरी तैयारी की। मन्त्री के

एक दर्जन ट्रक पकड़ लिए गये। मन्त्री जी की हालत खस्ता थी। मछली पानी के बाहर छटपटा रही थी। उन्होंने मैडम से सम्पर्क किया और गिड़गिड़ाते हुए सारा बाकया बता दिया। मैडम को जोर का झटका लगा। मानो उनकी नसों में खून का प्रवाह रुक गया हो। फिर उन्हें इसमें ठाकुरों-ब्राह्मणों की दलित सरकार के खिलाफ़ साजिश की बू आने लगी। उन्हें मन्त्री पर गुस्सा भी आया। उन्हें लगा कि यदि यह बात मीडिया में पहुँच गयी, तो विपक्षी दल सरकार का मुँह नोंच लेंगे और जनता सड़कों पर आ जाएगी। लोग शोषित दल को शोषक दल घोषित कर देंगे और पार्टी व सरकार हमेशा के लिए मिट जाएगी। ऐसी चिन्ता उन्हें काफ़ी सताने लगी। सहमते-सहमते बीस लाख रुपये में सौदा तय करके तत्काल मन्त्री को अभयदान देना उनकी मजबूरी थी। उनके मन में यह भी विचार आया कि इस ठाकुर को अगले फेर-बदल में कान पकड़कर निकाल देना है। न रहेगा बाँस, न बजेगी बाँसुरी। मगर गले में फँसी इस हड्डी को तत्काल निकालना ज़रूरी था वरना उसके साथ-साथ उनकी भी पोल खुल जाती और शायद सरकार ताश के पत्तों की तरह गिर न जाये। उनके चेहरे पर तनाव की मोटी पर्त जम गयी।

सो उन्होंने तुरन्त पुलिस महानिदेशक राही जी को फोन किया, ''सुन राही! राज्यमन्त्री, न्यास विभाग के परिवार वालों को सीमा पर उग्रवादियों ने घेर लिया है। शायद वे पड़ोसी देश के तख़्तापलट करने वाले लोग हैं। पुलिस और वन विभाग की वर्दी में दर्जनों की संख्या में उग्रवादी जलेबिया मोड़ पर डटे हैं। हमारी नाक का सवाल है सत्ता हमारी ही रहेगी। हम उनकी समानान्तर सरकार नहीं चलने देंगे। तू विभिन्न जिलों से और राज्य मुख्यालय से सैकड़ों सशस्त्र पुलिस जवान और अधिकारी भेजकर उनका सफाया कर। काम अधिक, बातें कम। सुबह मुझे खुशख़बरी खुद देना।''

पुलिस महानिदेशक मुख्यमन्त्री जी की फौरी कार्रवाई की शैली से वाकिफ़ थे और ना-नुकुर करने पर मुअत्तल हो जाते, सो वे तत्पर हो गये। उन्होंने एन्काउंटर के लिए मशहूर पुलिस अधिकारियों की टीम सीमावर्ती जलेबिया मोड़ के लिए रवाना कर दी और एक-एक मिनट का ब्यौरा मोबाइल पर लेने लगे। आख़िर सरकार की लम्बी नाक का सवाल था और बिना बारी की प्रोन्नति को याद करके पुलिस महानिदेशक सरकार की नाक का बाल बने रहना चाहते थे।

अगली सुबह पुलिस महानिदेशक के हवाले से एक ताज़ा ख़बर मुखपृष्ठ पर छपी, ''उग्रवादियों ने एक दर्जन वनकर्मियों की हत्या की। पुलिस ने जवाबी गोलाबारी की मगर उग्रवादी सीमा पार भागने में सफल हो गये।'' वर्दियाँ खून के छींटों के लगने के बावजूद मुस्करा रही थीं और अपनी-अपनी पीठ थपथपा

रही थीं।

मगर जलेबिया मोड़ के गाछ राज्यमन्त्री, मुख्यमन्त्री और पुलिस महानिदेशक के नाम पर बार-बार थूक रहे थे। मानो आसमान से थूक की बारिश हो रही थी।

धोखा

ब्लॉक परिसर में स्थित मेरे सरकारी आवास में कई तरह के गाछ थे। असल में यह पुराना ब्लॉक था। यहाँ की पोस्टिंग अच्छी मानी जाती थी क्योंकि सबको रहने की सुविधा मिल जाती थी, बच्चों को पढ़ने के लिए अच्छे स्कूल थे, सड़क और रेल यातायात की सुविधा थी और यहाँ का पानी अच्छा था। गाछों से हर साल अच्छा फ़ायदा होता था। मगर इस साल एक और फ़ायदा हमें मिला। कदम्ब के पेड़ की ऊँची डाल में मधुमक्खियों ने अपना छत्ता बनाया। मेरी पत्नी ने एक रात मुझसे धीरे-से कहा, "अपने यहाँ मधु का छत्ता लगा है।"

"तो इसमें धीरे-से कहने की क्या बात है?"

"श...श...श...आप तो बस माटी के लोंदा हैं। कहीं बीडीओ साहब को पता चल गया तो?"

"यह कोई चोरी थोड़े हैं? फिर बीडीओ साहब तो पी.सी. वाले हैं, "मैंने प्याज के छिलके की तरह बात को छीलते हुए कहा।

"उन्होंने नाजिर बाबू के हाते में फले आम को तुड़वाकर एक बोरा अपने घर अपनी जीप से भिजवा दिया था। कहीं इस छत्ते पर भी उनकी नज़र न लग जाए?"

"नाजिर बाबू के हाते में आम के दस पेड़ हैं—मिठउआ, खटउआ, सिन्दूरिया, बेलह्यिा, पेटभरवा और मालदहवा आदि। सो हर साल वे आम खाते हैं और बेचते भी हैं। हम मधु कहाँ बेचने वाले हैं?" मैंने आश्वस्त किया।

सो वह छत्ते के बढ़ने का बेसब्री से इन्तज़ार करने लगीं। एक माह बाद एक आदमी ने मेरा दरवाज़ा खटखटाया, 'हम मटरू नट हईं, हजूर? राउर गछिया में मधु माछी के छत्ता लागल बा। कहीं त काढ़ दी?"

मेरी पत्नी ने कहा, "अभी छत्ता बढ़ने दो।"

"मगर तब तक मधु माछिन मधुए चाट जइहन स, रउरा पछताये के पड़ी। हम मधु निकाल के आधा बाँट देब।" मेरी पत्नी की बाँछें खिल गयीं। नौ नकद

न तेरह उधार।

मटरू ने किरासन तेल, फटा बोरा, एक बाल्टी और एक चाकू माँगा। उसने बाँस की एक लम्बी लग्गी में बोरे का टुकड़ा लपेटकर उसमें किरासन तेल छिड़ककर और जलाकर छत्ते में आग लगा दी। फिर उसने छत्ते को बाल्टी में रोप लिया। थोड़ी देर में मैं घर से बाहर निकला। मेरी आँखें हज़ारों मरी मधुमक्खियों को देखकर रो पड़ीं, आँखों में अब भी संवेदना नहीं सूखी थी। "यह क्या हो गया?" ऐसा लगा जैसे कुरुक्षेत्र में हज़ारों योद्धा शहीद हो गये हों। कुरुक्षेत्र में भी शंख बजाकर ही बाकायदा युद्धघोष किया गया था। यहाँ युद्ध नहीं, जीवन का उत्सव था। फिर भी ऐसे परोपकारी जीवों को आग में भून दिया गया था। एक बार मन में आया कि इसके लिए पत्नी को डाँटा जाये, मगर उनके कर्कश स्वभाव को याद करके मैंने कुछ कहने की हिम्मत नहीं की, वरना सुबह-सुबह महाभारत शुरू हो जाता। थोड़ी देर में मटरू की पत्नी, माँ, भाई और बेटा सभी डेकची-कनस्तर लेकर पहुँच गये। मटरू बोला, "हजूर आगे रउरा मधु के ज़्यादा काम पड़ी, सभे मधु रउरा रख ली। एकरा बदले हमरा के पचास गो रूपिया दे दीं। ऊहे हमार मजूरी हो जाई।"

मेरी पत्नी घर से पैसा लाने अन्दर जाने लगीं तो मटरू ने उन्हें टोकते हुए कहा, "हजूर! एक काम आउर बा। हम ग्राहक खोजत बानी। रउरा ग्राहक सबसे बस एतने कहे के बा कि मटरू आपन हिस्सा के मधु बेचत बा। एतना कहे से ग्राहक के विश्वास होइ जाइ, आउर हमार काम बन जाइ।"

"तो तुम्हारे लिए मुझे झूठ बोलना पड़ेगा?"

"मालकिन! झूठ कवन पहाड़ बा। आपन देश के राजा-रानी झूठ के खान बाड़न स। ऊ बड़का-बड़का नेतवन सब 'एथी' के कसम खात-खात कबहूँ न थकलन स। बड़का-छोटका हाकिम लोग उहे सबे ताल-मा-ताल मिलावत बाड़न स। आपन बी.डी.ओ. जुम्मन बाबू के त रउरा जानते बानी। ओतना रंग त गिरगिटवो न बदली। हमनी के इन्दिरा आवास देवे के खातिर ऊ दू बरिस से दौड़ावत बा बाकिर कुच्छो न कइलस। हमनी के रोटी खात बानी कि माटी, एकर फिकिर उनका का बा? ऊ तो उतरे नेतवन के पीछे-पीछे दिन-रात घूमत रहेंले। मंडल बाबू मुख्यमन्त्री बनलन, बाकिर हम नट-चमार के सुधि न लिहलन। नौकरी-चाकरी में ऊ खूब जतियारी कइले। माने हमनी सब मानुस न हईं, जनावर हईं। आउर ऊ त जनावर के स्कूटर पर ढोइ-ढोइ क माला-माल होइ गइलन स। हमनी के त स्कूटर प चढ़े के सपनों में ना सोचनी। त मालकिन! जुग-जमाना बदल गइल। जइसन राजा, वइसने परजा।"

मेरी पत्नी को मटरू की बातें सोलहों आने सच लगीं। और मटरू अपने

भाई-बेटे के साथ ग्राहक खोजने निकल पड़ा। उसकी माँ और पत्नी वहीं गाछ के नीचे बैठ गयीं। थोड़ी देर में सात-आठ ग्राहक वहाँ आ गये। सभी ग्राहक मेरे घर पर यह दरयाफ्त करते कि क्या सचमुच छत्ते से ही मधु निकालकर मटरू बेच रहा है। मेरी पत्नी गोल-मटोल जवाब देती, ''मेरे पेड़ में मधु का छत्ता लगा था। इसने इस छत्ते से मधु निकाला । मैंने अपना मधु ले लिया। वह अपना मधु बेच रहा है।''

तभी मटरू ने कहा, ''रउरा देखीं मधु माछी के ढेर लागल बा कि ना? छत्ता के मोम इ बा कि ना?'' मटरू ने छत्ते की मोम को छितराकर लोगों को दिखाया।

और सचमुच एक लड़के ने चखा। उसने झटपट कहा, ''मीठा है मधु।'' मगर दूसरे-तीसरे ग्राहक को लड़के पर विश्वास नहीं हुआ। सो एक महिला ने उँगली से मधु निकालकर जीभ पर रखना चाहा, तभी उसके पति ने टोका, ''नहीं, नहीं। आज तुम्हारा गणेश चतुर्थी का निर्जली व्रत है।''

फिर बात मधु के दाम को लेकर शुरू हुई। मटरू ने बताया कि इसका बाज़ार भाव डेढ़ सौ रुपये किलो है, मगर मैं एक सौ पच्चीस रुपये किलो में देने को तैयार हूँ। मोल-भाव करते-करते एक सौ रुपये किलो में सौदा पट गया। मटरू ने दस किलो मधु बेच दी। मटरू खुश होकर कुछ बची मधु बेचने निकल गया। पत्नी को लगा कि उसने झूठ अपने स्वार्थ के लिए नहीं, मटरू के हित के लिए बोला था। जब शाम को दफ़्तर से मैं घर लौटा तो पत्नी ने हँसते-हँसते मुझे सारा वाकया बता दिया। तभी दरवाज़े पर किसी ने दस्तक दी। मैंने दरवाज़ा खोला, ''आइए, रमेश बाबू! बहुत दिनों बाद दर्शन दे रहे हैं। आप मासिक बैठक में आते हैं मगर आपसे मुलाकात नहीं होती।'' मैंने पुराने यार को उलाहना देते हुए उसका स्वागत किया।

''महेश बाबू, आपका आभार व्यक्त करने आया हूँ।'' उसने कहा।

''मेरा आभार?'' मैंने चौंककर पूछा।

''ऐसा है कि मैं आज अपने बेटे कमल को डॉक्टर आर.पी. शर्मा को दिखाने आया था। उन्होंने उसे चीनी खाने से मना कर दिया है। सिर्फ़ मधु से दवा लेने और मधु-नींबू का शर्बत पीने को कहा है। एक नट इधर ही मधु बेचने आया था। मैंने अपने बेटे को दो किलो मधु लेने भेजा था। दो सौ रुपये में ही दो किलो मधु उसने खरीद लिया। उसने दरयाफ़्त की तो भाभी जी ने सही गुणवत्ता भी बता दी। मैं बाज़ार में इसे खरीदने जाता तो दो सौ रुपये और गँवाकर आता।''

तब तक मेरी बेटी ने उन्हें चाय दे दी थी। रमेश बाबू ने हड़बड़ी में चाय पी और बोले, ''ग्रामसेवक की नौकरी क्षेत्र-भ्रमण की होती है। गेहूँ का फसल-

कटाई प्रयोग करना है। नये डी.एम. ने 'एथी' में डंडा कर रखा है। फिर कभी आऊँगा तो विस्तार से बातें होंगी।" जब वह हाथ मिलाने के बाद घर के बाहर जाने लगे तो मैं चीख पड़ा, "सुनिए!"

रमेश बाबू अचकचाकर मुड़े और मेरे पास दौड़कर आये। वह पूछने लगे, "क्या हुआ? कुछ छूट गया क्या? दवा की पर्ची तो नहीं छूटी?" लेकिन मैं कुछ बोल न सका। मेरे भीतर कुकुरकाट चल रही थी—एक ओर नट था और दूसरी ओर मेरी पत्नी थी। मैंने अगल-बगल झाँका, ऐसा लगा कि मैं अपना ही कुछ तोड़ रहा हूँ। पत्नी मेरी चीख सुनकर वहाँ आ गयी। मेरे बजाय उन्होंने ही जवाब दिया, "जो चीज़ आप बाज़ार में खरीदते हैं, उसे ठोंक-बजाकर लेते हैं कि नहीं? आप लोगों को जो अच्छा लगा और जितने में सौदा पटा, आपने खरीद लिया। इसमें हम लोगों का क्या दोष?"

"नहीं, भाभी जी! ऐसा क्यों कह रही हैं? हमने आपको दोषी कब कहा? भगवान आपका भला करे कि आपने ऐन मौके पर असली चीज हमें सस्ते में दिलवा दी। अच्छा चलता हूँ," यह कहकर रमेश बाबू तेज कदमों से चल दिये।

मुझे लग रहा था कि उसके जूते मेरी पीठ पर पड़ रहे हैं। मैं अपनी पीठ सहलाने लगा। तभी मेरी पत्नी ने मुझे घूरा, "जाने कैसे-कैसे लोगों से पाला पड़ता है? आपको अपना ख़याल नहीं, तो कम-से-कम परिवार का ख़याल रखना चाहिए। धोबी का कुत्ता न घर का, न घाट का।"

मैं चुप था मगर चुप्पी मेरे मन के भीतर खूब शोर मचा रही थी। मैं सीधे जाकर अपने बिस्तर पर बैठ गया। मैंने आँखें मूँद लीं। मगर रमेश बाबू बार-बार मुझे जगा देते थे, "आपको कुछ कहना है क्या?" मैं सोचने लगा कि मटरू न आया होता मधु निकालने तो ऐसा पाप हमारे परिवार से न होता। फिर यदि मेरी पत्नी ने झूठ न बोला होता तो निर्दोष ग्राहक मिलावटी मधु नहीं खरीदते। यदि मैं रमेश बाबू से सच्चाई बयान कर दूँ तो वह अपने मरीज बेटे को मिलावटी मधु नहीं खिलाएँगे।

एक हफ़्ते बाद वेतन का दिन था। नाजिर बाबू वेतन बाँट रहे थे। मैंने वहाँ अचानक रमेश बाबू को देखा। मैंने दूसरी तरफ मुँह मोड़ लिया। मगर दूसरी बार मैं नज़र मिलने पर उनके पास गया और उनसे हाथ मिलाया। वह फिर जल्दी में थे। मैंने उनसे कहा कि कुछ ज़रूरी बातें करनी हैं। उन्होंने थोड़ी देर में आने को कहा। हमेशा उनके काम के आगे समय कम पड़ता था।

तभी मुझे किसी ने ख़बर दी कि बीडीओ साहब ने बुलाया है। मुझे लगा कि कहीं किसी ने उनसे मेरी शिकायत तो नहीं कर दी? कहीं मिलावटी मधु वाला मामला झमेला तो नहीं बन गया? मैं सकपकाते हुए उनके पास पहुँचा।

मैंने हाथ जोड़कर उन्हें प्रणाम किया। बीडीओ साहब ने आँखों की ओर से मुस्कराते हुए पूछा, "सुना है, आपके हाते में कदम्ब के पेड़ से दस किलो मधु निकली है और आपने उसे बेचकर अच्छी आमदनी की है।"

"जी नहीं, सर!" मैंने एक झटके में कहा।

"क्या मतलब? ग्रामसेवक रमेश बाबू ने दो किलो मधु आपके यहाँ से खरीदी है? दूसरों ने भी खरीदी है। मेरी माँ आयुर्वेदिक दवा खाती हैं गठिया रोग के लिए। सो उनके लिए एक किलो मधु भिजवा देना।" बीडीओ ने आदेश दिया। 'जी, सर!' इसके अलावा मेरे पास कोई जवाब देने को नहीं था। मैं उन्हें नहीं कह सका कि मैंने नहीं, मटरू नट ने मधु बेचा है। किसी तरह उनसे पिंड छुड़ाना था।

मैं तुरन्त भागते हुए घर गया। पत्नी से मैंने हकलाते हुए सारी बातें बतायीं। वह आगबबूला हो गयीं, "जब माँ के दूध से बच्चे का पेट नहीं भरेगा तो बाप के 'एथी' से कैसे भरेगा? बीडीओ साहब नहीं, उनकी मेम साहब की खुराफ़ात है यह। पता नहीं, कितना बड़ा पेट है उनका? उनका बस चले तो वह ब्लॉक की मिट्टी भी खुदवाकर अपने घर भेज दें।"

मगर आदेश आदेश होता है। उनमें नुक्स देखने की न तो फुर्सत होती है और न हिम्मत। मैं फटाफट मधु देने के लिए पत्नी पर दबाव देने लगा। वह रसोई में गयी और एक पुराने प्लास्टिक के डिब्बे में मधु पलटने लगी और थोड़ा मधु अपने डिब्बे में बचा लिया। मुझे लग रहा था कि बीडीओ साहब ने एक किलो मधु लाने को कहा है और मैंने हामी भी भर ली है तो कम मधु देना अनुचित होगा। मगर पत्नी का कहना था, "दान की बछिया के दाँत नहीं देखे जाते। बीडीओ साहब को तौलने की फुर्सत है क्या?" पत्नी ऐसे मामलों की व्यावहारिक विशेषज्ञ थीं। बीडीओ साहब के यहाँ रोज़ जुटती ठेकेदारों-इंजीनियरों-नेताओं की भीड़ से वह बहुत कुछ सीख गयी थीं। अपनी पत्नी की आदत और शऊर से मैं अच्छी तरह परिचित था। मैंने मधु का डिब्बा उठा लिया और चुपचाप मेम साहब को दे आया।

करीब दो माह के बाद रमेश बाबू से ब्लॉक ऑफिस में फिर मुलाकात हो गयी। दूर से ही मैंने उन्हें पुकारा। वह चींटी की-सी चाल चलकर मेरे नज़दीक आये। मैंने पूछा, "रमेश बाबू! आपने सिर क्यों मुड़वा लिया बरसात में?"

"पूछिये मत, महेश बाबू। मैं कहीं का नहीं रहा। मेरा इकलौता बेटा गुजर गया। डॉक्टर ने बताया कि उसके खून में चीनी बहुत ज़्यादा बढ़ गयी थी और वह उसकी ज़िन्दगी पर भारी पड़ गयी। हम लोग उसे चीनी छूने भी नहीं देते थे। बस, आपके यहाँ का शुद्ध मधु और नींबू का शर्बत ही वह रोज़ाना पीता

था। मगर भगवान ने मेरी कुटिया उजाड़ दी। अब मैं क्या करूँ?'' यह कहते-कहते वह जोर-जोर से रोने लगे। मुझ जैसे कठकरेजी की आँखों से भी लोर लुढ़कने लगे। मैं उनका हाथ पकड़कर वहीं ज़मीन पर बैठ गया। मुझे लगा कि मैं किसी कब्र में नीचे धँस रहा हूँ। किसने धोखा दिया, यह सोच-सोचकर मैं अपने बाल नोंचने लगा और रमेश बाबू मुझे सान्त्वना देने लगे।

संगसार

जब मरीना खातून की शादी बेगमपुर में हुई, तो वह मुश्किल से सोलह साल की थी और उसका शौहर जुबैर महज अठारह का। उसके अब्बाजान की माली हालत अच्छी न थी मगर शौहर का घर-कुनबा ठीक-ठाक था। उसके पास कास्त जमीन के कुछ बीघों के अलावा एक किराने की दुकान भी थी जिस पर उसके ससुर बैठा करते। उसका शौहर इकलौती सन्तान था, इसलिए ससुराल में उसका मान-सम्मान काफ़ी हुआ। घर का माहौल ऐसा बना था कि बच्चे चाहे जितने हों कोई फ़र्क़ न पड़े, सब खुदा की मेहरबानी से उस दुनिया में जन्म लेते हैं, तो फिर इन्सान की क्या बिसात जो इसमें दखल दे? सो न चाहते हुए भी मरीना तीन बच्चे जनती गयी। तीन बच्चों के बाद उसने अपनी सास से कहा कि वह ऑपरेशन कराना चाहती है मगर सास ने ऐसी घुड़की लगायी कि दुबारा कहने की हिम्मत नहीं हुई। उसकी सास ठहरी मजहब की पक्की। पाँचों वक्त की नमाज़ अदा करती, रसोई साफ रखती, बच्चों की देखभाल करती, बड़े मियाँ को वक्त पर खाना देती, सो मरीना भी पाँचों वक्त की नमाज़ अदा करने लगी, रोजा रखती, जकात देती, फकीरों की सेवा करती। उसके ससुर सिर्फ़ जुम्मे की नमाज़ अदा करने जाते। उसकी सास में एक ही ऐब था कि वह पान-सुपाड़ी की बेहद शौकीन थी। पनडब्बा उसकी छाया की तरह साथ रहता, उठते-बैठते, खाते-पीते, आते-जाते, सब जगह।

फिर मरीना के दो बच्चे और हुए। यानी पाँच। उन्हें सम्भालना मुश्किल होने लगा क्योंकि उसकी सास को कैंसर हो गया और वह अचानक चल बसी। उसका शौहर जुबैर खेती के काम करता और ससुर दुकान सँभालता। बच्चे आपस में लड़ते-झगड़ते। उनकी सफाई, खान-पान कपड़ा-लत्ता, नहाना-धोना वगैरह के काम मरीना की कमर तोड़ देते। घर में चौका-बर्तन, सफाई आदि भी उसे ही करना पड़ता।

कुछ समय बाद उसके ससुर मन्नान मियाँ बच्चों का विशेष ख्याल करने

लगे। पहले तो वह बच्चों पर तवज्जो ही नहीं देते थे। अचानक यह क्या हो गया? सोचा तो मरीना का माथा ठनका, पर उसने इस पर कोई खास ध्यान नहीं दिया, लगा कि सास के मर जाने के कारण शायद ससुर उसकी भरपायी कर रहे हैं। जो भी हो, उसे एक सहारा-सा मिला। एक दिन मन्नान मियाँ मरीना के लिए कपड़े भी खरीदकर लाये। कपड़ों के साथ क्रीम, पाउडर, लिपस्टिक, नेल-पालिश भी थी। उसे तब और अचरज हुआ जब एक दिन मन्नान ने उसे सोने की एक अँगूठी लाकर दी। उसकी खुशी का ठिकाना न रहा। पर वह उस वहशत से अनजान ही रही जो उस पर मुसीबत बन कर टूटने वाला था। एक दिन वह लम्हा भी आ ही धमका। जुबैर एक रिश्तेदारी में गया था। मन्नान मियाँ उस दिन दोपहर को ही अपनी दुकान से घर आ गये। बच्चे स्कूल में थे। उन्हें आया देखकर मरीना को अचरज हुआ। मगर जब तक वह कुछ पूछ पाती, मन्नान ने खुद ही समस्या सामने रख दी, ''मेरी एक कुंजी खो गयी है। वह दुकान के बक्से की कुंजी है। उसके बिना मेरा हर्जा हो रहा है। जरा तुम भी खोजो।'' सो दोनों कुंजी ढूँढ़ने में लग गये। मगर मरीना ने वह कुंजी कभी देखी हो तब तो जाने कि वह कैसी है, सो वह उसके हाथ नहीं लगी। वह अपने कमरे में चली गयी। काफी देर तक इधर-उधर की परख के बाद मन्नान मरीना के कमरे में घुसा और बिस्तर के नीचे, कोने अँतरे में कुंजी खोजने लगा मगर वहाँ कुंजी हो तब तो मिले। करीब दस मिनट के बाद मरीना की देह पर बाघ की तरह झपट्टा मारकर मन्नान ने उसे काबू में करना चाहा। वह चीख कर उछल पड़ी, बहुत कोशिश भी की कि उसके चंगुल से निकल भागे, पर उसके ताकतवर पंजों से निकलकर भागना मुमकिन न हुआ। डर, दबाव तथा धमकी के बीच मन्नान ने मरीना के आँसुओं को सुखा दिया और उसके जिस्म को आटे की लोई की तरह रौंद डाला। छटपटाती और कराहती मरीना लाख कोशिश करती रही पर उस नर पिशाच से छूट न सकी और अपने को तबाह होते देखा। जाते समय वह धमकी देता हुआ निकला, ''किसी से इस बाबत साँस नहीं लेना। यदि किसी को इसकी भनक तक लगी तो हलाल होगी। और यदि मेरी बीवी बनकर रहोगी तो जन्नत इसी दुनिया में पाओगी।'' यह कहकर वह तूफान की तरह बाहर चला गया। मरीना मौत-सी पसरी खामोशी में खो-सी गयी। उसके सामने सिर्फ़ अँधेरा था।

उसे अपनी अम्मी द्वारा बचपन में दी गयी सीख याद आयी, ''अपनी देह और मन सिर्फ़ एक इन्सान को सौंपना, शौहर को। औरत किसी गली की कुतिया नहीं होती जिस पर कोई भी मुँह मार दे।'' मगर आज वह गली की कुतिया बना दी गयी। उसके अब्बा सरीखे ससुर ने ही उसके मुँह पर कालिख पोत दी।

उस दिन उससे कुछ भी खाया नहीं गया। अगले दिन उसका शौहर लौटा।

वह बुत की मानिंद पड़ी रही। उसका शौहर कुछ नहीं समझ पाया। बहुत कोशिशों के बाद उसने वाकया बयान किया। उसका शौहर भौंचक था। उसे काजी में आशा की किरण दिखायी दी। उसने मस्जिद के काजी को सारा वाकया बताया। काजी ने जुबैर से कहा, "इस वाकये में ससुर-पतोहू दोनों गुनाहगार हैं। अब तुम मरीना को अपनी बीवी नहीं मान सकते।" मरीना अपने शौहर पर थाने में रपट करने का दबाव बनाती रही। उसने यहाँ तक कहा कि 'अब मेरे लेखे तुम भी मर चुके हो, क्योंकि मेरा साथ नहीं दे रहे।" यह सुनकर जुबैर की रगों में जमे खून के थक्के पिघलने लगे थे। खैर...उसने थाने में रपट लिखवायी, कैसे रपट लिखी जा सकी, यह बताना न ज़रूरी है, न इसका कोई मतलब है।

जब अगले दिन बड़े दारोगा छुट्टी से वापस आये तो मुंशी ने उनके सामने वह रपट पेश कर दी। उनके मुँह का नक्शा बिगड़-सा गया। बहुत खोजबीन के बाद भी मन्नान मियाँ का अता-पता न चला। केस दर्ज़ कराने में पहले ही चार दिनों का विलम्ब हो चुका था, थाने में भी एक दिन का विलम्ब हो चुका था। बड़े दारोगा चाहते थे कि मन्नान मियाँ से मिलकर मामला सलटा लें, मगर उसकी भी गुंजाइश ख़त्म हो गयी थी। सो कोई विकल्प न देखकर बड़े दारोगा ने केस दर्ज़ कर लिया। उसके बाद यह घटना सार्वजनिक हो गयी। दर्जनों अख़बार, टीवी वाले इस स्टोरी को कवर करने के लिए बेताब थे। कोई मरीना का बयान पूछता, फोटो खींचता, कोई उसके शौहर जुबैर से पूछता, पर कोई मरीना की हालत का जायज़ा न लेता कि वह क्या से क्या बना दी गयी? है। दारोगा को डाँट मिली। छोटे दारोगा ने सिपाही को डाँट पिलायी और सिपाही ने चौकीदार को डाँट पिलायी। सो अन्ततः अगले दिन मन्नान मियाँ पुलिस की हिरासत में आ गया और मजिस्ट्रेट के सामने उसे हाज़िर करके जेल भेज दिया गया।

मगर इस पर बात न बनी। कुरान का हवाला देते हुए कहा गया, "अब मरीना जुबैर की अम्मी बन गयी है। इसलिए वह जुबैर के साथ नहीं रह सकती। ऐसा करना हराम है। शरीयत के मुताबिक मरीना और मन्नान दोनों गुनाहगार हैं। अरब देशों में दोनों को संगसार (पत्थर से मार-मारकर जान लेना) की सजा मिलती है। मगर हिन्दुस्तान में ऐसा कायदा-कानून नहीं है।"

मरीना हार मानने वाली न थी। वह गरजी, "मुझे मेरी हालत पर छोड़ दो। मैं शरीयत को मान लूँगी।" लेकिन जब चारों ओर घुप्प अँधेरा छा गया तो अचानक आसमान में जोर से बिजली कड़की। सूबे की शिक्षित मुस्लिम महिलाएँ

इकट्ठा हुईं। सबने 'महिला मुस्लिम पर्सनल ला बोर्ड' बनाया और कुरान शरीफ की नयी व्याख्या की और कहा, 'मरीना गुनाहगार नहीं है। असली गुनाहगार है उसका ससुर मन्नान मियाँ जिसे भारतीय संविधान के अधीन बने कानूनों के तहत कड़ी-से-कड़ी सजा मिलनी चाहिए। वैसे तो उसे संगसार किया जाना चाहिए मगर भारत के संविधान में ऐसा कायदा नहीं है। मरीना और जुबैर को पहले की तरह मियाँ-बीवी का रिश्ता कायम करने का पूरा हक है। दोनों की निजी ज़िन्दगी में शरीयत के नाम पर या कुरान शरीफ के नाम पर मुफ्ती लोग फ़तवा जारी करके दखल देने के हकदार नहीं हैं। हम ऐसे फ़तवों का खुलकर विरोध करते हैं और आगे करते रहेंगे।'' मरीना को लगा कि उसे अलादीन का असली चिराग मिल गया है। अब वह अपनी ज़िन्दगी उसकी रोशनी में जी लेगी। अचानक उसके शरीर में झुरझुरी हुई, माथा ऊर्जान्वित हुआ, कलेजा जुड़ाया। उसे महसूस हुआ, कि उसके बदन के फोड़े से काँटा बाहर निकल गया है। मगर वह खुशनुमा एहसास क्षणिक था। कुछ मुसलिम लड़कों का झुंड मरीना के घर के सामने आ पहुँचा। एक लड़के ने भीड़ का नेतृत्व करते हुए कहा, ''बाहर निकल कमबख़्त! जो इस कौम की ओर उँगली उठाएगा उसके हाथ काट लिए जाएँगे और गर्दन छोटी कर दी जाएगी।''

मरीना जान तजकर अन्दर से निकली और धर्म को कलंकित करने वाले दलीलों पर से पर्दा उठा दिया, ''ये मुफ्ती और उलेमा अपनी बहन-बेटियों के साथ ऐसा जुर्म होने पर बर्दाश्त करेंगे? क्या वे तब गुनाहगार को सजा देने की बजाय पीड़ित औरत को सजा देंगे? और हाँ, खुदा न करे, अगर आपके घर में ऐसा वाकया हो जाये, तो क्या आप लोग अपनी बहन, बेटी, बीवी को ही गुनाहगार मानेंगे? क्या कोई हिन्दुस्तानी औरत अपने दामन पर ऐसा बदनुमा दाग खुद लगाएगी जो उसके शरीर, मन और रूह में हमेशा के लिए अमिट हो जाये?''

मरीना ने पूरे जोश से कहा। बदनुमा भीड़ का जोश और गुस्सा अंगारे से राख में तब्दील हो गया। वे कुछ भी नहीं कह पा रहे थे। वे शर्मसार थे। उनके सिर नीचे झुक गये। फिर वे इस घटना पर मौन रहते हुए जहाँ से आये थे, वहीं वापस लौट गये।

मरीना को पहली बार एहसास हुआ कि उसने समाज में अपनी लड़ाई का पहला चरण जीत लिया है। तब उसे इतने पर सन्तोष नहीं हुआ। उसने अपने शौहर जुबैर से कहा, ''हम 'मुस्लिम महिला पर्सनल ला बोर्ड' की मदद लेंगे

और उसे अदालत से कड़ी-से-कड़ी सजा दिलवा कर ही दम लेंगे।'' जुबैर ने सिर हिलाकर हामी भर दी। फिर मरीना ने कुछ सोचते हुए कहा, ''अब मैं इस काले कफ़न को नहीं पहनूँगी। इस कफ़न ने ही आज तक मेरे भीतर की आग को ठंडा करने का काम किया है। अब मैं अपने भीतर की आग की लौ को और तेज करना चाहती हूँ। मैं इस्लाम को मानूँगी मगर फतवा को नहीं। मैं इस्लाम को मानूँगी मगर बुर्का को नहीं। मैं यह भी ज़ाहिर कर देती हूँ कि औरत मर्द के बराबर है, सभी मजहब, जाति और नस्ल बराबर हैं, जिसमें सभी इलाके और जबान के लोग बराबर हैं। संविधान सबको सम्मान से जीने का हक देता है। इसे न इस्लाम छीन सकता है, न हिन्दुत्व। मजहब हरेक का निजी मामला है जो संविधान की रोशनी में ही अपनाया जा सकता है।'' कहते हुए मरीना को लगा कि अब तक की मौत से वह उबर गयी है और वह देखेगी कि कैसे उसकी ज़िन्दगी पर कोई फैसला लेता है और क्यों लेता है? आसमान साफ था और उस दिन चारों तरफ उसकी आभा पसर गयी थी।

जुबैर ने मरीना का गुस्से से काँपता हाथ मजबूती से थाम लिया था।

अपने लोग

उन दिनों मेरी मुसीबत बढ़ गयी थी। उस साल भदई फसल बाढ़ से नष्ट हो गयी थी और रबी की फसल कुहरा और पाला से। मेरा घर निम्न मध्यमवर्गीय किसान परिवार का है। घर में मेरे सास-ससुर के अलावा जेठ, पति और देवर हैं। कुल मिलाकर छोटे-बड़े बारह लोग कुनबे में हैं और कोई बाहरी आमदनी नहीं। इस तरह संयुक्त परिवार की गाड़ी वर्षा-आधारित खेती की तरह भगवान भरोसे चल रही है। इसी बीच पता नहीं, किसी का शाप था या होनी थी कि मेरे बड़े लड़के की तबियत खराब हो गयी। मेरी सास ने बहुत दौड़-धूप की। गाँव-जवार का कोई ओझा-सोखा नहीं बचा था जिसके पास वह न गयी हो। ओझा-गुनी के यहाँ जाना केवल विश्वास नहीं, आर्थिक मजबूरी भी थी। वैसे मेरे पति ने एक-दो देसी वैद्य को भी दिखाया, मगर कोई फ़ायदा नहीं हुआ। सो मैं उन दिनों अपने बेटे के रोग को लेकर बहुत चिन्तित थी।

उस दिन मैं दाल छाँट रही थी। मेरी देवरानी ने मुझसे कहा, ''दीदी! आपके छोटे भैया मुम्बई में रहते हैं। वहाँ बड़े-बड़े अस्पताल हैं। क्यों नहीं आप उनसे मदद लेतीं?'' मुझे यह सुनकर अपने आप पर दुख हुआ। अब तक यह बात मेरे मन में क्यों नहीं आयी थी?

मुझे अपने भैया पर मुम्बई जैसे महानगर में कमाने-खाने के कारण गर्व हुआ, पर मैं सकुचा रही थी कि बाबूजी के इन्तकाल के बाद मायके का सहारा मजबूत होगा या नहीं। मैंने माँ की बात अपने आँचल के टोंग में गठिया ली थी। मैं कई दिनों तक सोचती-बिसूरती रही। मेरे पिता के गुजरने के बाद माँ ने कहा था, ''बिटिया! अब अपनी ससुराल को ही अपना घर समझना। यहाँ से बिना किसी के कहे, अकेली मत आना वरना तुम्हारा मान घट जाएगा।'' माँ की यह सीख मुझे बार-बार याद आने लगी। जब-जब मेरी देवरानी मुम्बई जाने के लिए मुझे प्रोत्साहित करती तब-तब फौरन माँ की बात याद आ जाती।

कई दिनों तक सोच-विचार के बाद मैंने यह बात अपनी जेठानी से कही। उन्होंने बीच का एक रास्ता झट से निकाल दिया, "तुम चिट्ठी भेज कर मायके वालों की 'राय-चाह' ले लो। अगर उनका रुख़ नरम होगा तो चली जाना।" मुझे यह बात जँच गयी। करीब पन्द्रह दिनों बाद माँ की चिट्ठी आयी। माँ ने मुझे घुमा-फिरा कर ढाँढ़स बँधाया था कि जब मन कहे, गाँव चली आओ या मुम्बई चली जाओ। मुझे उस चिट्ठी में रोशनी की एक किरण नज़र आयी। मुझे अहसास हुआ कि वे सब अपने हैं। दो दिनों के बाद मैं अपने पति और बेटे को लेकर मायके चली गयी। मैंने सोचा था कि यदि माँ इस स्थिति में मुम्बई जाने के लिए कह देगी तो चली जाऊँगी। मायके पहुँचने पर माँ बहुत खुश हुई। हम सबका पूरे मन से सबने स्वागत किया। मुझे बहुत अच्छा लगा, पर इस बात की टीस मन से नहीं जा पा रही थी कि बाबूजी के मरने के बाद घर में कौन-सा ग्रह प्रवेश कर गया कि मेरे मायके को दो फाँक कर गया। रात में माँ ने धीरे-धीरे बताया 'बाबूजी के मरने के बाद बड़े भैया का दुश्मनों ने कत्ल कर दिया और डर के मारे हम सभी शहर में जाकर छिप गये। इस घटना ने परिवार को एकबारगी फिर हिला दिया। हम लोग सम्भले भी नहीं थे कि बड़को को मंझले भैया और मंझलो भाभी ने अलग कर दिया—तिनके की तरह घर उड़ने लगा। हाँ, छोटे भैया ने बड़को और उनकी दो बेटियों का आर्थिक भार सम्भाला। छोटे भैया घर बनवाने के लिए जो मनीऑर्डर भेजते उसका आधा मंझले भैया काम में खर्च करते किन्तु बाकी आधा अपने नाम से जमा करते जाते। धीरे-धीरे मंझले भैया आर्थिक रूप से मजबूत होते गये। इसके बाद मंझलो ने मंझले भैया को उकसाना शुरू किया। उनसे उसने माँ और बड़को की नाना प्रकार से बुराई की—यहाँ तक कि मुझ पर चोरी और बड़को पर चरित्र-भ्रष्ट होने का आरोप लगाया।" यह सुनते ही मेरा शरीर गिनगिना उठा मानो मैं निर्वस्त्र हो गयी। मुझे रात भर नींद नहीं आयी। बार-बार ऐसा लगता कि यदि बाबूजी ज़िन्दा होते तो यह घर कमजोर नहीं होता या यदि बड़े भैया जीवित होते तो वह भी इसे डगमगाने से रोक सकते थे। आख़िर बाबू जी के बूढ़े होने पर बड़े भैया ने ही सारा दारोमदार सम्भाला था। वे मंझले और छोटे भैया को अपने बेटे की तरह मानते थे। मैं रात भर घर, दीवारों और कड़ियों को निहारती रही। नींद आने का नाम नहीं ले रही थी और मेरी आँखें बार-बार गीली हो जाती थीं। पूरी रात जग कर ही मैंने गुजार दी। सुबह नित्य कर्म से निवृत्त होने के बाद मेरा शरीर अलसाया था। आँगन में दौड़-दौड़ कर दीपिका कुछ-न-कुछ कर रही थी। बड़ी प्यारी बच्ची है। बिल्कुल अपनी माँ पर गयी है। मैंने उसे पास बुला कर दुलारा और कहा, "बिटिया! ठंडी बहुत है, थोड़ी लकड़ी लाकर अलाव जलाओ।" दीपिका ने झटपट वैसा ही किया।

थोड़ी देर तक हाथ-पैर सेंकने के बाद मैंने दीपिका से कहा, ''दीपिका! जब मैं तुम्हारी तरह छोटी थी, तो चौका-बासन, दुआर-गोबर, झाड़ू-पोंछा मैं ही करती थी, मगर आजकल सर्दी बहुत पड़ रही है, इसलिए हाथ-पैर सेंक कर घाम निकलने के बाद ही ये काम करना।''

यह सुनकर वह हँसने लगी। कहा, ''बुआ! मुझसे काम डरता है। अपने काम में आलस्य कैसा? मैंने तो सारा काम निपटा लिया।''

''इतने सबेरे?'' मुझे आश्चर्य हुआ। इस घर-आँगन में मेरे बचपन के दिन ऐसे गुजरे थे। समय तो पंख लगाकर भागता है।

''बुआ! अपना चौका-बासन, दुआर-गोबर, झाड़ू-पोंछा कोई बड़ा काम है क्या?''

''तो बड़ा काम क्या है?''

''मेरे लिए तो बड़ा काम कुछ भी नहीं—लेकिन दूसरों का थोपा हुआ काम मुझसे नहीं होता।'' जब 'अपने लोग' काम कराने के लिए ही अपने होते हैं और प्यार-दुलार भूल जाते हैं तो छोटा काम भी पहाड़-सा लगता है और अजनबीपन की दरार बहुत बढ़ जाती है।''

''अरे पगली! यह तो बता, तुमसे दूसरे का काम करने के लिए किसने कहा?''

''नहीं, बुआ। मैंने दूसरों का भी काम किया है। मैं करीब दो साल चाचा के साथ मुम्बई में रह चुकी हूँ'', दीपिका ने कहा।

''पगली! चाचा का घर पराया थोड़े होता है। जैसे पिता, वैसे चाचा। जैसी माँ, वैसी चाची। मेरे काका-काकी मुझे बहुत मानते थे। मैं अपनी काकी के साथ ही खाना खाती थी। मेरे काका मेरे लिए बहुत बहुत सारे खिलौने, कपड़े, तेल-साबुन लाते थे।'' मेरी यह बात सुनते ही दीपिका रो पड़ी। फिर उसने सिसकते हुए कहा, ''बुआ! मैं झूठ बिल्कुल नहीं बोल रही हूँ। मैं चाचा को अपना पिता और चाची को माँ समझती हूँ, मगर उन लोगों ने मुझे कभी भी अपना नहीं समझा।''

''चुप दीपिका, ऐसा नहीं कहते बड़ों के बारे में। ऐसा कहोगी तो मैं तुम्हारी पिटाई कर दूँगी।'' मैंने बनावटी क्रोध दिखाते हुए कहा।

''हाँ बुआ, माँ की तरह तुम भी मुझे नहीं समझोगी? लेकिन जो कष्ट भोगता है वही दर्द जानता है। चाचा ने मुम्बई ले जाते हुए कहा था—'बेटी! वहाँ मैं तुम्हें पढ़ाने के लिए ले चल रहा हूँ। गाँव में तुम्हारी पढ़ाई नहीं हो सकती। खूब पढ़-लिखकर बड़ा आदमी बनना' पर हुआ क्या मेरे साथ? बड़ा आदमी? पढ़ना-लिखना? दिन-रात नौकरानी की तरह मैं वहाँ काम ही तो करती रही।

जानती हो बुआ? चाचा-चाची और उनके बच्चों के लिए मैं दिन-रात क्यों खटती थी? इसलिए कि वे हमारे अपने थे। उनके साथ रह कर मैं पढ़-लिख जाऊँगी और अपने पैरों पर खड़ी हो सकूँगी। तब पिता की कमी नहीं खलेगी, ऐसा सोचा था पर उन लोगों ने तो मुझे अपने लिए कुछ करने का भी समय मुझे नहीं दिया। मुम्बई में चाचा के यहाँ नल में पानी अक्सर रात में नौ बजे आता है और सुबह चार बजे चला जाता है। मैं रोज़ाना बारह बजे रात तक बर्तन माँजती, घर के बाहर गली में कपड़े धुलती, सारे घर में पोंछा लगाती। मुश्किल से पाँच घंटे सोती। पाँच बजे मुझे रोज़ाना उठना पड़ता। चाचा-चाची, उनके दोनों बच्चों, अंजना और अभिषेक, के लिए टिफिन बनाती। घर में झाड़ू-पोंछा करती। पूजाघर की सफाई करती। दो बिस्कुट खाकर अपने नगर निगम के स्कूल साढ़े छह बजे पैदल चली जाती। मेरा स्कूल चाचा के मकान से करीब तीन किलोमीटर दूर था। जाड़े में सुबह-सुबह जब कुहासा छाया रहता तो सड़क पर कुछ नहीं सूझता था। कई बार मैं मोटर गाड़ियों से टकराने से बची। कई बार बेवड़ों-मवालियों के चंगुल में फँसते-फँसते बची। जब मैंने इसके बारे में चाची को बताया तो वे उल्टे मुझे ही गाली देने लगीं। मेरी माँ की कोख को सरापने लगी। दोपहर को स्कूल से वापस आने पर फिर बर्तन माँजती, पोंछा लगाती और खाना बनाती। रात में सब काम करने के बाद मैं उनके पैर दबाती, उन्हें नींद आने पर ही मैं कथरी-गुदड़ी बिछाकर सोने की कोशिश करती और नींद की गोद में बहुत सारे सपने देखती।''

मेरा शरीर धीरे-धीरे सुन्न हो रहा था। मुझे लगा—काश! यह एक झूठी कहानी होती। तभी दीपिका ने मुझे झकझोरते हुए कहा, ''बुआ, मेरे हाथ देखो। मेरी उँगलियों में घेट्टे (निशान) पड़ गये हैं। मेरी हथेली एक मजदूरिन की तरह कड़ी हो गयी है। पहले मेरे पैर की उँगलियाँ सड़ गयी थीं। यहाँ आने पर माँ और दीदी ने घरेलू इलाज किया, तब ठीक हुईं। मुझे देह में लगाने के लिए अलग से साबुन नहीं मिला। बाल धुलने के लिए शैम्पू कभी भी नहीं मिला। एक बार मैंने चाची से शैम्पू माँगा तो वह उबल पड़ीं, ''गाँव में कभी तुमने शैम्पू लगाया था क्या? शैम्पू लगाने को बदा होता तो बचपन में बाप नहीं खा जाती। जब ससुराल जाना तो वहीं शैम्पू लगाना।'' तब मैं खून का घूँट पीकर रह गयी थी। सो कपड़े धोने के साबुन से मैं अपना सिर कभी-कभी धोती थी, मगर उससे बालों की लटें बन गयी थीं''....कहते-कहते वह अपने को जज्ब नहीं कर पायी। वह सुबक-सुबक कर रोने लगी।

''दीपिका, दूसरे से यह सब मत कहना। अपने घर की इज़्ज़त पर बाहर वाले धब्बा न लगाने पाएँ''—मैंने उसके घाव को सहलाते हुए कहा।

वह फिर रोने लगी, ''नहीं बुआ, ऐसा कभी नहीं होगा।'' उसके हाव-

भाव से लगा कि मैं उस पर विश्वास नहीं कर रही हूँ। थोड़ी देर बाद मेरे चुप होने पर वह अपनी राम-कहानी पुनः बताने लगी, "बुआ! तुम्हें मुझ पर विश्वास नहीं हो रहा है। मैं तरसती थी, मगर चाचा-चाची ने मुझे कभी 'दीपिका' तक कह कर नहीं पुकारा। सिर्फ़ 'दिपिया' कह कर पुकारते रहे। 'दिपिया पानी लाओ', 'दिपिया' कपड़े धुल दो', 'दिपिया पोंछा लगाओ', वगैरह-वगैरह। चाची कहती थीं —'दिपिया पर चर्बी चढ़ गई है।' चाचा, चाची की बातों में धीरे-धीरे, हुँकारी भरने लगे।"

"लेकिन अंजना और अभिषेक तो तुम्हारे साथ खेलते-कूदते रहे होंगे?" मैंने जानना चाहा।

"नहीं, बुआ! अंजना मुझे पीटती थी। अपने सारे कपड़े-लत्ते मुझसे धुलवाती। अभिषेक मुझसे छोटा है, मगर उसने मुझे दीदी कभी नहीं कहा। वह मेरी चोटी पकड़कर उखाड़ता था और तमाचे-घूँसे मारता था। जब मैं चाची से उसकी शिकायत करती तो वह उल्टे मुझ पर ही बरस पड़तीं।"

"लेकिन तुम्हारी पढ़ाई-लिखाई का बोझ तो वही लोग उठाते थे न?"

"बुआ! इसे बोझ कहोगी या खानापूरी। मैं म्युनिसियल स्कूल में हिन्दी माध्यम से पढ़ती थी जबकि अंजना दीदी और अभिषेक मशहूर अँग्रेजी स्कूल में पढ़ते थे। वे दोनों स्कूल की बस से आते-जाते थे। मैं पैदल स्कूल जाती थी। वे दोनों ट्यूशन भी पढ़ते थे। उन दोनों के पास नाना प्रकार के खिलौने थे—साँप-सीढ़ी का खेल, कैरम, जहाज, मोटर, गेंद वगैरह। घर के बर्तन ही मेरे खिलौने थे, घर की रसोई से लेकर बाहर की गली के नल तक मेरा खेलने का मैदान था। मैं मैदान में अक्सर डटी रहती। मैं मैदान छोड़कर कभी नहीं भागी। फिर भी मुझे दोषी करार दिया जाता। जब कभी चाचा मेरे लिए अभ्यास पुस्तिकाएँ वगैरह लाते तो चाची उनमें से कुछ निकाल लेतीं। जब मैं उसे माँगती, तो वे ऊलजलूल कहने लगतीं।"

"सो क्या छोटको कोई काम नहीं करती थी?" मैंने पूछा।

चाची टीवी देखने में माहिर हैं। वे रंगीन टीवी के सामने दिन-रात पड़ी रहती हैं। टीवी के पचासों चैनल थे उनके यहाँ। दिन-रात फिल्म और सीरियल देखना और बन-ठन कर शाम को बाहर घूमना उनका शौक था। उन्हें काम से कोई मतलब नहीं था। वे रोजाना फिल्म देखती थीं। उन्होंने अँग्रेजी नहीं पढ़ी है, मगर अँग्रेजी फिल्में ज़रूर देखती थीं। जब कभी मैं टीवी देखने के लिए उनके कमरे में बैठती तो वह कहने लगतीं, "दिपिया, अपना काम बटोर लो। ज़्यादा टीवी देखोगी तो आदत बिगड़ जाएगी।" तुम्हें सारी ज़िन्दगी गाँव में बितानी है।" उनके कहने के अन्दाज़ से लगता मानो मैं चिड़ियाघर से पकड़ कर लायी

गयी हूँ।''

इतना सुनते-सुनते मैं अर्द्धचेतनावस्था में पहुँच गयी थी। एक बारह-तेरह साल की छोकरी के मुँह से सयानी बातें सुनते-सुनते। सिर खुजलाते हुए मैंने पूछा, ''अच्छा, दीपिका! क्या तुम्हारे मामा तुमसे मिलने कभी नहीं आये?''

''आये थे, बुआ! मेरे मामा बहुत अच्छे हैं। वे मिठाई लाये थे। और मुझे अपने घर ले जाने के लिए कह रहे थे, मगर चाची ने बहाना बना दिया कि अभी इसकी परीक्षा होने वाली है। गर्मी की छुट्टी में मुझे भेजने की बोल दी। जाते समय मामा ने मुझे, अंजना दीदी और अभिषेक को पचास-पचास रुपये भी दिये थे, मगर चाची ने मुझसे वह नोट ले लिया...'' कहते-कहते वह फफक-फफक कर रोने लगीं।

मैंने छोटको को ऐसा कभी नहीं समझा था। वह हमेशा मीठी-मीठी बातें करती थी। उसकी बतरस में मैं उसके इरादों को परखना भूल गयी थी। तभी मुझे याद आया कि छोटको ने एक बार बताया था कि उसके घर में एक दाई काम करती थी। मैंने दीपिका से पूछा, ''वह दाई कहाँ गयी?''

''बुआ! वह दाई बूढ़ी थी। वह नीचे के कमरे में झाड़ू-पोंछा लगा देती थी, मगर ऊपरी कमरे के लिए लोहे की सीढ़ी पर नहीं चढ़ पाती थी। फिर वह रोजाना सुबह-शाम एक-एक घंटे ही काम करने आती थी, और इस पार्ट-टाइम काम के लिए उसे छह सौ रुपये महीने मिलते थे। मैं नीचे-ऊपर, रात-दिन फुल टाइम काम करती थी। मैं घर की थी और नहीं भी थी। मैं दाई थी और नहीं भी थी। मैं इन्सान थी और नहीं भी थी। मेरी परवरिश के लिए दादी और माँ पर चाचा-चाची का अहसान भी था। एक पंथ दो काज एक साथ हो रहा था...।''

''बस करो, दीपिका! मुझसे अब नहीं सुना जाता। मैं छोटे भैया को बहुत कुछ समझ रही थी। मैंने समझा था कि वह बड़े भैया के गुजरने के बाद घर को पूरी तरह सम्भाल लेंगे। उनका मुम्बई का कारोबार सुनकर मैं झूम उठती थी, मगर जब तुम जैसी बच्ची की हिफ़ाजत वह नहीं कर सके तो किसी और के लिए क्या करेंगे? जो लोग यह कहते नहीं थकते कि खून पानी से ज़्यादा गाढ़ा होता है, वे भूल जाते हैं कि कीचड़ पानी से ज़्यादा गाढ़ा होता है, कि खून में आजकल विष घुल गया है, कि अपना खून अब खूनी हो गया है। मान गयी दीपिका! तुम मुझसे ज़्यादा समझदार हो। तुम बारह साल में जितना ज़्यादा कष्ट भोग चुकी हो, मैं उसका आधा भी नहीं भोग सकी हूँ। पता नहीं क्यों, भगवान ने हम बेटियों को भोगने के लिए ही बनाया है, सिर से पैर तक। और हमें भोगने में अपने लोग भी पीछे नहीं रहते।'' ये शब्द मेरे आँसुओं से पूरी तरह लथपथ हो गये थे। अपने बीमार बेटे की चिन्ता मुझे कम होने लगी और अब बड़को,

उनकी दोनों बेटियों और माँ की चिन्ता सताने लगी। फिर भी मैंने उन सबको सान्त्वना दी। मैंने प्लान बनाया कि सुबह-सुबह मैं अपने पति और बेटे के साथ अपने घर चली जाऊँगी। मुझे विश्वास हो गया था कि मेरा बेटा पराये घर में और ज़्यादा बीमार हो जाएगा और शायद मैं भी बीमार हो जाऊँगी। बच्चे नयी परिस्थितियों में जितना ढल जाते हैं उतना सयाने नहीं।

शाम को डाकिया आया और माँ को नमस्ते कहते हुए एक खत दे गया। मैंने उसे पढ़ना शुरू किया। चिट्ठी की शुरुआत 'बड़ों को प्रणाम और छोटों को आशीर्वाद' से हुई थी। उसके आगे जो कुछ लिखा था, उसे मैं पढ़ नहीं पा रही थी। माँ ने आगे पढ़ने के लिए बार-बार कहा। फिर मैंने पढ़ना शुरू किया, ''माँ, अंजना का मुँह फूल गया। शायद गलसुआ हो गया है। अभिषेक ने क्रिकेट खेलने में अपनी बायीं टाँग तोड़ ली है। एक दिन अंजना की मम्मी ऊपर के कमरे से नीचे उतर रही थी तो उसकी टाँग सीढ़ी पर फिसल गयी और गिरने से उसकी कमर की हड्डी टूट गयी। वह अस्तपताल में भर्ती है। घर में कोई दाना-पानी देने वाला नहीं है। दीपिका को किसी के साथ जल्द भेज दो वरना सारा घर-परिवार बर्बाद हो जाएगा। मेरी गलती को क्षमा करना। आपका बेटा—अंजनी।''

माँ की आँखों से लोर लुढ़कने लगे।'' मेरा अंजनी', 'मेरा अभिषेक', 'मेरी अंजना', 'मेरी छोटको' आदि कह कर वह रोने लगीं। मैं माँ को चुप कराने लगी। दीपिका गुमसुम खड़ी थी। बड़को दीपिका की ओर देखते हुए किंकर्तव्यविमूढ़ थी। मेरे मुँह से निकला, ''जो जैसा करेगा वैसा भरेगा। आज उन लोगों को दीपिका की याद आ रही है जिसे वे दुश्मन समझ कर रोटी का टुकड़ा देकर पूरी तरह से गुलाम बना लिए थे। अभी छोटको की कमर टूटी है, कल उसकी देह में कीड़े पड़ेंगे। उसने हमारी फूल जैसी दीपिका को झाँखर बना दिया। भगवान के घर देर है, अँधेर नहीं। वज्र गिर पड़ेगा उन सब पर एक दिन। मेरा रोआँ-रोआँ उन्हें सराप रहा है...।''

तभी दीपिका ने दौड़ कर मेरे मुँह पर अपनी दायीं हथेली रख दी। वह चीख पड़ी, ''ऐसा मत कहो, बुआ! भगवान दुश्मन पर भी वज्र नहीं गिरायें। उन्होंने मेरे साथ बदसलूकी की तो क्या मैं भी उनके साथ वैसा ही करूँ? तब हममें और उनमें फ़र्क़ ही क्या? तब उनके अच्छे दिन थे और सारी दुनिया उनके साथ थी। आज उनके बुरे दिन हैं और कोई भी साथ नहीं है। एक कहावत मैंने सुनी है, ''जब तुम हँसते हो तो सारी दुनिया तुम्हारे साथ हँसती है मगर जब तुम रोते हो, तो अकेले रोते हो। वे जो भी हैं, अपने हैं। आदमी हमेशा एक-सा व्यवहार नहीं करता। वह समय देख कर व्यवहार करता है। इसलिए मैं अपनी परवाह नहीं करके उनकी सेवा करने ज़रूर जाऊँगी। दादी! शायद पश्चिम टोले

के घुरहू बाबा मुम्बई जाने वाले हैं, मुझे उनके साथ जल्दी भेज दो। शायद इस बार ठोकर खाने के बाद वे मुझे बेहतर समझेंगे, अपना समझेंगे। दादी कहती थीं, लाठी से मारने पर तालाब की काई नहीं फटती।''

यह कहते-कहते वह भोंकार छोड़ कर रोने लगी। माँ भी रोने लगी। बड़को भाभी और मैं भी रोने लगी। मुझे लगा कि दीपिका ने ऐसा कह कर अपने बड़ों के चिकने गालों पर करारा तमाचा भी मारा और मुसीबत के दिनों में उन्हें गिरने पर उठा भी लिया।

चौथा खम्भा

'अख़बार नहीं, आन्दोलन', 'लोकतन्त्र का प्रहरी', 'पत्र नहीं, मित्र भी', 'हिन्दी का सर्वाधिक लोकप्रिय दैनिक', 'सबसे पहले और निर्भीक।'

ये सूक्तियाँ मैं अपने अख़बार 'भारत-भारती टाइम्स' के कार्यालय में रोज़ाना पढ़ता था। मुझे गर्व होता था कि मैं ऐसे अख़बार, नहीं, नहीं, आन्दोलन से जुड़ा हूँ। जबसे दिवाकर जी इनके सम्पादक बने हैं, इसकी प्रसार संख्या में कई गुना इजाफ़ा हुआ है। क्यों न हो, दिवाकर जी ने प्रतिभा को सम्मान दिया, पैरवी और पैसे को नहीं। मुझे भी इस बात का फख्र है कि मेरा चयन भी मेधा के आधार पर हुआ। वरना मेरे जैसा गरीब एवं बिना पहुँच वाला आदमी, बिना किसी सिफ़ारिश के यहाँ खड़ा भी नहीं हो सकता था। हमारे महानगर से तीन राष्ट्रीय स्तर के दैनिक छपते थे। मगर 'भारत-भारती टाइम्स' की पहचान बिल्कुल अलग थी। कितने सरकारी घोटालों, प्रशासनिक बदहालियों, फर्जी नियुक्तियों, अवैध लाइसेंसों, मिलावटी वस्तुओं, अनुपयुक्त दवाओं की खरीद-बिक्री, वगैरह के मामलों का पर्दाफाश 'भारत-भारती टाइम्स' ने ही किया था।

मगर इसका मतलब यह नहीं था कि सनसनी ख़बरें बिना पुष्ट किये ही छापी जाती थीं। यही इस अख़बार की ख़ासियत थी। बिना पूरी छानबीन किये और संवाददताओं से साक्ष्य प्राप्त किये दिवाकर जी कुछ भी नहीं छपने देते थे। फिर भी यदि किसी संवाददाता की असावधानी से गलत ख़बर छप जाती तो उसका खंडन प्राप्त होने पर उसे विधिवत् छापा जाता था। यहाँ दूसरे अख़बारों की तरह गैर-ज़िम्मेदार संवाद-सूत्र नहीं थे। सम्पादक को उनके सहयोगी श्रीश भरपूर मदद करते थे। सम्पादक का श्रीश पर पूरा विश्वास था। जिस किसी ख़बर की तह में सम्पादक को जाना होता, वह श्रीश को ही स्थल-अध्ययन के लिए भेजते थे। फर्जी पुलिस मुठभेड़ों का मामला हो या कांग्रेसी मुख्यमन्त्री के परिवार द्वारा दलित महिलाओं का उत्पीड़न, डी.एस.पी. कारूराम द्वारा एक महिला सिपाही का बलात्कार का मामला हो या स्वयंसेवी संस्था 'मानव विकास संगठन' द्वारा विदेशी पैसे का दुरुपयोग, बंगलादेशियों की सीमावर्ती जिलों में अवैध घुसपैठ हो या पाकिस्तानी

खुफिया एजेंसी आई.एस.आई. द्वारा सार्वजनिक स्थानों पर विस्फोट का मामला, इन सबमें सम्पादक ने श्रीश को भेजकर जाँच-पड़ताल कराई थी। इन सभी ख़बरों के मुख्य पृष्ठ पर छपने से शासन की कुर्सी रबड़ के खिलौनों की तरह काँप उठी थी। सो श्रीश की पूरे पत्रकारिता-जगत में अच्छी पहचान बन गयी थी। जनता में भी उनकी लोकप्रियता बढ़ गयी थी। जितने खत सम्पादक के नाम आते, लगभग उतने ही श्रीश के नाम से अलग से आते थे। सूबे की सरकार ने उन्हें कई बार सुख-सुविधाओं की सौगात भेजी मगर उन्होंने एक भी स्वीकार नहीं की। उन्हें सस्ती दर पर महानगर में आवासीय भूखंड देने की पेशकश भी सरकार ने की, मगर उन्होंने लात मार दिया, "मेरे लिए यह अतिरिक्त उदारता क्यों? यदि सभी पत्रकारों के लिए सरकार भूखंड आबंटित करने की नीति बनाये, तो अलग बात है। पत्रकारों के लिए पेंशन की नीति बने, तो ठीक है। मगर मैं अपने लिए सरकार से कोई चीज़ कृपा के रूप में नहीं प्राप्त करूँगा।"

"श्रीशजी! सभी पत्रकारों के लिए भूखंड अभी उपलब्ध नहीं हैं। नीति भी बनेगी मगर समय लगेगा। देखिए न, हाईकोर्ट के चीफ जस्टिस महेश्वरन जी को एक भूखंड आबंटित हो रहा है।" सरकार के प्रतिनिधि ने कहा।

"जी नहीं, मैं किसी से अपनी तुलना नहीं करता। मैं इसे विनम्रतापूर्वक अस्वीकार करता हूँ।"

"ठीक है, श्रीशजी! यदि आपकी ज़मीर को यह मान्य नहीं है, तो कोई बात नहीं। मगर इस प्रकरण को अख़बार की सुखियों में मत लाइएगा।" उस प्रतिनिधि ने अनुरोध किया।

"इस बाबत आप निश्‍चिंत रहिए। मैं अपनी प्रशंसा में कोई ख़बर नहीं छपने दूँगा। यह भी अप्रत्यक्ष रूप से पीत पत्रकारिता होगी।"

मगर कोई ख़बर छपे बिना भी यह बात पत्रकारों के बीच फैल गयी। मैंने श्रीशजी को इस नाते बधाई दी। मुझे उनकी दाढ़ी में उनकी प्रतिभा और ईमानदारी झलकती दिखायी दी। जब वह किसी प्रकरण पर अपनी दाढ़ी खुजलाते, तो मुझे एक विशेष प्रकार की प्रेरणा मिलती। मेरे अन्दर भी यह इच्छा बलवती होने लगी कि मैं भी दाढ़ी बढ़ाऊँ और उनके ही पदचिन्हों पर चलूँ। मैंने उस दिन अपनी ज़िन्दगी का आदर्श पुरुष मन-ही-मन चुन लिया।

इसी बीच 'श्रमजीवी पत्रकार संघ' का चुनाव नज़दीक आ गया। कुछ मित्रों के कहने पर मैंने भी महासचिव पद के लिए पर्चा भर दिया। मगर कुछ वरीय पत्रकारों ने श्रीश पर दबाव डाला। पर उन्होंने इनकार कर दिया। फिर कुछ पत्रकारों ने सम्पादक से इस बाबत मुलाकात की। और सम्पादक के बार-बार कहने पर श्रीश ने पर्चा भर दिया। मैंने तथा दूसरे अख़बार के रतन जी ने अपना-अपना पर्चा वापस ले लिया। श्रीश जी के लिए अपना परचा वापस लेने में भी मुझे गर्व का

एहसास हुआ। सो श्रीश निर्विरोध महासचिव चुन लिए गये।

श्रीश जी के निर्विरोध चुने जाने पर जश्न मनाया गया। खुद सम्पादक दिवाकर ने उसमें शिरकत की। उन्होंने खुले दिल से श्रीश की प्रशंसा की और नये पत्रकारों से अपील की कि वे श्रीश से आदर्श पत्रकारिता के गुण सीखें। बाद में तमाम सहकर्मियों ने श्रीश जी से 'दो शब्द' कहने का अनुरोध किया। थोड़े संकोच के बाद उन्होंने बोलना शुरू किया, ''सम्पादक महोदय और साथियो! हम लोगों ने अपने विद्यार्थी जीवन में 'प्रगतिशील छात्र संघ' का गठन किया था। हम लोगों ने दक्षिणपंथी, जातिवादी, धर्मोन्मादी तथा संकीर्णतावादी मानसिकता के खिलाफ़ संघर्ष शुरू किया था। करीब तीन वर्षों के परिश्रम से बुर्जुआ और सामन्ती ताकतों के गढ़ में हम लोगों ने सेंधमारी की थी। विद्याथियों के हित में ही नहीं, समग्र शिक्षा—नीति, रोज़गार-सृजन और समतामूलक समाज की स्थापना के लिए हम लोगों ने आन्दोलन चलाया था। फिर क्या था? अगले साल से हमारे संगठन से अध्यक्ष या महासचिव का उम्मीदवार विश्वविद्यालय छात्र संघ के चुनाव में लगातार जीतता रहा। हम लोगों ने कई सांस्कृतिक कार्यक्रम किये और नये समाज के निर्माण हेतु जनचेतना विकसित की। मगर दुर्भाग्य यह रहा कि मेरे विभागाध्यक्ष को यह रास नहीं आया कि मैं विश्वविद्यालय संघ का अध्यक्ष पद सँभालूँ। उन्हें डर था कि डीन के रूप में जो गलत कार्य किये हैं, उनका पर्दाफाश मेरे द्वारा हो जाएगा। सो उन्होंने मेरे खिलाफ विश्वविद्यालय प्रशासन के कान भर दिये। उच्च प्रशासक कान के कच्चे निकले और उन्होंने अनुशासनहीनता के आरोप में मुझे विश्वविद्यालय से निकाल दिया। उसी समय मैंने अपनी डिग्रियाँ फाड़ दीं। अलग से कुछ करने को ठान ली। इसलिए मैं कोशिश कर रहा हूँ कि अपने तन-मन से इस पिछड़े समाज को कुछ जगाऊँ, इसे उद्वेलित करूँ और इसकी हौसला अफ़जाई भी करूँ।...'' यह कहते-कहते उनकी आँखें डबडबा आयीं। वे अपने रूमाल से अपनी आँखें पोंछने लगे। मैं भी अपनी आँखें मलने लगा। एक बार फिर मेरा विश्वास दृढ़ हो गया कि श्रीश जी अन्दर और बाहर, शब्द और कर्म से एक हैं। सो मैं उनकी एक-एक गतिविधि पर निगाह रखने लगा जिससे मैं भी उनके आदर्श अपने जीवन में उतार सकूँ।

वहाँ मौजूद सभी लोगों ने तालियों से उन्हें बधाई दी। धीरे-धीरे समय बीतता गया। श्रीश के कार्यों में और निखार आता गया। उन्होंने प्रदेश के पिछड़ेपन पर एक किताब भी लिखी। मगर उसके छपने में दो साल लग गये। उनकी इच्छा थी कि वह किताब दिल्ली के सबसे प्रतिष्ठित प्रकाशन संस्था में छपे। पर वहाँ पांडुलिपियों की भरमार थी। सो उनकी पांडुलिपि स्वीकार हो जाने के बाद भी छपने में दो साल लग गये।

इस बीच सूबे की चुनाव राजनीति में भूचाल आ गया। कांग्रेस की सरकार बुरी

तरह हार गयी और जनतान्त्रिक मोर्चे में पुराने समाजवादी, लोहियावादी, गाँधीवादी जे.पी. वादी, वगैरह सभी शामिल थे। अधिकांश नेता युवा थे और अन्य पिछड़े वर्ग से आते थे। सो सामाजिक न्याय की शक्तिशाली सरकार बन गयी। सरकार बनने के एक माह बाद श्रीश की पुस्तक छपकर आ गयी। उसका शीर्षक था 'विकास की राह में काँटे'। प्रकाशक की सोच थी कि यदि इसका नये मुख्यमन्त्री महोदय विमोचन कर दें, तो उसकी अच्छी-खासी बिक्री हो जाएगी। श्रीश ने भी अनमने ढंग से हुँकारी भर दी। सो अन्तर्राष्ट्रीय श्रम दिवस, यानी पहली मई को पुस्तक का विमोचन कार्यक्रम तय हो गया। सम्पादक के अलावा तीन अन्य वक्ताओं ने उस किताब की भूरि-भूरि प्रशंसा की। उस किताब में सच और साहस भरपूर था। एक खोजी पत्रकार के रूप में श्रीश ने नदियों, बाँधों, जंगलों, सड़कों, बिजली, जलापूर्ति, शिक्षा और स्वास्थ्य जैसे गम्भीर मुद्दों को बखूबी उठाया था। विमोचन के बाद मुख्यमन्त्री ने भी जोशीला भाषण दिया। फिर उसके पन्ने पलटने शुरू किये। अचानक समर्पण पृष्ठ पर उनकी निगाह टिक गयी। मुख्यमन्त्री ने उसे पढ़कर श्रीश की ओर देखा। फिर दुबारा पढ़ा, ''अपने पूज्य पिताजी श्री रघुवर चौधरी को समर्पित, जिन्होंने आजीवन सामाजिक न्याय के लिए संघर्ष किया।''

मुख्यमन्त्री ने इधर-उधर देखा। श्रीश के चेहरे को पढ़ा। मगर उनसे नहीं रहा गया। फिर वे श्रीश की ओर देखकर बोले, ''मैंने तो आपको गोरा चित्ता देखकर भूराबाल समझा था मगर आप तो अपनी बिरादरी के निकले। चलिए, आज से हमारा ख्याल कीजिए। मैं आपका ख्याल करूँगा।''

परन्तु श्रीश इस प्रसंग से लजा-से गये। उन्होंने कभी जात-पात की बात नहीं की थी। समारोह खत्म होने पर श्रोताओं में कानाफूसी होने लगी। अब तक कोई नहीं जानता था कि श्रीश की सामाजिक पृष्ठभूमि क्या है। वह सभी लोगों से आत्मीयता से व्यवहार करते थे। मैं खुद दूसरी जाति का होते हुए भी उनके काफी नज़दीक था।

सूबे में पंचायतों का चुनाव लम्बे अरसे से नहीं हो रहा था। सो सम्पादक के नाम सैकड़ों पत्र आया करते थे। एक दिन सम्पादक ने श्रीश से कहा कि वह इस पर मुख्यमन्त्री का साक्षात्कार लें। श्रीश ने मुख्यमन्त्री सचिवालय से समय माँगा। मुख्यमन्त्री ने अपने नाटकीय अन्दाज़ में कहा, ''ऐसे समय कैसे मिलेगा? कुछ दिन मेरे साथ टूर पर चलिए। जनता का उत्साह देखिए। उस पर रिपोर्टिंग कीजिए, तब इंटरव्यू लीजिए।''

श्रीश ने सम्पादक को सारी बातें बता दीं। वह खुद टूर पर नहीं जाना चाहते थे। सम्पादक ने टूर पर जाने की सहमति दे दी। उन्होंने पहली बार हेलीकाप्टर से एक सीमावर्ती जिले का दौरा किया। यात्रा के दौरान उन्हें काजू-बादाम दिया

गया, मगर श्रीश ने मना कर दिया। गन्तव्य पर पहुँचने के बाद मुख्यमन्त्री ने पूरे उत्साह के साथ भाषण दिया, "अब हम चालीस साल तक शासन करेंगे। कांग्रेस ने चालीस साल तक शासन किया है, तो हम उनसे कम नहीं हैं।" यह कहते हुए वे अपने साथ कुर्सी उठाये हुए माइक के पास तक गये। फिर बोले, "मनुवादी ताकतों ने देश का बंटाढार कर दिया है। लेकिन अब भूरा बाल साफ करना है। इसका मतलब नहीं समझे? भूमिहार, राजपूत, ब्राह्मण, लाला। यही हैं सवर्ण अपने सूबे में। बाकी तो दबले-कुचले, गरीब-गुरबा हैं। उनको संगठित कर शासन करना है। लोहिया और जे.पी. के सपनों को साकार करना है।"

अगले दिन अख़बारों के मुख्य पृष्ठों पर उस कार्यक्रम का 'आँखों देखा हाल' छपा। 'भारत-भारती टाइम्स' में उनकी रंगीन तस्वीर भी छपी। सकारात्मक रिपोर्टिंग देखकर मुख्यमन्त्री बाग-बाग हो गये। सो उन्होंने श्रीश को फोन कराकर साक्षात्कार का समय नियत कर दिया। उस दिन शाम को पर्यटन विभाग और 'युवा मारवाड़ी संघ' के संयुक्त तत्वावधान में एक फैशन शो आयोजित था। श्रीश ऐसे कार्यक्रमों में पहले कभी नहीं गये थे। मगर 'युवा मारवाड़ी संघ' के कुछ युवकों के अनुरोध पर वे इसमें शरीक हुए। पुरस्कार वितरण के बाद उन्हें भी मंच पर 'दो शब्द' बोलने के लिए बुलाया गया। मैं भी वहाँ गया था। उन्होंने फैशन शो को उपभोक्तावादी संस्कृति से जोड़ते हुए गम्भीर भाषण दिया।

अगले दिन साक्षात्कार निर्धारित था। इसी बीच जापान से एक शिष्टमंडल आ गया। मुख्यमन्त्री उस शिष्टमंडल से देर तक बतियाते रहे। इसमें एक घंटे का समय जाया हुआ। श्रीश को अन्दर से कोफ़्त हो रही थी कि पहले से समय नियत रहने का क्या मतलब है? तभी अन्दर से बुलावा आया। श्रीश अन्दर चले गये। जापान का शिष्टमंडल चाय की चुस्कियाँ ले रहा था। मुख्यमन्त्री ने श्रीश का परिचय शिष्टमंडल से कराया और कहा, "यह भी यहाँ से आपके यहाँ सांस्कृतिक विनिमय कार्यक्रम के तहत जाएँगे। ये पत्रकार तो हैं ही, बड़े लेखक भी हैं।" शिष्टमंडल के सदस्यों ने बारी-बारी से हाथ मिलाकर उन्हें जापान आने का आमन्त्रण दिया। शिष्टमंडल के प्रमुख ने अपना विजिटिंग कार्ड भी उन्हें थमा दिया। शिष्टमंडल के जाने के बाद मुख्यमन्त्री श्रीश को अन्दर के कमरे में ले गये। उन्हें अपने परिवार वालों से मिलाया। फिर अपने सरकारी आवास की खेती-बाड़ी भी उन्हें दिखायी। उसके बाद बोले, "श्रीश जी! आपकी रिपोर्टिंग लाजवाब थी। आख़िर हैं तो आप अपने आदमी न? दूसरा कोई सवर्ण होता, तो पता नहीं क्या ऊटपटांग लिखता। आपने जो यह लिख दिया था "एक नेता ने मंच से कहा—भूरा बाल साफ करो" यह मेरे लिए सकारात्मक निकला वरना सारे विरोधी मेरा मुँह नोंच लेते। यह तो जनतन्त्र है। यहाँ सबको बोलने, लिखने, विरोध करने

का अधिकार है। यदि मैं हिटलर होता, तो इन सवर्णों के विरोध को कानून से रोक देता, मगर दुनिया बहुत आगे बढ़ गयी है। आप तो भोगे होंगे सवर्णों का शोषण? आपसे ज़्यादा आपके पिताजी भोगे होंगे जिन्हें चारपाई पर गोड़वानी बैठाया जाता रहा होगा, ऊँची जातियों के लोगों के आने पर चारपाई से उठना पड़ता रहा होगा, वगैरह-वगैरह...''

''पंचायत चुनाव पर क्या कहना है आपको?'' श्रीश ने बीच में उन्हें टोक दिया। ''देखिये, पूरी पृष्ठभूमि समझनी होगी। कांग्रेस के शासन काल में पंचायतों का चुनाव हुआ था, तो उसमें तीन दर्जन लोग मरे थे। अभी चुनाव करा देंगे, तो सैकड़ों लोग मर जाएँगे...''

''यानी सरकार विधि-व्यवस्था बनाये रखने में सफल नहीं होगी?''

''नहीं ऐसी सीधी बात नहीं है। हमारी सरकार ने दलितों, पिछड़ों और अल्पसंख्यकों को ही अधिकतर जिलों में डी.एस.पी. तैनात किया है। हमें उन्हीं पर भरोसा है मगर अभी उनमें आत्मविश्वास की कुछ कमी है...''

''तो क्या अच्छे अफसरों की कमी हो गयी अचानक?''

''नहीं, नहीं! जो सबसे महत्त्वपूर्ण बात है, वह यह है कि विधान सभा का चुनााव हाल में हुआ था। बड़ी मुश्किल से हम लोगों ने रात-दिन एक करके दलितों-पिछड़ों-अल्पसंख्यकों को गोलबन्द किया है। किन्तु पंचायत-स्तर पर चुनाव होने पर फिर ये लोग बिखर जाएँगे और आपस में ही लड़-मरेंगे।''

''तो क्या यही कारण अख़बार में दिखाया जाये?''

''देखिये श्रीशजी! आप तो अपने हैं। सुलझे हुए हैं। ऐसा क्यों नहीं लिखते कि जनसंख्या वृद्धि के कारण जिलों, ब्लॉकों और पंचायतों का पुनर्गठन किया जाना है। हमारी कोशिश रहेगी कि सत्ता का विकेन्द्रीकरण भौगोलिक आधार पर भी हो। जिस गाँव में दलितों-पिछड़ों का बाहुल्य होगा, उसे ही पंचायत का मुख्यालय बनाया जाएगा। अब तो हमें दूर तक सोचना-समझना है। यह सब नहीं करेंगे, तो दीर्घकाल तक कैसे राज करेंगे?''—कहते हुए मुख्यमन्त्री खिलखिलाकर हँराने लगे।

और श्रीश ने अगने दिन साक्षात्कार छाप दिया। मुख्यमन्त्री के पक्ष को जिस चतुराई और दृष्टि-सम्पन्नता से पेश किया गया था, वह चौक-चौराहों पर चर्चा का विषय था। सम्पादक और मेरे अलावा शायद ही किसी को मालूम हो पाया कि किसने साक्षात्कार लिया था। श्रीश के नाम की जगह 'भारत-भारती टाइम्स टीम' लिखा था। वह साक्षात्कार हिन्दी साहित्य की धरोहर कहा गया।

इस साक्षात्कार के छपने के बाद मुख्यमन्त्री आवास से श्रीश को फोन आया। वहाँ उन्हें बुलाया गया था। वे बिना किसी से कुछ कहे वहाँ चले गये। मुख्यमन्त्री

ने उन्हें सीधे अन्दर बुलाया और शाबाशी दी, ''गजब की भाषा-शैली है आपकी। मैंने सपने में भी नहीं सोचा था कि सत्तारूढ़ दल और सरकार के बारे में इतने कोणों से सोचा-समझा जा सकता है। आख़िर अपना खून अपना ही होता है। अंग्रेजी में कहते हैं, ''ब्लड इज़ थिकर दैन वाटर। श्रीश जी! क्यों नहीं आप मेरे प्रेस सलाहकार बन जाते?''

''जी नहीं। ऐसा करने से मेरी प्रतिष्ठा पर आँच आ जाएगी।''

''नहीं, नहीं। मैं आपकी प्रतिष्ठा पर आँच नहीं आने दूँगा।''

और अन्त में विदाई। श्रीश ने कुछ लेने से मना कर दिया। मगर मान-मनौवल के बाद एक थैले में मिठाई और मेवे का उपहार उन्हें दे दिया गया। वहाँ से वे सीधे घर आये। घर में बच्चे इस उपहार से प्रसन्न हुए। यह श्रीश के घर में पहला उपहार था।

इसी बीच सम्पादक दिवाकर का वहाँ से ट्रांसफर हो गया। विरोधी पक्ष ने इस अख़बार समूह के मालिक को बताया था कि वह सरकार के नक्शे-कदम पर चल रहे हैं। सबूत के लिए मुख्यमन्त्री का साक्षात्कार पर्याप्त था। मगर श्रीश की निकटता उन अधिकारियों से बढ़ती गयी जो मुख्यमन्त्री के करीब थे। गोपनीय ख़बरें श्रीश ही उन अधिकारियों के ज़रिये पाते थे।

इसके बाद कोई सम्पादक वहाँ नहीं आया। एक अधेड़ उम्र के नलिन जी प्रभारी की हैसियत से काम करने लगे। वे इस कार्य में दक्ष नहीं थे। प्रबन्धन हावी था। सो सम्पादकीय की जगह 'प्रबन्धकीय' छपने लगा। सरकार-विरोधी ख़बरें कम-से-कम छपने लगीं। श्रीश के अन्दर की बेचैनी, जज़्बात और प्रतिबद्धता कुन्द होने लगी। यदि कोई सहकर्मी नयी ख़बरें खोज-खाज कर लाता, तो उसे हतोत्साहित किया जाता। उस ख़बर को बासी बना दिया जाता। बिक्री पक्ष पर भी ध्यान नहीं दिया जाता। स्थानीय प्रबन्धन को मुख्यालय से संकेत मिल गया था कि प्रादेशिक संस्करण बन्द करना है। तो उसने भी प्रबन्धन ढीला कर दिया। प्रसार-संख्या घटने पर पत्रकारों की छँटनी होने लगी। तो बेचैन हुए सभी सहकर्मी श्रीश के पास गये। पहले वह झुँझलाये। मगर समवेत स्वर में प्रबंधन का विरोध सुनकर उन्होंने श्रम विभाग से दखल देने का वादा किया। जब कुछ लोगों ने शिष्टमंडल ले जाने का सुझाव दिया, तो वह ख़फा हो गये, ''आप लोगों को मुझ पर विश्वास नहीं है, तो मैं पत्रकार यूनियन से त्याग पत्र दे देता हूँ।''—वह गुस्से से काँप रहे थे। उनके मुखड़े पर पसीना चुहचुहा आया था।

कुछ वरीय सहकर्मियों ने उन्हें समझाया-बुझाया। उन्हें मालूम हो गया था कि आजकल श्रीश की पूछ सत्ता के गलियारे में है, कि श्रीश की मुख्यमन्त्री से खूब छनती है, कि सिर्फ़ वही मुख्यमन्त्री के बेडरूम तक जा सकते हैं, कि यदि

वे मुख्यमन्त्री को कनविंस कर दें, तो प्रबन्धन छँटनी नहीं कर पाएगा, वगैरह-वगैरह।

मगर श्रीश को 'चैम्बर ऑफ कामर्स' के द्वारा आयोजित मेले में शरीक होना था। वे वहाँ 'प्रदेश के औद्योगिक विकास' पर बतौर मुख्य वक्ता भाषण दे रहे थे। उसमें वह पिछले शासन की धज्जियाँ उड़ा रहे थे और बार-बार दुहराते थे, "नयी सत्ता को काम करने के लिए पूरा समय देना होगा। उद्योगपतियों को सरकार से कन्धा से कन्धा मिलाकर चलना होगा। तभी गरीबों के हित में सही विकास सम्भव होगा।" इस पर मुझे कुछ असमंजस-सा हुआ।

और अन्ततः श्रीश मुख्यमन्त्री से मिलने गये। मुख्यमन्त्री के यहाँ उस अख़बार के स्थानीय महाप्रबन्धक पहले से ही उनका इन्तज़ार कर रहे थे। दोनों ने गर्मजोशी से हाथ मिलाया। मुख्यमन्त्री ने दोनों की बातें सुनीं। मुख्यमन्त्री की बातों से लगा कि उन्हें सिर्फ़ एक ही चिन्ता थी कि श्रीश का क्या होगा? मगर महाप्रबन्धक ने कच्ची गोली नहीं खायी थी। उसने वैकल्पिक चिकित्सा का नुस्खा पेश किया, "सर, श्रीश को चीफ रिपोर्टर के पद पर दिल्ली में नियुक्त कर लिया जाएगा। इतना मेरा वादा है।" "मगर इन्होंने इतने सालों तक इस अख़बार में काम किया। आपके अख़बार को शोहरत दिलायी। ये तो बेचारे गरीबी में पले-बढ़े हैं, किसी जमींदार परिवार के थोड़े हैं। घर में भरा-पूरा परिवार है। उन सबको सँभालना है। अभी तक इन्होंने अपना घर-बार नहीं बनाया है। आपको इतने बड़े बोझ से मुक्ति दिला रहे हैं, तो इनका भी आगा-पीछा देखिये। अपने सहकर्मियों के बीच बदनाम होंगे और फूटी कौड़ी भी नहीं मिलेगी, तो 'माया मिली न राम' वाली कहावत सिद्ध होगी। न घर के रहेंगे, न घाट के।"—मुख्यमन्त्री ने श्रीश के लिए पुरजोर वकालत की। श्रीश की समझ में नहीं आ रहा था कि वह क्या कहें।

महाप्रबन्धक कुछ देर तक चुप रहे। मुख्यमन्त्री ने उन्हें फिर प्रोत्साहित किया। महाप्रबन्धक उठकर खड़े हो गये। थोड़ी दूर हटकर उन्होंने मोबाइल पर मुख्यालय से बात की। फिर मुख्यमन्त्री की ओर मुखातिब होकर कहा, "सर! हम श्रीश जी का पूरा ख्याल रखेंगे। हम रिश्तों को पूरी तरजीह देते हैं।"

"यह हुई न काम की बात। क्यों श्रीश जी?" मुख्यमन्त्री ने कहा।

"सर! मैं अभी कुछ कहने की स्थिति में नहीं हूँ। समूह, व्यक्ति से बड़ा होता है। मेरी पूरी यूनियन का सवाल—बड़ा सवाल है। मैं अपने लिए कुछ ले लूँ और हमारे सहकर्मी खाक छानते फिरें, यह बात मेरे गले नहीं उतरती।" श्रीश ने जवाब दिया। "देखिए श्रीश जी! आदमी को व्यावहारिक होना चाहिए। सिद्धान्त अच्छी चीज़ है, मगर सरज़मीन पर हमेशा उसे नहीं उतारा जा सकता। दुनिया तेजी से बदल रही है। 'दुनिया के मजदूरो एक हो' का नारा देने वाला सोवियत

रूस फूटी ककड़ी-सा छितरा गया है। अपने ही देश के बड़े-बड़े नेता, कामरेड, वगैरह गैताल में चले गये हैं। अब तो जो अपना और अगली पीढ़ी का पूरा ख्याल करके आगे बढ़ेगा, वही सफल होगा। आजकल बुढ़ापे में असहाय बाप को बेटा भी नहीं पहचानता। फिर आपके अख़बार में सामन्ती तत्त्व भरे पड़े हैं। वहाँ से भी भूरा बाल साफ हो जाएगा, तो कौन-सी आफ़त आ जाएगी?'' मुख्यमन्त्री ने पूरी दृढ़ता से कहा।

श्रीश को ये बातें अच्छी नहीं लगीं। वे अब तक सफलता के बजाय सार्थकता को महत्त्व देते रहे थे। श्रीश ने बाद में निर्णय लेने को कहा और वह चल दिए। महाप्रबन्धक भी मुख्यमन्त्री से दुआ-सलाम करके प्रस्थान कर गये। दफ्तर लौटकर श्रीश ने बताया कि अभी कोई फैसला नहीं हुआ है। लेबर कमिश्नर के यहाँ कल बातचीत होगी।

लेबर कमिश्नर को मुख्यमन्त्री ने पहले ही फोन से ब्रीफ कर दिया था कि वह प्रबन्धन के द्वारा तालाबन्दी करने की अनुमति दे देंगे। श्रीश सहित पाँच पत्रकार यूनियन की ओर से लेबर कमिश्नर के यहाँ पहुँचे। महाप्रबन्धक और प्रबन्धक वहाँ पहले से मौजूद थे। त्रिपक्षीय वार्ता असफल रही। दो घंटे की गहमागहमी के बाद लेबर कमिश्नर ने प्रबन्धन को तालाबन्दी की अनुमति दे दी। पत्रकार यूनियन के कई सदस्यों, जिसमें मैं भी था, ने इसका प्रतिरोध किया और लेबर कमिश्नर पर सीधे-सीधे आरोप लगाया, ''आप प्रबन्धन से मिल गये हैं।''

लेबर कमिश्नर ने श्रीश की ओर देखा। वह अपनी नज़रें नीचे गड़ाये हुए थे। मानो खुद अपनी नज़रों से अपने को छिपा रहे हों। यूनियन ने अपने महासचिव से कुछ नहीं पूछा।

अगले दिन 'भारत-भारती टाइम्स' में तालाबन्दी हो गयी। दूसरे अख़बारों ने इस ख़बर को मुख्य पृष्ठ पर छापा। लेकिन जो ख़बर नहीं छपी थी, उसे तीन ही लोग जानते थे—मुख्यमन्त्री, महाप्रबन्धक और श्रीश।

अब उस अख़बार के सारे पत्रकार पस्त थे। नाना प्रकार की प्रतिक्रियाएँ व्यक्त हो रही थीं। कई लोगों को श्रीश की चुप्पी पर आश्चर्य हो रहा था मगर किसी को उनकी ईमानदारी पर अभी तक शक नहीं था। हालाँकि वे लोग इतना जान गये थे कि नये मुख्यमन्त्री के सत्ता सम्भालने के बाद से वह वर्ग-चेतना के बजाय बिरादरी-चेतना पर उतर आये थे। मगर इससे ज़्यादा शंका की गुंजाइश हम लोगों को नहीं थी।

एक माह बाद जब सारे पत्रकार नये रोज़गार की तलाश में दर-दर की ठोकरें खाते हुए भटक रहे थे, तो श्रीश को दिल्ली में चीफ रिपोर्टर के पद पर नियुक्ति-पत्र मिला। उनकी खुशी का ठिकाना न था। उनके परिवार की डूबती नाव किनारे

लग गयी थी। वह फौरन ज्वाइन करने चले गये और पत्नी-बच्चों को उसी किराये के मकान में छोड़ गये। नयी जगह पर उन पर ज़्यादा बड़ी ज़िम्मेदारी थी क्योंकि अब सिर्फ़ एक प्रदेश का नहीं, समूचे देश का सवाल था। अब तक वह एक सूबे के विकास पर चिन्तित थे, अब सौ करोड़ जनता उनके सामने खड़ी थी।

एक माह बाद जब वे अपने पुराने महानगर में पहुँचे, तो उनके पास काफी सामान था। वे सामानों से लदे थे। नहीं, नहीं, सामान उन पर लदा था। वे मन-ही-मन सोच रहे थे कि आज उनकी पत्नी और बच्चे खुशी से झूम उठेंगे। उन्होंने जादुई अन्दाज़ में कहा, ''जानती हो, इससे क्या-क्या है?'' पत्नी ने कहा, ''मेरे और बच्चों के लिए कपड़े होंगे।'' लेकिन नहीं। उनकी अटैची में चारमीनार अपार्टमेंट के फ्लैट नम्बर दो के आबंटन कागजात और कुंजियों का गुच्छा था! उसे देखकर उनकी पत्नी चौंक पड़ी, ''आजकल पत्रकारिता कर रहे हो या और कुछ?'' श्रीश का मुँह खुला-का-खुला ही रह गया।

उभरती लकीरें

सारा रामपुर गाँव एकदम चुप था, मगर रामदुलारी दहाड़ रही थी, 'नहीं-नहीं, मैं वेश्या नहीं हूँ।'' दरअसल रामदुलारी के माँ-बाप बचपन में ही मर गये थे। उस समय उसकी उम्र सिर्फ़ तीन साल की थी। उसके बप्पा को टीबी हो गयी थी। इलाज न करा पाने और अच्छी खुराक न मिलने के करण साल भर में वह गुजर गया। बाद में रंड़ापा खेते-खेते माँ भी मर गयी। गाँव के झोला छाप डॉक्टर ने बताया था कि उसको डबल निमोनिया हो गया था। गाँव वाले कहते थे—नीमर पड़िया छत्तीस रोग। सो, वह भी रामदुलारी की सुध-बुध भूलकर सीधे स्वर्ग चली गयी। अब बची रामदुलारी। कोई भाई-बहन भी नहीं। काका-काकी थे और उनके बाल-बच्चे भी थे। काका-काकी ने रामदुलारी को पाला-पोसा और उसने भी घर से लेकर खेती-बारी तक का काम किया। रोज़ाना घर बुहारती, चौका-बर्तन करती, दुआर-गोबर करती और काका-काकी का कलेवा लेकर खेत तक जाती। काका-काकी भी उसकी मेहनत से खुश रहते थे। इसीलिए उसे कभी स्कूल नहीं भेजते थे। आख़िर वह उनकी आँखों की तारा थी न। काका-काकी के बच्चे घर और खेती के काम में ना-नुकुर करते रहते, जिसके कारण काका उन्हें डाँटते और उसी बहाने काकी को भी फटकारते कि वह बच्चों को बिगाड़ रही है। उनके बिगड़ने का एक कारण और भी था। वे सभी स्कूल जाते थे। काका का स्पष्ट मानना था कि जो भी स्कूल जाता है, वह अपनी बोली-बानी भूलकर गिटपिट-गिटपिट करने लगता है। इतना ही नहीं, वह अपने हाथों में मेहँदी लगाकर बैठ जाता है। वह न खेती करता है, न बादल का मुँह देखता है। खेती के कामों में उसका अंग नहीं निहुरता, झकझक कपड़ों में माटी लग जाएगी न। सो, काका अपने स्कूल जाने वाले बच्चों से दूर ही रहते थे। वैसे बच्चे भी इतने चालाक हो गये थे, कि वे अपने बाप से कन्नी काटते रहते थे, वरना वह किसी अढ़वा-टढ़वा के लिए उन्हें टोक देंगे।

घर और बाहर का काम करते-करते रामदुलारी सयानी होने लगी। काका

और काकी ने उसकी शादी की बात चला दी, मगर तिलक-दहेज़ का रेट आसमान पर था। जहाँ-जहाँ जाते, माँगों की लम्बी सूची सुनकर अवाक् रह जाते। काका की चिन्ता बढ़ते देखकर काकी ने अपने रिश्तेदारों से भी चर्चा चलाई। उसके नन्दोई ने एक रिश्ता बताया, मगर उस गाँव में रामपुर गाँव की कोई लड़की पहले से ब्याही नहीं थी। काकी को यही डर लग रहा था कि रामदुलारी को हारे-गाढ़े अपने गाँव-जवार की लड़कियों से भेंट-मुलाकात करने का भी मौका नहीं मिलेगा। फिर गाँव के पट्टीदार कहेंगे कि बिना माँ-बाप की बेटी को विदेश में ब्याह दिया, परन्तु काका ने इस पर खूब माथा-पच्ची की और आख़िर में निर्णय के अन्दाज़ में काकी से कहा—'सुनती हो! पट्टीदार के ताने की फिक्र मत करो। कौआ क्या खाता नहीं, पट्टीदार क्या कहता नहीं? देवी-देवताओं के पट्टीदार भी उनका हित नहीं सधने देते। नारद किसी पट्टीदार से कम थे क्या? मेरी मानो, रामदुलारी की शादी मलिकपुर के धनेसर से कर दो। शादी अपने माफिक है। बहनोई की बिचवई से सस्ते में बेड़ा पार हो जाएगा। फिर हर साल एक-एक बच्चे की शादी कर अपने सिर का बोझ हल्का करना है।'' और काकी ने हुँकारी भर दी। फिर क्या था? काका ने अपनी सामर्थ्य भर तिलक-दहेज़ दिया और रामदुलारी रोती-गाती अपनी ससुराल चली गयी।

ससुराल में सास थी और पति था। छोटा-सा घर। कच्चा घर। सास बूढ़ी हो गयी थी, इसलिए झाड़ू लगाना, लीपना, पोतना आदि न के बराबर था। रामदुलारी को घर की भीतरी गन्दगी पसन्द नहीं थी। जब वह कोनिया वाले भित्तर में रखी डेहरी से अनाज निकालने जाती तो वहाँ की गन्दगी से उसे छींक-पर-छींक आने लगती और साँस फूलने लगती। सो, उसने एक दिन लालटेन जलाकर कोने-अँतरे तक जाकर पग-पग चमका दिया। उसका घर थोड़े ही दिनों में चमकने लगा मानो चन्दन छिड़कने से महक रहा हो। उसकी सास खुशी से झूम उठी। पति ने भी एहसास किया कि ठीक ही कहा गया है—बिनु घरनी घर भूत का डेरा। सचमुच वह कच्चा मकान अब घर बन गया था। रामदुलारी की सोहरत चारों ओर फैल गयी। समूचे गाँव की औरतें अपनी बहुओं को रामदुलारी जैसी बनने की सीख देने लगीं।

मगर डाह करने वाली औरतें भी उसी गाँव में थीं। रामदुलारी को ससुराल आये एक साल बीत गया मगर उसे कोई बच्चा क्यों नहीं हुआ? उसे अन्दरूनी बीमारी है। उसकी बच्चेदानी छोटी है, उसे महीना ही ठीक से नहीं होता, वह ठिगनी है, इसलिए उसके भीतरी अंग विकसित नहीं हुए, वगैरह-वगैरह। मगर धनेसर अब तक इस पर ध्यान नहीं दे रहा था। करीब दो साल बीत गये तो उसकी एक पड़ोसिन भाभी ने उसे ऐसा ताना मारा कि वह शर्मिन्दा हो गया। उसने कई

दिनों तक यह बात अपने पेट में रखी मगर अपच होने लगी तो उसने पत्नी से यह बात कह दी। रामदुलारी लजाकर भीतर चली गयी, लेकिन धनेसर उसे कई दिनों तक छेड़ता रहा। रामदुलारी ने आख़िर में हँसते हुए कहा, "अभी तो मैं छोटी हूँ। मेरे गाँव में मेरी उम्र की कई लड़कियाँ अभी तक अनब्याही हैं। फिर लड़का-बच्चा अपने वश में थोड़े रहता है, जब बदा होगा, तो भगवान आपके अरमान पूरे कर देंगे। मेरी काकी कहती है—समय से पहले और तकदीर से ज़्यादा किसी को कुछ कभी नहीं मिलता। और धनेसर फिलहाल उस जवाब से सन्तुष्ट हो गया। उसे लगा कि रामदुलारी पढ़ी-लिखी भले नहीं है, मगर कढ़ी है। शरीर से न सही, मगर दिमाग से पक्की सयानी लगती है। सो, धनेसर का पुनः खेती-बाड़ी में मन लगने लगा। उसके उदास चेहरे पर रौनक लौटने लगी।

लेकिन विधि का विधान कुछ और था। पड़ोसिन भाभी ने धनेसर को बातों-बातों में बता दिया कि रामदुलारी तो ठांठ है, ठांठ। उसे कितना भी अच्छा चारा खिलाओ, आराम कराओ, घुमाओ-फिराओ, मगर वह लरकोर नहीं हो सकती। वह कहने लगीं, "बबूल के पेड़ में आम का फल कैसे लगेगा? उसमें से तो काँटा-ही-काँटा निकलेगा। जरा दिमाग लगाओ, कभी पत्थर पर दूब उगी है? सूखे खेत में कितनी बार भी छान मारो, मछली नहीं मिल सकती। सूरज पूरब के बजाय पश्चिम से उगने लगे तो भी रामदुलारी की कोख हरी नहीं होगी। एक औरत ही औरत की अंदरूनी बात जान सकती है।"

और वहाँ से लौटते ही धनेसर धम्म से चारपाई पर गिर पड़ा। कहने को उसकी साँसें चल रही थीं, मगर लगता था, शरीर में जान ही नहीं है। जिससे वह इतना प्यार करता था, वही खोटा है। खोटे सिक्के को कितना भी रगड़ो, वह किसी काम का नहीं हो सकता। वह बिना कुछ खाये-पिये सिरदर्द का बहाना कर चारपाई पर पड़ा रहा। रातभर इस मसले पर सोचता रहा। कभी सोचता कि कुछ समय और इन्तजार कर लेना ठीक होगा, मगर दूसरे क्षण पड़ोसिन भाभी की बातें सवालिया निशान बनकर उसका मुँह बन्द करा देतीं। फिर उसके दिमाग में एक बात आयी कि वह दूसरी औरत को ब्याहकर लाएगा और इसी घर में रामदुलारी के साथ-साथ वह भी रहेगी।

अगले दिन उसने मन की बात अपनी माँ से कह दी। माँ भी उदास हो गयी। उसकी झुर्रियों से एक उपाय निकला कि पहले दक्खिन बाबा से झाड़-फूँक करा लें। बहुत फलित होती है उनकी भभूत। सो, वह उसी दिन दक्खिन बाबा के पास गई और उनसे भभूत माँग कर ले आयी। कुछ रामदुलारी को खिला दी, कुछ उसके शरीर पर लगा दी, कुछ उसके कमरे में छिड़क दी और बाकी

एक कपड़े में बाँधकर उसकी बाँह में बाँध दी। दक्खिन बाबा ने छः माह का समय दिया।

घर वाले इन्तजार करने लगे। पाँचवे माह रामदुलारी को महीना नहीं हुआ। सास को मालूम हुआ तो वह खुशी से झूम उठी और दक्खिन बाबा का गुणगान करने लगीं। अगले दिन वह फिर दक्खिन बाबा के यहाँ गयी और मनौती कर आयीं कि पोता होगा तो उन्हें खूब बढ़िया कुर्ता-धोती और प्रसाद बाँटेगी। धनेसर को भी खुशी हुई। यह खुशख़बरी सुनकर गाँव की औरतों में रामदुलारी के पक्ष में बतकही होने लगी।

लेकिन इसी बीच अनहोनी घटी यानी रामदुलारी मासिक धर्म से गुजरी। घर में हहान-खहान पड़ गया। उसकी सास और पति के सारे अरमान आसमान से टूटकर भद्द से गिर पड़े। अगले दिन पड़ोसिन भाभी टिपोरी बोलीं, "भैया, मैं तो पहले से ही कह रही थी। इस निगोड़ी की कोख हरी नहीं होने वाली। रेगिस्तान में बाग उगने लगें तो दुनिया का आठवाँ आश्चर्य हो जाएगा। फूटी कौड़ी से सम्पत्ति में बरकत नहीं होती। मेरी मानो, कुछ दूसरा जतन करो वरना...का वर्षा जब कृषि सुखाने।" और धनेसर महरा का माथा चकरा गया जैसे पृथ्वी सूरज के चारों ओर घूमती है, वैसे ही वह रामदुलारी के चारों ओर घूम-सा गया। थोड़ी देर में चुपचाप घर लौट पड़ा। घर आते ही रामदुलारी पर बरस पड़ा। उसकी गलती सिर्फ़ इतनी थी कि वह इन्हें आते देखकर पानी लेकर आ गई। लाल-पीली आँखों में गुस्सा अदहन-सा उबल रहा था। उसने पानी भरा गिलास फेंकते हुए कहा, "साली कुतिया, हरामजादी, बाँझ, तेरे हाथ का पानी पीना भी पाप है। तू निकल जा मेरे घर से। तू यहाँ रहेगी तो हमारे कुल का सत्यानाश हो जाएगा।"

रामदुलारी बुक्का फाड़कर रोने लगी। और धनेसर महरा लात-घूसों से उसकी मरम्मत करता रहा। मार सुनकर उसकी सास दौड़कर आयी और बहू के शरीर पर ढाल बनकर झुक गई। तब वह मार से बच सकी। उसे मायके की आस बिल्कुल नहीं थी। क्योंकि उसके काका इस दुनिया से चल बसे थे और काकी अपने बेटों के अधीन थीं। चचेरे भाई शुरू से ही उससे नफ़रत करते थे। इधर कुआँ उधर खाई थी।

अगल-बगल के तमाशबीन मजा लेते रहे। कोई धनेसर को डाँटने वाला न था। हाँ, उसकी माँ उसे झिड़की दे रही थी, "मैंने तुम्हें कभी दूब से भी नहीं मारा। तुम हमारी फूल-सी बहू की जान क्यों ले रहे हो? ऊपर वाले के ऊपर किसका जोर है? जो हमारी तकदीर में लिखा है, वही होगा। इसमें बहू क्या कर सकती है?"

कुछ दिन बीत गये। धनेसर कुछ उपाय सोचता रहा। एक दिन उसे गंजहा

बाबा मिल गये, जब वह दिशा-फराकत होकर लौट रहा था। गंजहा बाबा रसिक थे अपने ज़माने के मगर अब जुगाड़ ही नहीं बैठता। तब भी वह शिकार तलाशते रहते हैं। जाने कब कोई मछली कंटिया में फँस जाए। सो, चटखारे लेकर वह धनेसर को कुरेदते रहे और वह पत्नी को ओटका-पुरान बकता रहा। आखिर में गंजहा बाबा ने उसके कान में कुछ कहा। धनेसर का मुरझाया चेहरा मानो पानी पाते ही खिल गया।

अगले दिन धनेसर ने अपनी माँ से कह दिया कि वह नानी को देखने ननिहाल जा रहा है। रात में घुप्प अँधेरा था। गाँव और बिजली एक-दूसरे के पक्के दुश्मन। दोनों एक-दूसरे को नहीं सुहाते। सास के कहने पर रामदुलारी ने लालटेन बुझा दी और आँगन में अपनी चारपाई पर सोने चली गई। अचानक रात में चार लड़के घर की डंड़वारी कूदकर अन्दर घुस गये और रामदुलारी को दबोच लिये। वह जोर से चीख पड़ी। ओसारे में लेटी उसकी सास ने गोहार लगाया तो आस-पास के लोग जुट गये। वे चारों भागने में सफल हो गये। धनेसर भी आ गया, क्योंकि वह ननिहाल न जाकर किसी के घर रुक गया था। वह जोर-जोर से चिल्लाने लगा, "यह सत्तर भतार करने वाली हरजाई है। अभी तक इसका कुकर्म खुला नहीं था। आज दूध-का-दूध और पानी-का-पानी हो गया। इसका चाल-चलन मुझे पहले से ही ठीक नहीं लगता था। मैंने भी ननिहाल जाने का बहाना बना लिया था, जिससे यह निश्चिन्त होकर अपने यारों से आँखें लड़ा सके। मैं इसे दागकर घर से निकाल दूँगा।"

भीड़ में नाना प्रकार की बातें हो रही थीं। अन्त में बड़े बुजुर्गों ने यह कहकर अपना पल्लू झाड़ लिया कि वह धनेसर का निजी मामला है और निजी मामले में दखल देना ठीक नहीं है। रामदुलारी कहती रही, "नहीं पंचो, मैं वेश्या नहीं हूँ। मैं दूसरे मरद की परछाईं से भी परहेज करती हूँ। यदि गलत होती तो गुहार क्यों लगाती?"

उसकी सास अब उसके पक्ष में नहीं थी। उसे लगा कि उसकी नाक कट गयी है। अब वह अपने बेटे के साथ थी, परन्तु चाल नहीं जानती थी। सारे नर-नारी कान में तेल डाले वापस चले गये। और धनेसर ने सचमुच उसे गरम तेल भरी कलछुल से पुट्ठे पर दाग दिया। वह कराहती रही, रोती रही, सरापती रही और ज़्यादा मार के डर से घिसट-घिसट कर बाहर की ओर चलती रही। रात बीत नहीं रही थी। वह गाँव के सिवाने पर बरगद के पेड़ के नीचे रात भर काँखती रही। उसकी कराह से रात भी बेचैन हो गयी थी और आसमान के तारे शोक मनाते-मनाते बादलों की चादर ओढ़कर सो गये।

रामदुलारी को याद आया कि राम ने भी एक धोबी के कहने पर सीता पर

लांछन लगाकर अग्नि-परीक्षा ली थी और धरती को दुःखं बर्दाश्त नहीं हुआ था। सो, धरती फट गयी और सीता उसमें समा गयी। रामदुलारी धरती को घूरती रही मगर धरती नहीं फटी और वह उसमें नहीं समाई। सुबह पश्चिम टोले की रमा उसके पास गयी। उसके घावों को सहलाया। अपने फ्रॉक की कोर से उसके आँसू पोंछे। उसे बहुत समझाया-बुझाया और सहारा देकर उसे अपने घर ले गयी। रमा को शुरू से ही उस पर दया आती थी। उसके भोलेपन पर। उसकी मेहनत पर। उसकी निश्छलता पर, मगर उसके माँ-बाप और भाई-भाभी उसे देखकर भन्ना उठे। सहसा उसका भाई चीख पड़ा, ''इस रंडी का साथ क्यों दे रही हो रमा? लगता है इसके साथ पहले से ही कुछ साठ-गाँठ है? जब समूचे गाँव ने उसे बहिया दिया तो इस कलमुँही को तूने क्यों छुआ? क्यों लायी अपने घर? हमारी भी बदनामी हो गयी। जो कोई सुनेगा कि तुम उसकी हमदर्द हो, वह हमारी इज़्ज़त में बट्टा लगा देगा। तुमने हमारी मर्यादा का ख्याल नहीं रखा। सो, आज से तुम भी यहाँ से निकल जाओ। जाओ उसके साथ। वही तुम्हारी माँ-बहन है। आज से तुम्हारे लिए इस घर का दरवाज़ा हमेशा के लिए बन्द हो गया।'' भाई ने रमा की चोटी पकड़ कर घर से बाहर कर दिया। माँ को विश्वास नहीं था कि उसका भाई रमा को निकाल बाहर करेगा। उसका विरोध कारगर नहीं हुआ।

और रमा रामदुलारी के साथ गाँव छोड़कर चली गयी। जाते समय रामदुलारी ने लाख कहा, ''रमा तू लौट जा। मेरे साथ तू क्यों आफत मोल ले रही है? मैं तो बदनाम हो ही चुकी हूँ। मैं कहीं गुजारा कर लूँगी। लौट जा मेरी बहन! लौट जा।'' कहते हुए रामदुलारी की आँखों से आँसू गिर रहे थे। रमा को काठ मार गया था। वह कुछ बोल नहीं रही थी। वे दोनों किसी अनजाने ऊबड़-खाबड़ रास्ते पर चली जा रही थीं। बहुत देर बार रमा ने रामदुलारी से कहा, ''मेरा भाई नहीं है वह। बिल्कुल शैतान है। आज तुम्हारी जो गति हुई है, उसका ज़िम्मेदार वही शैतान है। उन चार लोगों में एक मेरा भाई भी है। मैंने छुप कर उन लोगों की सारी बातें सुन ली थीं। मुझे अपने से दूर न करो भाभी!'' कहकर वह फफक पड़ी।

''रो मत बेटी! सचमुच आज से तू ही मेरी बेटी है। विपत की संगिनी, सब कुछ है तू मेरी। आज हम सिर्फ़ दो एक ओर हैं और दूसरी ओर पूरा गाँव है, लेकिन अन्याय की मियाद ज़्यादा नहीं होती। हम सरकार और समाज से लड़कर अपना हक लेंगे। समय धीरे-धीरे करवट लेगा और हमारा साथ देगा।'' कहती हुई रामदुलारी आगे बढ़ने लगी। उसकी ताकत और चाल रमा का सहारा मिलने से बढ़ गयी थी।

ध्वंस

''और पंडीजी कैसे हैं?'' बैठकखाने के बीच बहस में मेरे मुँह से अचानक निकला यह सवाल मुझे गमगीन कर गया। दो-एक दिन के लिए जब भी छुट्टी मिलती है, मैं गाँव जाता हूँ। मेरा मानना है कि पूरे विश्व में अपने गाँव जैसा आत्मीय संसार दूसरा कहीं नहीं है। पर दो-एक दिन के प्रवास में गाँव की चौहद्दी तक फैली चीज़ों का साक्षात्कार कर लेना सम्भव नहीं हो पाता। इसलिए बैठकखाने में परिवार के सदस्यों से पूछ-ताछकर सन्तोष कर लेना पड़ता है।

''कौन पंडी जी?'' मेरे छोटे भाई ने पूछा।

''अरे पागल! वही जो तुम्हारी अकल पर पत्थर पड़े रहने के कारण स्कूल में खूब धुनाई करते थे।'' मैंने कुरेदने के खयाल से छोटे भाइे को लगने वाली बात कह दी। कुछ देर तक सोचने के बाद उसने अनमने ढंग से कहा, ''पंडी जी चले गये।''

''कहाँ गये? किस स्कूल में?''

''बड़के स्कूल में।'' उसने फिर पहेली बुझाई।

''मिडिल स्कूल में कि हाई स्कूल में?'' मैंने जिरह की।

''उससे भी बड़े स्कूल में।''

''अरे उल्लू के पट्ठे साफ-साफ क्यों नहीं बताता कि किस डिग्री कॉलेज में उनकी बदली हो गयी?''

''भैया, वे इस दुनिया के नहीं, स्वर्ग की दुनिया के स्कूल में चले गये।''

''क्या? नहीं, नहीं। पंडी जी अभी कैसे मर सकते हैं? ऐसी कौन-सी लाइलाज बीमारी थी जिससे वह नहीं बच सके? सही-सही बताओ?'' यह कहते-कहते मैं बेचैन हो गया। ऐसा क्या कारण था कि सर्वनाश हो गया। मैं परदेश से कमाकर लौटा था। इस बार मैंने मन बनाया था कि पंडी जी से भेंट करूँगा और गुरु-दक्षिणा भी देने की कोशिश करूँगा। पंडी जी की मृत्यु का दुखद समाचार सुनने के बाद बातों से मेरा मन उचट गया। मैं उठकर अन्दर चला गया। उस रात मैंने खाना भी नहीं खाया। उनके साथ बचपन के गुजरे दिन फिल्म की

रील की तरह एक के बाद एक आँखों के सामने तैरने लगे...

मैं राजकीय प्राइमरी पाठशाला रोशनपुर में पढ़ता था। मैं करीब पाँच साल का था, तो मेरी माँ ने एक दिन मेरे दादा जी से मेरी पढ़ाई के बारे में चिन्ता प्रकट की। माँ ने स्कूल का मुँह नहीं देखा था, सो वह मुझे खुद अक्षर ज्ञान नहीं करा सकती थी। दादा जी ने तुरन्त जवाब दिया, ''वह किसान का बच्चा है, बड़ा होकर किसानी करेगा। पढ़-लिखकर कौन मुंशी बनना है?'' माँ ने कुछ जवाब नहीं दिया। जीवन में मेरी माँ को शायद दादा जी की बात पहली बार सबसे बुरी लगी थी। अपने दुख को उसने मर्यादा के भीतर जज्ब कर लिया था और सोचकर सन्तोष कर लिया था कि जब ऊपरवाले ने लिखा होगा, तब मेरा बेटा पढ़ेगा। मगर कुछ दिनों के बाद मेरा हमजोली विवेक स्कूल जाने लगा। उसके पिता रोडवेज में मुलाजिम थे। सो वह पढ़ाई का मूल्य जानते थे। दूसरी ओर मेरे दादा जी हर चीज़ की क़ीमत तो जानते थे मगर पढ़ाई का मूल्य नहीं जानते थे। मेरे दादा जी इतने भी बुरे नहीं थे। औपचारिक शिक्षा भले नहीं पायी थी, मगर जीवन में कढ़े थे। मगर एक माह बाद मेरे साथी विवेक को तख्ती-झोला लेकर स्कूल जाते देखकर मेरी माँ का मातृत्व भाव प्रबल हो उठा। उन्होंने दादा जी से पुनः मेरा नाम स्कूल में लिखाने के लिए चिरौरी-विनती की। इस बार दादा जी दस साल पुराना एक मुकदमा जीतने के बाद खुश थे। उन्होंने बजरंगबली के नाम पर प्रसाद बाँटा था। सो उसी दिन वह मुझे लेकर गाँव के राजकीय प्राइमरी पाठशाला पहुँच गये। वह खुशी के कारण या माँ की पीड़ा समझ गये थे या कचहरी में जाते-जाते पनही घिसने से पढ़ाई का मूल्य समझ गये थे, ठीक-ठीक उनके हृदय की थाह पाना सम्भव नहीं था। पर मुझे तैयार कर स्कूल में दाखिला कराने वे चल पड़े। दादा जी ने 'पंडी जी', जो स्कूल के सबसे पुराने और बड़े शिक्षक थे, से मेरा नाम लिखने के लिए कहा। पंडी जी ने मुझसे कहा, ''दाहिने हाथ से बायाँ कान मूड़ के ऊपर से पकड़ा, बच्चा!'' मैंने सहमते-सहमते अपना बायाँ कान पकड़ लिया। सो पंडी जी ने फैसला सुना दिया, ''ई लरिका छोटी जमात मा पढ़ै लायक अहै।'' उसके बाद उन्होंने मेरे दादा जी से मेरा नाम पूछा। दादा जी हँसकर कहने लगे, ''एकर बाल बड़ा-बड़ा अहै, तो एकर नाग बालेसरलाल।'' सो पंडी जी ने मेरा नाम बालेसर लाल लिख लिया। दादा जी को मेरी उम्र की पूरी जानकारी नहीं थी, जन्मतिथि की कौन कहे? सो पंडी जी ने अन्दाज से मेरी उम्र पाँच वर्ष और जन्मतिथि 20 अगस्त दर्ज़ कर ली।

और अगले दिन से मैं स्कूल जाने लगा। मेरे पास तख़्ती नहीं थी। सो माँ पड़ोसी के घर से एक तख्ती 'मँगनी' माँग लायी। मैंने दो-तीन जगहों से कड़ाहों से कालिख छुड़ा-छुड़ाकर 'कज्जी' बटोर ली। फिर अपनी तख़्ती पर पानी डालकर कज्जी से पुताई की, उसे सुखाया। फिर रद्दी कागज से मैंने कज्जी साफ

की और उसके बाद एक शीशी को 'घोटन्ना' बनाया। उसकी पेंदी से तख़्ती को मैंने कई बार घोटारा, जिससे वह चमकने लगी।

मगर मैं कैसे लिखता मुझे क से कबूतर मालूम था मगर क लिखना नहीं आता था। घर में कोई बतानेवाला नहीं था। दादा जी, दादी, माँ और दीदी सभी 'लिख लोढ़ा पढ़ पत्थर' थे। आज जब मैं सोचता हूँ तो लगता है मेरे जीवन के वे सबसे कठिन दिन थे जब मैं अक्षर-ज्ञान पा रहा था। पिताजी लापता थे। सो पुनः माँ ने पड़ोसी के बच्चे से मुझे अक्षर-ज्ञान कराने के लिए चिरौरी की। लड़के में सम्मान-भाव जागा कि विद्यार्थी होकर भी वह शिक्षक की तरह मुझे पढ़ाएगा।'' सो उसने दुद्धी से क ख ग से लेकर क्ष त्र ज्ञ तक मेरी तख़्ती पर लिख दिया। मुझे आश्चर्य हो रहा था कि वह कितनी जल्दी-जल्दी क में गोला, पाई आदि बनाकर आगे फटाफट बढ़ रहा था। मगर उन अक्षरों की भिन्नता, आकार-प्रकार, क्रम आदि समझने में मेरी छिछनी चटक रही थी। माँ ने 'सेंठा' छीलकर क़लम तथा दुद्धी घोलकर एक दवात तैयार कर दी थी। मेरा हाथ ऐसे काँप रहा था मानो स्प्रिंग की कोई गुड़िया हो। इस समय भी माँ ने हिम्मत नहीं छोड़ी। मगर मैं अक्षरों से बुरी तरह डर गया था। फिर माँ के कहने पर उस पड़ोसी लड़के ने 'सेंठा' की कलम मुझे पकड़ाई और खुद भी पकड़कर कच्ची दुद्धी से लिखे गये अक्षरों के ऊपर मुझसे गीली दुद्धी से दुहराने का काम करने लगा। कुछ देर की मशक़्क़त के बाद मैंने सारे अक्षरों के ऊपर गीली दुद्धी चढ़ा दी। मेरा हाथ काँपने से कई अक्षर टेढ़े-मेढ़े हो गये थे। मगर ऐसा लगता था कि मैंने माड़वगढ़ की लड़ाई जीत ली है। मगर मेरी माँ मुझसे भी ज़्यादा खुश थी। दादा जी हँसने लगे, ''ई ससुरा मास्टर-मुंशी थोड़े बने, किसान का लरिका तो किसान ही रहे। मुदा दुई अच्छर पढ़ लेहे से मर-मुकदमा के तारीख-वारीख जानि ले। चला ठीके अहै। जवन कुछ भगवान करिहें, ठीक करिहें।''

अगले दिन मैं विवेक के साथ स्कूल चला गया। स्कूल में पहुँचते ही सभी बच्चों से स्कूल की सफाई कराई गयी। मेरे स्कूल में स्कूल के नाम पर बस एक पेड़ और खुला मैदान था। पेड़ से पत्तियाँ गिरी थीं, उन्हें चुनकर हम लोगों ने एक गड्ढे में डाल दिया। उसके बाद प्रार्थना शुरू हुई। हर जमात के लिए अलग-अलग लाइन बनी थी। मैं भी विवेक के साथ छोटी जमात वाली लाइन में सबसे पीछे खड़ा हो गया। मुझे वह प्रार्थना याद नहीं थी। मगर विवेक के मुँह से बार-बार सुनते-सुनते उसकी पहली लाइन मुझे याद थी, ''तू ही राम है, तू ही रहीम है...'' मैंने वह लाइन दुहराई और आगे न जानने से चुपचाप खड़ा रहा। उसके बाद प्रार्थना करानेवाले लड़के ने कहा, ''भारत देश?'' सभी बच्चों ने कहा, ''हमारा देश।'' फिर उस लड़के ने कहा, ''इसकी रक्षा?'' सभी ने कहा, ''हम करेंगे, हम करेंगे, हम करेंगे।'' फिर उस लड़के ने कहा, ''भारत माता की।''

सभी बच्चों ने जोर से कहा, "जय।" मैं भी जोर से चिल्लाया, "जय।" जब दादा जी मुकदमा जीतकर आये थे और बजरंगबली का प्रसाद उन्होंने बाँटा था, तब भी हम सभी बच्चों ने जोर से "जय" कहा था। मुझे जय कहना अच्छा लगा।

थोड़ी देर के बाद छोटी जमात वाले अपनी-अपनी तख़्ती लेकर पंडीजी के पास खड़े हो गये। उनकी बड़ी-बड़ी मूँछों और भारी-भरकम शरीर से मुझे डर लगता था। थोड़ी देर बाद वह मेरे सामने खड़े हो गये और मैंने उन्हें अपनी तख़्ती दिखा दी। वह खुश होकर मेरी पीठ थपथपाने लगे। मैं भी मुस्कराने लगा। अक्षर के महत्त्व का पहली बार अहसास हुआ। मैंने पहली बार दुनिया जीती थी। वैसे पंडी जी का डील-डौल देखते ही बच्चों में डर समा जाता था। धीरे-धीरे मैं पंडी जी को जीतने लगा था। उन्होंने मुझे खूब लिखने के लिए कहा, चाहे जितनी गलतियाँ हों।

फिर हम लोग स्कूल के पिछवाड़े वाले तारा पर तख़्ती धोने और कज्जी लगाने के लिए गये। वहाँ मैंने विवेक से पूछा, "ई भारत देश कवन चीज होय? अउर भारत माता के अहीं?" लेकिन विवेक को भी नहीं पता था जबकि वह मुझसे पहले से पढ़ता था। दूसरे बच्चे भी नहीं बता पाये। फिर तख़्ती पुतारकर कज्जी लगाकर हम सबने अपनी-अपनी तख़्ती सुखाई। फिर काग़ज़ से साफ कर घोटन्ना से घोटारी। अब समस्या थी कि मेरी तख़्ती पर कौन लिखे? मैंने विवेक से मदद माँगी। उसने अपना काम पहले करने का बहाना किया। मैं अपनी माँ को याद करने लगा। मगर बीच-बीच में 'भारत-देश' और 'भारत माता' टपक पड़ते जैसे दाल-भात खाते समय कंकड़। मैं कुछ देर तक बैठा रहा। मैं अपनी तख़्ती को फिर से घोटारने लगा। बैठे से बेगार भली। थोड़ी देर में पंडी जी ने सूरजरूपी घड़ी की ओर ऊपर ताका और कहा, "खाने की छुट्टी।" सारे बच्चे हल्ला मचाते और 'खाने की छुट्टी बजाने की ढोल' कहते हुए फुर्र हो गये। मैं भी अपने मुँह से पीं-पीं की सीटी बजाता हुआ घर पहुँचा। पहुँचते ही माँ ने मुझे अपनी गोद में उठा लिया और मेरा माथा चूमने लगी। मैंने देखा कि उसके हाथ में भी कज्जी लग गयी, तब मैंने अपना चेहरा 'शीशे' में देखा। मैं तो कज्जी लगने से बन्दर बन गया था। फिर माँ ने मेरा हाथ-मुँह धोया और मुझे खाना परोसा। उन दिन मुझे खूब भूख लगी थी। मैंने छककर खाया। फिर मैंने माँ को अपनी कठिनाई बतायी कि मेरी तख़्ती पर दुद्धी से दुबारा किसी ने नहीं लिखा। माँ ने शाम को लौटने पर फिर लिखवाने का वादा किया।

खाना खाने के बाद मैं फिर बच्चों के झुंड के साथ गाते हुए स्कूल चला गया। थोड़ी देर में पंडी जी भी अपने घर से खाना खाकर आ गये। उसके बाद

वह हमारी छोटी जमात के पास पहुँचे। मेरी खाली तख़्ती देखकर वह सारा मजमून भाँप गये। उन्होंने बड़ी जमात के एक लड़के से मेरी तख़्ती पर कच्ची दुद्धी से लिखने को कहा। उसने लिख दिया। मैं उसके ऊपर दवात में घुली दुद्धी से लिखने लगा। मगर मेरा हाथ काँप रहा था। सो उस लड़के ने मेरी उँगली पकड़कर मुझसे लिखवाया। वह लड़का मेरे लिए उस समय भगवान था क्योंकि भीतर से मैं डर रहा था कि यदि मैं नहीं लिख सका तो पंडी जी मुझे मोटी छड़ी से पीटेंगे। उनके बारे में सुना था कि वह मोटी अकल वालों को मोटी छड़ी से पीटते हैं जिससे उनकी बन्द अकल खुल जाये। मुझे पेशाब लगी। मैं देर तक दबाये बैठा रहा मगर सहन नहीं हो पा रहा था सो मैंने पंडी जी के सामने खड़े होकर एक उँगली दिखायी और उन्होंने मुझे जाने की आज्ञा दे दी। मैं मन-ही-मन सोच रहा था कि यदि मेरी फूल जैसी कोमल हथेलियों पर उनकी छड़ी पड़ी, तो मेरी हथेली टूटकर जमीन पर गिर पड़ेगी और मैं लूला हो जाऊँगा। मगर नहीं, पंडी जी मेरे लिए यमराज नहीं, भोले शंकर सिद्ध हुए। इसी प्रकार मैंने दूसरे दिन की लड़ाई भी किसी तरह जीत ली। फिर पंडी जी ने भारत देश और भारत माता के बारे में बताया।

फिर क्या था? माँ ने हौसला बढ़ाया और मैं फटाफट लिखने लगा। अक्षर-ज्ञान के बाद मैंने गिनती भी सीख ली। पहले मुँहजबानी सौ तक गिनती याद कर ली। फिर तख़्ती की दूसरी ओर गिनती भी लिखने लगा। जैसे नया बाछा कुछ दिनों की हैरानी के बाद हल खींचने में सफल हो जाता है, वैसे मैं भी लिखने-पढ़ने में सफल हो गया। मैं आसमान में उड़ने लगा। पहली सीढ़ी की सफलता से आगे की सीढ़ियाँ आसान लगने लगती हैं। मैं सरल शब्द भी लिखने लगा। फिर छोटे-छोटे वाक्य लिखा, जैसे, ''दिन निकला। मीरा पानी भर लायी।'' मैंने माँ का नाम लिखा। माँ ने उसे अपनी उँगलियों से टटोला, फिर मेरी ओर देखा और उसने मुझे अपने कलेजे में समा लिया। उसकी आँखों से आँसू गिरने लगे। ऐसा लगा, माँ ने खुद अपना नाम लिखा हो।

अन्ततः मैं छोटी जमात से पास होकर बड़ी जमात में चला गया। मैं खुश, माँ खुश और दादा-दादी भी खुश। दीदी भी खुश थी मगर ऐसा लगता था जैसे उसका हक परिवार ने छीन लिया हो। तभी तो वह माँ को कोस रही थी कि उसे क्यों नहीं पढ़ाया? माँ ने बार-बार यही कहा था, ''तू पराया धन है और तुझे चौका-बर्तन सम्भालना है।''

बड़ी जमात में पंडी जी नहीं पढ़ाते थे, वहाँ मुंशी जी पढ़ाते थे। मुंशी जी शिव नारायण लाल। वह बड़े नेक इनसान थे। मगर वह अवधी नहीं, खड़ी हिन्दी

बोलते थे। हम बच्चे अवधी में ही उनसे बातें करते थे। मुंशी जी कभी छड़ी नहीं छूते थे। वे बच्चों का कान पकड़कर हल्के से उमेठ देते थे। वह पढ़ाने में ध्यान देते थे। वह हम लोगों को पहाड़ा याद करवाते थे। फिर बच्चों के अलग-अलग समूह बनाकर आपस में प्रतियोगिता करवाते थे। फिर खेल-कूद पर उनका ज़्यादा जोर रहता था। लम्बी दौड़, लँगड़ी, खो-खो, चोर-सिपाही, चूहा-बिल्ली आदि खेल होते थे। वह खूब हँसते थे और खूब हँसाते थे। रोते बच्चों को भी वह हँसा देते थे। पंडी जी से जितना ज़्यादा बच्चे डरते थे, मुंशी जी से उतना ही ज़्यादा घुलते-मिलते थे। मुंशी जी ने पढ़ा-लिखाकर मेरी नैया पार लगा दी। यानी मैं कक्षा दो का विद्यार्थी बन गया। मुझे ऐसा लगने लगा कि मैं छोटी और बड़ी जमात के ऊपर हूँ। उन जमातों के बच्चे कभी-कभी मुझसे कुछ पूछने आते थे और मैं सहर्ष बतला देता था।

कक्षा दो में, 'मास्टर जी' हमें पढ़ाते थे। राम प्रसाद नये-नये मास्टर हुए थे। उन्होंने पहले ही दिन हम लोगों को कहानी सुनायी। वाह क्या राजा था, क्या रानी थी। मेरा मन करता था कि काश, मैं भी उस राजा-रानी के घर पैदा हुआ होता और अच्छी-अच्छी चीज़ें खाता-पीता, पहनता-ओढ़ता और सैर-सपाटे करता। मास्टर जी कहानी सुनाकर बोले, "तुम लोग भी ऐसे बन सकते हो। खूब मेहनत करो।" धीरे-धीरे उनसे भी दोस्ती हो गयी। मगर उनमें एक ख़राबी थी, वह बीड़ी पीते थे। हालाँकि पंडी जी के डर से वह स्कूल में बीड़ी नहीं पीते थे, मगर थोड़ी दूर जाकर खूब धुआँ-धड़ाका करते थे। एक बार मैंने अपनी आँखों से उन्हें बीड़ी पीते हुए देखा था। मेरा मन हुआ कि पंडी जी से मास्टर साहब की शिकायत कर दें मगर मेरी घिग्घी बँध गयी। एक तो पंडी जी का डर था, दूसरे राम प्रसाद मास्टर जी का। खैर...वे बहुत अच्छा पढ़ाते थे। कई बार समझाते थे। जो बच्चे नहीं समझते थे, उन्हें डाँटने-फटकारने के बजाय पुचकारकर समझाते थे। इमला, हिसाब, सुलेख वगैरह पर खूब ध्यान देते थे। वे महीने में एक-दो दिन अंत्याक्षरी कराते थे। फिर नाटक, गीत, लोकगीत, राष्ट्रीय गीत आदि कार्यक्रम कराते थे जिनमें हम बच्चों को खूब मज़ा आता था। अब हम कॉपी पर लिखने लगे थे। इसलिए कज्जी की जगह स्याही ने ले ली थी। अब तख़्ती पुतारने, और घोटारने का झंझट बिल्कुल नहीं था। उन्होंने एक काम और कराया था—पी.टी.। इसमें मुझे मज़ा आता था। उन्होंने मुझे अगुआ बना दिया। सो मैं सीटी बजा-बजाकर तरह-तरह से पी.टी. कराता था। पर मुझसे कभी भी शीर्षासन नहीं हुआ। मैं डरता था कि कहीं गिरने से शरीर का कोई अंग भंग न हो जाये। स्कूल के खुले मैदान में आसन करना धूल का फूल बनना था। मगर पी.टी. टूर्नामेंट के

बहाने हमें नये-नये पैंट-बनियान अपने-अपने घर से मिले। पी.टी. में पूरे ब्लॉक में हमारी टीम दूसरे नम्बर पर आयी थी, इसलिए मास्टर जी ने मुझे खूब शाबाशी दी। बाद में मुझे पता चला कि चूँकि मुझसे शीर्षासन नहीं हो पाता था, इसलिए मुझे अगुआ बना दिया गया था। मगर इस सफलता से पंडी जी गद्गद् थे। पूरे ब्लॉक में उनके विद्यालय का नाम रौशन हुआ था।

फिर दूसरी, तीसरी, चौथी जमात भी मैं पास कर गया। पाँचवीं कक्षा में फिर पंडी जी से पाला पड़ा। पाँचवाँ दर्ज़ा वैतरणी नदी माना जाता था—जो फेल हो गया, वह ज़िन्दगी भर नरक में रहेगा। पंडी जी अब एक कुबरी लेकर स्कूल आते थे। जिस बच्चे को सवाल नहीं आता था, उसके गले में कुबरी फँसाकर अपनी ओर खींच लेते थे। परन्तु मारने से ज़्यादा डराते थे। एक दिन पंडी जी ने अपने छोटे भाई आशुतोष को एक रजिस्टर लाने के लिए अपने घर भेजा। उसने आकर बताया कि वह रजिस्टर वहाँ नहीं है। पंडी जी को गुस्सा आ गया। उन्होंने उसे अपनी कुबरी से खींच लिया और दर्जनों बार पीटते रहे। तभी मुंशी जी और मास्टर जी ने दौड़कर पंडी जी की कुबरी पकड़ ली। दरअसल पंडी जी अनुशासन के मामले में बड़े कठोर थे। उनका मानना था कि जीवन की सफलता के लिए हर हालत में अनुशासन का पालन होना चाहिए पर उनका हृदय मोम से भी कोमल था। उस घटना के बाद उनका भाई आशुतोष बीमार होकर बिछावन पर पड़ा रहा। आशुतोष जल्दी ठीक हो जाये, इसके लिए मीलों पैदल जाकर वैद्य जी से दवा वह लाते। देर हो जाती तो बिना खाना खाये समय पर स्कूल चले आते।

खैर, अब हम लोगों को बीस तक पहाड़ा, अद्धा, पवन्ना, सवैया, ड्योढ़ा, अढ़ैया आदि याद करने पड़ते थे। जैसे मेरा तोता कहता था 'सीताराम-सीताराम', वैसे हम लोग इन चीज़ों का रट्टा मारते थे। एक के गड़बड़ाने पर सब गड़बड़ा जाता था। हमारी कक्षा में तीन लड़के हरेराम, छोटेलाल और छेदी को ये सब चीज़ें बिलकुल याद नहीं होती थीं। हरेराम का पिता हलवाहा था, छोटेलाल का पिता किसान था और छेदी का पिता दुकानदार था। मगर यह भी सही था कि वे हम लोगों की औसत आयु से दो-तीन साल बड़े थे।

और हाँ, पंडी जी पढ़ाने के अलावा दैनिक जीवन के बारे में भी बतलाते थे और वैसा करते भी थे—जैसा कथनी, वैसी करनी। वह कहते थे—

मूड़ मुड़ाये छः नफा,
कन्धा मोटा, सिर सफा;
तेल, साबुन, कंघी, शीशा बचा।।

वह प्रायः सिर मुड़ाए रहते थे। उन्होंने कभी कंघी से बाल नहीं झाड़ा।

उन दिनों पाँचवीं कक्षा की परीक्षा ब्लॉक स्तर पर होती थी। हमारे स्कूल का परीक्षाफल हमेशा अच्छा होता था। परीक्षा शुरू होने के एक महीने पहले पंडी जी का फरमान जारी हुआ कि कक्षा पाँच के सभी बच्चे सूर्यास्त तक खा-पीकर उनके घर पर पढ़ने के लिए पहुँच जाएँ। अपना-अपना ओढ़ना-बिछौना प्रत्येक बच्चे को ले जाना पड़ता था और पढ़ाई के लिए सबको एक-एक बोतल किरासन तेल चन्दा के रूप में देना पड़ता था। अब तक हम लोग दिन में ही बाघ के सामने रहते थे, अब रात में भी उसके सामने रहना पड़ता था। पंडी जी रात में दो बार चेक करते थे कि कोई सो तो नहीं रहा है। सोनेवाले के ऊपर वे एक लोटा पानी डाल देते थे। फिर उसकी नींद हिरन हो जाती थी। जब परीक्षा हुई तो सारे सवाल हम लोगों को सरल लगे। सो सभी अच्छे अंकों से पास हो गये। सिर्फ़ तीन लड़के फेल हुए—हरेराम, छोटेलाल और छेदी। फेल तो अशुतोष भी हो जाता क्योंकि पढ़ाई-लिखाई से अधिक उसका मन खेलने में लगा रहता था पर पंडी जी की एक ही साध थी कि उनका छोटा भाई पढ़-लिखकर अफ़सर बने। पंडी जी ने पहले ही ऐलान कर दिया था कि यदि वे तीनों पास हो जाएँगे, तो उनकी गदोरी में बाल उग आएँगे। सो न बाल उगे, और न वे तीनों पास हुए। परीक्षाफल के बाद हमें अहसास हुआ कि पंडी जी हम लोगों की भलाई के लिए ही कड़ाई करते थे। मेरे दादा जी कहा करते थे कि बिना डंडे की मार खाये अकल का ताला नहीं खुलता सो वे अपने कथन की पुष्टि हमारे परीक्षाफल में देखते थे।

जुलाई में हम लोगों ने अपने विद्यालय से टी.सी. लेकर एक कोस दूर कक्षा छह में नाम लिखा लिया। वह स्कूल भी इलाके में नामी था। तीन दिन के बाद पढ़ाई शुरू हो गयी। पहले दिन की पहली कक्षा में शिक्षक ने पूछा, "ग्रामरक्षा दल क्या होता है?" सभी लड़के चुप। सबकी सीटी-संपाटी गुल। मगर मेरे गाँव के पाँचों लड़कों ने हाथ उठा लिए। मैंने सही जवाब दिया। इतना सुनते ही उस शिक्षक ने मेरी पीठ थपथपायी और पूछा, "किस स्कूल से पढ़कर आये हो?"

"राजकीय प्राइमरी पाठशाला, रोशनपुर।" मैंने जवाब दिया।

यह सुनकर उस शिक्षक ने मेरे स्कूल को बधाई दी और कहा, "रोशनपुर स्कूल में पंडी जी के पढ़ाये विद्यार्थी दूर से ही पहचान लिए जाते हैं। जिसकी नींव पंडी जी ने डाली हो, उसे भला कौन हरा सकता है?"

पर जीवन के भीतर रहस्य का एक और संसार चलता रहता है जिसकी गति जानना आसान नहीं है। मेरे कुरेदने पर मेरे छोटे भाई ने जो बयान किया, वह इस प्रकार है—

जिस दिन पंडी जी रिटायर हो रहे थे उस दिन रोशनपुर के प्राइमरी स्कूल

में मजमा लगा हुआ था। ऐसे बहुत से परिवार थे जिनकी दो पीढ़ियों को पंडी जी ने पढ़ाया था। उनके चेले किसी व्यवसाय, नौकरी या दूसरे काम में लगे थे। सरकारी कार्यालयों की तरह उन्हें गाँववालों ने कोई 'मानपत्र' नहीं दिया लेकिन निर्जीव कागजी मानपत्र के बजाय उनके आँसुओं के कतरे आभारपूर्ण विदाई के जीवन्त प्रमाण थे। पंडी जी को देने के लिए कई लोग कई तरह के उपहार वहाँ लाये थे मगर पंडी जी ने एक भी उपहार नहीं लिया। उन्होंने हाथ जोड़कर कहा, "प्रेम को इन क्षयशील चीज़ों से मत तौलिये। मैं आप लोगों के दिल की भाषा भीतर-ही-भीतर समझ रहा हूँ। दिल के विशाल दरिया को गाँव की पोखरी मत बनने दीजिए। मैं सरकारी नौकरी से रिटायर हुआ हूँ, मगर जीवन से नहीं। अभी तो मुक्त ज़िन्दगी की शुरुआत करनी है और सोने के पहले सैकड़ों मील चलना है..." यह कहते हुए उनका गला रुँध गया।

अगले दिन पंडी जी दस बजे सरकारी प्राथमिक विद्यालय पर पहुँच गये। विद्यार्थी अवाक्। दूसरे शिक्षक आये ही नहीं थे। उन्हें यह दृश्य देखकर बेहद तकलीफ हुई। मगर उन्होंने बच्चों से विद्यालय परिसर की सफाई कराकर प्रार्थना करवा दी। थोड़ी देर बाद दोनों शिक्षक आ गये। वे हाथ जोड़कर बोले, "पंडी जी, आपने हमारी नाक बचा ली वरना गाँव-जवार के लोग कहते कि पंडी जी के रिटायर होते ही मास्टरों ने 'मास्टरी' करनी शुरू कर दी।" यह सुनते ही पंडी जी चौंके और उन्हें अहसास हुआ कि वे सचमुच सरकारी सेवा से रिटायर हो गये हैं और पढ़ाने के लिए अब विद्यालय नहीं आना है। वह कुछ सकुचाते, कुछ मुस्कराते हुए वापस चले गये।

समय टाटा सुमो, वर्षा, स्कॉर्पियो, मार्शल जैसी गाड़ियों से भी ज़्यादा तेजी से भागता गया। एक शिक्षक यानी मास्टर साहब को राजनीति में कूदने का भूत सवार हुआ। वह अपने विद्यालय के बजाय शिक्षक संघ के कार्यालय, बेसिक शिक्षा अधिकारी के कार्यालय और चौक-चौराहों पर नियमित कक्षाएँ लेने लगे। बच्चों की जगह बड़े-बूढ़ों को 'पढ़ाने' में उन्हें बेसी मज़ा आने लगा। सर्वत्र न्याय के घोड़े दौड़ रहे थे, सो उन्होंने भी बहती गंगा में डुबकी लगा ली। अन्ततः वे जिला प्राथमिक शिक्षक संघ के महासचिव चुन लिए गये। उनका जिला स्तर पर अभिनन्दन हुआ। फूल-मालाओं से लदे वह बेसिक शिक्षा अधिकारी के कार्यालय में पहुँचे नये दल-बल के साथ। नये बेसिक शिक्षा अधिकारी मोहम्मद मुस्तफा अंसारी ने हाथ मिलाकर और गले लगाकर पूरी गर्मजोशी से उनका स्वागत किया। मास्टर साहब की बाँछें खिल गयीं। जिस शख्स का नाम सुनकर वह कोसों दूर रहकर भी सिहर उठते थे, उससे गले मिलने से उनकी हीन ग्रन्थि काफ़ूर हो गयी और लगभग बराबरी का-सा अहसास उन्हें होने लगा। आख़िर क्यों

न हो, जहाँ हज़ारों शिक्षकों के वह महावत बन गये हों वहाँ डर किस बात का? एक बार उनके दिमाग़ में यह बात कौंधी कि यदि पंडी जी न रहे होते तो वे पहले ही महासचिव बन गये होते। फिर वही सरकारी दफ़्तर उनका अड्डा बन गया। बेसिक शिक्षा अधिकारी उदारता दिखाते हुए शिक्षकों के आवेदन उनकी मार्फत प्राप्त होने पर हाथों-हाथ निपटाने लगे। तुरन्त दान महाकल्याण! मास्टर जी यानी महासचिव अपनी बिरादरी से चन्दा स्वरूप कुछ राशि हर आवेदन पर लेने लगे। सो उनकी दुकान चलने लगी। अंसारी साहब कलाकन्द मिठाई और मगही पान के बेहद शौकीन थे। सो महासचिव ने उन्हें पीतल का एक बेशकीमती पनडब्बा उपहार में दिया। अंसारी साहब फूले नहीं समाये।

दूसरी तरफ मेरे गाँव का विद्यालय दिन दूनी रात चौगुनी गति से अवनति करने लगा। मुंशी जी पाँच कक्षाओं को भेड़ की तरह खेत में 'दोबते' थे। हर कक्षा के मॉनीटर से कुछ इमला, गाणित के कुछ सवाल आदि दिलवाकर अपने काम की इतिश्री समझ लेते थे। मगर सब्र का बाँध धीरे-धीरे टूट गया। वे झल्लाने लगे और 'बारह बजे लेट नहीं, तीन बजे भेंट नहीं' का सिद्धान्त मन, वचन और कर्म से निभाने लगे। नतीज़ा यह हुआ कि बहुत सारे विद्यार्थी साल के अन्त में 'फेलार्थी' होने लगे। मगर एक दिन अचानक अध्यापक रूपी बिल्ली के भाग्य से सिकहर टूटा। सूबे की सरकार ने एक शासनादेश निकला, "वंचित बच्चों की साक्षरता दर बढ़ाने और ठहराव सुनिश्चित करने के लिए अब कोई भी विद्यार्थी कक्षा पाँच तक फेल नहीं किया जाएगा अर्थात् एक कक्षा से दूसरी कक्षा में हर विद्यार्थी प्रोन्नत होगा।" मुंशी जी यह जानकर बेहद खुश हुए कि अब पास-फेल का चक्कर नहीं होगा। उन्हें महसूस हुआ कि पंडी जी भले रिटायर हो गये वरना इस आदेश के बावजूद पढ़ाना पड़ता। मगर एक बात उनके लिए गठिया का दर्द बन गयी थी कि परीक्षा का प्रपंच खत्म क्यों नहीं कर दिया जाता। ऐसा होने पर एक बार नाम लिखने के बाद कक्षा पाँच पास होने पर टी.सी. ही लेने विद्यार्थियों-अभिभावकों को स्कूल आना पड़ेगा। मगर उनका वश कहाँ।

मगर विद्यालय में पढ़ाई न होने से पंडी जी चिन्तित होने लगे। कुछ लोगों से सलाह-मशविरा करके सुबह-शाम अपनी बगिया में बच्चों को मुफ़्त में पढ़ाना शुरू कर दिया। करीब पच्चीस बच्चे सुबह और पच्चीस बच्चे शाम को उनके पास पढ़ने के लिए जाने लगे। वह अपनी पेंशन से कुछ पैसे गरीब बच्चों के लिए कॉपी-किताब पर भी खर्च करने लगे। सो बभनौटी के अलावा चमरौटी से भी कई बच्चे नियमित रूप से वहाँ पढ़ने लगे। कक्षा पाँच के बच्चों पर वह विशेष ध्यान देते। जिससे वे अगली कक्षा में जाने के पहले मूलभूत चीज़ों को अच्छी तरह सीख लें। वह भाषा और गणित पर विशेष जोर देते थे। गाँधी जी

की तरह वह भी मानते थे कि प्राथमिक स्तर पर मातृभाषा ही शिक्षा का माध्यम हो और कक्षा छह से ही अँग्रेजी की पढ़ाई शुरू की जाए। मगर विधि का विधान कुछ और था।

एक दिन ग्राम पंचायत के प्रधान मुख़्तार चौधरी ने अँग्रेज़ी माध्यम से पढ़ाने के लिए रोशनपुर गाँव में ही सेंट जॉन पब्लिक स्कूल खोल दिया। शुरू में शीट की छत बना दी गयी। उसका उद्‌घाटन स्थानीय नेता माननीय विधायक नथुनी राम के कर-कमलों से सम्पन्न हुआ। सजावट, खान-पान आदि का पूरा इन्तजाम था। उन्होंने गद्‌गद् भाव से कहा, ''भाइयो और बहनो! अब इस इंग्लिश स्कूल में हम गरीब-गुरबों के बच्चे भी अँग्रेजी पढ़ सकेंगे। उनको अँग्रेजी पढ़ने के लिए शहर जाने का कष्ट नहीं उठाना पड़ेगा। एक सोची-समझी सामन्ती साजिश के तहत गरीबों को सरकारी विद्यालयों में पढ़ने के लिए मजबूर किया जाता रहा है, जहाँ अँग्रेजी सीखने का कोई बन्दोबस्त नहीं होता जिससे वे बड़े-बड़े कम्पटिशन से बाहर हो जाते हैं। यदि बाबा साहब अँग्रेजी नहीं पढ़ते तो विदेश नहीं जाते और फिर संविधान नहीं बनाते। अब वक्त ने करवट ली है, तेजी से सामाजिक बदलाव हो रहा है। इसके लिए मैं यहाँ के ग्राम प्रधान जी को साधुवाद देता हूँ जिन्होंने देर से ही सही, मगर गरीबों की सुधि ली। और हाँ, इस विद्यालय के भवन-निर्माण के लिए मैं अपने विधायक कोष से पाँच लाख रुपये देने की सहर्ष घोषणा कर रहा हूँ। इसके अलावा यहाँ के डी.एम. और एस. डी. एम. से मैं बात करूँगा कि इस विद्यालय को दस एकड़ सरकारी ज़मीन आवंटित की जाये।'' इस घोषणा पर श्रोताओं ने काफी देर तक जमकर तालियाँ बजाईं। ग्राम प्रधान चौधरी जी इतने खुश हुए कि विधायक जी के चरणों में लोटने लगे। वैसे उन्होंने अपने आप ग्राम समाज की करीब पाँच एकड़ ज़मीन इस पावन कार्य हेतु पहले से ही हथिया ली थी, बस उस पर हाकिम-हुक्कामों की मुहर लगनी थी। चौधरी जी ने अपने खानदान और रिश्ते के सभी बेरोज़गार नवयुवकों को शिक्षक के रूप में बहाल कर दिया और विभिन्न राजनैतिक दलों से नकद, ईंटें, सीमेंट आदि का दान लिया। अगले दिन से माननीय विधायक के सम्माननीय पुत्र ने शिक्षक के रूप में नये स्कूल में योगदान दिया। उन्हें कक्षा एक, दो, तीन, और चार में अँग्रेज़ी पढ़ाने का दायित्व दिया गया क्योंकि वह हाई स्कूल फेल थे। गाँव-जवार के युवक तो यहाँ तक कहते हैं कि उन्हें अँग्रेज़ी में ज़ीरो अंक मिले थे और जब उनके बूढ़े दादा ने पूछा कि तुम्हें गोला क्यों मिला तो उसने कहा था, ''खेती-बारी का काम करने से स्कूल पहुँचने में देर हो गयी थी। तब तक सारे बच्चे पूरे नम्बर लूट लिए थे।'' दादा जी ने इसे बच्चे की तकदीर मान ली थी। इस पब्लिक स्कूल के खुलने पर जो बच्चे पंडी जी के यहाँ पढ़ने जाते थे, उन लोगों ने वहाँ जाना छोड़ दिया और एक सौ रुपये मासिक फीस देकर सेंट जॉन पब्लिक स्कूल

में नाम लिखवा लिया। दो हफ़्ते में ही इस पब्लिक स्कूल के छात्र-छात्राओं की संख्या एक सौ से ऊपर हो गयी और सरकारी प्राथमिक विद्यालय में मात्र तेरह बच्चे बचे रह गये। पब्लिक स्कूल के प्रबन्धक और शिक्षक प्रायः व्यंग्य कसते, ''सरकारी स्कूल में तेरह बच्चे सरकारी शिक्षा का श्राद्ध करने के लिए बचे हैं।''

यह सुनकर पंडी जी बहुत दुखी हुए। उन्हें रात में नींद नहीं आयी। उनसे बैठे नहीं रहा गया। वे एक ठेगना के सहारे आसपास के टोले-मुहल्ले में घूमने लगे। सबसे चिरौरी-विनती करते, ''अपने बच्चों को सरकारी विद्यालय में न सही, तो मेरे विद्यालय में भेजो। कोई फीस नहीं लगेगी। जो गरीब माँ-बाप किताब-कॉपी नहीं खरीद पाएँगे, उनके बच्चों को मैं अपनी पेंशन से किताब-कॉपी दूँगा। अँग्रेजी का नाम देना और अँग्रेज़ी माध्यम से पढ़ना उत्तम विधि नहीं है। अँग्रेज़ चले गये मगर अँग्रेजियत नहीं गयी। अँग्रेजी रानी हिन्दी को दासी बना रही है। अधकचरे मास्टर लोग अधकचरा ज्ञान देंगे जिससे बच्चों की नींव हमेशा के लिए कमजोर ही रहेगी। यह नकली-पब्लिक स्कूल असल में प्राइवेट हैं। इसकी चमक-धमक, तड़क-भड़क में धोखा है, धोखा। सभी चमकने वाली चीज़ें सोना नहीं होतीं। अपनी बोली-बानी में बच्चे जल्दी और आसानी से सीखते हैं। गाँधी जी कहा करते थे...।'' ''मगर पंडी जी, ज़माना बदल गया है। अँग्रेज़ी अन्तर्राष्ट्रीय भाषा है जिसे बचपन में सीखने पर ही अच्छी तरह बोलना आता है।'' एक अभिभावक ने कहा।

''यह भी बात है कि अब तक आप लोग सनातनी चीज़ें पढ़ाते रहे। अब हज़ारों बच्चे नयी-नयी आधुनिक चीज़ें भी तो सीखें देश के बाहर से, वरना वे कूप-मंडूक ही रह जाएँगे।'' दूसरे ने कहा।

''जैसे पुराने कपड़ों को फेंक दिया जाता है, वैसे ही इस ज़हरीली और सड़ी-गली शिक्षा-व्यवस्था को भी फेंकना ज़रूरी है।'' तीसरे ने कहा।

पंडी जी कसमसाने लगे। उनका माथा भारी होने लगा। उन्होंने अपना सिर झुका लिया। मगर वहाँ खड़े बच्चे पंडी जी की ओर टुकुर-टुकुर निहार रहे थे। उनकी ओर से एक बूढ़े ने कहा, ''पंडी जी के यहाँ सूरज की रोशनी मुफ़्त में मिलती है और नये प्राइवेट स्कूल में जुगनू की रोशनी भी महँगी मिलती है। जुगनू की रोशनी से ज़िन्दगी का बेड़ा पार नहीं होगा।'' कुछ बच्चे कानाफूसी करने लगे कि पंडी जी के यहाँ पढ़ाई जारी रखी जाए मगर वे अपने अभिभावकों से बग़ावत करने की हिम्मत नहीं जुटा पा रहे थे। एक बच्चे ने कुछ संकेत किया तो एक महिला ने उस बग़ावत पर लगाम लगाते हुए उसे घूरा। उसकी सीटी-सपाटी गुल, चुरकी सोंटा हो गयी। ज़्यादातर बड़ों का समर्थन न पाकर पंडी जी उदास हो गये। उन्हें लगा कि बिना पढ़ाये वे कैसे साँस लेंगे? कभी आँखें फड़फड़ातीं', कभी भुजाएँ, उनकी आँखों से आँसू बहने लगे। निराशा और हताशा

के बादल उन्हें घेरने लगे। वे कुछ दूर गये थे कि कुछ बच्चों ने कहना शुरू किया, ''पंडी जी बहुत अच्छा पढ़ाते हैं, हमें वहीं पढ़ने जाने दीजिए।'' इस पर उस युवती ने उन्हें डाँटते हुए कहा, ''क्या? इस लँगड़े-अन्धे और अनकल्चर्ड आदमी से क्या पढ़ोगे? जो लकड़ी के ठेगना के सहारे चल रहा है, वह तुम्हें क्या सहारा देगा? मैं तीन साल अपने हस्बैंड के साथ दिल्ली में रहकर आयी हूँ। वहाँ अँग्रेज़ी स्कूलों का बोलबाला है। वहाँ पढ़नेवाले बच्चे फर्राटेदार अँग्रेज़ी बोलते हैं, टाई लगाते हैं। वे टी.वी. में अँग्रेज़ी वाले प्रोग्राम देखते हैं, इंग्लिश न्यूज़ पेपर पढ़ते हैं और बड़े होकर बड़ी-बड़ी नौकरियाँ हासिल करते हैं। दूसरी ओर गाँव के हिन्दी-चिन्दी माध्यम वाले सरकारी स्कूलों के बच्चे गोबर-गणेश होने के नाते रोजगार की तलाश करते-करते बूढ़े हो जाते हैं। यह कम्पटिशन और भूमंडलीकरण का युग है। लाखों में एक सफल होगा, सो बिना अँग्रेज़ी के कोई व्यक्ति अच्छी नौकरी का पाला भी नहीं छू सकता।''

जाते-जाते पंडी जी ने ये बातें सुन लीं। उन्होंने कई बार मुड़-मुड़कर पीछे देखा मगर उन्हें ऐसा आभास हुआ कि अधिकतर गाँववाले कह रहे हैं, 'मुड़-मुड़ के न देख, मुड़-मुड़ के।'' 'अनकल्चर्ड' शब्द उनकी पीठ में छुरे की तरह घुस गया। अँग्रेजी ज्ञान की सूचक इस गाली ने उनकी लम्बी तपस्या को भंग कर दिया था। उनकी जबान ने काम करना बन्द कर दिया। सो वह अपनी पगडंडी पकड़े आहिस्ता-आहिस्ता अपने घर चले गये।

घर पहुँचने पर आधुनिकतावाद की मार से पंडी जी निखरहरी खटिया पर ऐसे गिरे मानो आसमान से गिरे हों। वह कभी सीने पर हाथ रखते, तो कभी माथे पर। कभी अपने दोनों हाथ से अपनी आँखें मलते, तो कभी आसमान की ओर कातर दृष्टि से निहारते और मुक्ति के लिए ईश्वर से प्रार्थना करते। उनके माथे पर पसीने की बूँदें आ गयी थीं। फिर पंडी जी एक महीने तक शर-शय्या पर पड़े भीष्म पितामह की तरह आह भरते मौत का बेसब्री से इन्तज़ार करते रहे। उनका एक चेला गोवर्द्धन बार-बार कहता था, 'स्कूल के मायने पंडी जी। पंडी जी के मायने स्कूल। सरकारी स्कूल ठीक हो जाये, तो शायद पंडी जी ठीक हो जाएँगे।''

मगर सरकारी स्कूल को ठीक करना दुनिया का आठवाँ आश्चर्य होता। सेंट जॉन पब्लिक स्कूल के पीछे गाँव-जवार के अधिकतर पढ़े-लिखों तथा ग्राम प्रधान से लेकर विधायक जैसे जन-प्रतिनिधियों तक ने पूरी ताकत झोंक दी थी। सो पंडी जी खटिया से उठ नहीं सके बल्कि खटिया ने ही उन्हें एक दिन चुपचाप ऊपर उठा लिया।

अब गाँव, गाँव न रहा, ऐसा मैं सोचने को विवश था।

पुजारी काका

पुजारी काका पूरे गाँव के काका थे और उनका ठिकाना काली माँ का नीम-चौरा था। गर्मी में चौरा धोने के बाद वे नीम के पेड़ से कुछ पत्तियाँ तोड़ते और खा जाते काली माई के नाम पर।

उनके दालान पर उस दोपहर कई लोग जुटे थे। एक बूढ़ा आदमी अपनी फरियाद कर रहा था, "पुजारी काका! मेरे घर में जिन्न घुस गया है। रात में खाने-पीने के बाद मेरे आँगन में ढेले-पत्थर गिरते हैं। कई बार मेरी पतोहू चोट खा गयी। कुछ उपाय करिये, काका!"

"ठीक है, अब ढेला नहीं गिरेगा।" काका ने ऐसे सहज ढंग से कहा जैसे कोई पैर में गड़े काँटे को निकाल ले। बिसुन को याद आया कि सोमन ओझा ने उससे मुर्गा, नींबू, मिठाई और डेढ़ सौ रुपये नकद पहले लिया था मगर कुछ असर नहीं हुआ। शाम में वह किसुन के दालान में जाकर उसकी चारपाई पर बैठ गये। क्रोध से उसकी आँखें लाल थीं। वे बिगड़े, "आपके लाड़ले को कुछ-कुछ हो गया है। कुछ लोगों पर भूत चढ़ता है मगर उस पर जवानी का जिन्न चढ़ा है और वह बिसुन की पतोहू के पीछे हाथ धोकर पड़ा है। रात में वह उसके घर में ढेला-पत्थर फेंकता है। आज आपकी बिरादरी का बिसुन मेरे यहाँ विचरवाने आया था। काली माई ने आपके लड़के का नाम लिया है और कहा है कि यदि एक हफ़्ते में वह नहीं सुधरा तो उसकी जान ले लूँगी।"

"राम-राम! मेरे घर में यह राक्षस कैसे पैदा हो गया? पराई औरत को मैंने हमेशा अपनी माँ-बहन समझा। काली माई ने एक हफ़्ते का समय दिया है। वह बहुत ज़्यादा है। आज के बाद अगर शिकायत मिले तो अपनी मूँछें मुड़वा दूँगा।"

और सचमुच बिसुन के घर में तब से कभी ढेला-पत्थर नहीं गिरा। इससे पुजारी काका की शोहरत चारों ओर फैल गयी।

मगर जिस मिज़ाज के काका थे, उस मिज़ाज के उनके परिवार के लोग नहीं थे । उनका बड़ा लड़का फौज में था। उन्होंने दो गायें और दो भैंसें पाल

रखी थीं। वह उनकी वैसी ही सेवा करते थे जैसे गोपाल वृन्दावन बिहारी लाल। वह सुबह 'चाह' पीकर जानवर चराने निकल जाते थे और शाम को ही आते थे।

'चाह' के अलावा जितना दूध बचता, उसे फौजी के आधा दर्जन बेटे-बेटियाँ और उसकी धर्मपत्नी खाती थीं। काकी को दूध सूँघने को भी नहीं मिलता था। पुजारी काका को यह बात मालूम हुई तो उन्होंने बहू को समझाया-बुझाया। मगर उस घमंडी औरत पर कोई असर नहीं हुआ। पुजारी काका ने अपने फौजी बेटे को एक खत लिखा। वह धर्मसंकट में पड़ गया। एक उँगली कटे तो भी दर्द, दूसरी कटे तो भी दर्द। मगर तीसरे दिन फौजी को उसकी धर्मपत्नी का खत मिला। उसकी धर्मपत्नी ने पुजारी काका की ढेर सारी बुराइयाँ की थीं। हर बात में नमक-मिर्च लगी होती। अन्ततः फौजी ने एक कड़ा जवाब लिखा, "मैंने इतने बड़े परिवार की नैया अब तक खेई मगर अब मेरी ज़िम्मेदारी बढ़ गयी है। बड़ी बेटी शादी करने के लायक हो गयी है। बड़े बेटे को बारहवीं पास करने के बाद शहर के किसी अच्छे डिग्री कॉलेज में भर्ती कराना है। और आपके तीन बेटे हैं। अब मँझले बेटे का सहारा लीजिए। छोटे बेटे को घर पर रखकर आपने बिगाड़ दिया है। वह चार साल से दसवीं में फेल हो रहा है और आप उसे कहीं बाहर कमाने के लिए नहीं भेज रहे हैं। सो मैंने सोच-समझकर निर्णय ले लिया है कि अब आपका चूल्हा अलग जले।"

काका ने वह चिट्ठी दो-तीन बार पढ़ी, 'मंझले बेटे का सहारा लीजिए', 'छोटे बेटे को घर पर रखकर आपने बिगाड़ दिया है', 'अब आपका चूल्हा अलग जले'। वे जितनी बार चिट्ठी पढ़ते, उतनी बार उनके माथे पर जूते की-सी चोट पड़ती और वे अपना माथा सहलाने लगे। ऐसे में मंझले बेटे की याद उन्हें आयी। मगर वह तो बीस साल से घर ही नहीं आया, चिट्ठी-चौपाती, पैसा-कौड़ी की बात कौन कहे? छोटे बेटे को किसने बिगाड़ा? सच है कि उसे कॉपी-किताब कभी नहीं मिली। फीस समय पर न जमा करने के कारण कितनी बार उसका नाम काटा गया और बाद में चिरौरी करने पर लिखा गया। फिर घर पर कोई हलवाहा नहीं है। छोटू ही हलवाहे का काम करता—हल जोतने से लेकर कटाई-मड़ाई तक। बाहर का आदमी होता तो रोज़ाना चालीस-पचास रुपये मजूरी ले लेता। मगर उसे न तो ठीक से खाना मिला, न कपड़ा। फिर गाँव-जवार में हल चलाने से जो बदनामी हुई, सो अलग। सब कहते, "ब्राह्मण होकर हल चलाता है।" यही वजह है कि अब तक कोई भी उसकी शादी करने के लिए नहीं आया। यदि कोई गरीब कन्या पक्ष वाला शादी के लिए उस गाँव में आता भी तो गाँव वाले छोटू को 'छोटुआ हरवाह' सम्बोधित करके उसे बिदका देते। ऐसा बार-

बार होने पर एक बार छोटू ने काका से कह दिया था, "काका! मेरी शादी कब होगी? यदि आप कोई जतन नहीं करेंगे तो बुढ़ापे में कोई मुझे एक रोटी बनाकर देने वाला भी नहीं रहेगा।"

यह सुनते ही काका की छाती फट गयी थी। वे अन्दर से भभकते दीये की तरह काँप गये थे। तब से वह छोटू की शादी कराने के लिए कोशिश करते रहे हैं मगर अभी तक उन्हें सफलता नहीं मिली। और अब अलगौझी से स्थिति और नाजुक हो जाएगी। मगर करें तो करें क्या? चारों ओर अँधेरा छा गया है। अरे, कुछ भी नहीं दिख रहा है। अपना हाथ, कान, मुँह, पैर, कुछ भी नहीं। उन्होंने अपनी उँगलियों से कई बार अपनी आँखें मलीं मगर कोई फ़ायदा नहीं।

उधर दो-तीन दिनों के बाद ही बड़ी पतोहू ने भोंदू नामक एक बारहवर्षीय बालक को चरवाहा रख लिया। दोनों वक्त का भोजन और दो सौ रुपये महीने पगार पर। अब भोंदू सब काम करने लगा। पुजारी काका की मानो ज़रूरत नहीं रही। भोंदू सुबह आठ बजे आता और सूरज डूबने के समय चला जाता। जबकि काका इतने समय के पहले और बाद में भी खटते थे। मगर उन्हें दोनों वक्त भरपेट भोजन नहीं मिलता था, पगार की कौन कहे। एक बार उन्होंने सोचा कि वह भोंदू को बहकायें कि अभी खेलने-खाने की उम्र में क्यों अपनी गर्दन फँसा रहा है? लेकिन दूसरे क्षण उन्होंने सोचा कि ऐसा करने पर गाय-भैंस भूखों मर जाएँगी।

काका, काकी और छोटू ने मजबूरन अपना अलग चूल्हा जलाया। खाना बनाने के नाम पर चार रोटियाँ सेंकी गयीं। काकी ने काका और छोटू को डेढ़-डेढ़ रोटियों के साथ थोड़ा नमक और एक हरी मिर्च परोस दी। काका ने छोटू से कहा, "छोटू! खाना कैसा लग रहा है?"

"बहुत नीक।"

"हरी मिर्च बहुत तीती है?"

"तिताई में भी बहुत मिठास है।"

बाप-बेटे में संवाद चल रहा था। दोनों एक-दूसरे की थाली निहार रहे थे कि तभी काकी ने पूछा, 'छोटू के बाबूजी! और रोटी लेंगे?"

"नहीं, नहीं। पेट भर गया है।" कहते हुए काका एक लोटा ठंडा पानी गटागट पी गये। छोटू ने भी एक लोटा पानी पिया।

"पानी कैसा है, छोटू?" काका ने पूछा।

"आज का पानी अमृत-सा लग रहा है, बाबूजी!" छोटू ने जवाब दिया। खाना खाने के बाद दोनों ने अपना-अपना पेट सोहराया, डकार ली और अपने-अपने बिस्तर पर चले गये। फिर काकी ने भी अपने हिस्से की एक रोटी धीरे-

धीरे चबाकर खा ली और एक लोटा पानी पीकर सो गयी। लेकिन अगली सुबह काकी ने काका से कहा, "ज़िन्दगी बहुत लम्बी है। मैं आपकी और छोटू की खुराक जानती हूँ। क्यों नहीं मँझले बेटे के यहाँ जाकर कुछ पैसा-कौड़ी लाते और कहते कि छोटू की कोई छोटी-मोटी नौकरी लगवा दे?"

काका सहमत नहीं हुए। मगर छोटू ने भी जब दबाव दिया तो मँझले बेटे के यहाँ जाने को तैयार हो गये। काकी ने पड़ोसियों से चावल, तिल, गुड़ आदि उधार लिया और तिलवा, ढूँढ़ी आदि की एक छोटी-मोटी गठरी तैयार कर दी। भारी मन से काका ने गाड़ी पकड़ ली। चालू डिब्बा भीड़ के भार और बदबू से गरुवा रहा था। उसमें भी रैली वालों की रेलम-पेल ने साँस लेना दूभर कर दिया था। कई घंटे तक वे शौचालय के बगल में कभी खड़े, कभी बैठे गाजियाबाद पहुँच गये। महलनुमा घर के दरवाजे की घंटी बजाई। बहू ने दरवाजा खोला पर खुश नहीं हुई। उन्हें देखकर भुनभुनाती हुई अन्दर चली गयी। काका भी अन्दर गये। तभी उनका मंझला बेटा वहाँ आया। काका ने कहा, "रामबदन! तुम्हारा काम कैसा चल रहा है?"

सुनकर रामबदन ने कुछ नहीं कहा। मगर रामबदन का बेटा अंकुर बीच में टपक पड़ा, "पापा, ये कौन हैं? कहाँ से आये हैं?"

"मैं तुम्हारा दादा हूँ, मुन्ना? मैं बहादुरपुर गाँव से आया हूँ। लो, तुम्हारी दादी ने तुम्हारे लिए कुछ भेजा है।" अंकुर ने उसे लेने के लिए हाथ बढ़ाया ही था कि उसे पापा ने डपट दिया, "नहीं अंकुर, गन्दी चीज़ से इन्फेक्शन हो जाएगा।" काका इन्फेक्शन का अर्थ तो नहीं जानते थे, मगर डपट के भाव से वह यह ज़रूर जान गये कि हमारी सौगात को वह गन्दी चीज़ बतला रहा है। सो वह उबल पड़े, "मुन्ना! तुम्हारे पापा का जन्म बहादुरपुर गाँव में ही हुआ था। गाँववालों का रसोईघर उतना ही साफ रहता है जितना उनका दिल।"

"आप क्या फालतू बातें कर रहे हैं? एक तो रात के दस-ग्यारह बजे बिना किसी ख़बर के आ धमके और दूसरे...।"

"तुम्हारे जैसे वंश के होते हुए भी मैं अपने को निर्वंश मानता हूँ। मैंने कर्ज़ लेकर तुम्हें पढ़ाया था और आज तुम मेरी औकात और सीमा बता रहे हो। मैं खुद लज्जित महसूस कर रहा हूँ।"

"जब मुझ पर आपको शर्म आती है तो मेरे घर आए क्यों?" रामबदन ने शब्दों की कुल्हाड़ी उनके सिर पर पटक दी।

"मैं आया नहीं हूँ, तुम्हारी माँ ने भेजा है। मैं गोमांस खाऊँ यदि यहाँ पल भर रुकूँ।" यह कहकर काँपते हुए काका उठ खड़े हुए।

काका किंवाड़ खोलकर बाहर निकलने ही वाले थे कि एक बूढ़ी आवाज़ आयी, "बाबूजी! मैंने सब कुछ सुन लिया है। आपने मेरी आँखें खोल दीं। मैंने

अपनी जवानी कोठे की बदबू और सीलन में बितायी। मैं कोठे से खारिज कर दी गयी तो मैंने इसे सहारा दिया। अपना बेटा समझा। मगर जिसने खून का रिश्ता भुला दिया, वह पैसे के रिश्ते को कभी भी भुना और भुला सकता है। मेरी आपसे विनती है कि आप इस गरीबखाने में अपनी ज़िन्दगी अच्छी तरह गुजारिये। खुदा कसम! मैं आपके धर्म-निर्वाह में कोई कोर-कसर नहीं छोड़ूँगी..."

"नहीं-नहीं, मुझे शहर की ज़िन्दगी पसन्द नहीं। जब अपना सिक्का ही खोटा हो तो परखने वाले का क्या दोष?" और काका तेज कदमों से घर से बाहर निकल पड़े। काका का इस तरह जाना बुढ़िया बर्दाश्त नहीं कर सकी। उसने चीखते हुए कहा, "रामबदन! अब तुम इस घर के सदस्य नहीं रहे। घर सिर्फ़ दीवारों का नाम नहीं होता। मैंने तुम्हें अपना बेटा मानकर सब कुछ तुम्हारे हवाले कर दिया था। मगर आज से मैं तुम्हारी माँ नहीं और तुम मेरे बेटे नहीं। तुमने झूठ कहा था कि तुम्हारे माँ-बाप नहीं हैं। इसी वक्त तुम लोग मेरे घर से बाहर निकल जाओ। मैं अपनी सारी जायदाद एक ट्रस्ट को सौंपकर बाकी ज़िन्दगी नये ढंग से जिऊँगी।" इस कथन में कँपकँपी और बेशुमार दर्द भरा था।

यह सुनते ही रामबदन, उसकी पत्नी और उनके बच्चों पर पहाड़ टूट पड़ा मगर बुढ़िया उनकी चिरौरी पर टस से मस नहीं हुई। आधे घंटे के अन्दर रामबदन को परिवार सहित उस घर से बाहर निकलना पड़ा। बुढ़िया ने अन्दर से किंवाड़ बन्द करते हुए बाहर की बत्ती भी बन्द कर दी। रामबदन, उसकी पत्नी और उनके बच्चे अँधेरी गुफा में धँसे चले जा रहे थे। वे एक-दूसरे को टटोल रहे थे। रामबदन रो रहा था, उसकी-पत्नी रो रही थी और उनके बच्चे भी रो रहे थे। पत्नी शायद पिछले बीस सालों में पहली बार रो रही थी। वह पुजारी काका को सरापने लगी, "अरे, वह कलमुँहा! उसके यहाँ आने के कारण ही यह गति हम लोगों की हुई। काली माई की पूजा का ढोंग करता है। मरने पर कीड़े पड़ेंगे उसकी लोथ में।" दूसरे क्षण उस बुढ़िया को कोसने लगी, "सत्तर भतार कर चुकी हरजाई। एक से न अघाने वाली, दिल की काली है निगोड़ी। ज़िन्दगी भर कुकर्म किया और अब धरम-करम की बात करती है। कौआ गंगा नहाने से सफेद नहीं हो सकता। सीधे नरक जायेगी रंडी!" उनकी रोने-कोसने की आवाज़ सुनकर गली के कुत्ते जोर-जोर से भौंकने लगे। रामबदन को पहली बार लग रहा था कि उन लोगों की ज़िन्दगी से कुत्तों की ज़िन्दगी बेहतर है जिनकी कम-से-कम एक गली तो निश्चित है। वह धीरे-धीरे पैदल चलकर सुनसान चौराहे पर पहुँचे। वहाँ आधी रात को कोई इनसान उनकी ज़िन्दगी की ढहती दीवारों को देखने-सुनने वाला नहीं था। रामबदन और उसकी पत्नी को लग रहा था कि उन्हें बीच चौराहे पर पिता ने निर्वस्त्र कर दिया है।

खोज

उस दिन जमा भीड़ बढ़ती ही जा रही थी। तरह-तरह के लोग और तरह-तरह की टिप्पणियाँ। कोई कहता, 'मेहनती था।' कोई कहता, 'बहुत ईमानदार था।' नेकदिल इनसान और सीधा-साधा कहने वाले लोग भी थे। आगे की भीड़ के कारण टेम्पो वाले ने मुझे वहीं पतुरिया के बाग में सड़क पर उतार दिया। मेरा घर वहाँ से करीब तीन सौ गज की दूरी पर था। पत्नी और बच्चों को सामान के साथ सड़क के किनारे बिठाकर मैं आगे बढ़ गया। भीड़ में सरसों फेंकने तक की जगह नहीं थी। किसी तरह धक्का-मुक्की सहते-सहते मैं भीतर तक गया। मैंने देखा कि एक बूढ़ा अधनंगा ही ज़मीन पर पड़ा है। लोगों के 'बोलकारने' पर बड़बड़ाने लगता है। कभी कपड़े फाड़ने लगता है, कभी बाल नोंचने लगता है, तो कभी सफेद दाढ़ी खुजलाने लगता है। उसके गालों पर गाँव की सड़क की तरह गड्ढे हो गये हैं, आँखें आसमान के तारों की तरह प्रकाश देने पर भी भीतर की ओर बैठ गई हैं, चेहरा आम की तरह चुचुक गया है, नाखून घास की तरह बहुत बढ़ गये हैं, बालों में रस्सी की तरह गाँठें पड़ गयी हैं, पेट अल्युमिनियम के बर्तन की तरह पिचक गया है, सीना पोखर-सा सूख गया है, उसके सिर और माथे पर चोट के निशान मौजूद हैं और खून के धब्बे भी हैं। एक बार देखने पर मैं पहचान न सका तथा 'और भी ग़म हैं जमाने में,' गुनगुनाता हुआ वापस होने को हुआ। मगर दिल नहीं माना। सो मैंने उसे फिर घूरकर देखा, "अरे, यह तो अपने चाचा हैं। सुलेमान चाचा! हे भगवान! हुआ क्या इन्हें। दादा-रे-दादा।" मुझे चाचा के साथ अतीत में घटी घटनायें याद आने लगीं।

उस दिन किसी बड़े नेता की मौत हो गयी थी और इसीलिए सभी स्कूलों की तरह मेरे स्कूल में भी छुट्टी हो गयी थी। मैं दर्जा दो का विद्यार्थी था। नेता की मौत माने छुट्टी। छुट्टी माने मौज-मस्ती। हम बच्चे ऐसी छुट्टी बार-बार मिलने की कामना करने लगे। हम लोग छुट्टी होते ही घर की ओर फुर्र हो गये। रास्ते में एक पुराना तालाब पड़ता था। उसमें कुछ लोग कंटिया लगाकर

मछली मार रहे थे। हम लोग इस दृश्य को देखने लगे। थोड़ी देर में बंसी में एक मछली फँसी। मछली मारने वाले ने पूरी ताकत के साथ डंडी खींची जिससे सूत की डोरी टूट गयी और मछली पानी में ही गुम हो गयी। मुझे अच्छा लगा कि हमारी तरह मछली की भी छुट्टी हो गयी। हमें लगा कि आज का दिन अच्छा है। मछली मारने वाले ने वहाँ खड़े बच्चों से कहा, ''जो सूत की डोरी लाएगा, उसे एक रुपया इनाम मिलेगा।'' दादा-रे-दादा, एक रुपया!

तब हम लोगों के लिए एक रुपया का इनाम बहुत बड़ा इनाम था। घर से तब जल्दी पैसे नहीं मिलते थे। हम लोगों ने वहीं निश्चय कर लिया कि जल्दी से कहीं से एक डोरी लाई जाये। हम लोग आगे बढ़े। रामू के घर के बाहर दढ़ियल चाचा अपनी दुकान फैलाये बैठे थे। वह अपनी साइकिल में पीछे कैरियर पर एक बड़ी टोकरी रखते थे और आगे हैंडिल में दो बड़े झोले लटकाते थे। सो वहाँ तमाम औरतें और लड़कियाँ जुटी हुई थीं। अपनी-अपनी रुचि के सामान वे देख-परख रही थीं। हम लोग वहीं खड़े हो गये। वहाँ फैली चीज़ों को देखकर मेरा मन ललचा गया। मेरा साथी रामू मुझसे बहुत ज़्यादा होशियार था। कौवे की तरह उसकी निगाह सूत की डोरी पर थी। इस तरह की निशानेबाजी में वह अर्जुन से कम नहीं था—वह अपने लक्ष्य को ही देखता था। वह किसी तरह भीड़ के अन्दर तक पहुँचा और डोरी पर हाथ साफ करके नौ-दौ-ग्यारह हो गया। किसी ने भी उसे नहीं देखा। मगर मेरी निगाह पहले से ही रामू पर थी। साझा उद्देश्य होने के कारण मैंने उसे डोरी चुराते हुए देखा। लेकिन मैंने अपने सहपाठीनुमा दोस्त की शिकायत चाचा से नहीं की। बल्कि मैंने उनका ध्यान बँटाने के लिए आगे बढ़कर कहा, ''चाचा, सलाम!'' चाचा ने कहा, ''खुश रहो, बेटा!'' मुझे ऐसा लगा कि चाचा मेरे ऊपर खुश हो गये हैं और अब वह मुझ पर इस प्रकार की शंका नहीं करेंगे। रामू की सफलता ने मेरे भीतर ईर्ष्या का भाव पैदा कर दिया। मैंने सोचा कि यदि रामू, कक्षा दो का छात्र, सूत की डोरी का पूरा बंडल बिना चुटकी बजाये यूँ गायब कर सकता है, तो क्या उसका दोस्त प्यारेलाल एक छोटा-सा सामान भी गुम नहीं कर सकता। मैंने यह बात कई बार सोची। अन्तरात्मा की आवाज़ को पिंजड़े में कैद करके अन्ततः मैंने अपने को तैयार कर लिया। सो मैंने अपनी निगाह वहाँ फैले नाना प्रकार के सामानों पर दौड़ाई। दादा-रे-दादा, मेरी निगाह प्लास्टिक की एक सीटी पर पड़ गयी। मेरी कक्षा का गोलू रोज़ फुर्र-फुर्र सीटी बजाता था।

खैर, गोलू की सीटी मेरी आँखों की किरकिरी बन गयी थी। उसने मुझे अपनी सीटी कभी छूने तक नहीं दी, बजाने की कौन कहे। जब मैं कभी-कभार उससे सीटी बजाने के लिए माँगता तो वह सीधे कहता, ''बाप से कह दो कि

सीटी खरीद दे।'' मुझे यह बात बार-बार अन्दर तक चुभती रहती थी। सो चाचा की दुकान में सीटी देखकर मेरे मुँह में पानी आ गया। मगर मुझे गुरुजी की बात याद आयी, ''चोरी करना पाप है, नदी किनारे साँप है।'' सो मैं पीछे हट गया। लेकिन मुझे याद आया कि अभी-अभी रामू, कक्षा दो, ने सूत की डोरी चुराई है। उसे भी तो गुरु जी शिक्षा देते हैं कि चोरी मत करो। तभी थोड़ी दूर से मुझे गोलू की सीटी की फुर्र-फुर्र आवाज़ सुनाई दी। मुझसे नहीं रहा गया। सो मेरे आगे जो औरतें बैठकर चाचा की दुकान से सामान खरीद रही थीं, उनकी पीठ के पीछे से मैंने अपना हाथ धीरे-धीरे टोकरी की ओर बढ़ाया और सीटी उठा ली। मेरे हाथ की सफाई का कमाल यह था कि किसी ने मुझे सीटी उठाते हुए नहीं देखा। मजे की बात तो यह थी कि खुद चाचा ने नहीं देखा। वह खरीदारों को अपने सामानों की विशेषताओं और कीमतों के बारे में बतलाने में व्यस्त थे। मुझे अपनी सफलता पर सन्तोष हुआ। लेकिन एक लड़के ने आग में घी डालते हुए मेरे पिता जी से कह दिया, ''इसने किसी के घर से सीटी चुरा ली होगी।''

यह सुनते ही पिता जी ने मेरे दोनों गालों पर तड़ाक-तड़ाक कई थप्पड़ जड़ दिये। उनका इरादा और ज़्यादा मारने का था, मगर तब तक फरिश्ता बनी मेरी दादी ने उन्हें अपनी कसम खिला दी और मैं तमाचों की मार से बच गया।

लेकिन मेरा बेड़ा गर्क होने वाला था सो अचानक मेरा विरोधी गोलू वहाँ आ धमका। उसे इस बात से तकलीफ थी कि पूरे गाँव और स्कूल में उसके सिवा दूसरे के पास सीटी कैसे आ गयी? सो उसने मेरे पिताजी के कान भरे, 'मनतोरा के दुआर पर सुलेमान चाचा अपनी दुकान लगाये बैठे हैं। प्यारेलाल वहाँ गया था शायद, वहीं से इसने सीटी चुराई होगी।' सो पिताजी मेरा हाथ पकड़े हुए मनतोरा के दुआर तक मुझे ले गये। वहाँ चाचा खरीदारों से घिरे थे। पिताजी ने जोर से कहा, ''चाचा! सही-सही बताओ कि ये सीटी आपकी है या नहीं?''

चाचा ने वह सीटी अपने हाथ में ले ली और उसकी परख बारीकी से करने लगे। फिर उन्होंने मेरी ओर ताका। उसके बाद फिर सीटी पर वह हाथ फेरने लगे। फिर मेरी और उन्होंने निहारा।

''प्यारे लाल, यह सीटी तुम्हें कहाँ मिली?''

''...'' मेरे मुँह से कोई शब्द नहीं निकला।

''सच-सच बताओ, बेटा! डरो मत।''

''मुझे स्कूल के रास्ते में 'परुआ' मिली,'' मैंने सिसकते हुए कहा।

''परुआ? स्कूल के रास्ते में?''

''हाँ, चाचा। सच्ची-मुच्ची, ''मैंने इतने धीरे से कहा मानो मेरे प्राण सूख

रहे हों।

''प्यारेलाल ठीक कह रहा है। इसे यह सीटी कहीं परुआ मिली होगी...'' चाचा ने फैसला सुना दिया।

''मगर चाचा! गोलू कह रहा है कि इस गाँव में किसी लड़के के पास कोई सीटी नहीं है,'' मेरे पिताजी ने वकीलों की तरह ज़िरह की। मेरे पिताजी मेरे पीछे ही पड़े थे जैसे पुलिस निर्दोष लोगों के साथ मुठभेड़ करती है।

''दरअसल, मेरी दुकान में सीटी थी ही नहीं। हो सकता है, बगल के गाँव के बच्चों के पास से यह सीटी स्कूल जाते समय गिर गयी हो। प्यारेलाल तो शरीफ लड़का है। जब भी मुझे देखता है, 'सलाम चाचा' कहता है,'' चाचा ने गाँव की समूची कचहरी के सामने अपना दो टूक फैसला सुना दिया।

घर पहुँचने पर दादी ने पूरा हाल पूछा। मेरे द्वारा चाचा के फैसले की सूचना देने पर वह पिताजी पर बरसने लगीं, ''रघुवीर कसाई है। मगर कसाई भी आदमी को नहीं, जानवरों को मारता है। मेरे निर्दोष पोते को बिना वजह मार रहा था।'' यह कहते-कहते उनकी आँखों में आँसू आ गये। उन्होंने मुझे छाती से लगा लिया और मेरा सिर, गाल, पीठ आदि चूमने-सहलाने लगीं। मुझे लगा कि मैं कालिख की कोठरी से बेदाग निकल आया हूँ।

करीब एक महीने बाद शाम को मैं अकेले अपने गन्ने के खेत की ओर जा रहा था। रास्ते में मुझे चाचा मिल गये थे। मैंने उनसे दुआ-सलाम की। हाल-चाल पूछने के बाद उन्होंने एक थैला मेरे कन्धे पर लटका दिया। थोड़ी दूर जाने के बाद उन्होंने अपना थैला मेरे कन्धे से उतारकर पुनः साइकिल के हैंडिल में लटका लिया। मैंने कहा, ''चाचा, उस सीटी का क्या दाम है? मैं अपनी दादी से पैसे माँगकर लाऊँगा और आपको अगली बार दे दूँगा।''

''देखो प्यारेलाल! बचपन में गलती सबसे होती है। मुझसे भी हुई थी और तुम्हारे पिताजी से भी हुई होगी। गलती होना बिल्कुल स्वाभाविक है मगर ज़्यादा ज़रूरी यह है कि हम अपनी गलती सुधार लें। फिर मैंने तुमसे अपना झोला थोड़ी दूर तक ढुलवाया है तो उसका दाम वसूल हो गया। यदि तुम सीटी की चोरी नहीं भूलोगे तो यह भी नहीं भूलोगे कि तुमने मेरा झोला ढोया है। तुम मुझे भगवान समझकर मेरी पूजा मत करो, न पुलिस समझकर मुझसे डरो। मुझे सिर्फ़ इनसान समझकर लोगों पर विश्वास करो, बस,'' चाचा ने मुझे समझाते हुए कहा। मैंने उसी समय उनके सामने चोरी न करने के लिए अपने कान पकड़े थे। आज मैं चाचा को बताना चाहता था कि सीटी की चोरी मेरी पहली और अन्तिम चोरी थी। मैं उन्हें यह भी बताना चाहता था कि मैंने गलतियाँ कीं मगर कभी उन्हें दुहराया नहीं और उसी की बदौलत मुझे नौकरी में बराबर तरक्की

मिलती गयी। मगर पता नहीं, क्यों चाचा ने मुझे पहचाना ही नहीं। नौकरी मिलने के बाद मेरे मन में कई बार इच्छा हुई थी कि चाचा से भेंट कर लूँ, मगर घर-परिवार और नौकरीपेशा की झंझटों से कभी फुर्सत नहीं मिली।

थोड़ी देर में माँ ने रहस्य से पर्दा उठाते हुए बताया कि चाचा का बेटा एक छोटी-मोटी दुकान चला रहा था। चाचा ने बुढ़ापे में भी अपनी साइकिल पर लदी चलती-फिरती दुकान बन्द नहीं की। मगर कस्बे के एक दुकानदार ने पुरानी कसर निकालने के लिए उनके बेटे पर चोरी का झूठा मुकदमा दर्ज़ करा दिया। पुलिस थाना की उसने बाकायदा पूजा की। सो दारोगा ने उसके इकलौते बेटे को गिरफ़्तार कर जेल भेज दिया। वह थानेदार के पास फरियाद करने गये थे मगर कुछ सुनने के बजाय उन लोगों ने उनके साथ बदसलूकी करते हुए उन्हें धकेल दिया। बाद में अदालत में मामला शुरू हुआ। मुकदमे में दो गवाह चाचा के गाँव के भी थे। उन दोनों ने उन्हें पूरा आश्वासन दिया कि वह सही गवाही देंगे। चाचा पूरी तरह आश्वस्त हो गये थे कि अब अदालत में दूध-का-दूध, पानी-का-पानी हो जाएगा। मगर गवाही के दौरान दोनों गवाहों ने बयान दिया कि सुलेमान के लड़के कल्लू मियाँ ने मुद्दई को जान से मारने की कोशिश की थी। ज़िरह के दौरान बचाव पक्ष के वकील ने गवाह से पूछा, ''कल्लू मियाँ किस तरह के लोगों के साथ उठता-बैठता है?'' वह गवाह भड़क उठा, ''वह तो आतंकवादियों के साथ रहता है। उसके यहाँ बाहरी लोग बराबर उठते-बैठते हैं।'' तभी सरकारी वकील ने अन्तिम वार किया, ''हुजूर! इन्हीं देशद्रोहियों के लिए नया कानून बना है पोटा। पोटा के तहत ऐसे लोगों को सजा मिलेगी, तभी यह देश बचेगा।''

इस मुकदमे में चाचा की दुकान ख़त्म हो गयी। एक दिन फास्ट ट्रैक अदालत ने कल्लू मियाँ को आजीवन कारावास की सजा सुना दी। चाचा बहुत सारे मन्त्रियों, अफसरों, प्रधानमन्त्री, राष्ट्रपति आदि से गुहार लगाते रहे और दरगाहों में मन्नतें मानते रहे, मगर किसी की कृपा उन्हें नहीं मिली। सो वह विक्षिप्त हो गये।

पूरी कहानी सुनकर मैं अवाक् रह गया। मैं बेचैन हो उठा। मुझसे खाना नहीं खाया गया। उस रात मैंने सपने में देखा कि मैं एक अन्धी सुरंग में हूँ। थोड़ी देर में चाचा अपनी दुकान फैला देते हैं और सभी बच्चों को सीटियाँ बाँट देते हैं। सारे बच्चे सीटियाँ बजाने लगते हैं, सीटियों में गीत गाकर नाचने लगते हैं। मैं भी जोर से फुर्र-फुर्र अपनी सीटी बजाता हूँ। अचानक सीटियाँ गाछ का आकार ले लेती हैं और उनमें फल लगे होते हैं। तभी माँ ने मेरे माथे पर अपना हाथ रख दिया। मैं अचकचाकर उठ बैठा। मगर न तो वहाँ मेरी सीटी थी और न मेरे चाचा थे। सो थोड़ी देर में हाथ-मुँह धोकर मैं चाचा की खोज में निकल पड़ा।

दंश

राजन लद्धड़ प्रवृत्ति का नौजवान था। अनायास किसी-न-किसी से झगड़ा कर लेना, किसी विद्यार्थी की कलम चुरा लेना, लड़कियों पर बेतुकी टिप्पणियाँ करना, अध्यापकों से बहाना बनाना, माँ-बाप से झूठ बोलना, सहपाठियों पर रोब झाड़ना आदि उसकी फितरत थी। बावजूद इसके कछुवा गति से वह बी.ए. पास हो ही गया। हालाँकि इसमें उसे पाँच साल लगे और श्रेणी मिली तो तीसरी। लेकिन डिग्री के इस रुतबे ने उसके व्यक्तित्व में निखार ला दिया था। अब वह अधिकारपूर्वक टाई लगाने लगा था। जब वह इंटर में पढ़ता था, तो दूसरे की टाई मँगनी माँगकर कभी-कभार पहनता था। कहते हैं एक दिन उसको किसी ने टोक दिया था कि टाई पहनने से पहले ग्रेजुएट होना चाहिए। उसके बी.ए. करने में यह ताना बड़ा सहायक हुआ। अब वह सबसे कहता है कि वह ग्रेजुएट हो गया है, इसलिए टाई लगाता है। उसे भी महसूस हुआ, टोकने वाला आदमी सचमुच सही था। टाई लगाने के लिए तो सचमुच ही ग्रेजुएट होना चाहिए।

आँखों पर चश्मा, गले में टाई और सुन्दर गठीला बदन देखकर उसे अक्सर अनजाने लोग अफ़सर समझने लगते थे। उसने कई बार फार्म भरे, मगर किस्मत ने साथ नहीं दिया। सो उसने एक प्राइवेट स्कूल में शिक्षक के पद पर नियुक्ति हेतु आवेदन दिया। किस्मत ने करवट ली। उसका चयन हो गया। मगर उसके चयन का भांडा बाद में फूटा—दरअसल उसने जाली सर्टिफिकेट जमा की थीं। अवध विश्वविद्यालय के थर्ड डिवीजन बी.ए. पास के बजाय उसने इलाहाबाद विश्वविद्यालय से फर्स्ट डिवीजन बी.ए. पास की जाली सर्टिफिकेट बना ली थी। मगर उसका प्रबन्धक ऊँची खोपड़ी का था, सो उसने उस सर्टिफिकेट को इलाहाबाद विश्वविद्यालय को सत्यापन हेतु भेज दिया। मगर जानकारी मिलते ही सत्यापन प्रतिवेदन आने से पहले ही राजन वहाँ से इस्तीफ़ा देकर चला गया।

सो मुकदमे से तो वह किसी तरह बच गया, लेकिन बेरोज़गारों की स्थायी सूची में वह दर्ज़ हो गया। फिर भी उसने हार नहीं मानी। उसके गाँव में यू.पी.

रोडवेज का एक कंडक्टर रहता था। उसे विभाग के लोग धुरंधर जी कहा करते थे। वह छप्पन सीटों में से छत्तीस सवारियों को डब्ल्यू.टी. ले जाता था। वह कितनी बार पकड़ा गया और मुअत्तल हुआ, यह उसे भी याद नहीं। विभागीय सहकर्मी उसे हिकना, थेथर वगैरह उपाधियों से नवाज़ते हैं। फिर भी 'हिकना के चूतड़ पर रूख जमा, उसके लिए छाँह' वाली कहावत सिद्ध हो रही थी।

बेरोज़गार राजन ने इन्हीं धुरंधर जी की चेलाही स्वीकार कर ली। वह उनके कुछ निजी काम कर देता था। बदले में वे उसे अपने साथ मुफ़्त में कानपुर, लखनऊ, गोरखपुर, बनारस आदि की सैर कराते थे। इस सैर सपाटे में अपने गुरु से 'सफलता' के गुर भी सीख रहा था। इसी अन्दाज़ में वह धुरंधर का बेटा बनकर एक बार दिल्ली भी घूम आया। दिल्ली की ऊँची-ऊँची इमारतें, चौड़ी और समतल सड़कें, बड़े-बड़े स्टोर और प्रतिष्ठान आदि उसे भा गये। वह वहाँ गाँव के नौकरी करने वालों के यहाँ दस दिनों तक मेहमानबाजी करता रहा। जब वह घर लौटा तो उसका घर पर मन ही नहीं लगा। सो उसने गाँव-जवार से दिल्ली-कानपुर-गोरखपुर-बनारस आदि जाने वालों से सीधे सम्पर्क साधा और बड़ी सफाई से उन जगहों के लिए पास बनाने लगा। वे सवारियाँ भी खुश थीं कि उन्हें पूरे किराये के बदले आधा किराया देना पड़ता था। सो धुरंधर जी जैसे गुरु गुड़ ही रह गये और उनका चेला चीनी हो गया। यह रहस्य जान लेने पर धुरंधर जी ने अपना माथा पीट लिया कि उन्हें पहले यह युक्ति क्यों नहीं सूझी। और धुरन्धर जी ने अपनी हार मान ली। अब वह राजन को 'गुरु' कहने लगे। राजन की कमाई काफ़ी अच्छी तरह चल रही थी। मगर विनाश काले विपरीत बुद्धिः। वह खुद कानपुर से बनारस तक रोडवेज की बस में अपने बनाये पास पर यात्रा कर रहा था। एक बार रास्ते में बस की चेकिंग हो गयी। चेक करने वाले ने पूछा, "कहाँ का पास है?"

"गोरखपुर डिपो का?"

"किसका हस्ताक्षर है?"

"ए.आर.एम. का।"

"ए.आर.एम. का क्या नाम है?"

"श्री ए.पी. श्रीवास्तव।"

"क्या यह उन्हीं का हस्ताक्षर है?"

"जी, जी।"

"नामसेंस। वह जोर से गरजा, मैं ए.पी. श्रीवास्तव का हस्ताक्षर पहचानता हूँ। वह मेरी बैच के हैं।"

तब राजन की चुरकी सोंटा हो गयी। वह काँपने लगा। तभी उसके सामने

धुरंधर जी का चेहरा झिलमिलाने लगा। उसके मुँह से निकल पड़ा, "सर! माफ कर दीजिए। मैं धुरन्धर जी कंडक्टर का भतीजा हूँ। आइन्दा ऐसी गलती नहीं होगी।"

चेकिंग टीम के सदस्यों ने आपस में सलाह-मशविरा किया और विभागीय कर्मचारी के परिवार का होने के नाते उसे सिर्फ़ कड़ी चेतावनी देकर छोड़ दिया।

फिर उसने वह काम करना छोड़ दिया। अब तक उसके पास इतने पैसे हो गये थे कि उसकी जीवन शैली बदल गयी थी। उसने एक मोबाइल फोन रख लिया था जिससे उसके व्यक्तित्व में और चमक आ गयी थी। उसने अब विकास कार्यों में रुचि लेनी शुरू कर दी। उत्तर प्रदेश जल निगम से सभी सुदूर गाँवों में हैंडपम्प लगाने का अभियान शुरू हो गया था। उसने वहाँ के एक स्टाफ से सम्पर्क साधा और आकर्षक कमीशन पर काम बन गया। चूँकि वह अभी भी अपने गाँव वालों से प्यार करता था, इसलिए उसने यह शुभ कार्य अपने गाँव से ही शुरू किया। इंडिया मार्का दो की एक हैंडपम्प गड़वाने में दरअसल दस हजार रुपये का खर्च पड़ा था, मगर सरकारी विभाग ने उसका इस्टीमेट बीस हजार रुपये का बनाया था। इस प्रकार शेष दस हजार में बन्दरबाँट का खेल होता था। राजन को इसमें से कुछ नहीं मिलता था और उसे इसकी आशा भी नहीं थी। मगर वह बालू से इत्र निकालने की कला सीख चुका था। उसने अपने गाँव के दस लोगों से सम्पर्क किया जो इंडिया मार्का दो का बड़ा हैंडपम्प लगवाना चाहते थे। गाँववालों को यह अच्छी तरह मालूम था कि इसमें करीब दस हजार रुपये लगते थे। सो उसने गाँव के भाइयों पर कृपा करते हुए कहा, "मैं दस के बदले पाँच हजार रुपये में ही वह हैंडपम्प लगवा दूँगा। उसे चाहे घर के बाहर लगवाइये या भीतर।"

गाँववालों को यह सौदा मुनाफ़े का लगा। सो राजन ने दो महीने के अन्दर दस हैंडपम्प लगवा दिये। फिर माँग बढ़ती गयी, हैंडपम्प घटती गयी। सो राजन ने एक सुलझे हुए व्यवसायी की तरह अपना रेट सात हजार रुपये कर दिया। उधर प्रदेश जल निगम के इंजीनियर, जूनियर इंजीनियर, क्लर्क वगैरह भी खुश थे कि उन्हें अपने विभागीय कमीशन के अतिरिक्त कुछ सुविधा-शुल्क मिल रहा था।

कुछ महीनों के बाद राजन का यश चारों ओर फैल गया। उसे पड़ोसी गाँवों से हैंडपम्प लगवाने का काम मिलने लगा। उसे सारी रकम एडवांस मिलने लगी। सो उसने एक दर्जन लोगों से सात हजार रुपये की दर से राशि वसूल ली। जब दफ़्तर में काम की गति धीमी हो गयी, तो उसने सारी राशि हड़पने का जुगाड़ किया और ऐलान कर दिया कि निगम वाले सारा पैसा खा गये हैं। मगर उसने

उन लोगों से काम करवाने का अपना संकल्प घोषित किया। दुर्भाग्य से वहाँ के इंजीनियर और जूनियर इंजीनियर का तबादला हो गया और राजन ने अपने ग्राहकों को दो-दो हजार रुपये वापस किये। शेष इंजीनियरों के मत्थे मढ़ दिया। चूँकि अब राजन जिला मुख्यालय में रहने लगा था, इसलिए उसके ग्राहक उससे बार-बार तगादा नहीं कर सकते थे। फिर वह हीरो होंडा लेकर इधर-उधर घूमता रहता था। सो उससे मिलना टेढ़ी खीर थी।

मगर उसने अपने खास लोगों को अभी तक धोखा नहीं दिया था। उसकी छोटी बहन की ससुराल में पीने के पानी की बहुत दिक्कत थी। सो उसने उनसे पाँच हज़ार रुपये मात्र लेकर इंडिया मार्का दो हैंडपम्प लगवा दी थी। वे लोग काफी खुश थे, क्योंकि पड़ोसी की हैंडपम्प से पानी लाने में कई बार झगड़ा हो गया था। बहन की ससुराल वालों का उस पर विश्वास बढ़ता गया। कुछ हफ़्ते पहले उसने अपनी पड़ोसिन भाभी को आँगनबाड़ी सेविका के रूप में नियुक्त करवा दिया था। उसके एवज़ में चचेरे भाई ने पाँच हज़ार रुपये देने को कहा था। मगर बेरोज़गार आदमी पैसे कहाँ से दे? उसने दो हज़ार रुपये ही दिये। इसके लिए भी गहना बेचना पड़ा। बाकी तीन हज़ार के लिए चचेरे भाई ने लम्बा वादा किया, "हर महीने इसकी तनख्वाह से तीन सौ रुपये चुका देंगे।" राजन को दूसरा रास्ता सूझा। एक रात उसने पड़ोसिन भाभी की देह से बकाया वसूलने की भरपूर कोशिश की। मगर अँधेरे में वह चिल्ला पड़ी और उसके पति ने उसे एक लाठी मारी। किसी तरह वह वहाँ से भाग निकला। इज़्ज़त का ख्याल करके पड़ोसिन और उसके पति ने यह वाकया किसी से नहीं बताया। वे हर महीने तीन सौ रुपये चुकाते गये।

इसी बीच उसकी बहन के ससुर ने बेटे की बेरोज़गारी दूर करने की चर्चा राजन से की। राजन ने चुटकी बजाते हुए नौकरी दिलाने की बात कही। जब ग्रामसेवक की भर्ती की ख़बर अख़बारों में निकली, तो उसके बहनोई उसके शहर वाले क्वार्टर पर आ धमके। राजन ने दो-तीन दिनों की दौड़-धूप की। फिर अपने जीजा को हिसाब-किताब समझाया-'सौदा महँगा ज़रूर है मगर पक्का है। एक लाख रुपये लगेंगे। भगवान ने चाहा तो आप साल भर में ही पचास हजार कमा लेंगे। अभी स्वर्णजयन्ती रोज़गार योजना, प्रधानमन्त्री सड़क योजना, अन्त्योदय योजना, अन्नपूर्णा योजना, दीनदयाल योजना, अम्बेडकर योजना, इन्दिरा आवास योजना, शौचालय योजना वगैरह ग्रामीण क्षेत्रों में चल रही हैं जिनमें रसोई स्वत: परोसी रहती है। आपको सिर्फ़ मुँह में कौर डालना है और भगवान की कृपा से आपके हाथ-पैर दुरुस्त हैं। 'लेकिन इतनी बड़ी रकम यकायक कहाँ से मिलेगी? हम लोग तो छोटे-मोटे किसान हैं।"

"घबड़ाने की बिल्कुल ज़रूरत नहीं है। पहली किस्त में पच्चीस हज़ार से काम चल जाएगा। फिर साक्षात्कार के पहले पचास हज़ार और उसके बाद अन्तिम किस्त पच्चीस हज़ार देनी होगी। वैसे बाजार का रेट सवा लाख है, लेकिन आप तो मेहमान हैं। इसमें मेरी बहन की भलाई का भी सवाल है न?" और राजन का बहनोई रुपये की व्यवस्था करने के लिए निकल पड़ा। जब वह घर पहुँचा, तो चारों ओर समस्या का पहाड़ नज़र आने लगा। उसके पिता ने खेत बेचने पर सहमति देने से इनकार कर दिया। घर के सारे जेवरात से सिर्फ़ पहली किस्त की व्यवस्था हो सकी। सो ग्रामसेवक का फार्म भर दिया गया। साक्षात्कार अगले महीने था। महाजन ने पचास हज़ार रुपये उधार देने से मना कर दिया, मगर उसने खेत खरीदने के बदले उतनी राशि देने पर सहमति दी। उसके पिता के न चाहने के बावजूद पूरब तरफ का एक बीघा खेत बेच दिया गया और साक्षात्कार के एक पखवाड़ा पहले राजन को वह रकम दे दी गयी। कुछ ही दिनों में राजन ने साक्षात्कार का पत्र लाकर उसे दे दिया।

साक्षात्कार के एक दिन पहले राजन का जीजा उसके शहरी क्वार्टर पर आ गया था। राजन ने अपने मोबाइल फोन पर कई लोगों से बातें कीं जो साक्षात्कार से सम्बन्धित थीं। उसने अपने जीजा को बता दिया, "आप तो जानते ही हैं कि यह कलियुग है। सुविधा शुल्क के बिना कोई सुविधा नहीं मिलती। सभी लोगों से बातें हो गयी हैं। उनका रुख सकारात्मक है। सबको मैंने एडवांस भी दे दिया है। अब आप अपने को उसी तरह सुरक्षित समझिये जिस तरह कोटा वाले सुरक्षित होते हैं। बल्कि उससे भी ज़्यादा। आपका नाम सूची में ग्यारहवें नम्बर पर है।"

"राजन भैया! सब कुछ आपकी मुट्ठी में है। मुझे भी प्रसाद दे दीजिए।"

और उसका साक्षात्कार निपट गया। साक्षात्कार इतना असन्तोषजनक था कि किसी भी तरह कोई उम्मीद नहीं बनती थी। फिर भी राजन ने उसे नौकरी मिलने का पूरा आश्वासन दे दिया था।

करीब एक माह बाद रिजल्ट निकल गया। सूची में बबलू श्रीवास्तव का नाम था। सो बबलू के घर वाले खुशी से झूम उठे। राजन ने साक्षात्कार बोर्ड को हृदय से धन्यवाद दिया और भगवान भास्कर को आभार के साथ प्रणाम किया। उसने अपने जीजा से उसी दिन अन्तिम किस्त के रूप में पच्चीस हज़ार रुपये ले लिए। उसकी बहन के ससुराल वालों ने राजन के साथ-साथ उसकी बहन को भी सगुनही घोषित किया। इसके पहले उस पर शनि की साढ़े साती सवार होने का आरोप लगता था। बहन ने भगवान के साथ-साथ अपने बड़े भाई का शुक्रिया अदा किया।

पन्द्रह दिन बीत गये मगर बबलू को कोई नियुक्ति पत्र नहीं मिला। सो

उसने राजन से सम्पर्क किया। राजन ने मोबाइल घुमाकर खोज-ख़बर ली और उसे बताया कि डाक में विलम्ब हो रहा है। मगर जब ज्वाइन करने की तिथि तक नियुक्ति पत्र नहीं मिला, तो राजन के जीजा और उसके पिता राजन के शहरी क्वार्टर पर पहुँचे। मगर वहाँ अलीगढ़ का बना एक बड़ा-सा ताला उसकी गैरहाजिरी दर्ज़ कर रहा था। वे लोग सारी रात उस क्वार्टर के बाहर बैठे रहे। परन्तु राजन नहीं आया।

सुबह-सुबह वे लोग राजन के मूल गाँव शहादतपुर पहुँच गये। राजन के पिता यह सब सुनकर अवाक् थे। मगर अपना मुँह जिलाने के लिए उन्होंने कहा, "समधी जी! आपने मुझसे कभी चूँ तक नहीं किया और इतनी बड़ी रकम उस लौंडे-लफाड़ी को कैसे दे दी? वह मुझसे कई सालों से अलग रहता है। न लेना एक, न देना दो।" वह सिर पर हाथ रखकर शोकमग्न हो गये। "मगर आप कुछ कीजिए। उसका पता लगाइये। उसने इतनी बड़ी रकम खर्च नहीं की होगी। दस-पाँच हज़ार भले खर्च किया हो। जो मिल जाये, वही सही। भागे भूत की लँगोटी ही सही..." यह कहकर समधी रोने लगे। उन्हें लगा कि उनका घर-परिवार उजड़ गया है।

परन्तु कोई रास्ता नहीं निकला। समधी अपने बेटे के साथ अपने गाँव वापस चले गये। जब समधिन ने यह सब सुना, तो उनका पारा गरम हो गया, "वह मुँहझौंसी पूरे कुल का नाश कर देगी। धोबी के कुत्ते की तरह न घर की है, न घाट की। इसी ने अपने ठग भाई से ऐसा करवाया होगा। मैं इसका बदला लेकर रहूँगी।"

राजन के जीजा ने भी अपनी माँ की बातों में हामी भरी। और देखते-देखते राजन की बहन शिमला को जलती कलछुल से माथे, पीठ, पेट, हाथ और पैर में दाग दिया गया। वह कराहती रही। कोई मदद करने नहीं आया। मगर उसकी सास को इतने से भी सन्तोष नहीं हुआ। सो उसने मुँगरी से पीटते हुए उसे रात में ही घर से निकाल दिया।

शिमला ज़्यादा पढ़ी-लिखी नहीं थी। गाँव वालों ने उसकी ससुरालवालों के डर से उसे आश्रय नहीं दिया। उसने एक पेड़ के नीचे कराहते-कराहते रात बिता दी।

उसका साथ सिर्फ़ अँधेरी रात दे रही थी। भोर से पहले वह गाँव के बाहर निकल गयी। दो दिनों बाद वह किसी तरह अपने मायके पहुँची। उसकी दुर्दशा देखकर उसकी भाभी को लगा कि ननद की खुशी हमेशा के लिए छिन गयी है। वह मन-ही-मन अपने पति को सरापने लगी "भगवान! उस बेईमान को इस अधर्म की कड़ी सजा दो। मेरी जैसी सधवा से विधवा ही भली।"

राजन की खोज उसके पिता और पड़ोसियों ने कई जगहों पर की। मगर पता चला कि वह जिला मुख्यालय छोड़कर राजधानी लखनऊ चला गया है। लखनऊ में इस गाँव के धुरंधर जी ने उसे कार में जाते हुए एक बार देखा था। राजन को अपनी पत्नी, माँ, बहन और पिता की परवाह बिल्कुल नहीं थी। वह अपनी धुन में मस्त था। राजन के पिता की आँखों से नींद गायब हो गयी। सारी रात वे चिन्ता में डूबे रहते। उठते-बैठते उन्हें बेटी का दुःख सालता रहता। वे सोचने लगते कहीं बेटी स्थायी भार न हो जाये। फिर उनकी ज़िन्दगी कितने दिनों की बची है? करीब एक हफ़्ता बीता था, राजन की पत्नी अपने मायके अपनी बीमार माँ को देखने चली गयी।

उसकी माँ ने उसे बताया, ''बेटी! पिछले दिनों इस इलाके में चोरी-डकैती की कई घटनाएँ घटीं। सो सुरक्षा के ख्याल से पाहुन ने मेरी हंसुली, हार, कंगन, करधन, पाजेब, टीका, पायल, वगैरह बैंक के लाकर में रखने के लिए एक पखवाड़े पहले ले लिया था। उसका काग़ज़ वापस लाकर देने को कहा था, मगर अभी तक नहीं दिये। तुम्हारे बाबूजी को बिना कुछ बताये ऐसा मैंने चुपके से किया था। मैं उसी को लेकर अब चिन्तित हूँ।''

फिर वह रोने लगी। राजन की पत्नी भी माँ से लिपटकर रोने लगी। सिसकते हुए उसके मुँह से सिर्फ़ इतना ही निकला, ''माँ! उस अपराधी पर तुमने क्यों विश्वास किया? जो अपनी बहन का न हुआ, वह ससुराल का क्या होगा?'' उधर राजन के पिता के ऊपर मुसीबत ने पहाड़ का रूप ले लिया। शिमला के ससुराल वालों ने उन्हें सन्देश भेजा था, ''यदि एक महीने में हमारे सारे रुपये वापस नहीं मिलेंगे, तो बहू को हमेशा के लिए छोड़ दिया जाएगा।'' और देखते-देखते एक महीना बीत गया। अब शिमला अपने मायके पर एक बोझ-सी मानी जाने लगी। गाँव-घर की औरतें कहने लगीं, ''बहू रखना आसान है, बेटी नहीं।'' सो शिमला के पिता उसकी दूसरी शादी के लिए दर-दर भटक रहे हैं। वे जहाँ जाते हैं, उन लोगों को राजन का कच्चा चिट्ठा पहले से ही मालूम रहता है। सो अक्सर उन्हें दुत्कार-फटकार मिलती हैं। कुँआरों की तो बात ही नहीं, विधुर भी दूसरी शादी के बाबत उन्हें घास नहीं डालते।

और उस दिन डाकिया ने एक चिट्ठी दी। राजन के पिता ने सोचा कि शायद इसमें राजन के घर वापस आने की सूचना हो या शायद शिमला के ससुराल वालों ने अपने फैसले पर पुनर्विचार किया हो। लिफाफा के ऊपर भेजने वाले का नाम नहीं था, वे झटपट लिफाफा फाड़कर पढ़ने लगे, ''बाबूजी! प्रणाम। मैं आपके घर की बहू थी। जितना मुझसे बन सका मैंने घर-गृहस्थी सँभाली। मगर जब आपका बेटा अपने गाँव-घर को ठगा, अपनी बहन के सुहाग को नहीं बख्शा,

मेरी माँ के गहनों को बेच दिया, तो मैं ऐसे अपराधी पति के साथ कैसे रह सकती हूँ? आज से मैं आपका घर हमेशा के लिए छोड़ रही हूँ। अब मैं अपने पैरों पर खड़ी होकर जीवन-निर्वाह करूँगी। सम्मान की एक सूखी रोटी अपमान के अकूत वैभव से लाख गुना बेहतर है। आपकी अभागिन बहू।''

और चिट्ठी पढ़ते ही राजन के पिता ज़मीन पर धम्म से गिर पड़े जैसे आसमान से तारा टूटकर गिरता है। वंश ही दंश बनकर सता रहा था। इससे तो अच्छा निर्वंश होना था।

छाया

वीरनपुर के साधुराम तिवारी की शादी उसके पिता ने महज बारह वर्ष की उम्र में ही कर दी थी। इसके तीन साल बाद उसका गौना हुआ था। तब तक साधुराम ने दसवीं कक्षा पास कर ली थी और उधर उसकी पत्नी से गौने के तीस-पैंतीस साल बीत जाने पर उन्हें कोई सन्तान नहीं हुई। उसने ओझा-सोखा की झाड़-फूँक तथा हकीम-वैद्य से लेकर अँग्रेज़ी डॉक्टरों तक से इलाज़ करवाया मगर कोई असर नहीं हुआ। किसमें दोष था, वह जानना मुश्किल था। कुछ हमउम्र वालों और रिश्ते-नातेदारों ने उसे दूसरी शादी करने के लिए बार-बार कहा, मगर उसने साफ इनकार कर दिया। वह एक ही जवाब देता, ''मैं अपनी पत्नी से बेवफ़ाई नहीं करूँगा। हम साथ जिएँगे, साथ मरेंगे यदि कोई दूसरी औरत इस घर में आ जाएगी तो कलह का जंगल उग आएगा घर में।''

एक बार उसकी बूढ़ी माँ बहुत बीमार पड़ी। उसकी माँ कम उम्र में विधवा हो गयी थी। मगर बुढ़ापे तक रँड़ापा खेती रही साधुराम के सहारे। सो लायक बेटे के रूप में साधुराम ने एक पुश्तैनी खेत गिरवी रखकर अपनी माँ का महँगा इलाज कराया। माँ धीरे-धीरे ठीक हो गयी। मगर दमा की बीमारी जड़ से नहीं गयी। फिर भी उसकी माँ अपने बेटे-पतोहू की सेवा से सन्तुष्ट थी। एक दिन उसने दोनों से कहा, ''मेरी सेवा जिस तरह तुम दोनों कर रहे हो, वैसे ही तुम्हारी भी सेवा बुढ़ापे में होगी।'' और इतना कहकर माँ ने हमेशा के लिए आँखें मूँद लीं।

साधुराम ने माँ का क्रिया-कर्म विधिवत् निपटाया। दसवें दिन 'शुद्ध' के समय तीन पीढ़ी के पितरों का श्राद्ध किया और तेरहवें दिन ब्रह्मभोज सम्पन्न हुआ। सब कुछ करने के बाद उसे अपनी माँ की कही हुई बात याद आयी। फिर उसके मन में दूसरी आवाज़ उठी कि जब उसको आद-औलाद ही नहीं है, तो उसका श्राद्ध कौन करेगा? हाँ, यदि उसकी पत्नी की मृत्यु सुहागिन के रूप में हो जाएगी, तो वह उसका श्राद्ध विधिवत् करेगा। मगर यदि वह खुद पहले गुजर जाएगा,

तो उसका और उसकी पत्नी का श्राद्ध कौन करेगा? थोड़ी देर में उसके अन्दर से एक आवाज़ आयी, ''जिसने ज़िन्दगी दी है उस पर भरोसा करो। समाज में नेक इन्सान दूसरों की मुसीबत में खड़े मिलते हैं। दसई मिसिर और भुल्लू कहार भी तो बेऔलाद मरे थे मगर गाँव वालों ने उनका श्राद्ध कर्म किया कि नहीं।'' सो उसकी जिज्ञासा फूस की आग की तरह जल्दी ही शान्त हो गयी।

साधुराम के बगल में ही धीरजराम का घर था जो उसका चेचरा भाई था। धीरजराम पास के कस्बे के मिडिल स्कूल में हेडमास्टर थे। उनकी पूरे 'पवस्त' में अच्छी शोहरत थी। अब सरकारी अध्यापकों की तनख़्वाह अच्छी-खासी हो गयी थी। सो उन्होंने पुश्तैनी ज़मीन के अलावा पाँच बीघे ज़मीन और खरीद ली थी। मगर उनके छः बेटों और तीन बेटियों के परिवार का खर्चा ज़्यादा था। मास्टराइन तेज दिमाग़ की थी। उनका नैहर सदोखापुर था जहाँ की लड़कियाँ जादू-टोना करके अपने-अपने मर्द को अपने वश में कर लेती हैं, ऐसी गँवई कहकूत थी। सो सीधे मास्टर को टेढ़ी मास्टराइन अपनी उँगली पर नचाती थी। वही खेती-बारी, घर, दुआर, हित-नात, अड़ोस-पड़ोस के मामले निपटाती थीं और मास्टर साहब के ज़िम्मे सिर्फ़ स्कूल के बच्चों को सम्भालना था। तो मास्टराइन साधुराम को 'बड़े देवर' कहकर सम्मान देतीं और गुड़ से लेकर नये पकवान तक समय समय पर उन्हें खिलाती रहतीं। अपनी आदत से लाचार होने के कारण वह साधुराम के खेतों में क्या-क्या बोया जाएगा, इस पर भी बिना माँगे सलाह-मशविरा देती रहतीं। तीज-त्योहार से लेकर शादी-ब्याह के मौकों पर वह साधुराम के घर 'बायन' भेजती रहतीं और उसके बावत हफ़्तों चर्चा यूँ करतीं, ''गौने का लड्डू बहुत अच्छा था। तिलक में चढ़े फल मीठे और ताज़े थे।'' ऐसा कहकर वह साधुराम के नगीच बनी रहना चाहती थीं।

इसे सुनकर साधुराम उन चीज़ों का बखान शुरू कर देते और इससे मास्टराइन का जी जुड़ा जाता था। मगर साधुराम के बाल भी धूप में नहीं पके थे। चूँकि उनके वहाँ शादी, छट्ठी वगैरह के अवसरों का सवाल ही नहीं उठता था, सो वह मास्टराइन के बच्चों के लिए खाने या खेलने की कोई न कोई चीज़ हाट-बाजार से ज़रूर लाकर देते थे। मास्टराइन के छोटे बेटे मोहन से उसका विशेष लगाव हो गया था। गाँव में वह कहावत खूब प्रचलित थी 'खाई मीठ कि माई मीठ'—यानी मोहन के लिए साधुराम अपने माँ-बाप से ज़्यादा करीब थे।

साधुराम के गाँव वीरनपुर में कई जातियों के परिवार सदियों से रहते थे। उनमें माला के मनकों की तरह आपस में मेल-मिलाप था। एक दूसरे के हारे-गाढ़े कभी भी जाति आड़े नहीं आती थी। शादी को छोड़कर बाकी सारे कामों में सभी लोग साझीदार होते थे।

वीरनपुर के पश्चिम टोले में महेश कहार सबसे गरीब था। गरीबी के कारण उसकी शादी समय पर नहीं हुई। मगर पिछले साल एक ऐसी लड़की से उसकी शादी हुई जो पहली शादी के एक साल के अन्दर ही विधवा हो गयी थी। महेश के घर में उसके सिवाय कोई नहीं था। पत्नी के आ जाने से घर की ज़रूरतें बढ़ गयी थीं। गाँव में सूखा पड़ने के कारण खेती का काम उसे नहीं मिला। साधुराम अक्सर सुबह-शाम गाते, "सूखा पड़ा देश माँ बरसा नाहीं पनिया, किसनिया गड़बड़ानी सजनी।" उनके इस गाने में गाँव-जवार के सारे लोगों का रोना शामिल था। महेश की आँखों के आँसू सूख गये थे। कभी वह आसमान को निहारता, तो कभी अपनी मेहरारू का चाँद-सा मुखड़ा। मगर पता नहीं, सूखे से या और किसी वजह से उसकी पत्नी को कै होने लगी। महेश के पास दवा खरीदकर लाने के पैसे नहीं थे। सूखे में कोई कमाई नहीं थी। उसके पास पुश्तैनी ज़मीन एक इंच भी नहीं थी। सो शादी के तीन महीने बाद वह कमाने के लिए दिल्ली चला गया। जब कोई नौकरी नहीं मिली, तो वह अनाज मंडी में दिहाड़ी मजदूर का काम करने लगा।

कुछ दिनों के बाद महेश की औरत के पास जमुनी गयी थी अपने बेटे की तिलक में गीत गाने का न्यौता देने। महेश की बहू ने कै की बात उसे बतायी। जमुनी ने भाँप लिया कि वह पेट से है। फिर समूचे वीरनपुर में यह बात हवा के झोंके से फैल गयी कि महेश की बहू का पाँव भारी है। गाँव के जवान और बूढ़े, स्त्री और मर्द, प्रायः सभी की एक राय थी कि महेश की बहू किसी गैर-मर्द से फँस गयी है, तभी तो महेश के न रहने पर भी वह पेट से है। मगर महेश के परदेश चले जाने पर और वहाँ से कोई चिट्ठी-पत्री, रुपया-पैसा न आने के कारण महेश की औरत गाँव-जवार में कुछ फुटकर काम-काज करने लगी। कुछ दिनों तक सरकारी ग्रामीण रोज़गार योजना के तहत उसने कच्ची सड़क बनाने में मिट्टी का काम किया। मगर वह काम ख़त्म हो जाने के कारण उसके सामने विपत्ति का पहाड़ खड़ा हो गया। सो वह पड़ोस के गाँव मोचनपुर में काम करने जाने लगी। मोचनपुर के मुखिया के पास प्राइवेट ट्यूबवेल थी सो उनकी खेती का काम बराबर चल रहा था। मगर झूरा के कारण काम कम मिलने और मजूरों की संख्या ज़्यादा होने का फ़ायदा मुखिया ने उठाया तथा उसने मजूरी की दर कम कर दी। फिर भी महेश की औरत उतने पर भी काम करती रही। मगर बाद में पेट ज़्यादा भारी हो जाने के कारण उसे काम बन्द कर देना पड़ा।

एक दिन जमुनी की देखरेख में उसे पुत्र-रत्न की प्राप्ति हुई। मगर गाँव की सारी औरतें उससे नाराज़ थीं, सो बेटा जनने पर भी उसके घर सोहर गीत नहीं गाये गये, ढोल-मजीरा नहीं बजे और औरतों की नाच नहीं हुई। मगर बच्चे

का क्या दोष? जमुनी के दिल में यह बात खटकी मगर वह छोटे मुँह बड़ी बात नहीं कह सकती थी। फिर भी जमुनी ने एक सोहर के बोल शुरू कर दिये, ''कौनी कोखी राम जन्मेन कौनी कोखी लछिमन हो, रामा कौनी कौखी भरत भुआल अजोधिया माँ जन्मेन हो।''

लेकिन मास्टराइन को बिल्कुल अच्छा नहीं लगा। वह ऐसी अन्धेर बर्दाश्त नहीं कर पायी। वह पूरे गाँव में घूम-घूमकर महेश की औरत के चाल-चलन पर छींटाकशी करती, ''इन छिनार ने तो पूरे गाँव की नाक कटवा ली। मैं तो पहले ही कहती थी कि जो दुबारा ब्याह कर आती है, वह ऊपर से नीचे तक देह खोलने में नहीं शरमाती। जिसे समाज की लाज और डर नहीं, वह मैला भी खा सकती है। जो सौ-सौ घाट का पानी पी सकती है, वह कुच्छो कर सकती है। मेरा वश चले तो उसे मौत के घाट उतार दूँ। जो समूचे समाज को, पूरी औरत जाति को मार रहा है, उसे मारने से पाप नहीं लगेगा।''

यह सुनकर आस-पास की औरतें उसकी हाँ-में-हाँ मिलाने लगीं। एक तो मास्टराइन जैसी धाकड़ औरत के सामने बाकी औरतें बौनी बनी रहती थीं और दूसरे उन्हें यह भी डर था कि एक कुलटा का पक्ष लेने पर खुद उनके चरित्र पर भी आँच आ सकती थी।

खैर...गाँव में यह भी बात फैल गयी कि महेश ने अपनी पत्नी को छिनाल होने के नाते छोड़ दिया है और अब वह बेचारा अपना मुँह दिखाने कभी भी वीरनपुर गाँव में नहीं आएगा। मगर किसी ने यह समझदारी नहीं दिखायी कि तहकीक़ात कर ले कि क्या महेश ने इस बावत कोई चिट्ठी-चौपाती लिखी है या गाँव-जवार के किसी आदमी के सामने इतना बड़ा फैसला लिया है। मगर वीरनपुर गाँव ऐसा है कि पिसान की जगह समूचा जाँत ही उड़ा देता है।

उधर महेश की औरत के मायके वाले इतने गरीब थे कि वे उसको चार-छः दिन की भी शरण नहीं दे सकते थे। वैसे भी उसके माँ-बाप बूढ़े हो गये थे और घर में उनके सिवाय कोई न था। खेती-बारी सुई की नोक के बराबर भी नहीं थी। ऐसे में महेश की औरत ने अपने को नैहर, ससुराल, पड़ोस, और समाज सबसे अलग-थलग महसूस किया।

उधर साधुराम को माँ के मरने के बाद घर काटने को दौड़ता था। वह बुझे मन से काफ़ी देर तक घर के बाहर समय काटने लगे। नहीं, नहीं, समय उन्हें काट रहा था। लम्बे समय तक ऐसा होने पर एक दिन पत्नी ने चुप्पी तोड़ने के लिए उसे टोका, ''आप आजकल उदास, निराश और हताश क्यों रहते हैं?''
''क्या बताऊँ? लगता है, यह घर मरघट बन गया है। माँ इस घर की कस्तूरी थीं। उनके स्वर्ग सिधारने के बाद अँधेरा-ही-अँधेरा दिखता है। मैं अब किसके

लिए जिऊँ।''

यह सुनते ही उसकी पत्नी रोने लगी। उसे लगा कि क्या मेरे पति मेरे लिए नहीं जी सकते। फिर सोचने लगी कि शायद वह औलाद के न होने का संकेत कर रहे हैं। उसे भी एक दिन बात-बात में एक औरत ने कहा था कि निरवंश होना पृथ्वी पर भार होना है। सो रात भर दोनों प्राणी बेचैन रहे। उन्हें नींद बिल्कुल नहीं आयी।

सुबह साधुराम की पत्नी अपना अरहर का खेत देखने गयी थीं। बगल में मेड़ पर चीथड़े में लिपटा एक बच्चा रो रहा था। उसे देखते ही वह चिल्लायी, ''किसका बच्चा रो रहा है? दौड़ो, दौड़ो।'' यह कहते हुए उसने बच्चे को उठा लिया। बच्चे ने टट्टी-पेशाब कर दी थी। मगर उसे घिन नहीं लगी। मुखिया के खेत में काम कर रही महेश की औरत हाँफते हुए आयी। साधुराम की पत्नी ने उससे कहा, ''तू कैसी माँ है? इसे अकेला क्यों छोड़ दिया? इस बच्चे को ठंड लग रही है। यह काँप रहा है। इसे डॉक्टर के पास ले जाओ, वरना गड़बड़ हो जाएगी।''

''मगर मेरे पास पैसे नहीं हैं...'' यह कहकर महेश की औरत रोने लगी।

''चलो, मैं तुम्हें कुछ रुपये-पैसे देती हूँ। साधुराम की पत्नी ने दिलासा देते हुए कहा, ''बच्चे से बढ़कर कुछ भी नहीं। मगर तुम्हें कुछ मजूरी-धतूरी भी तो मिली होगी?''

''मुखिया पूरी मजूरी नहीं देता। उसने अधूरी मजूरी दी थी सो उससे कुछ दाना-पानी खरीद लिया। बाकी मजूरी माँगने पर आनाकानी करते हुए कहता है, ''पैसा पेड़ में नहीं फलता।''

साधुराम की पत्नी ने अपने दोनों हाथों की उँगलियाँ फोड़ते हुए मुखिया को सरापा और तेजी से अपने घर की ओर बढ़ गयी। उसने महेश की औरत को पचास रुपये दिये। महेश की औरत रुपये लेकर चौराहे के कस्बे की ओर भागी। वहाँ उसने एक डॉक्टर को दिखाया। डॉक्टर बोला, ''इसे डबल निमोनिया हो गया है। इसे महँगी सुई-दवा देनी होगी इसमें तीन सौ रुपये लगेंगे।'' उसने पचास रुपये कम्पाउंडर को देते हुए चिरौरी की कि बाकी पैसे बाद में दे देगी। मगर काइयाँ कम्पाउंडर उखड़ गया, ''यहाँ उधार-बाढ़ी में इलाज नहीं होता। ऐसे दस मरीज़ आ जाएँ तो हमारी दुकान चौपट हो जाएगी। यह कोई सरकारी अस्तपाल नहीं है कि खैरात में दवा-दारू होगी। इस हाथ पैसा दो, उस हाथ दवा लो।''

उसने डॉक्टर से भी चिरौरी की मगर वह भी टस से मस नहीं हुआ। उसने सिर्फ़ इतना कहा, ''अब तक इसको कुछ दवा देकर बचाते हैं, मगर तुम फौरन

जाकर पैसे का बन्दोबस्त करो। तीन सौ रुपये लेकर आना, तो बच्चे को वापस ले जाना।'' सो महेश की औरत सिर पर पाँव रखकर वहाँ से भागी।

हाँपते हुए वह अपने गाँव पहुँची। उसने सारी गाथा साधुराम और उसकी पत्नी को सुना दी। साधुराम का रुख ठंडा हो गया था क्योंकि उसे लग रहा था कि महेश की औरत तीन सौ रुपये कभी नहीं चुका पायेगी। मगर उसकी पत्नी ने बिना सोचे अन्दर से तीन सौ रुपये लाकर उसे दिये। महेश की औरत की आँखों में आँसू छलछला आये। वह साधुराम की औरत के पैरों पर गिरकर दुआ देने लगी। तभी उसने कहा, ''आप भी इसके साथ डॉक्टर के पास चले जाइये। यह बेचारी अनजान है। और हाँ, जब तक बच्चा ठीक न हो जाये, आप छोड़कर मत आना।''

सो साधुराम और महेश की औरत भागते हुए डॉक्टर के पास चले गये। पैसे पाकर डॉक्टर ने पूरी संजीदगी और गम्भीरता से किशन का इलाज किया। करीब आधी रात को किशन की हालत में सुधार हुआ। वह चैन की नींद सोने लगा। उसकी माँ ने उसका माथा चूमा और मन से साक्षात् देवी के रूप में मिली साधुराम की पत्नी को याद किया। डॉक्टर ने बताया कि अब वह ख़तरे से पूरी तरह बाहर है। सुबह तक ठीक हो जाएगा। साधुराम भी वहाँ पूरी रात किशन की सेवा में लगे रहे। उन्हें लगने लगा कि एक बच्चे की सेवा करके वह पुण्य कमा रहे हैं। उनके मन में ज़िन्दगी को लेकर नाना प्रकार के विचार आने लगे।

सुबह किशन बिल्कुल ठीक हो गया। डॉक्टर ने किशन की माँ को हिदायत दी, ''बच्चे को कमज़ोरी है। इसलिए एक टॉनिक पिलाना पड़ेगा महीने भर। दूसरे, आपको प्रोटीन वाली चीज़ें जैसे दाल वगैरह खानी है जिससे ज़्यादा दूध बने। माँ का दूध अमृत समान होता है, इससे कई बीमारियों से बचाव भी होता है। तीसरे, बच्चों की सफाई पर ध्यान दो ठंड, नमी आदि से बचाओ।''

सो साधुराम और किशन की माँ किशन को लेकर चलने लगे। रास्ते में किशन की माँ डॉक्टर के द्वारा बतायी गयी तीनों बातों पर एक-एक करके गौर करने लगी। वह जितना सोचती, उतना ही ज़्यादा बेचैन हो जाती कि कैसे दाल का जुगाड़ करेगी, कैसे रजाई-कम्बल की व्यवस्था करेगी और कैसे किशन को बचाएगी। वे तीनों साधुराम के दरवाज़े पर पहुँच गये।

किशन की हालत में सुधार देखकर साधुराम की पत्नी खुशी के मारे उछल पड़ी। साधुराम ने उसे डॉक्टर द्वारा बतायी गयी तीनों हिदायतों को बता दिया। साधुराम की पत्नी ने किशन की माँ की बेचैनी भाँपते हुए कहा, ''तो इसमें परेशान होने की क्या बात है? बच्चे के हित में तुम मेरे घर पर ही कुछ दिन रह जाओ। जब किशन पूरी तरह स्वस्थ हो जाएगा, तब अपने घर चली जाना।''

सो ज़िन्दगी के सारे रास्ते और गलियाँ बन्द पाकर किशन और उसकी माँ

वहीं रहने लगे। साधुराम की पत्नी ने उन दोनों का पूरा ख़्याल रखा। किशन की माँ चौका बर्तन, झाड़ू आदि करती, फिर खाना बनाती। उधर साधुराम की पत्नी किशन की तेल-बुकवा से मालिश करती। उसे अपने बिस्तर पर सुलाती। उसे खिलाती-पिलाती रहती। उसका मातृत्व इस कदर छलछला रहा था मानो किशन उसकी कोख से ही जन्मा हो। उस घर में पसरा अन्धकार प्रकाश की चकाचौंध में छूमन्तर हो गया।

कई दिनों के बाद साधुराम की पत्नी ने उससे शाम के समय कहा, ''क्यों जी, किशन और उसकी माँ को हम नहीं जाने देंगे। उनके आने से घर में सूरज उग आया है।''

''तो भेज कौन रहा है उन्हें? क्या वे लोग जाना चाहते हैं?'' साधुराम ने जानना चाहा।

''नहीं, आपने मेरी बात समझी नहीं। मैं किशन को हमेशा के लिए अपने पास रखना चाहती हूँ। उसके बिना मैं रह ही नहीं सकती। मैं उसे गोद लेना चाहती हूँ।'' पत्नी ने कहा।

''पर यह कैसे हो सकता है? वह हमारी जाति का नहीं है। फिर दुनिया क्या कहेगी?'' साधुराम ने अपनी पत्नी को समझाया। पर उसकी पत्नी पर इसका कुछ भी प्रभाव नहीं पड़ा। वह तपाक् से बोली, ''मुझे हर हालत में किशन ही चाहिए।''

''तब तो सारी दुनिया मुझ पर यही तोहमत लगाएगी कि ज़रूर किशन की माँ से मेरे अनैतिक सम्बन्ध हैं। पहले भी कुछ लोगों ने दबी जुबान से ऐसा ही कुछ कहा है। अब तो लोग सरेआम कहेंगे।''—साधुराम ने तर्क दिया। थोड़ी देर बाद पत्नी का हठ सुनकर उन्होंने टालने के अन्दाज़ में कहा, ''क्या किशन की माँ इसके लिए तैयार होगी?''

''उसे क्या आपत्ति हो सकती है भला।''

''क्यों, क्या कोई अपनी औलाद किसी को यूँ ही दे देता है?''

''वह गुझ गर छोड़ो। किशन की माँ मेरी बहन जैसी है, उसे मैं मना लूँगी।''

इसी बीच किशन की माँ वहाँ आ गयी। उसे देखकर साधुराम की पत्नी ने कहा, ''क्यों नगीना, क्या तुम मुझे किशन को गोद लेने दोगी।''

''क्या?'' वह चौंकी।

''बुरा मत मानो, नगीना।'' साधुराम की पत्नी को लगा कि नगीना को वह बात जँची नहीं।

''नहीं, नहीं, मैं बिल्कुल बुरा नहीं मान रही हूँ। इससे तो किशन की तकदीर खुल जाएगी। उसे छाया मिल जाएगी। उसका भविष्य सँवर जाएगा और मेरी

कोख भी तर जाएगी। छतनार गाछ ही किसी व्याकुल राही को छाया दे सकता है।'' नगीना ने शालीनता से कहा।

उस रात साधुराम और उसकी पत्नी रातभर नाना प्रकार की योजनाएँ बनाते रहे। सुबह होते ही साधुराम की पत्नी ने किशन को गोद में उठा लिया। उसे दुलारने लगी। फिर वह सोहर गाने लगीं। किशन भी धीरे-धीरे उन्हें परक गया था। उनकी गोद से दूसरे की गोद में नहीं जाता था। इसी बीच मास्टराइन आ गयीं। वह अपने मायके से कल रात ही लौटी थी। उसने साधुराम को फटकारने के लहजे में कहा, ''देवरजी, आपने इस कलमुँही को इस घर में पनाह क्यों दे दी। इसे तो इसकी बिरादरी ने भी टाट—बाहर कर दिया है। दुनिया थूक रही है। आपके चलते हमारा भी नेवता-हंकारी बभनौटी में बन्द हो जाएगा।''

साधुराम के मुँह पर ताला लगा था। वह बड़ी भाभी का अदब-लिहाज भले कर रहे थे मगर उनके शब्दबाण उन्हें छेद रहे थे। यह सुनकर साधुराम की पत्नी किशन को लेकर बाहर आयीं और अपनी जेठानी को जवाब देने लगी, ''दुनिया कौन होती है हम पर थूकने वाली? हमने कोई अनैतिक या अवैध काम नहीं किया। इसे बहन के रूप में अपनाया है। यदि ऐसा करना गुनाह है, पाप है, तो हम लोग ऐसा बार-बार करते रहेंगे।''

''अच्छा-अच्छा! इस चाल में तुम्हारी भी सहमति है। बाद में तुम पछताओगी। चार दिन की चाँदनी, फिर अँधेरी रात।''

''नहीं, अब हमारी ज़िन्दगी से अँधेरा मिट गया है। हम किशन को गोद ले लेंगे। जहाँ किशन रहेगा, वहाँ अँधेरा पास नहीं फटकेगा। और उसकी माँ नगीना इनके हाथ की कलाई में राखी बाँधेगी। वह उनकी मुँह बोली बहन होगी।'' पत्नी की सोच सुनकर साधुराम का मुरझाया चेहरा अचानक खिल गया। उन्हें लगा कि बिना बदनामी के उन्हें किशन की छाया मिल गयी है। किशन एक दिन छतनार गाछ की तरह छाया देगा, ऐसा साधुराम सोचने लगे। नगीना की खुशी उसकी आँखों से छलकने लगी। उसे एक बड़ा भाई मिल गया था। अब उसे मायके का सुख भी मिलने वाला था। उसने भी मास्टराइन की आँखों से आँखें मिलायीं। मास्टराइन ने आँखें नीची कर लीं। उसका सपना चूर-चूर हो गया था। वह मुँह बनाये तेज कदमों से वापस लौट गयी। उधर किशन किलकारियाँ मार रहा था। तभी वहाँ जमुनी आ गयी। उसने माहौल देखकर फिर एक सोहर गाना शुरू कर दिया, ''कौनी कोखी किशन जन्मेन, कौनी कोखी दाऊ हो...।'' तभी नगीना ने साधुराम की दायीं कलाई में धागे की राखी बाँध दी। आसमान में खिला इन्द्रधनुष खुले दिल से अपना सौन्दर्य सबको लुटा रहा था।

चक्रव्यूह

और बमबम बाबा ने अपनी बड़ी बेटी चम्पा की शादी तय कर ली। इसकी शादी के लिए वे तीन बरसों से भटक रहे थे। वे दर्जनों दरवाज़ों पर दस्तक दे चुके थे। कुछ के द्वारा दहेज़ की बाबत कुत्ते की तरह दुत्कार दिये गये थे। उस लड़की की तकदीर को कल तक घर और बाहर वाले बराबर कोसते रहते थे। कोई कहता, "मंगली होगी। मंगली लड़की जिस घर में जाएगी उसका भला नहीं हो सकता।" कोई कहता, "जन्म से कुलच्छिनी है।" कोई कहता, "साँवली है।" कोई कहता, "नाटे कद की है," कुछ दिनों बाद गाँव में यह भी कानाफूसी होने लगी कि उसमें ज़रूर कोई ऐब है। जितने मुँह, उतनी बातें। विडम्बना यह कि जो लड़कियाँ काली और कुरूप थीं, वास्तव में कुलच्छिनी थीं, मंगली थीं—वे भी चम्पा को कोसती थीं और भरसक एक दूरी बनाये रखती, जिससे चम्पा की काली छाया न पड़ जाये। उसके बारे में जो कहा-सुना जाता, उसे सुनकर वह भीतर-ही-भीतर भुरकुस होती थी, मगर उसने अपनी ओर से किसी को कभी भी कुछ नहीं कहा। माँ-बाप, भाई-बहन, सगे-सम्बन्धियों की मर्यादा का ख्याल करके चुप रहती। पर इस तरह के अपमानजनक व्यवहार से तंग आकर कई बार उसके मन में यह भी ख़्याल आता कि क्यों नहीं कुएँ में डूबकर अपनी जान दे दे। सो शादी तय होने से घर-गाँव के अधिकतर लोग खुश हुए।

उन लोगों की तरह वह उतनी खुश तो नहीं हुई पर दुःख ज़रूर कम हुआ। यह सोचकर कि माता-पिता के लिए अब वह बोझ नहीं समझी जाएगी। बाबा छोटे-मोटे किसान थे। खेती-गृहस्थी ही मुख्य जीविका थी। बड़ा लड़का पढ़-लिखकर बेरोज़गार था। प्राय: बाबा सोचते—जवान लड़की का विवाह कर पिता के कलेजे का बोझ हल्का हो जाता है, मगर बेकार लड़का ऐसा बोझ होता है जो पिता की चिता के साथ ही ख़त्म होता है। पर अभी यह सोचने का समय नहीं था। लड़की की शादी सम्पन्न होने तक चिन्ता बनी हुई थी। उन्होंने किसी

तरह कुछ कर्ज़ लेकर शादी का इन्तज़ाम किया। राशन तो घर का था पर और तमाम इन्तज़ाम तेल-फुलेल, साग-सब्जी, टेंट, लाइट, कपड़े-लत्ते, बर्तन-बासन, मिठाई, वगैरह में ही बुढ़ऊ को छट्ठी का दूध याद आने लगा।

जो हो बारात का स्वागत बाबा ने औकात भर किया। उस रात पुराने रस्म-रिवाजों को निभाते हुए विधिवत् शादी सम्पन्न हो गयी यानी आग के सामने सात फेरे, कन्यादान वगैरह पूरा हो गया। लेकिन सुबह खिचड़ी की रस्म-अदायगी में सब कुछ गड़बड़ हो गया, जिसकी थोड़ी बहुत आशंका पहले से ही थी। खिचड़ी परोसने के बाद दूल्हे ने कौर नहीं उठाया। चंपा के बड़े भाई ने दूल्हे के दायें हाथ की ऊँगली में सोने की अँगूठी पहना दी। दूल्हे की नाराज़गी अपनी जगह बनी रही। बुढ़ऊ ने भरसक तरह-तरह के मीठे बोल से दूल्हे को मनाने की कोशिश की, परन्तु कोई फ़ायदा नहीं हुआ। बरामदे में बैठ कर औरतें गा रही थीं—

"दूल्हा खिचड़ी न खइहें, भगाव मड़वे से, डोलाव मड़वे से..."

बमबम बाबा ने हार मानकर करुण-भाव से समधी से अनुरोध किया कि वे दूल्हा को खिचड़ी खा लेने के लिए कहें। परन्तु समधी जी की आँखों में आग के अँगारों का ताप साफ झलक रहा था, मानो इस अवसर पर उनकी प्रतिष्ठा के अनुरूप उन्हें कुछ नहीं मिला हो। उनकी भृकुटि तनी हुई थी। कहा, "बमबम जी! आपने होंडा देने के लिए कहा था। कहा था कि नहीं? कहाँ है होंडा?" इतना सुनते ही बाबा ने एक कुशल अभिनेता की तरह मुद्रा बनायी। आँगन से दौड़े-दौड़े एक कोठरी में गये। वहाँ से पीतल का चमचमाता एक हंडा लाकर समधी के सामने रख दिया। फिर क्षमा-प्रार्थना करने के अन्दाज़ में कहा, "समधी जी! सचमुच मैं भूल गया था। आपने हंडा माँगा था, सो लीजिए।"

यह सब देख सुन समधी जी का पारा सातवें आसमान पर पहुँच गया। वह अब दहाड़ने लगे, "मैं पूछता हूँ, आप आदमी हैं कि पाजामा? आपसे बड़ा धोखेबाज मैंने दुनिया में नहीं देखा। कहाँ चालीस हज़ार की हीरो होंडा मोटर साइकिल और कहाँ यह पीतल का हंडा। मैं नहीं जानता कि आप बहरे हैं। आप सरासर चालाकी पर उतर आये हैं। मैं खूब समझता हूँ। पर याद रखिये आप जैसे चालाकों को पानी पिलाना मुझे आता है।" समधी की डाँट सुनकर बाबा हकलाने लगे। वे समधी के पैर पर धम्म से गिर गये। कहने लगे, "समधी जी! नाराज़ मत होइये। मैं सच कहता हूँ। मुझसे गलती हो गयी। मैंने आपसे कोई धोखा-धड़ी नहीं की। मोटर साइकिल देने की औकात ही कहाँ थी कि उसका वचन देता। सो मेरे बड़े लड़के की शादी में यही हंडा मिला था। मेरी पगड़ी की लाज रख लीजिए। बड़े पेड़ों की छाया में ही राही आराम पाते हैं।" परन्तु बाबा ने

जो पैर पकड़ा था, उसे समधी जी ने झटके से खींच लिया। वे औंधे मुँह गिरे। इसके पहले कि समधी जी शान्त होते, आग में घी का काम किया दूल्हे के चाचा ने। वह चिल्ला-चिल्ला कर बोलने लगे—हे बम भोले। कान खोल कर सुन लीजिए—यह शंकर जी की बारात नहीं है कि हम लोग भांग-धतूरे से खुश हो जाएँगे। यदि आप हीरो होंडा नहीं देंगे, तो आपकी लड़की की विदाई हरगिज नहीं कराएँगे। आपने बर्रे के छत्ते को छेड़ा है तो उसका अंज़ाम भी भुगतने के लिए तैयार हो जाइये...

देखते-ही-देखते बाराती नरमदल और गरमदल में बँट गये। गनीमत यह थी कि बारात में भी कुछ शरीफ़ लोग थे। दूल्हे के मामा ने आगे बढ़ कर अत्यन्त विनम्र भाव से दूल्हे के पिता को समझाने की कोशिश की, "जीजा जी! अपनी इज़्ज़त अपने हाथ। हम लोगों ने अब तो बमबम जी के यहाँ अच्छा या बुरा रिश्ता जोड़ लिया है, इसलिए किसी रिश्तेदार की बेइज़्ज़ती अपनी बेइज़्ज़ती है। वर-वधू ने कल रात पवित्र अग्नि के चारों ओर सात फेरे ले लिए हैं। हिन्दू परम्परा में विवाह पूरा हो चुका है। जब बमबम जी की औकात हीरो होंडा देने की नहीं है, तो वे बेचारे क्या करें?" इस कथन का प्रभाव अच्छा पड़ा। समधी जी कुछ नरम हुए। मगर दूल्हे का चाचा बिगड़ैल घोड़े की तरह बिदका हुआ था। उसने दूल्हे को आँख से इशारा किया। सो वह बिना खिचड़ी खाये उठ खड़ा हुआ। उसने घड़ी और अँगूठी मंडप में ही छोड़ दी। एक घराती को मौका मिला। उसने जले पर नमक छिड़कते हुए कह डाला, "दहेज़ माँगना कानूनन जुर्म है। हमलोग मुकदमा दायर कर दें तो ये सीधे जेल ही जाएँगे। यहाँ की खिचड़ी नहीं खाते, वहाँ खानी पड़ेगी।" अब जोर-शोर से तू-तू मैं-मैं होने लगी। थोड़ी देर तक अराजकता का माहौल बना रहा। मगर दूल्हे के मामा और चम्पा के बड़े भाई ने बीच-बचाव किया। एक-डेढ़ घंटे बाद किसी तरह समझा-बुझाकर चम्पा की विदाई कर दी गयी। चम्पा रोते-रोते चली गयी। उसे अपने परिवार से दूर जाने का दुःख तो था ही, जिस परिवार में जा रही थी, उस बिगड़ैल परिवार में अपने भविष्य की दुश्चिन्ता को लेकर भी लगातार रोये चली जा रही थी...

जैसे-तैसे बारात विदा हुई। बाबा का बोझ हल्का हुआ। जो बोझ उनके सिर पर पच्चीस बरसों से था, उसके उतरने से उन्हें थोड़ा सुकून मिला। वह मन-ही-मन आश्वस्त हुए, "हो गया ब्याह, तो क्या करेगा शाह?"

परन्तु लड़की की विदाई के बाद चम्पा की माँ बिल्कुल चुप-न भूख, न प्यास न नींद, न चैन। भविष्य की आशंका से थर-थर काँपती। कभी घर, कभी आँगन। वह सोचने लगतीं, "ससुर तो बिगड़ैल है, अगर उसकी सास भी वैसी ही होगी तब क्या होगा? ननद पर भी बाप का असर पड़ सकता है। जैसा बीज,

वैसा फल। माटी की सास सतावै, काठ की ननद चिढ़ावै। फिर बड़ा परिवार है। सबके लिए काम करना होगा। अभी घर में उसकी कोई देवरानी नहीं है, सो सारा भीतरी भार उसी के सिर पर पड़ेगा। फिर उस गाँव में इस गाँव की लड़कियाँ पहले से नहीं हैं जिनसे वह अपना दुःख-दर्द बाँट सके...''

चम्पा रोती सुबुकती बाबुल की देहरी छोड़ चली। ऊबड़-खाबड़ सड़क पर जीप दौड़ती चली जा रही थी। हिचकोलों से उसका शरीर जीप में बैठे पति की ओर झुक जाता। वह तुरन्त सँभल जाती। निगोड़े ड्राइवर ने ऐसे समय में एक कैसेट लगा रखा था। इस गीत को तो उसने सुना कई बार था, पर आज जैसे कसक की अनुभूति इसके पहले कभी नहीं हुई थी। वह बहुत ध्यान से गीत सुनने लगी—

''काहे को ब्याहे बिदेस, अरे लखिया बाबुल मोरे,
हम तो बाबुल तोरे पिंजड़े की चिड़िया कुहुक रोती जाये,
भैया को बाबुल दियो महला-दुमहला हमको दिये परदेस...''

करीब चार-पाँच घंटे के सफर के बाद ससुराल की देहरी पर जब जीप खड़ी हुई तो वह संज्ञा-शून्य थी। किसी ने उसे सहारा देकर जीप से उतारा। धरती पर पहला पग रखते ही उसे लगा पाँव बिल्कुल बेजान हो गये हैं। शायद झुनझुनी पकड़ ली है। साँस उखड़ी जा रही है। घूँघट के भीतर उसकी आँखें ठीक-ठीक कुछ भी नहीं देख पा रही थीं। एक बार तो वह गिरते-गिरते बची। उसने मन की आँखों से देखा—ससुराल में किसी के पास नयी दुल्हन के स्वागत का वैसा उत्साह नहीं था जो आम तौर पर उसके गाँव में होता है। मायके में उसने सुन रखा था कि उसकी दो ननदें हैं। पर द्वार-छेंकाई के अवसर पर कोई सामने नहीं थी।

उसकी सास ने सूप, जल, अक्षत आदि से उसकी परछन नहीं की। पहली बार घर में प्रवेश के अवसर पर औरतें जो गीत गाती हैं, वह गीत नहीं गाया गया। दरअसल ससुराल वालों को दुल्हन से ज़्यादा हीरो होंडा का इन्तज़ार था। पूरा परिवार चाहता था कि सूरज भले बेरोज़गार है, मगर उसकी मोटर साइकिल की इच्छा दहेज़ से पूरी हो जाये। सो सास तथा अन्य औरतों ने मुँहदेखायी भी चम्पा को नहीं दी। उसे सीधे एक कोठरी में ले जाकर बैठा दिया गया। वहाँ पर मात्र एक दरी बिछी थी। उसके पास कोई आ-जा नहीं रहा था। वह अकेली गुमसुम पड़ी थी। उसे खूब प्यास लगी थी। समस्या थी कि किससे पानी माँगे? वह भीतर-भीतर काँपने लगी। डर के मारे उसका बुरा हाल था। वह पसीने से लथपथ कब सो गयी, कुछ पता नहीं।

कल रात क्या हुआ, उसे कुछ पता नहीं। धीरे-धीरे जब होश आया, तो अपने पास एक बूढ़ी औरत को पंखा झलते पाया। वह अचकचा कर उठ गयी। उस औरत ने बताया कि वह सूरज की बुआ है। वह कल रात बेहोश हो गयी थी।

फिर बड़ी ननद कोठरी में आयी। उसने आते ही व्यंग्य का पहला वाण फेंका, ''मेरे भैया की तकदीर ही फूट गयी जो तुम्हें ब्याह लाये। कहाँ बी.ए. पास मेरे भैया लाखों में एक और कहाँ तुम मिडिलची। कहाँ सूरज, कहाँ काली अमावस। नाम चम्पा रानी, गन्ध महुआ की।'' मुँह बिचकाते वह चली गयी। ननद से पहली मुलाकात की तिलमिलाहट अभी ख़त्म नहीं हुई थी, तभी सास आ धमकी। उनके हाथ में एक प्लेट था उसमें दो सूखी रोटियाँ और नमक-मिर्च भी था। जैसे घरेलू कुत्ते के आगे रोटी फेंकी जाती है, उसी अन्दाज़ में उसके सामने प्लेट रख दिया गया। 'खा लेना' कहकर डपटती हुई जिस झटके से वह आयी थी, वैसे ही बाहर चली गयी। उसे अपने मिठुआ कुत्ते की याद आयी जिसे वह कौरा देती थी।

रोटी उसी तरह सामने पड़ी थी। खाने की इच्छा मर गयी। सो उसने अपने बक्से में रखी मिठाई का एक टुकड़ा मुँह में डाला और एक लोटा पानी पी गयी। थोड़ी देर बाद उस कोठरी में किसी के आने की आवाज़ आयी। उसने सर उठाकर देखा। सास खड़ी थी। उसे अपनी माँ की बात याद आयी। वह उठी और सास के चरणों को छूकर माथे पर हाथ लगाया। किन्तु आशीर्वाद के बदले उसने मुँह बिचकाया, ''चल हट। बड़ी आयी चरण छूनेवाली, तुम लोगों ने तो मिलकर मेरे घर में आग लगा दी। तुम्हारा कोई दोष नहीं। हमलोगों का भाग्य ही खोटा था। तुम्हारे बाप ने हम सबको धोखा दिया, उसे नरक में भी जगह नहीं मिलेगी। खाना मैंने रख दिया है। खाना हो तो खा।'' आयी थीं सास आँधी की तरह और तूफान की तरह चली गयी।

वह सास की बातों का अर्थ जोड़ने लगी। इस धरती पर उसके होने का क्या मतलब? ईश्वर ने उसे क्यों जन्म दिया? मायके से लेकर ससुराल तक उपेक्षा और केवल उपेक्षा ही उसे मिलती रही। वहीं माँ ही थी जो उसकी खोज-ख़बर लिया करती थी। सबकी यही इच्छा थी कि किसी तरह इसकी शादी सम्पन्न हो जाए तो उसकी छोटी बहन कृष्णा की शादी हो। वह छोटी बहन की शादी की अड़चन बनी थी। अब रास्ता साफ है।

तभी बाहर छोटी ननद की आवाज़ सुनायी पड़ी, ''उठिए महारानी जी! कब तक बैठे-बैठे पाँव तोड़ते रहिएगा। खाना-वाना तो बनाना आता होगा या वह भी नहीं। माँ ने ख़बर करने के लिए कह दिया था। आज ही चौका छुवावन

की रस्म अदायगी है। मुझे कहना था सो कह दिया। मुझे क्या? तीन दिन हो गये, पता नहीं भैया कहाँ चले गये। बाबूजी चिन्ता से मरे जा रहे हैं। आपको क्या? आप तो पलंग तोड़ रही हैं।'' आसमान को सुनाते हुए वह चली गयी। उसे देखने का अवकाश नहीं था कि उसने फेंकी हुई रोटी खायी या नहीं।

मायके में वही तो रसोई की जिम्मेवारी निभाती चली आयी है। यह काम उसके लिए न तो पहाड़ था और न दुखदायी। किन्तु अन्तिम वाक्य सुनकर उसका कलेजा धड़कने लगा था। इसका मतलब है कि वे यहाँ नहीं हैं। कहाँ गये? क्यों गये? किससे पूछे, कौन बताये। क्या वे भी बिना मुझसे मिले नाराज़ हो गये। उनके घर से जाने का कारण क्या मैं हूँ? नहीं-नहीं।

कुछ सोचती हुई वह चौके में गयी। अपने पाक विद्या के अनुभवों का उपयोग किया। तरह-तरह के व्यंजन बनाये। देर रात तक घर के लोग और मेहमान नयी दुल्हन के बनाये व्यंजनों का आनन्द लेते रहे, खा-पीकर सभी लोग सोने चले गये। वह भी खाकर सोने चली गयी। किसी के पैरों की आहट उसे महसूस हुई। वह बोली, ''कौन?'' पर कोई जवाब नहीं मिला। थोड़ी रात और ढली। उसे भी नींद आ गयी।

चम्पा की सास और ननद ने पूरा इन्तज़ाम कर लिया था। सुबह चारों ओर हल्ला हो गया कि सूरज की बहू रात में डिबरी ऊपर गिर जाने से जलकर मर गयी।

सप्ताह भर बाद बमबम बाबा को चम्पा की मौत की ख़बर मिली। सुनते ही वे आग-बबूले हो गये। अपने बड़े बेटे के साथ चम्पा की ससुराल पहुँचे। मौत को लेकर लोगों के भिन्न-भिन्न मत थे। पर बमबम बाबा को पूरा विश्वास था कि चम्पा स्वत: नहीं जली बल्कि उसे जलाकर मार दिया गया है। वे थाना पर रिपोर्ट दर्ज़ कराने गये। थानेदार उनकी बात अनसुनी कर डपटने लगा, ''मुझे मालूम है, वे लोग रईस और खानदानी हैं। उनके खानदान में ऐसा कभी नहीं हुआ कि किसी बहू को प्रताड़ित किया गया हो। मुझे तो लगता है कि आपकी बेटी ही किसी से प्यार-वार करती रही होगी। उससे आप लोगों ने इसकी शादी नहीं की सो उसने आत्महत्या कर ली होगी। देखिए, बमबमजी! इसे आप तिल का ताड़ बना रहे हैं। बता देता हूँ कि आप किसी के बहकावे में मत आइये वरना पछताना पड़ेगा। मैं आपके दु:ख को समझ सकता हूँ। लेकिन बिना ठोस सबूत के किसी पर आरोप लगाना ठीक नहीं है। इससे ये लोग बेवजह परेशान हो जाएँगे।''

थाने में इस अधिकारी की बात सुनकर बाबा का माथा चकराया। लगा कि वे पुलिस थाने में नहीं, धर्म सभा के किसी प्रकांड विद्वान सन्त के प्रवचन सुन

रहे हैं। इसके पहले कि वे सहज होते, वहाँ मौजूद एक तोंदियल आदमी ने बेरुखी से रास्ता नापने के लिए कह दिया। बाबा तमतमाकर चल पड़े।

थाना से निराश लौटते समय रह-रहकर एक ही बात मन में आ रही थी। किसी तरह बदला लेना है जो उस दिन लड़के के पिता और चाचा ने सब लोगों के बीच खिल्ली उड़ाई थी। पूरे समाज में बेइज़्ज़ती की थी। इससे अच्छा अवसर उन्हें नहीं मिल सकता है। घर लौटते हुए उन्हें अपने इलाके के एक पत्रकार रमण की याद आयी। वे चौराहे पर उतर कर सीधे रिक्शे पर बैठे पत्रकार के यहाँ चल पड़े। स्थिति सुनायी। रमण जी ने सलाह दी कि एक जीप रिज़र्व करके अगले दिन पत्रकारों को वहाँ लेकर पूरे मामले की छानबीन कराई जाये, जिससे सभी अख़बारों में यह मामला उछले। बमबम बाबा ने वैसा ही किया। पत्रकारों ने जीप से स्थल-भ्रमण किया, कस्बे के सबसे बड़े होटल में नाश्ता आदि किया और अपने कर्तव्य का निर्वाह करते हुए बड़े-बड़े अक्षरों में 'दरिंदों की शिकार हुई एक और चम्पा' नामक शीर्षक छापा। बमबम बाबा के अलावा चम्पा की ससुराल के घर का चित्र भी बाकायदा छप गया। पुलिस कप्तान ने थानेदार को फोन किया थानेदार पहले दर्ज़े का खुर्राट था। उसने बहाना बना लिया कि वह उस समय बाज़ार में ट्रक उलटने के मामले में तहकीकात करने गया था। पुलिस कप्तान ने जिला पुलिस की बदनामी सभी अख़बारों में पढ़ी और उनका मन हुआ कि वे उस थानेदार को तुरन्त मुअत्तल करके लाइन हाजिर कर दें, मगर उसी समय उनकी मेमसाब वहाँ आ गयी और उन्होंने फोन का चोंगा अपने हाथ में ले लिया। कप्तान साहब के गुस्से को शान्त करते हुए वह बोली, "जो बदनामी होनी थी, वह हो गयी। वहाँ अपना आदमी ही थानेदार है। अब उसी थानेदार को आरोपियों को पकड़ने के लिए कहिए। इससे पुलिस की बिगड़ी छवि सुधर जाएगी।" और कप्तान साहब ने 'गृहमन्त्री' के आदेश का पालन किया।

अगले दिन पुलिस ने सूरज और उसके पिता को गिरफ़्तार कर जेल भेज दिया। अख़बारों में पुलिस कप्तान ने डैमेज कंट्रोल किया। मुख्य न्यायिक दंडाधिकारी की अदालत से सूरज और उसके माँ-पिता व बहन की ज़मानत अर्जी ख़ारिज़ हो गयी। उनकी अर्जी सत्र न्यायाधीश और हाईकोर्ट ने भी ख़ारिज़ कर दी। बमबम बाबा को खुशी हुई कि आख़िरकार चम्पा के ससुराल वालों की गिरफ्तारी हो गयी।

कुछ महीने बाद मुकदमे का पेशी शुरू हुई। बमबम बाबा को यह धुन सवार थी कि चम्पा की हत्या करनेवालों को फाँसी की सजा ज़रूर मिले। इसके लिए वह सरकारी वकील के पास हर तारीख को जाते थे और सबूत वगैरह पुख़्ता करने की व्यवस्था करते थे। सरकारी वकील ने उन्हें आश्वस्त किया था कि

अभियुक्तों को सजा मिलेगी। उन्होंने चम्पा की ससुराल के गाँव के कुछ लोगों की किसी तरह अपनी तरफ कर लिया था और अभियुक्त पक्ष की कमजोरियों का ब्योरा वे सरकारी वकील को समय-समय पर बतलाते थे।

एक दिन जब बमबम बाबा सरकारी वकील के निवास पर सुबह आठ बजे पहुँचे, तो वहाँ बचाव पक्ष के वकील रणेन्द्र शर्मा को देखकर उन्हें काटो तो खून नहीं। उनका हलक सूखने लगा। सरकारी वकील इसे ताड़ गये और कहने लगे, ''बमबम जी! वकील-वकील मौसरे भाई होते हैं। शर्मा जी एक-दूसरे मामले में मुझसे सलाह-मशविरा लेने आये थे।''

मगर बमबम बाबा की खोपड़ी पुरानी थी। कई सेर शुद्ध सरसों का तेल सोख चुकी थी। सो वह सोचने लगे कि यदि सरकारी वकील की बात सही मान ली जाये, तो भी यह गुंजाइश है कि सूरज का वकील सरकारी वकील से सलाह-मशविरा के नाम पर मिल जाएगा। एक बार उनके दिमाग में आया कि अदालत से प्रार्थना करके दूसरे सरकारी वकील तैनात करायें। मगर उनके बेटे ने इसलिए मना कर दिया कि बिना मतलब एक और दुश्मन तैयार हो जाएगा।

अगली पेशी पर भी बमबम बाबा सरकारी वकील के पास समय पर पहुँच गये। सरकारी वकील ने उन्हें चाय पिलायी और घर का हाल-चाल पूछा। बाबा ने कहा, ''क्या कहूँ, वकील साहब! बड़ी बेटी की हत्या ने हमें अन्दर तक हिला दिया है। अभी दूसरी बेटी भी ब्याहनी है। मगर दिमाग काम नहीं करता।''

''बमबम जी! भगवान ने जिसका मुँह चीरा है, उसे अन्न भी देता है। जब आपको उसने बेटी दी है, तो उसके विवाह का इन्तज़ाम भी वही करेगी। इन्सान को मौके पर चूकना नहीं चाहिए।'' सरकारी वकील ने सहानुभूति का छींटा मारा। बमबम बाबा फिर धार्मिक उपदेश सुनकर चौंके—कहीं कुछ अनर्थ तो नहीं होनेवाला है? फिर उनके अन्दर 'मौके पर नहीं चूकना चाहिए' गूँजने लगा। सरकारी वकील की खोपड़ी कानूनों से सीझी हुई थी। सो उसने पासा फेंका, ''बमबम जी! आप बिल्कुल फिक्र मत करें। उन लोगों को फाँसी होकर ही रहेगी। लेकिन एक बात बताइये। आप चौथेपन में हैं। भगवान में विश्वास ज़रूर करते होंगे। सो क्या उनको फाँसी देने से आपकी बेटी चम्पा आपको वापस मिल जाएगी?''

''वकील साहब! यह सब मुझसे क्यों कह रहे हैं? जानता हूँ, मेरी मरी बेटी अब वापस नहीं आएगी मगर उनको सजा मिलने से मेरी अन्तरात्मा और चम्पा की आत्मा को शान्ति ज़रूर मिलेगी। फिर ऐसा घोर अपराध करनेवालों को धक्का लगेगा। समाज में दहेज़ रूपी दानव की नाक में कानून की नकेल लग जाएगी...''

"सो तो ठीक है, मगर आपको कोई खास फ़ायदा नहीं होगा। फिर आपका नाम बेटी की ससुराल वालों को तंग करनेवालों में शुमार हो जाएगा और आपको दूसरी बेटी की शादी करना दूभर हो जाएगा। सो सोच-समझकर कदम उठाइये, तो ज़्यादा बेहतर होगा।"

बमबम बाबा की अक्ल पर ताला लग गया। वे सरकारी वकील के कानूनी चक्रव्यूह में ऐसे चकराये कि माथा थामकर बैठ गये। सरकारी वकील ने जब बमबम बाबा को समझाया कि आप सूरज के पिता से समझौता कर लें, इसी में आप दोनों की भलाई है तो बाबा को इस बारे में फिर से सोचने पर मजबूर होना पड़ा। सरकारी वकील के आवास पर दोनों पक्षों के बीच एक घंटे तक बात हुई।

आज अदालत खचाखच भरी थी। करीब डेढ़ घंटे बाद इस मुकदमे की गवाही शुरू हुई। सरकारी पक्ष की ओर से बमबम बाबा पहले गवाह के रूप में पेश हुए। सरकारी वकील के दो वाक्यों के बाद जब बचाव पक्ष के वकील रणेन्द्र शर्मा ने जिरह में उनसे पूछा, "चम्पा की हत्या कब हुई?" तो बमबम ने एक बार सरकारी वकील को देखा, दूसरी बार मजिस्ट्रेट को देखा और तीसरी बार ऊपर की ओर देखा। पर चुप खड़े रहे। "तो चम्पा की हत्या कब हुई? जल्दी बताइये, अदालत का समय बर्बाद मत कीजिए।" रणेन्द्र शर्मा ने डपटकर पूछा। "मुझे याद नहीं।"—बमबम ने कहा

"क्या उस समय आप घटनास्थल पर थे?'

"नहीं"

"किस आधार पर आपने थाने में रिपोर्ट दर्ज़ कराई?"

"मुझे बेटी की ससुराल वाले गाँव के कुछ लोगों ने बताया था कि मेरी बेटी की हत्या सूरज और उसके परिवार वालों ने की।"

"तो आपने सुनी-सुनायी बात पर थाने में रिपोर्ट दर्ज़ कराई?"

"जी हाँ।"

"क्या मरने के पहले चम्पा ने आपको कोई खत दिया था कि उसकी ससुराल वाले उसे तंग-ओ-तबाह कर रहे हैं?"

"बिल्कुल नहीं।"

"तो क्या चम्पा की मौत के लिए उसके पति-ससुर, सास-ननद को आप दोषी मानते हैं?"

"दोहाई भगवान की। मैंने ऐसा करते उन्हें नहीं देखा, तो इस बुढ़ापे में झूठ कैसे बोलूँ?"

इसके बाद मजिस्ट्रेट ने बमबम बाबा से सम्पुष्ट किया कि वे जो कुछ कह रहे हैं, स्वेच्छा से कह रहे हैं अथवा उन पर कोई दबाव है। बमबम बाबा ने

किसी के दबाव होने से इनकार किया। वे बार-बार कहते रहे, ''अब मेरी बेटी इस दुनिया में लौटकर नहीं आएगी। मैं किसी को बेवजह नहीं फँसाना चाहता।'' तो अन्तत: अदालत ने सूरज और उसके परिवार वालों को बरी कर दिया। जेल से छूटने की खुशी में उन लोगों ने जश्न मनाया। दोनों पक्षों ने हाथ मिला लिया। करीब छ: माह के बाद बमबम बाबा ने अपने बेटे और पतोहू से मशविरा किया, ''कृष्णा की शादी में देर हो रही है। पिछली बार की तरह कई सालों तक भटकने का समय नहीं है। क्यों न उसकी शादी सूरज से कर दी जाये?'' उनका बेटा कुछ सोच रहा था! उसका मन इस प्रस्ताव के पक्ष में नहीं था। मगर बहू ने घूँघट की ओट से कहा, ''बाबू जी। आपका अन्तिमपन चल रहा हैं। बेटी तो गाय होती है जिसके गले में एक पगहा ज़रूर होना चाहिए।''

थोड़ी देर तक विचार-विमर्श करने के बाद इस प्रस्ताव पर तीनों की सहमति हो गयी। घर के अन्दर बैठी कृष्णा सब सुन रही थी। उसने अपनी भाभी से इसका विरोध भी किया मगर भाभी ने उसे झिड़क दिया। माँ के मरने के बाद भाभी के अलावा वह किसी और से अपने मन की बात कह ही नहीं सकती थी। बमबम बाबा तो शुरू से ही बिगड़ैल और हावी पिता थे जिनकी हर इच्छा बच्चों के लिए आज्ञा होती थी।

कुछ दिनों बाद अपने बेटे को लेकर वह सूरज के यहाँ पहुँच गये। पैलगी-अशीष हुई और चाय-नाश्ता भी। फिर बमबम बाबा धीरे से बोले, ''समधी जी! आपको अपनी बात तो याद ही होगी? मेरी छोटी बेटी का हाथ थाम लेने के लिए आप लोगों ने वादा किया था। मेरी विनती है कि किसी पंडित-पुरोहित को बुलाकर बैसाख में शादी की तिथि पक्की कर ली जाये।'' मगर उनके समधी-दामाद चुप रहे। बमबम बाबा से नहीं रहा गया। उन्होंने फिर अपनी बात जारी रखी, ''मेरी ज़िन्दगी का क्या ठिकाना? कृष्णा के हाथ पीले कर दूँ, तो मेरी चिन्ता दूर हो जाएगी। मेरी कृष्णा बहुत काम-काज करती है। इस घर के लिए वह लक्ष्मी साबित होगी।''

सूरज ने कुछ नहीं कहा। मगर समधी जी खड़े होकर दहाड़ने लगे, ''इस घर में एक काली के आने से ही सब कुछ तहस-नहस हो गया, अब यहाँ किसी लक्ष्मी की ज़रूरत नहीं है। फिर कैसा वादा? आपने तो सजा दिलवाने में कोई कोर-कसर उठा नहीं रखी। हमने क्षमा माँगी, पर आप नहीं पसीजे। आपने बदले में पच्चीस हज़ार रुपये का सौदा किया। दूसरी लड़की की शादी सूरज से करने का वादा कराया। मरता क्या नहीं करता? हमने आपकी सारी बातें मान लीं—पच्चीस हज़ार नगद चुका दिया। जो बाप पहली बेटी की लाश तक का सौदा कर सकता है, वह दूसरी बेटी का शुभचिन्तक कैसे हो सकता है? हम लोग

दुबारा जेल नहीं जाना चाहते।" बमबम बाबा का सिर कोंहड़ा की तरह लटक गया। वह दूसरी बेटी की शादी के लिए ही वह पैसा लिये थे। अपने लिए नहीं, बमबम ने इस पैसे की बात घर में नहीं बतायी थी। उन्होंने सोचा था कि चम्पा की शादी में जो मुख्य खर्चे थे लगभग वही खर्चे फिर करने के लिए पच्चीस हज़ार से कम नहीं लगेंगे। बेटा अभी तक बेरोज़गार है।

इतनी ज़्यादा जमीन-जायदाद नहीं कि उसे बेचकर कृष्णा की शादी करें। मगर ये सब बातें वे सबके सामने कैसे करते? वे अवाक् थे किन्तु विचारों के चक्रव्यूह में बुरी तरह फँस गये थे। इतना सुनते ही बमबम बाबा को उनके अपने बेटे ने दाँत पीसते हुए एकटक घूरा और फिर उनके पास ही थूककर तेजी से वहाँ से दफा हो गया।

लाक्षागृह

गोवर्धन पंडित के यहाँ एक सप्ताह से श्रीमद्‌भागवत का पाठ चल रहा था। वजह यह कि गोवर्धन के माँ-बाप बद्रीनाथ-केदारनाथ की तीर्थयात्रा करके सकुशल लौट आये थे। सातवें दिन हवन आदि हुआ तो आठवें दिन ब्रह्मभोज का कार्यक्रम रखा गया। चूँकि रबी की बुआई का वक्त था, इसलिए गाँव-जवार के लोग अलस्सुबह खेती के काम के लिए निकल जाते। गोवर्धन ने दस क्विंटल आटा पिसवाया था। सो गाँववालों ने तय किया कि रात में ही पूड़ी बना ली जाए। गाँव की औरतें और लड़कियाँ पूड़ी बेलने के लिए बुलाई गयीं। रात भर पूड़ी बेलने और वनस्पति में उसे छानने का काम होता रहा। सुबह जब कटहर गोंसाई विषहरी तलैया के पास गा रहा था, 'जानकी न लौलऽ पिया, आफ़त लौलऽ जानकी', तब औरतों और लड़कियों का हुजूम घर-वापसी के लिए निकला। उसमें से एक लड़की मीनू अपने घर नहीं पहुँची। लोगों ने कहा, "उसका अपहरण हो गया।" सब जानते थे, वह जानती नहीं थी। पर गाँव पर आफ़त-विपत का खेल तभी से शुरू हुआ। पूरा गाँव 'लाख का घर' बन गया। हर-हर जलने लगा केशोपुर मझगाँव का सिवाना। वारदातों का सिलसिला शुरू हुआ तो चितोखर पुलिया का वज्र हृदय भी हहर गया।

उस दिन की वारदात के बारे में जमुनिया मुसहर की बात मान ली जाय तो हुआ यूँ कि छिनरभुतही थान के पास कुछ लड़के घात लगाये बैठे थे। मीनू वहाँ पहुँची तो उसे दबोच लिया। पास की कोठरी में ले गये। वह छटपटाती रही। फिर भी उन तीनों ने बारी-बारी से उसके साथ मुँह काला किया। चौदह वर्षीया मीनू निश्चेत हो गयी। दो लड़के उसकी हालत देखकर चम्पत हो गये। मगर रीतेश कहाँ भागकर जाता? यह कोठरी उसके घर का एक भाग थी, जिसमें वह लिखता-पढ़ता और रात में सोता था।

मीनू अपने ननिहाल में रहकर पढ़ती थी। जब दिन चढ़ आया और मीनू नहीं लौटी तो मीनू के नाना-नानी उसे तलाशने निकले। पर कहीं पता नहीं चला।

पूरे गाँव में हहोर हो गया कि मीनू नामक लड़की लापता हो गयी है। पर शाम तक उसका अता-पता नहीं चला। सो थाने में गुमशुदगी की रिपोर्ट दर्ज़ करा दी गयी। थाने वाले ने कहा, ''सुबह-सुबह कोई अपराधी गैंग वहाँ थोड़ी गया होगा? आसपास तलाश कीजिए, मिल जाएगी।'' यह कहकर थाने वालों ने अपना पल्लू झाड़ लिया। गाँव में यह चर्चा जोरों पर थी कि थाना बिना पूजा लिये अपना हथियार नहीं उठाता। मीनू के नाना-नानी की हालत अच्छी नहीं थी। नाना प्राइमरी स्कूल के शिक्षक के पद से रिटायर हुए थे। मीनू के मामा दिल्ली में एक दुकान पर छोटी-मोटी नौकरी करते थे।

शाम होते-होते यह अफ़वाह फैल गयी कि मुखिया के भतीजे ने उसे गायब किया है। एक-दो लड़कों ने यह भी बताया कि वह मीनू का पीछा बहुत पहले से कर रहा था। मगर कहीं कोई खास जगह के बारे में सूचना नहीं मिली। सुबह हुई तो मालूम हुआ कि मुखिया के भतीजे रीतेश ने आत्महत्या कर ली। गाँव के लोग उसे देखने के लिए पहुँचे। मुखिया के परिवार ने यह शिगूफा छोड़ा कि उसे साँप ने डँस लिया है। मगर गाँव के अधिकतर लोगों ने इस पर विश्वास नहीं किया। मुखिया के परिवार वालों ने आनन-फानन में मृतक को गोमती के घाट पर ले जाकर जला दिया। गाँववालों की नज़र में रीतेश पर मीनू को गायब करने का आरोप पुष्ट हो गया। मगर मीनू कहीं मिली नहीं। रीतेश के मरने पर गाँव वालों ने शोकसभा में शामिल होना मुनासिब नहीं समझा। मुखिया को करारा धक्का लगा। मगर मुखिया का परिवार इतनी आसानी से हार मानने वाला नहीं था। वह रात काली रात साबित हुई। मुखिया के परिवार वालों ने अचेत मीनू को बिजली का करंट छुआकर मार दिया। वे इतने से भी सन्तुष्ट नहीं थे। उन्होंने उसे उस कुएँ में डाल दिया जो मीनू के ननिहाल का था। मगर हैंडपम्प लग जाने के कारण कोई उससे पानी नहीं भरता था। जब दूसरे दिन किसी ने उस कुएँ में झाँका तो लाश दिखायी पड़ी। लोगों ने लाश को बाहर निकाला। पुलिस को इसकी सूचना दी गयी। पुलिस ने पोस्टमार्टम के लिए उसे जिला अस्पताल में भेज दिया। मुखिया का परिवार चूकने वाला नहीं था। उसने कई ग्राहक तैयार कर लिए जो ग्राम समाज की जमीन का पट्टा पाने, चकरोड पर मिट्टी डलवाने और खड़ंजा कराने की लालच में थैलियाँ सौंपते गये। पूँजी एकत्रित हो गयी तो मुखिया ने जिला अस्पताल के सिविल सर्जन डॉक्टर चौधरी का पैर पकड़ा। डॉक्टर चौधरी ने चिकित्सा कार्य में कम और सुविधा-शुल्क में ज़्यादा नाम कमाया था। कई बार फेल होकर किसी तरह उन्होंने मेडिकल की पढ़ाई पूरी की थी। मगर उनकी किस्मत रंग लायी। आरक्षण नीति ने उन्हें नौकरी दे दी और बाद में तीन प्रमोशन भी। मगर उनकी निजी प्रैक्टिस कभी नहीं चली। उनके बारे में यह बात

चारों ओर फैल गयी थी कि जिसका उन्होंने खुद इलाज किया, उसे सीधे मोक्ष मिला!

सो ऐसी शोहरत वाले सिविल सर्जन ने, मुखिया को भी उपकृत किया, मात्र पच्चीस हज़ार रुपये में। पोस्टमार्टम रिपोर्ट में यह अच्छी तरह लिख दिया गया, "उक्त लड़की के शरीर पर घाव का कोई निशान नहीं पाया गया। मृत्यु का कारण पानी में डूबना है।"

मगर थानेवाले मुखिया को यूँ ही छोड़ने वाले नहीं थे। उन्हें घटना-स्थल से मीनू के कान की एक बाली मुखिया के द्वार पर गिरी मिल गयी थी। इसके अलावा उसकी चप्पल मुखिया के घर के सामने तालाब के किनारे मिली थी। उधर मुखिया ने धूप में बाल नहीं सुखाये थे। उन्होंने पिछले विधान सभा चुनाव में जिस उम्मीदवार की मदद की थी, वह आज विधायक है। यह भी संयोग है कि जिले का एस. पी. भी उन्हीं की जाति का है। सो सोच-समझकर मुखिया विधायक जी के दरबार में पहुँचे। चूँकि विधायक जी पहली बार विधायक बने थे और अभी भी उन पर राजनीति का रंग पूरी तरह नहीं चढ़ा था, सो वे मुखिया को पहचान गये। मुखिया की बात सुनकर उन्होंने एस.पी. को फोन किया। एस. पी. ने मामला देख लेने की बात कही।

उधर धीरे-धीरे गाँव से यह बात शहर तक पहुँच गयी। एक दिन पत्रकारों का एक दल मझगाँव पहुँच गया। उन्होंने फोटो खींची, गाँववालों से बातचीत की। अगले दिन कई अख़बारों के बीच के पृष्ठों पर 'मझगाँव में एक लड़की की हत्या' की ख़बर विस्तार से छपी। चारों ओर तहलका मच गया। पुलिस की गाड़ियाँ मझगाँव में दौड़ने लगीं। इंस्पेक्टर के अलावा डी. एस. पी. भी वहाँ स्थल निरीक्षण करने गये। गाँववालों को लगा कि मुखिया की अब नहीं चलेगी। पर डी.एस.पी. के बाल ही नहीं, खोपड़ी भी पकी थी। उसने मुखिया के लड़के के अलावा तीन अन्य लोगों को हिरासत में ले लिया। उधर मुखिया अपने गुप्त एजेंडे पर कार्य करता रहा। उसने थाना के एक दलाल से सम्पर्क साधा। दलाल थानेदार का खास आदमी था। जो मामला किसी से भी नहीं सलटता, उसे वह सलटा देता। सो थानेदार ने एक लाख रुपये में अपने को बेचने की बोली लगायी। मुखिया ने बहुत हाथ-पैर जोड़ा। मगर वह टस से मस नहीं हुआ और बोला, "मैं अपनी नौकरी दाँव पर लगा रहा हूँ। यदि प्रेसवाले मेरे पीछे पड़ गये, तो मेरी नौकरी चली जाएगी। भले इस वर्दी पर कुछ दाग लगे हों, मगर है तो वर्दी ही। यदि मेरी वर्दी उतर गयी, तो मैं जीते जी मर जाऊँगा। वर्दी सोने की चिड़िया है। मैं इससे रोज़-रोज़ सोने के अंडे पाता रहता हूँ।" और मुखिया ने काँखकर बात मान ली।

थानेदार सुलझा हुआ था। उसने डी.एस.पी. का कड़ा रुख जान लिया था। सो वह सीधे एस.पी. साहब के पास पहुँचा और उन्हें सारी कहानी बता दी। एस.पी. को याद आया कि इसी मामले में विधायक ने भी फोन किया था। सो उन्होंने थानेदार के माध्यम से पचहत्तर हज़ार रुपये ले लिए और विधायक को फोन करके बता दिया कि उनका काम कठिन ज़रूर है, मगर उनकी इज़्ज़त करते हुए वे उसे सम्भाल लेंगे। सो एस.पी. ने आदेश दिया कि इस मामले का पर्यवेक्षण वे खुद करेंगे, डी.एस.पी. नहीं। आख़िर उन्हें भी तो अठारह वर्षों तक डी.एस.पी. के रूप में कार्य करने का अनुभव था।

दूसरी तरफ थाने ने एक नयी रणनीति बनायी। थानेदार ने सबसे पहले गोवर्धन पंडित को बुलवाया। फिर मीनू के मामा के लड़के को। उसके बाद मुखिया के लड़के और अन्य दो लड़कों को हाजत में बन्द कर दिया। यह ख़बर सुनकर गाँववाले परेशान हो गये। आख़िर गोवर्धन पंडित का क्या दोष था? क्या मीनू का ममेरा भाई उसके साथ दुष्कर्म करेगा? फिर वह हत्या क्यों करेगा? यदि वह किसी चरण में शामिल होता, तो घरवालों को ज़रूर कुछ-न-कुछ बताता। मगर थाना का मानना था कि यदि गोवर्धन पंडित ने रात में पूड़ी बनवाने का कार्यक्रम नहीं किया होता, तो यह जघन्य अपराध ही नहीं होता। गोवर्धन पंडित ने बताया कि पिछले साल उसी गाँव में तीन-चार आयोजनों में रात में ही पूड़ी बनवायी गयी थी और कोई शिकायत नहीं हुई थी। पुलिस ने मीनू के ममेरे भाई को गाली ही नहीं दी बल्कि चार-छः बेंत भी मारे। पुलिस की रणनीति उसके ज़रिये मीनू के ननिहाल वालों पर दबाव डालना था। अब वे लोग मीनू के हत्यारों को सजा दिलाने और इन्साफ पाने की बात भूल से गये, बल्कि उस लड़के को छुड़ाने में लग गये। बेटी मरी सो मरी, बेटा बच जाये। तीन दिनों तक पुलिस गोवर्धन पंडित तथा अन्य लोगों को थाने पर रखे रही। जब मामला शान्त हो गया, तो उसने सभी को छोड़ दिया।

दूसरी ओर गाँववाले मुखिया की फितरत पर आश्चर्यचकित थे। यद्यपि अविवाहित लड़के या लड़की का श्राद्धकर्म उस इलाके में नहीं किया जाता था, फिर भी मुखिया ने ऐसा आयोजन किया और लोगों को तेरही ब्रह्मभोज में आमन्त्रित किया। इस मसले पर गाँव के अधिकतर परिवारों ने उसका बहिष्कार किया। मुखिया को यह बहिष्कार नागवार गुजरा। उसने इसका बदला लेने की ठान ली। और उसे गेहूँ की सिंचाई के वक्त मौका भी मिल गया। पूरब दिशा में उसकी एक निजी ट्यूबवेल थी और वहाँ सिंचाई का कोई दूसरा साधन नहीं था। गाँववाले जब सिंचाई हेतु पानी माँगने जाते, तो वह उसके खराब होने का बहाना बना देता था। गेहूँ के पौधे पीले होने लगे। कई परिवारों की जीविका एकमात्र खेती थी।

लोग मजबूरी में टूटने लगे। 'जीव जाय जीविका न जाये' वाली बात थी। सो छोटे किसानों ने मुखिया से चिरौरी-विनती की। मुखिया थोड़ा पिघला, मगर जलकर भी रस्सी की तरह ऐंठा रहा। उसने सिंचाई की दर एक सौ रुपये प्रति घंटा से बढ़ाकर चार सौ रुपये प्रति घंटा कर दी। किसानों की कमर टूट गयी। धीरे-धीरे बहिष्कार करने वालों में से कई परिवारों ने मुखिया से मेज-जोल, खाना-पानी, बर-बरात का सम्बन्ध स्थापित कर लिया।

करीब एक साल बीत गया। मुखिया का कारोबार बढ़ता गया। उसके परिवार के कुछ बच्चे चोरी-डकैती भी करने लगे। एक बारात में किराये पर आये जेनरेटर की चोरी उसके परिवार के लड़कों ने कर ली। गाँव से दो टी.वी. लापता हो गये। एक दिन गाँव-जवार के तीन-चार लोग मुखिया के दरवाज़े पर गये और ज़मीन पट्टा करने अथवा उनका पैसा वापस करने पर जोर देने लगे। मुखिया ने उन्हें जोर से डाँटा क्योंकि उनके मेहमान के सामने वे लोग पैसे माँग रहे थे। मुखिया ने कहा, "कैसा पैसा? कोई काग़ज़-पत्र हो तो दिखाओ।" इतना कहने पर एक नौजवान ने मुखिया की गर्दन पकड़कर उठा लिया और ज़मीन पर पटक दिया। मुखिया के लड़के किसी गुप्त एजेंडे को अंजाम देने के लिए बाहर गये थे। इधर मुखिया की इज़्ज़त मिट्टी में मिल गयी। मुखिया को अस्पताल में भर्ती कराया गया। उनकी पीठ की हड्डी टूट गयी थी। डॉक्टरों ने स्पष्ट बता दिया कि ढलती उम्र में वह नहीं जुड़ेगी।

अब मुखिया एक बैसाखी के सहारे ही दिशा-मैदान जा सकते थे। मगर उन्हें अपनी गलती का अहसास था, सो उन्होंने थाने में केस दर्ज़ नहीं करवाया। एक दिन शाम में वह दिशा-मैदान से नहीं लौटे, तो उनके घरवाले उन्हें खोजने निकले। ललमुँहा चौर के पास वे एक खेत में औंधे मुँह गिरे हुए पाये गये। मुखिया की पत्नी को लगने लगा, यह आदमी-आदमजात का कारनामा नहीं। वह ओझा-सोखा की शरण में भागी। हटेंगर ओझा ने मामले को पूरी तरह सुना-गुना, फिर बताया कि चुड़ैल का कारनामा है। तिस पर नाम भी खोल दिया—मीनू चुड़ैल! आगाह कर दिया, बिना जान लिए नहीं छोड़ेगी। मुखिया की पत्नी जोर-जोर से रोने लगी। ओझा ने जीव के बदले जीव देने, पताका, मिठाई, हवन—सारा विधि-विधान 'फारे-फार' बता दिया। अगले दिन सारा कर्मकांड सम्पन्न हो गया। लेकिन मुखिया मर गया, ऐन उसी रात को।

गाँव में यह बात तेजी से फैल गयी। फलतः लाश जलाने के उपक्रम में अधिकांश ग्रामीण शामिल नहीं हुए। वजह वही, मीनू चुड़ैल! उपद्रव थमा नहीं उसका। अगले महीने मुखिया का मंझला लड़का कुएँ में गिरकर मर गया। दहशत ने और जोर पकड़ा। वैसे जमुनिया मुसहर की बात मानें तो मामला दूसरा था।

लड़का चोरी करने गया था। लौट रहा था तो बिजली के तार में पैर फँस गया और 'स्पाटे पर' टें बोल गया। पर एक मुँह की बात पर किसको यकीन होता। गाँव के सारे मुँह में एक ही बात ठूँसी हुई थी—बदला ले रही है मीनू चुड़ैल।

एक महीने के अन्दर आधा दर्जन लोग 'सीरियस' हुए, अस्पताल गये, मरके बचे। सबको ओझा ने एक ही बात बतायी, "छोड़ेगी थोड़े। सबको ले मरेगी मीनू चुड़ैल।" सबने नाना तरह के कर्मकांड करवाये। पर शान्ति नहीं मिली। लपट थाने तक पहुँच गयी। वही घूसखोर दारोगा। सिर्फ़ वही काहे, पूरा थाना। एक महीने तैयारी चली उसके लिए। ओझा ने जैसा कहा, वैसा किया गया। भाद्रपद की अमावस्या को अनुष्ठान पूरा हुआ। लोगों ने देखा, मसान में छः फुटा 'लाख का घर' जल रहा है—'अनमन' थाने की इमारत के आकार का! अब पता नहीं उसमें कितने 'लाख के आदम-आदमजात' थे!

वजूद

और चाची दहाड़ मार कर रोने लगीं। उनकी आँखों से खून सने आँसू झरने की तरह निकलने लगे। धीरे-धीरे पूरा गाँव रोने लगा। मैं भी उसमें शामिल हो गया। मुझे लगा कि पूरी धरती रो रही है और रोते-रोते धरती सूरज का चक्कर लगाना ही भूल गयी। कभी-कभी एक आदमी के मरने पर ऐसा लगता है कि पूरा गाँव मर गया है। सो चारों ओर अँधेरा छा गया। मुझे उस समय कुछ भी नहीं सूझ रहा था। अब्दुल चाचा को खोकर चाची ने कितना कुछ खो दिया, इसका तो कोई अनुमान ही लगा सकता है किन्तु मैंने जो पारस पत्थर खो दिया, उसके एक-एक कण का मैं प्रत्यक्ष अनुभव कर रहा हूँ। मुझे इसलिए भी दुख हो रहा था कि इस दुख का पहाड़ ढोने के लिए खुदा ने मुझे ही क्यों ख़बरिया बनाया? अब्दुल चाचा थे ही ऐसे। दुनिया के तमाम लोग उनके अपने थे। कहाँ था किसी से उनका भेद? पराया शब्द उनके दौलतखाने में नहीं था। अपने को जब कोई खोता है तो उस समय उससे पहली बार मिलने वाली कोई घटना मन में कौंधती है।

हुआ यूँ कि परदेश से घर लौटे अब्दुल चाचा से जो मेरा दुआ-सलाम हुआ, तो वह फौरन मेरे रोज़गार के बारे में पूछ बैठे। उस समय मेरे चेहरे पर चुप्पी और मायूसी छा गयी। वह मेरी मायूसी का मतलब ताड़ गये। फिर तपाक से बोले, ''रजाक! घबड़ाओ मत। इस बार तुम मेरे साथ चलो। मैं अपने कारखाने के मालिक से तुम्हारी रोज़ी-रोटी के बारे में बात करूँगा। खुदा ने चाहा तो ज़रूर तुम्हारी रोज़ी-रोटी का प्रबन्ध हो जाएगा।'' उनके अलफाज़ ने मेरे भीतर रोशनी की किरण पहुँचा दी। मुझे कुछ तसल्ली मिली। सो करीब एक महीने के बाद मैं अब्दुल चाचा के साथ कर्णनगर के लिए चल दिया। मेरे पास तो अधेला भी नहीं था। चाचा ने ही सारा जुगाड़ किया। सो जुमे की नमाज़ अदा करके शाम में हम दोनों धक्का-मुक्की, तू-तू, मैं-मैं और चिल्लपों के बीच किसी तरह रेलगाड़ी के चालू डिब्बे में घुस गये। जब गाड़ी चल दी, तो जो लोग हमारे

चढ़ने पर गरम होकर एतराज़ कर रहे थे, वे थोड़ा नरम हो गये। जब अपने स्टेशन पर हम लोग ट्रेन पर चढ़े, उस समय भीड़ में सरसों डालने की भी जगह नहीं थी। आगे भी पूरे सफर में वही हाल रहा, प्रायः हर स्टेशन पर नये मुसाफिर चढ़ जाते और रेंगते हुए गाड़ी चल पड़ती। चालू डिब्बे में कभी उकड़ूँ बैठे, कभी खड़े होकर और कभी किसी के पाँवों तले दबकर हम सफ़र करते रहे। रास्ते में हम मुफ़्त में जितने धक्के खाये, उतनी गालियाँ भी। यह रेल का सफर मेरे लिए रोज़ा रखने और हज करने से कम नहीं था। मगर अब्दुल चाचा इसके अभ्यस्त हो गये थे। वह इसे इतमीनान से ले रहे थे। उनका मानना था कि अपने गाँव-घर से बाहर जाने पर सुख-चैन की आशा नहीं करनी चाहिए। यदि कोई ऐसी आशा करेगा, तो निराश होगा। वह बार-बार मुझे समझाते, ''बेटा! अपनी ज़रूरतें कम करो। अपने को थोड़े में समेटो। ताते पाँव पसारिये जाती लाम्बी सौर। फिर न कोई दिक्कत होगी और न कुछ अफ़सोस।'' मगर उनकी यह तकरीर मेरे गले उस समय नहीं उतरी।

खैर...दो दिन बाद हम कर्णनगर रेलवे स्टेशन पहुँचे। अब्दुल चाचा रेलवे स्टेशन से करीब पन्द्रह किलोमीटर दूर रहते थे। मगर उस दूरी को कमतर मानते हुए पन्द्रह किलोमीटर को वह महज़ पाँच कोस कहते। वहाँ से राजपुर तक का सरकारी बस का किराया दस रुपये था। आँटो रिक्शा का किराया पन्द्रह रुपये था। इसलिए उन्होंने मन-ही-मन गणित लगाया कि अगर 'छकड़ा' में बैठें, तो दस रुपये में हम दोनों राजपुर पहुँच जाएँगे। स्टेशन से दो सौ गज की दूरी पर स्थित, अपनी गठरी-मोटरी लिए, हम लोग छकड़ा स्टैंड पहुँचे। अब्दुल चाचा ने बताया कि 'छकड़ा' खरीदना ऑटो रिक्शा से सस्ता पड़ता है। इसका किराया कम होने के कारण तकलीफ सहकर भी इधर के लोग इसी से सफ़र करते हैं। छकड़े की सवारी हम लोगों की तरफ नहीं है। मैंने देखा कि उसका आगे का हिस्सा मोटर साइकिल की तरह था—हैंडिल, पेट्रोल टंकी वगैरह। लेकिन पिछला हिस्सा सामान लादने वाले खुले ऑटो रिक्शा की तरह था जिसमें लोहे की पटरी सीट का काम करती थी और उसके ऊपर कोई मुलायम रबड़ या रूई की गद्दी नहीं थी। मगर चारों कोनों पर और बीच में हाथ से पकड़ने के लिए रॉड लगे थे। ऊपर की ओर अन्तहीन खुले आसमान की छत थी जिससे मुसाफिरों को धूप और बारिस का खुले आम मुकाबला करना पड़ता था।

जो भी हो, छकड़ा देखना मेरा नया अनुभव था। सो मैं उस पर बैठकर काफी खुश था। मैं इसलिए भी खुश था कि उसमें रेलगाड़ी की तरह धक्का-मुक्की नहीं थी। चूँकि वह डीजल से चलता था, इसलिए उसमें धुआँ और आवाज़ अधिक थी। अब्दुल चाचा ने बताया कि हालाँकि उसमें बैठने की इजाज़त छः

लोगों की है मगर ड्राइवर लोग उसमें बारह-चौदह सवारियों को बैठाते हैं। भाड़ा कम होने के कारण मजदूर ज़मात के लोग उसमें ज़्यादा बैठते है। हम लोगों के साथ भी दस लोग बैठे थे। इसका बाँकापन मुझे खूब अच्छा लगा भले वह ज़्यादा प्रदूषण फैला रहा था। जिसकी जैसी औकात होती है, वह वैसी ही मोटरगाड़ी में सफर करता है। इधर भी कुछ लोग बसों और जीपों की छतों पर बैठे थे, हम लोगों के इलाके में भी जो कम किराया देते है या जिन्हें अपनी ज़िन्दगी की परवाह नहीं होती, इसी तरह सफ़र किया करते है। ऐसा जाना-पहचाना दृश्य जल्दी ही याददाश्त में आ गया। इसलिए तसल्ली हुई कि कानून तोड़ने वाले सब जगह हैं और दुनिया मिला-जुला कर सचमुच गोल है। और हमारा महान भारत तो निश्चित रूप से एक है।

खैर...हम लोग राजपुर स्टैंड पर उतरकर पैदल चलकर थोड़ी देर में शाहपुरा बस्ती में अब्दुल चाचा के क्वार्टर पर पहुँच गये। चाचा ने एक जंग लगे ताले में अपनी चाबी लगायी और दरवाज़ा खोला। अन्दर से धूलभरी गैस का भभका निकला। टिन के पतरों, बाँस और बरसाती से बने क्वार्टर में अकूत जाला लगा था। बाहर की नाली से निरन्तर निकल रही बदबू मेरे नथुनों में भर गयी। बगल में कूड़े का ढेर गँजा था, जिसमें कई सुअर कुछ खाने की चीज़ें पूरे दमखम से खोज रहे थे। उसी छः फीट लम्बे और पाँच फीट चौड़े कमरे में खाना बनाना, उसी में नहाना, उसी में बर्तन और कपड़े धोने पड़ते थे। उसी में राशन-पानी रखना और उसी में सोना पड़ता था। बदबू से मेरे दोनों नथुने भर गये और झोपड़ी की बेतरतीबी से मेरा दिल बैठ गया। मगर कोई विकल्प नहीं था। सो भीतर-ही-भीतर घुटकर, मैंने चाचा के साथ मिलकर उस क्वार्टर की सफाई की। बारी-बारी से हम दोनों ने नहाया-धोया। फिर हम दोनों ने खिचड़ी बनाकर खाये। तब जाकर कहीं आराम करने की बारी आयी। यहाँ भी एक समस्या उठ खड़ी हुई। इस घर में एक भी खाट नहीं थी। अब्दुल चाचा ने फटाफट यह तय किया कि जमीन पर बिछे टाट-पट्टी और कथरी पर लेटने पर जिधर अब्दुल चाचा का पैर हो, उधर मेरा सिरहाना रहे जिससे दोनों को लेटने में कुछ सहूलियत हो। ऐसा ही किया गया। अब्दुल चाचा ने बायीं करवट ली और दस मिनट में वह खर्राटें भरने लगे। मुझे अपने गाँव में सुनी हुई यह कहावत फौरन याद आयी, ''नींद न जाने टूटी खाट। भूख न जाने सूखा भात।''

मगर मेरी आँखों में नींद कहाँ? मैं ऊपर बनी छत में लगी बाँस की बल्लियाँ गिनने लगा। तभी मेरे मुँह, आँख और सिर पर घुन की धूल गिरी और मैं अचकचाकर उठ बैठा। मैं अपना शरीर झाड़ने लगा मगर घुन की धूल साफ होने

की बजाय बढ़ती जा रही थी। यह क्या? मेरे सिर से अकूत घुन की धूल निकल रही थी। फिर आँख, कान, नाक और मुँह से भी धूल निकलने लगी मानो देह के भीतर तक घुन की सुरंग चली गई हो। फिर हाथ, पैर, छाती, पेट से भी घुन की धूल निकली। मेरा पूरा शरीर और दिमाग घुन का भीटा बनता गया। इस समय मेरी बेचैनी देखने लायक थी, पर देखे कौन? चाचा की झोपड़ी में आईना भी नहीं था। अब्दुल चाचा तो मानो घोड़े बेचकर सो रहे थे।

करीब एक घंटे बाद अब्दुल चाचा सोकर उठे और गहरी नींद के लिए उन्होंने खुदा का शुक्र अदा किया। मैं ऐसा खुशकिस्मत नहीं था। मैंने घुन के बारे में चाचा को बताना चाहा मगर उनके शरीर पर गिरे किलो भर घुन को देखकर और उस पर उनकी कोई खास प्रतिक्रिया न सुनकर मैंने वह प्रसंग शुरू नहीं किया। वे घुन को बहुत हल्के से ले रहे थे जैसे कोई बच्चा माटी में खेलने में मज़ा लेता है। मुझे अपने दोस्त श्रीराम के पिता की याद आने लगी। वह एक पौराणिक कहानी सुनाया करते थे : एक राजा था। उसका राजपाट बहुत बड़ा था। धन-दौलत और परिवार से वह बड़ा आदमी था। मगर एक बार यमराज उसका प्राण हरने पहुँचे, तो उसकी पत्नी और बच्चों ने बहुत चिरौरी की। मगर यमराज ने बताया कि राजा को मरकर सुअर की योनि में पुनः जन्म लेना है। उनकी पत्नी और बच्चों ने विनती की कि सुअर की ज़िन्दगी से राजा को जल्दी निज़ात पाने का कोई उपाय बताएँ, महाराज! इस पर यमराज ने कहा, ''एक ही शर्त है। राजा के सुअर के रूप में जन्म पाने के बाद उसे मार दिया जाय। सुअर के मरने के बाद राजा को पुनः मनुष्य की योनि मिल जाएगी। इसके सिवाय दूसरा कोई उपाय नहीं है।'' बड़े राजकुमार ने यमराज से प्रार्थना की, ''उस समय आप मुझे इशारे से बता देने की कृपा करेंगे कि किस सुअर के रूप में मेरे पिता जी हैं, मैं उनके उद्धार के लिए उन्हें तीर से मार दूँगा और तब उन्हें नया जन्म मिल जाएगा''। यमराज ने राजा से इस बावत पूछा। राजा ने इस युक्ति पर सहमति दे दी। कुछ समय के बाद राजा का देहान्त हो गया। फिर कुछ समय बीत जाने पर यमराज ने राजकुमार को इशारे से बताया, ''वह सुअर जो इस समय गन्दी नाली में बार-बार लोट रहा है, वही आपके पिता हैं।'' राजकुमार ने सुअर पर जैसे ही तीर का निशाना साधा, वह अकबका गया। उसने यमराज से प्रार्थना की, ''महाराज! मुझे मरने से बचा लीजिए। मुझे इस योनि में ही सच्चा सुख मिल रहा है।'' तब यमराज ने राजकुमार को उन पर तीर चलाने से मना कर दिया। इस कहानी को सुन लेने के बाद वह यह कहना नहीं भूले कि जो जिस महौल में रहता है, वही उसे अच्छा लगने लगता है। मगर यह तो सिर्फ़ सुनी हुई एक मिथकीय कहानी थी जबकि अब्दुल चाचा अपनी असली ज़िन्दगी कुछ इसी तरह जी रहे

थे। वह बड़े ही जीवट के व्यक्ति थे। आख़िर कथनी से करनी बड़ी होती है।

खैर...अब्दुल चाचा ने जागने पर चाय बनायी और चाय पीने से हम दोनों के शरीर में कुछ गरमाहट आयी। फिर चाचा मुझे लेकर काम के लिए बाहर निकल पड़े। करीब पाँच किलोमीटर पैदल चलकर वह अपने कारखाने पर पहुँचे। चाचा अपना पास दिखाकर कारखाने के अन्दर घुस गये। उनके लौट आने तक बाहर ही रहकर मुझे उनका इन्तज़ार करना था। दोपहर के डेढ़ बजे जब सायरन बजने पर लंच की छुट्टी हुई, तब चाचा बाहर निकले। उन्होंने कहा, "मुबारक हो, रजाक! अल्लाह ने चाहा, तो तुम्हारा काम हो जाएगा। साहब ने तुम्हारे कागजात को रखते हुए दो-चार दिन में खुशख़बरी देने को कहा है"। एक बार मुझे अपने कानों पर यकीन ही नहीं हुआ। जिस आदमी ने नौकरी के लिए पचासों दफ़े दरख़ास्त भेजी हो और कहीं से भी कोई खुशख़बरी नहीं मिली हो, उसे 'चट मँगनी, पट ब्याह' पर यकीन कैसे होगा? आधे घंटे तक इस खुशख़बरी को सुनकर मैं मन-ही-मन झूमता रहा। मैं तरह-तरह के सपने देखता रहा। मुझे लगा कि मैं उड़ रहा हूँ नीले आकाश में। मगर जैसे ही कारखाने का सायरन बजा, मैं ज़मीन पर आ गया। चाचा फिर कारखाने के भीतर काम करने चले गये। मैं इतना आशावादी बन गया कि मुझे लगा कि हो सकता है दयालु साहब आज ही कोई काम मुझे दे दें। सो मैं आसमान से फूल झरने का इन्तज़ार बेसब्री से करने लगा।

जब शाम के छः बजे चाचा कारखाने से बाहर निकले, तो मैं उनके साथ-साथ 'चाली' के लिए वापस चल पड़ा। वापस लौटते समय मेरे कदम जल्दी-जल्दी आगे बढ़ रहे थे। सुख की परछाईं भी मनुष्य का हौसला बढ़ा देती है। चाली पहुँचने पर हमने पाया कि बच्चों का झुंड, सुअरों का झुंड और मुर्गों का झुंड शोर मचाने में मशगूल है। मैंने चाचा से कहा, "इस माहौल में कैसे रहते हैं आप लोग?" चाचा ने कहा, "परिस्थितियाँ ज़िन्दगी जीने के नये तरीके सिखा देती हैं। जो सीख लेता है, वह धीरे-धीरे आगे बढ़ जाता है और फिर उनमें बदलाव लाता है। जो नहीं सीख पाता, वह ज़िन्दगी से भागने की कोशिश में तन-मन से पूरी तरह टूट जाता है।" और मैंने जीने के नये तरीके सीखने का फैसला झटपट ले लिया। फिर मैं चाचा के साथ कमरा साफ करने और खाना बनाने में जुट गया। जिस कारखाने में मुझे काम मिलने वाला था, उसकी बाबत पहली दफ़ा मैंने चाचा से पूछा। वह पल भर रुके। फिर चाचा काम करते जाते और अपने कारखाने व कारखाने के मालिक के बारे में मुझे बताते जाते। उस कारखाने में प्लास्टिक की बाल्टी, जग, मग, प्लेट, कनस्तर, पानी का टब वगैरह बनते थे। चाचा ने बीस साल के दौरान कई कारखानों में नौकरी की थी और तरह-तरह के मालिकों के साथ काम करने का अनुभव उन्हें हासिल हुआ था।

उनके बारे में वह बताते रहे। इस बीच प्रसंगवश मैं पूछ बैठा, ''मालिक लोग खूँखार होते हैं क्या?''

''देखो, रजाक! इस दुनिया में, दुनिया का मालिक एक ही होता है—खुदा! और खुदा तो खूँखार हो ही नहीं सकता! हाँ, आम इन्सान की ज़िन्दगी में तीन तरह के मालिक होते हैं। पहला, रोज़ी-रोटी देने वाला पेट का मालिक। दूसरा, किराये पर ठिकाना देने वाला मकान मालिक और तीसरा, आत्मा की देख-रेख के लिए ऊपर वाला मालिक। मगर ऊपर वाले से रोज़मर्रे की ज़िन्दगी में वास्ता नहीं पड़ता। पहले और दूसरे मालिकों में शायद ही कोई अच्छे होते हैं। जो मजदूरों का हक मारे, उससे क्या डरना? जो काम को इबादत समझे और निज़ी फ़ायदे के लिए कोई गलत काम न करे, उसे न नीचे वाले से डरने की ज़रूरत है और न ऊपर वाले से।'' चाचा ने अपना तजुर्बा बताया। मुझे लगा कि चाचा की खोपड़ी कोल्हू का तेल सोखकर अच्छी तरह पकी हुई है। खैर...रोटी-चोखा बनाकर और खा-पीकर हम लोग सो गये। अगले दिन जब मैं सुबह उठा तो अपने को तरोताज़ा महसूस कर रहा था। पिछली रात वाली ऊब और निराशा गायब हो चुकी थी। सफाई करके, नहा-धोकर और खा-पीकर चाचा ड्यूटी पर चले गये। मैं सारा दिन बस्ती से लेकर चौराहे तक यूँ ही घूमता-फिरता रहा। मैंने देखा कि चाली में कई प्रदेशों के लोग घुल-मिलकर रहते हैं। गरीबी और तंगी में भी हँसना उन्होंने अच्छी तरह सीख लिया है। मैं चौराहे पर बने छकड़ा स्टैंड पर खड़ा होकर जब यह सोच रहा था, उसी समय रेडियो पर ख़बर आयी कि महिषापुर में खड़ी रेल के एक डिब्बे में आग लगने से सात लोग मारे गये हैं। मरने वाले बनारस से तीर्थाटन करके लौट रहे थे। यह भी सुना कि मुसाफिरों और आस-पास के दंगाई लोगों के बीच खूब मारपीट हुई और उन्होंने स्टेशन को जला दिया। पुलिस के पहुँचने के बाद ही स्थिति पर काबू पाया गया। मेरा मन कैसे-कैसे होने लगा। मैं सीधे भागकर कमरे पर गया। कमरे का ताला खोलकर मैं अन्दर धुसा तथा डर के मारे अन्दर से ताला बन्द कर लिया—खुदा न ख़ास्ता हालात बिगड़ जाएँ। चाचा के बाहर होने की सोचकर मेरा दिल घबड़ाने लगा। सोच-सोच कर मैं इतना शंकित हो गया कि चाचा के लिए खुदा से दुआ करने लगा। बाद में मुझे दूसरों से मालूम हुआ कि जब वह कारखाने के पास पहुँचे, तो सायरन बज उठा। मगर अभी पाली शुरू होने में पन्द्रह मिनट बाकी थे। वह आस-पास घूर-घूर कर देखने लगे। तभी किसी ने उन्हें बताया कि 'धमाल' हो गया है। उनका माथा चकराया देखते-ही-देखते भगदड़ मच गयी। अब्दुल चाचा बूढ़े आदमी थे, सो वह दौड़ कर भाग नहीं सके। तभी दहाड़ते हुए नौजवानों का एक जत्था वहाँ आ धमका। उनके हाथों में बड़े-बड़े त्रिशूल थे। वे चीख रहे

थे, ''महिषापुर का बदला लेंगे, एक हिन्दू पर दस-दस लेंगे।'' अब्दुल चाचा को उस हरकत का कारण समझ में नहीं आया। मगर हवाओं का तूफानी और ज़हरीला रुख पहचान कर वह कारखाने के गेट पर बने सिक्योरिटी गार्ड वाले कमरे में घुस गये। वह जत्था आगे बढ़ता गया और सरकारी गाड़ियों को जलाता गया। अब्दुल चाचा ने खुदा को मन-ही-मन शुक्रिया अदा किया और जत्थे के चले जाने पर चैन की साँस ली। मगर अभी चैन कहाँ? सैकड़ों लोग जुलूस की शक्ल में आते फिर दिखायी पड़े। यह भीड़ खूँखार जानवरों में तब्दील हो गयी थी। उसने मुसलमान का कारखाना कहते हुए कारखाने में आग लगा दी। देखते-ही-देखते प्लास्टिक का यह कारखाना आग की लपटों में तब्दील हो गया और पूरा माहौल बदबू से भर गया। अब्दुल चाचा के साथ छिपे करीब दो दर्जन कामगार कारखाने से बाहर भागे, मगर भीड़ ने दिल और दिमाग को अपने शरीर से बाहर निकाल फेंक दिया था। उन लोगों ने वहाँ जान बचाकर छुपे दो दर्ज़न लोगों को एक-एक कर कारखाने के अग्नि-कुण्ड में झोंक दिया। वे मजदूर 'बचाओ-बचाओ' कहकर चिल्लाने लगे। पचास लाख की आबादी वाले कर्णनगर में कहीं से कोई बचाने वाला नहीं आया और कोई भी इस जघन्य कृत्य को रोकने के लिए सामने नहीं आया। पुलिस तो खुदा की तरह अदृश्य हो गयी। आसमान का चेहरा बिल्कुल काला पड़ गया। बड़े-बड़े जुल्मों के दौरान होने वाली कोई आकाशवाणी भी तब नहीं हुई। धीरे-धीरे जंगल की आग की तरह वह महानगर के चारों ओर फैल गयी। कारखाने से करीब पाँच सौ मीटर की दूरी पर पुलिस थाना स्थित था मगर थाने से कोई नहीं आया। थाना इंचार्ज इंस्पेक्टर मोदक की बेटी की शादी में कारखाना मालिक ने दो तोले की जंजीर तोहफ़े में दी थी। मगर वही मोदक वह कारखाना बचाने नहीं गया। उल्टे वह और उसके सहकर्मी धमाल करने वालों को पीछे से उकसाते रहे। पुलिस की शह पर कर्णनगर दंगाइयों के कब्ज़े में होकर एक 'प्रयोगशाला' बना दिया गया।

मोदक तो उस दिन शैतान ही बन गया था। वह धमाल करने वालों को उत्साहित करता था मानो भारत-पाक युद्ध हो रहा हो। हर मुस्लिम बस्ती को छोटे-छोटे पाकिस्तान की संज्ञा दी जा रही थी। मुस्लिम बस्ती में चारों ओर से घिरे लोगों को वह बाहर भागने भी नहीं देता था। इधर कुआँ, उधर खाई वाली कहावत सिद्ध हो रही थी। जैसा कि मुझे अपने पड़ोसियों से जानकारी मिली, ऐसा अपने देश के इतिहास में पहली बार हो रहा था जब धमाल के समय मुसलमान अधिकारियों को छुट्टी दे दी गयी। दरअसल सूबे की सरकार ने यह अलिखित आदेश दे दिया था कि अपनी सुरक्षा के लिए पुलिस-प्रशासन के मुस्लिम अधिकारी-कर्मचारी छुट्टी पर चले जाएँ। खुद पुलिस कमिश्नर सत्ता के डर से या किसी

बड़े लालच में हाथ-पर-हाथ धरे बैठे रहे जबकि पूरा शहर धमाल की आग में धूँ-धूँ जल रहा था जिस पर तवा रखकर सत्ताधारी दल बेफिक्र रोटियाँ सेंक रहा था। इन्सान को ईंधन की तरह इस्तेमाल करने पर जो रोटियाँ बन रही थीं, वे उसे देशी पिज़्ज़ा की मानिन्द जायकेदार लग रही थीं। इस मेहनत, सफ़ाई और हुनर के आगे पाँच सितारा होटलों के चेफ झख मारें। उनके आगे नीरो भी झख मारे। वर्दी पर खून के छींटे बार-बार पड़ रहे थे मगर वे उसे बदलते या छुपाते जा रहे थे। टिनोपाल या उजाला से वर्दी की झक धुलाई कराना सरकार के ज़िम्मे था और सरकार अपनी ज़िम्मेदारी से मुँह नहीं मोड़ रही थी। बाद में यह भी खुलासा हुआ कि तूफान मचाने वाले अधिकतर लोग भाड़े के थे। दूर-दूर के गाँवों से भाड़े पर दंगाई लाये गये थे। एक आदमी को जान से मारने के लिए एक हज़ार रुपये उन्हें इनाम में दिये जाते थे। एक घर-बार जलाने के लिए डेढ़ हज़ार रुपये और एक होटल-दुकान-कारखाना जलाने के लिए दो हज़ार रुपये दिये जाते थे। आदमी की जान सबसे सस्ती होने का ऐलान था। बाम्बे स्टॉक एक्सचेंज में सेंसेक्स तेजी से बढ़ने लगा। कई अर्थशास्त्री और पत्रकार इसे विकास का नया पैमाना कह रहे थे। दलाल स्ट्रीट पर धमाल के शेयर खूब बिक रहे थे। दस रुपये का शेयर पाँच सौ में बिक रहा था। जुआ खेलने वाले सट्टा लगा रहे थे कि अगले चुनाव में सत्ताधारी दल कितने बहुमत से जीतेगा। जब अल्पमत रूपी बाँस रहेगा ही नहीं, तो विरोध की बाँसुरी कैसे बजेगी?

मिल्लत और भाईचारे का भाव एकदम नीचे गिर गया। कोई इसका पुछत्तर नहीं था। मुफ्त में भी कोई नहीं ले रहा था इसे। एक तो ढोने का चक्कर, दूसरे रखने की जगह नहीं। सारे कुएँ और कूड़ेदान पहले ही लाशों से भर गये थे और नगर निगम नरक बन गया था और उसकी गाड़ियाँ लावारिस लाशों को ढोने में परेशान थीं। इन्सानों के साथ-साथ कौमी एकता की ऐतिहासिक मिसालें जैसे पीर बाबाओं की मजारें और मशहूर शायरों की दरगाहें भी पूरी तरह मटियामेट कर दी गयीं जिससे हिन्दू-मुस्लिम अपनी साझी विरासत भूल जायें। पक्की सड़क गिट्टी और कोलतार से बना दी गयी जिस पर दिन-रात मोटर गाड़ियाँ धक्-धक् करती सरपट दौड़ती रहती हैं। फिर पाँच सौ वर्ष पुरानी मस्जिद भी ढहा दी गयी जिससे इतिहास मिट जाये। फिर घरों-होटलों-कारखानों-दुकानों के रूप में उन्होंने वर्तमान को भी मिटा दिया।

मगर तूफान मचाने वालों ने भविष्य को भी नहीं बख्शा। अबोध नन्हे-नन्हे बच्चों को त्रिशूलों पर रखकर नचा दिया गया। माँ की कोख में पल रहे भ्रूणों को भी उन्होंने साग बना दिया। हे मालिक, तुम्हारे रहते हुए भी पहली बार भूत, वर्तमान और भविष्य को एक साथ मिटाने की मनौती पूरी की गयी।

जब कत्ले-आम पूरा हो गया, तो पुलिस आयुक्त द्वारा इत्मीनान से शहर में कर्फ्यू लगा दिया गया। पाँच दिनों तक अब्दुल चाचा का कोई पता नहीं चला। कर्फ्यू में छठवें दिन चार घंटे की ढील दी गयी, तो मैं अब्दुल चाचा की खोज में उनके कारखाने पर गया। कारखाने की ऊँची भव्य इमारत मलवे के ढेर में तब्दील हो गयी थी। मुझे अचानक ऐसा लगा कि इसी में मेरा भविष्य गड़ा है। मैं यहाँ कुछ देर तक बेसुध खड़ा रहा। कई लोग आये और गार्ड से पूछताछ किये। मैंने भी अब्दुल चाचा के बारे में उससे पूछा। उसने बताया कि अब्दुल चाचा का आधा शरीर जल गया था। उन्हें सदर अस्पताल में किसी ने भर्ती करा दिया था। मैं हाँफता हुआ भागा मगर मेरे पास अस्पताल का पूरा पता नहीं था, सो मैं चाचा को न खोज सका और चार घंटे की ढील का समय खत्म हो गया। मैं अब बाहर नहीं जा सकता था। दूसरे अस्पताल के एक कोने में बेहाल मरीज-सा मैं रात भर पड़ा रहा। मुझे लगा कि बीमारी सिर्फ़ किसी अन्दरूनी रोग के कारण ही नहीं, डर से भी होती है।

अब्दुल चाचा को खोजते-खोजते मैं परेशान हो गया। उनके साथ काम करने वाले किसुन चाचा के यहाँ मैं गया जो चाचा की बस्ती से दस किलोमीटर दूर मानपुरा में रहते थे। मैं पैदल गया क्योंकि मेरे पास किराये के पैसे नहीं थे और मुझे कोई उधार देने वाला भी नहीं था। मैंने चाचा के बारे में उनसे दरयाफ़्त की। वे खुद चाचा को तलाशने में परेशान थे। मुझमें और पैदल चलने की हिम्मत नहीं रह गयी थी। वहाँ कुछ क्षण के लिए सन्नाटा छा गया। थोड़ी देर बाद किसुन की पत्नी दो प्याली चाय लेकर आयी। मैं भूख-प्यास से बिलबिला रहा था सो मैंने बिना कुछ देखे एक घूँट चाय पी मगर मुँह बिचका कर कहा, ''इसमें चीनी नहीं?''

''मजबूरी का नाम महात्मा होता है। मैं पहले खालिस दूध और ज़्यादा चीनी की चाय पसन्द करता था। मगर परिस्थितियाँ नयी आदतें सिखा देती हैं। अब मैं बिना दूध और चीनी की चाय पीता हूँ। दूध की जगह पानी और चीनी की जगह नमक का इस्तेमाल करके चाय पीने की औपचारिकता पूरी करता हूँ।'' किसुन चाचा ने माथे पर हाथ रखकर धीमी आवाज़ में कहा। फिर उन्होंने बताया कि कारखाना जल जाने के बाद वह गल्ला मंडी में बोरे ढोने का काम करते हैं।

अब्दुल चाचा का पता नहीं चल पाया। उनके बिना इस अनाथ को पूछने वाला कौन है? फिर मैं अपने बारे में सोचने लगता हूँ। आख़िर मेरा कोई ज़रिया नहीं है। क्या मैं गल्ला मंडी में अपनी पीठ पर बोरे ढो सकता हूँ? मेरी सेहत इतना वजन उठाने की इजाज़त ही नहीं देती। घर-परिवार को खर्चा भेजना होगा

और अपनी खुराकी भी चलानी होगी। मैं साबुन-तेल लगाना बन्द कर दूँगा। मैं तेल-मसाला खाना बन्द कर दूँगा। सुबह और दोपहर का खाना बन्द कर दूँगा। मैं दाल खाना बन्द कर दूँगा। मैं सब्जी खाना बन्द कर दूँगा। मैं भात खाना बन्द कर दूँगा। मैं रोटी खाना बन्द कर दूँगा। अचानक मेरे दिमाग में कौंधा कि रोटी नहीं खाऊँगा, तो क्या खाऊँगा? सिर्फ़ हवा? सिर्फ़ पानी? सिर्फ़ धूप? रोज़ा रखने की अपनी आदत पुरानी है। कुछ दिन मैं उपवास रखूँगा। आख़िर गाँधी बाबा भी उपवास करते थे। मगर कब तक? दिन भर रोज़ा रखने से पहले अहले सुबह सहरी मिला करती थी और शाम को रोज़ा खोलने पर जायकेदार इफ़्तारी खाना। मगर लगातार कितने महीने तक मैं बिना खाये-पिये जिऊँगा? क्या? नहीं जी पाऊँगा? क्या खुदा मदद नहीं करेगा? यह सोचते-सोचते मैं बड़बड़ाने लगा और कुर्सी से नीचे लुढ़कने लगा। तभी किसुन चाचा ने मुझे सम्भालते हुए पूछा, "कहाँ खो गये हो, रजाक?" सचमुच मैं खो गया था। फिर किसुन से अब्दुल चाचा के बारे में मैं पूछने लगा, "और कहाँ गुंजाइश हो सकती है उनके मिलने की?" किसुन चाचा से भी नहीं रहा गया। वह मेरे साथ सदर अस्पताल तक गये। वहाँ लाशों का ढेर लगा था। डोम लोग नाक-मुँह पर पट्टी बाँधे लाशों को सफ़ेद कफ़न में लपेट रहे थे। वे लाशें मुस्कराने लगीं। हमें देर तक खड़े देखकर एक लाश ने मुझे इशारे से बुलाया और कहा, "आ जाओ हम लोगों के पास। क्यों परेशान हो रहे हो? तुम्हारी दुनिया में जाति-धरम का चक्कर है मगर हमारे यहाँ सब बराबर है। हम एक-दूसरे से नफ़रत नहीं करते। इस नापाक दुनिया में रहोगे, तो कभी भी मारे जा सकते हो। ज़मीन के ऊपर की दुनिया से बेहतर है जमीन के नीचे की दुनिया। हमारी दुनिया में बेरोज़गारी और भुखमरी नहीं है। हम सबको तुम्हारी दुनिया की चौकीदारी का काम मिला है। हम हिसाब लगाते रहते हैं कि तुम्हारी दुनिया में कब कौन किसको लंगी मार रहा है।" अचानक मुझे लगा कि अगली बारी मेरी है। मैं काँपने लगा। किसुन चाचा ने मेरा हाथ थाम लिया। बड़ी चिरौरी विनती करने पर दो घंटे में अब्दुल चाचा की लाश हमें मिली। उनका बदन जल गया था मगर चेहरे की शिनाख़्त आख़िर में हम दोनों ने कर ली। फिर हम दोनों चाचा की लाश एक ठेले पर लादकर ले गये कब्रिस्तान तक। वहाँ भी लाशों का मेला लगा हुआ था। ऐसा लगता था कि लाशें नाच रही हैं। मगर जगह कम पड़ती लग रही थी। मैंने खुद ज़मीन में एक गड्ढा खोदा और चाचा को सुपुर्दे-खाक कर दिया। चाचा की रूह की शान्ति के लिए मैंने खड़े होकर कुरान शरीफ की कुछ आयतें पढ़ीं। चाचा पाँचों वक्त की नमाज़ अदा करते थे, सो आयतें पढ़कर मुझे कुछ सुकून मिला। मैंने सोचा नेक दिल चाचा के साथ इस बेरुखी दुनिया ने सही सलूक नहीं किया था, कि चाचा ने किसी को हिन्दू-

मुसलमान नहीं, सिर्फ़ एक इन्सान समझा था। वहाँ से लौटकर मैं चाचा के क्वार्टर में आकर काफी देर तक बेसुध पड़ा रहा। मुझे रात भर नींद नहीं आयी। क्वार्टर के अन्दर अपने आप से, दीवारों से, बाँस की बल्लियों से और उनमें लगे घुन से मैं आहिस्ता-आहिस्ता बतियाता रहा। मगर थोड़ी देर में सूरज की रोने से सूजी और लाल हुई आँख दिखाई दी। तभी किसी ने किंवाड़ पर जोर से धक्का मारा। ऐसा लगा मानो किसी ने मेरे पेट पर जोर से एक लात मारी हो। मैंने अपना पेट एक हाथ से दबाये हुए झटपट किंवाड़ खोल दी।

पहले मैं बेराज़गार था, अब अनाथ भी हो गया था। किसी ने बाहर से जोर से आवाज़ दी, "अब्दुल भाई!" मैं बाहर निकला। उसने मुझसे पूछा कि अब्दुल भाई कहाँ है। मैंने डर और दुख के मिले-जुले भाव में कहा, "वह तो इस दुनिया से चले गये..."

"अच्छा हुआ! एक तो कम हुआ इस नरक में। मैं तो उसके मरने का इन्तज़ार कर रहा था।" वह तेजी से बोल गया।

"लेकिन ऐसा क्यों कह रहे हैं आप? लोग-बाग तो दूसरों के जीने की दुआ करते हैं।" मैंने उसे धीरे से टोका।

"तू कौन है जो इतनी बकवास कर रहा है?" उसने मुझे घुड़की दी।

"मैं उसका भतीजा हूँ"। मैंने लड़खड़ाती जुबान से कहा।

"इस क्वार्टर का किराया तीन महीने से बाकी है। लाओ दो किराया।" उसने अल्फाज़ के कोड़े से मुझे जोर से मारा।

"लेकिन मेरे पास पैसे बिल्कुल नहीं है..." मैं गिड़गिड़ाने लगा।

"मेरा नाम हकीम भाई रहीम भाई है। तू मुझे नहीं जानता?" कहते हुए उसने अपना आपा खो दिया और मुझे तथा चाचा को भद्दी-भद्दी गालियाँ देते हुए उसने धमकी दी, "अगर कल तक तुमने किराये के तीन सौ रुपये मुझे नहीं दिये, तो अब्दुल की तरह ही तुम्हारा भी हस्र हो जाएगा।" उसके जाने के बाद मैं घबड़ा गया। फिर भागकर गाँव जाने का निर्णय लिया। मैंने अब्दुल चाचा के टिन के एक बक्से को खोलकर देखा। दो पुराने कपड़े, जवानी के दिनों की अब्दुल चाचा की एक पुरानी तस्वीर और एक नोट बुक मिला। नोट बुक में अब्दुल चाचा के तीस रुपये थे और खर्च के हिसाब लिखे थे। मैंने ये सब ले लिए। फिर भागते हुए राजपुर चौराहे पर पहुँचा। वहाँ पहुँचकर मैंने राहत की साँस ली और सोचा कि जान बची तो लाखों पाये।

वहाँ कई छकड़े आये और स्टेशन-स्टेशन पुकारे। फिर मुसाफिरों को चढ़ाये और चले गये। मैं उन्हें निहारता ही रहा। मेरी जेब में पाँच रुपये भी नहीं थे

कि छकड़े में बैठकर स्टेशन चला जाऊँ। चाचा की नोट बुक में रखे तीस रुपये चाचा की अमानत थी। उसे खंर्च करने वाला भला मैं कौन हूँ? कोई चारा नहीं था। सो मैं पैदल ही स्टेशन के लिए चल दिया। स्टेशन पहुँचते-पहुँचते दोपहर हो गयी। मुझे बहुत भूख लगी थी। चारों ओर छोटी-बड़ी दुकानें सामानों से भरी थीं। मगर मेरे पास न कोई खाने का सामान था, न सामान खरीदने के पैसे थे। मैंने स्टेशन पर खूब पानी पिया लेकिन कुछ न खाने से मेरे पेट में धक्का लगा। मैं प्लेटफार्म पर लेट गया। आधे घंटे के बाद मुझे कुछ आराम महसूस हुआ। मैं शुद्ध हवा लेने के लिए स्टेशन से बाहर आकर टहलने लगा। थोड़ी दूरी पर मुझे एक गुरुद्वारा दिखायी दिया। मैं वहाँ धीरे-धीरे टहलते हुए चला गया। संयोग से दोपहर का लंगर शुरू हो गया था। मैं भी हाथ-मुँह धोकर पाँत में खाने के लिए बैठ गया। वहाँ का खाना मुझे बहुत अच्छा लगा। काश! मुझे ऐसा खाना रोज़ मिलता। मैंने छककर खाना खाया। ऐसा लगा कि मेरा पेट फट जाएगा। सो मैं गुरुद्वारा के अहाते में ही लेट गया। मुझे नींद आ गयी, थोड़ी देर बाद किसी आदमी ने मुझे जगाया। मैं उठ बैठा। फिर नल से मैंने पानी पिया। मैंने सोचा कि उसी लंगर में मुफ़्त खाना खाकर गुजारा किया जा सकता है। मगर सोएँगे कहाँ? फिर हकीम भाई रहीम भाई यहाँ आसानी से आ सकता है। तब क्या होगा? फिर धमाल हो सकता है, तब क्या होगा?

कुछ देर तक टहलने के बाद मैं फिर स्टेशन पर आ गया। थोड़ी देर बाद उत्तर भारत में मेरे इलाके को जाने वाली स्वराज एक्सप्रेस गाड़ी आ गयी। वह नाम मुझे अन्दर से मथने लगा। कहाँ है स्वराज? क्या मारने-काटने-जलाने की आज़ादी ही स्वराज है? क्या गाँधी का स्वराज का सपना यही था? मेरे पास टिकट खरीदने के पैसे नहीं थे। मैंने अभी तक बेटिकट सफर कभी नहीं किया था। मन ऊभ-चूभ कर रहा था। तभी रेलवे गार्ड ने सीटी बजा दी और रेलगाड़ी सरकने लगी। ख़ुदा का नाम लेकर मैं भी एक चालू डिब्बे में चढ़ गया। डिब्बे में समूची दुनिया खचाखच भरी हुई थी। गाली-गलौज, मार-पीट, धक्का-मुक्की का माहौल था। हिन्दी की कई बोलियाँ आपस में लड़-झगड़ रही थीं अपने-अपने वजूद के लिए। कोई कारखाना जलने, कोई होटल लूटने, कोई घर तोड़ने और कोई जान से मारे जाने की चर्चा कर रहा था। सबसे ज़्यादा दुर्दशा बच्चों, बूढ़ों और महिलाओं की थी। मगर इस भीड़ का एक फ़ायदा यह हुआ कि कोई टी.टी. टिकट चेक करने के लिए उस डिब्बे में न घुस सका। मैं एक सीट के नीचे छिपकर लेट गया था।

जब वह एक्सप्रेस गाड़ी करीब तीस घंटे बाद मुगलसराय पहुँची, तो इस

चालू डिब्बे के बहुत सारे मुसाफिर उतर गये। मैं सीट के नीचे से धीरे-धीरे निकल कर बाहर आया। मुझे ताज़ा हवा से कुछ सुकून मिला। मगर हम गरीबों की खुशी क्षण भर की होती है। जैसे गीदड़ की मौत आने पर वह शहर की ओर भागता है, वैसे ही बेटिकट यात्री की मौत आने पर वह दरवाज़े के पास की सीट पर बैठता है। ज्यों ही गाड़ी खुली, एक टी.टी. उस डिब्बे में घुस गया। उसका काला कोट देखते ही मेरी पतलून ढीली हो गयी और थोड़ी देर में गीली भी हो गयी। उसके द्वारा टिकट की माँग करने पर मैंने अपना सिर नीचे झुका लिया। वह सिंह की तरह गरजता रहा। फिर मैंने उसे बताया कि मैं अपने चाचा के साथ कर्णनगर गया था, वहीं उनकी मौत हो गयी और मैं उन्हें वहीं दफ़ना कर आ रहा हूँ। इस पर टी.टी. रावण की तरह जोर से हँसा और बोला, "तुम अच्छी कहानी गढ़ लेते हो। तुम्हारी कल्पनाशक्ति इतनी तेज है कि तुम एक अच्छे कहानीकार बन सकते हो। मगर कानून, कानून होता है। तीन सौ रुपये जुर्माना भरो, वरना एक हफ़्ते की जेल की सजा काटो।"

मेरे पास तीन सौ क्या, तीन रुपये भी नहीं थे। सो अगले स्टेशन पर मुझे उतारकर टी.टी और रेलवे पुलिस बल ने मेरे हाथों में हथकड़ी डालकर रेल थाना में कानून की काली कोठरी में बन्द कर दिया। शाम को मजिस्ट्रेट ने मुझे एक सप्ताह की कैद की सजा सुनायी। फिर मुझे जिला जेल में भेज दिया गया। वहाँ जाने पर शाम को खाना मिला। खाने की रोटी खिसक रही थी। मगर भूख में मैं तीनों रोटियाँ खा गया। फिर भी मेरा पेट नहीं भरा। पास के कैदी के घर से खाने का डिब्बा आया था सो उसने जेल में मिली अपनी रोटी-सब्ज़ी मेरे हवाले कर दी। मैंने सोचा—ऐसे ही बिल्ली की तकदीर से छींका टूटता है। किसी तरह मैंने वहाँ एक सप्ताह बिताया। मुझे जेल अधीक्षक ने रविवार को मुक्त करते हुए कहा, "भविष्य में कभी बेटिकट सफर मत करना, वरना फिर इस काल-कोठरी में आना पड़ेगा। जवान हो, कहीं कमाओ-खाओ। कानून को अपने हाथ में मत लो, उसके हाथ आदमी के हाथ से ज़्यादा लम्बे होते हैं।" मुझे यह सुनकर ऐसा लगा कि शैतान के हाथ मेरे हाथों से ज़्यादा लम्बे हैं।

उसकी चीख मेरे कानों में देर तक गूँजती रही। फिर मैं पैदल चलते हुए शाम को अँधेरा होते-होते अपने गाँव पहुँचा। मुझे सबसे पहले अब्दुल चाचा के घर जाना था। मुझे देखते ही चाची ने चाचा का हाल-चाल और घर का खर्चा भेजने के बारे में पूछा। फिर मेरी नौकरी के बारे में पूछा। मैं बिल्कुल चुप था। प्रिय पाठको, पूरा वाकया बयान करने की मुझमें अब हिम्मत ही नहीं बची थी और झूठी कहानी कहने की कला अभी तक मैं सीख नहीं पाया था। सो चाची

मेरे नजदीक आयी और मेरी बाँहें पकड़कर मुझे झकझोरने लगी। फिर मेरी चुप्पी की आवाज़ सुनने लगी। तब मेरे अन्दर थोड़ी चेतना लौट आयी। मैं आँसुओं की भाषा में उनसे सिर्फ़ इतना कह सका, ''चाची! सब कुछ गुजरात हो गया।''

कुदेशन*

तब टुममुनिया फूट-फूटकर रोने लगी। उसके मुँह से निकला, ''हाय भगवान! हाय रे करम! ई कोखी के कवन दोख भइल जे इ दिन देखइला भगवान! इ कोखिया, इ दुधवा के इहै मोल मिलल? पथरा गइल छाती फाटत काहे नइखे भगवान!'' यह कहते-कहते और आँसू पीते-पीते टुनमुनिया बेहोश हो गयी।

उस साल ऐसी घनघोर बारिश नेपाल और उसके सीमावर्ती इलाकों में हुई कि पूरा इलाका डूब गया। पानी में घर डूबा, खेती डूबी, तन डूबा, मन डूबा और डूब गया समूचा गाँव-जवार। सरकार ने राहत की घोषणा की। नेताओं और अफ़सरों ने अपनी-अपनी तोंद पर हाथ फेरा। उन्होंने अपने-अपने क्षेत्र तैयार किये, जैसे कोई पहलवान अपना अखाड़ा तैयार करता है—'अन्धा बाँटे रेवड़ी अपने को चीन्ह-चीन्ह' की तर्ज़ पर राहत सामग्री बाँटी गयी। कुछ लोगों को राहत के नाम पर कुछ धन मिला, कुछ को अनाज मिला, कुछ को कपड़ा मिला और कुछ को कुछ नहीं। कोई वाह कहता, तो कोई आह। बाढ़ के लिए कुछ लोग इन्द्र देवता को कोस रहे थे, कुछ नेपाल के मत्थे पर सारा दारोमदार डाल रहे थे, कुछ केन्द्र सरकार की अकर्मण्यता पर भड़ास निकाल रहे थे और कुछ सूबे की सरकार तथा प्रशासन को दोषी ठहरा रहे थे। भीखनपुर में कुछ पुराने ख़याल के लोग भी थे जो इस नरक के लिए अपनी तकदीर को कोस रहे थे। एक बुज़ुर्ग टेवाराम ने कहा, ''होइहें सोइ जो राम रचि राखा।''

सो दसवीं पास रामभरोसे एक दिन अपने साथियों से विचार-विमर्श करने लगा, ''पानी से घिरे हुए हम लोग क्या रास्ता अख़्तियार करें? कहाँ जायें, क्या करें?'' करीब एक हफ़्ते तक वे सभी सोचते-विचारते रहे। रामभरोसे ने ऊहापोह के बादलों को छाँटते हुए सलाह दी, ''अब हमें भीखनपुर से बाहर निकलना ही होगा। सिर्फ़ तक़दीर को कोसने से काम नहीं चलेगा। सरसों से अपने आप

* दूसरे प्रदेश की स्त्री।

तेल नहीं निकलता, उसे कोल्हू में पेरना पड़ता है। गोसाईं जी ने कहा है—कर्म प्रधान विश्व करि राखा।

सो कर्म करना होगा हमें।" उसके दिमाग में इससे ज़्यादा कुछ नहीं था। कोई धुँधला नक्शा नहीं था। कि किस ढर्रे से कहाँ जाये। लेकिन वह अपने जीवन में एक बार सूबे की राजधानी गया था। स्कूल की ओर से जादूघर, तारामंडल, चिड़ियाघर वग़ैरह देखने गया था। मगर जब रामभरोसे जैसे कुछ नौजवान एक कदम आगे बढ़ने की बात करते, तो बुढ़वन उसे दो पग पीछे खिसका देते। बाहरी दुनिया में फैले चोर-उचक्कों, छल-फ़रेब, दारू-नशा के गिरोह वग़ैरह के ऐसे क़िस्से वे सुनाते कि कई नौजवानों का दिल दहल जाता। फिर कोई जबर्दस्ती कोठों पर औरतों को खींच ले जाने की घटना बयान करता मानो उसकी आँखों के सामने वह घटना घटी हो। यह घोरमट्ठा दस दिनों तक चलता रहा मगर बाढ़ ने भागने का नाम नहीं लिया।

जब राहत की चीजें भी ख़त्म हो गयीं तो रामभरोसे की अगुवाई में कई नौजवानों ने बाहर जाने का फैसला एक दिन कर लिया। टुनमुनिया ने भी हुंकारी भरी और बोली, "हाथ पे हाथ धरे से काम ना चली। हम मानुष हई, जनावर ना।" जब रामभरोसे ने बाढ़ के पानी में बहते हुए टुनमुनिया को बाहर निकाला था, तब से वह उसे मन ही मन पूजती है। धीरे-धीरे पूजा से प्रेम की पंखुरी निकल आयी। टुनमुनिया के घर में उसकी बूढ़ी माँ के सिवाय कोई नहीं था। मगर दोनों की मुलाकात नज़रों तक सीमित थी। कुल पच्चीस औरत-मर्द-बच्चे रेलगाड़ी से किसी तरह सूबे की राजधानी पहुँच गये। उस समय सरकार बाढ़ से घिरे लोगों को बाहर पहुँचाने में मदद कर रही थी, सो टिकट लेने का झंझट नहीं था।

उन लोगों ने प्लेटफॉर्म नम्बर सात पर अपना डेरा जमा लिया। रामभरोसे पूछताछ कार्यालय में पता करने गया कि दिल्ली और पंजाब के लिए कोई गाड़ी कब आएगी। उसने अपनी पढ़ाई के दौरान जाना था कि पंजाब में उन्नत खेती होती है और वहाँ के किसान ख़ुशहाल हैं। वहाँ पता चला कि हावड़ा-अमृतसर एक्सप्रेस रात में डेढ़ बजे आएगी।

सो दाना-पानी कर लेने के बाद रामभरोसे ने सभी साथियों को सोच-विचारकर बताया, "एक आदमी का चालू किराया दो सौ रुपये है। कुल बारह मर्द, आठ औरतें और पाँच बच्चे हैं। बच्चों के लिए आधा टिकट हम नहीं लेंगे क्योंकि बीस सवारियों के पूरे टिकट में ही चार हजार रुपये लगेंगे। हमारे पास कुल मिलाकर केवल डेढ़ हजार रुपये हैं।" उसने जोड़ा-गाँठा। उसने कभी बाहर बेटिकट यात्रा नहीं की थी। सो उसे ऊभ-चूभ हो रही थी कि पकड़े जाने पर

क्या गति होगी? फिर उसने सोचा, 'जेल की चक्की पीसनी होगी। बदन पर कोड़े पड़ेंगे और सोने का जुगाड़ भी नहीं होगा। मगर रूखा-सूखा कुछ तो मिलेगा? भीखनपुर में तो सूखी रोटी भी नहीं मिल सकती।' आठों औरतों के लिए ही सोलह सौ रुपये की ज़रूरत थी। औरतों को बेटिकट ले जाना खतरे से खाली नहीं था। यदि वे बेटिकट पकड़ ली जाएँगी, तो जेल जाएँगी और उन पर न जाने क्या-क्या जुल्म होंगे। टुनमुनिया के पास एक पाई नहीं थी। मगर उसके गले में चाँदी की एक हल्की ज़ंजीर थी। वह रामभरोसे की पेशोपेश को ताड़ गयी और बोली, "हमार इ निहोरा बा। एकरा के मत ठुकराई। जिनगी रही त ढेर गहना बनी! गहना तो जिनगिये ख़ातिर होला। साँसत में जिनगी पड़ जाय त इ सब ना सोचल जाला।" रामभरोसे ने अन्दाज़ लगाया कि कम-से-कम तीन-चार सौ रुपये में वह बिक जाएगी। राजधानी होटल के पास पान की एक सजी हुई दुकान खुली थी। आधे घंटे तक मोल-तोल करने के बाद पान दुकानदार ने दौ सौ रुपये जंजीर के बदले उसे दे दिये।

अब रामभरोसे ने गणित लगाना शुरू किया, "यदि हम आठ टिकट ले लेंगे, तो सिर्फ़ एक सौ रुपये बचेंगे। तो रास्ते भर तीस-बत्तीस घंटों में हम लोग क्या खाएँगे? सो पाँच टिकट अमृतसर के लिए, तीन टिकट मुगलसराय के लिए और दस प्लेटफॉर्म टिकट ले लेते हैं।" करीब बारह सौ रुपये खर्च हो गये। डेढ़ बजे रात में धड़धड़ाती हुई हावड़ा-अमृतसर एक्सप्रेस प्लेटफॉर्म नम्बर सात पर आ गयी। सभी लोग अपनी-अपनी गठरी-मोटरी लेकर और बच्चों को पकड़े हुए चालू डिब्बे में घुस गये। कुछ मिनटों में ही गाड़ी सीटी बजाते हुए चल दी। एक साथी से वह फुसफुसाया, "ज़्यादा भीड़ न होने से टिकट चेकिंग में बेटिकट मुसाफिरों के पकड़े जाने की ज़्यादा गुंजाइश होती है। जरा सम्भलकर रहना।" तभी टुनमुनिया उसकी ओर ताकने लगी। दोनों की आँखें मिलीं, तो पलकें झपकने का नाम ही न ले रही थीं। मगर रात में उन्हें कहीं भी टी.टी. का दर्शन नहीं हुआ।

कई बार चढ़ते-उतरते वे लोग किसी तरह अमृतसर पहुँच गये। रास्ते में उन लोगों ने पेट की आग बुझाने के लिए भूँजा फाँककर पानी पी लिया था। प्लेटफॉर्म से बाहर निकलने के बाद टुनमुनिया को उल्टी होने लगी। रामभरोसे का जी घबरा गया। वहाँ उसकी दवा का इन्तज़ाम नहीं हो सकता था। मालूम हुआ कि उसने दो दिनों से कुछ नहीं खाया था, सिर्फ़ बार-बार पानी पी लेती थी। उसे ज़मीन पर लिटाकर हवा की जाने लगी। रामप्यारी ने उसका मुँह धोया। फिर उसने उसके सिर के बाल खोल दिये जिससे हवा लगे। थोड़ी देर में उसकी तबीयत ठीक हो गयी। फिर एक अधेड़ सरदार गुरिंदर सिंह ट्रैक्टर लेकर वहाँ

पहुँच गये। उन्होंने इन लोगों से पूछा, "क्यों जी, तुम लोग मेरे यहाँ काम करोगे?"

"बिलकुल मालिक!" रामभरोसे के मुँह से निकला। उसे लगा कि बिल्ली की भाग्य से सिकहर टूटने वाला है।

"मजदूरी क्या लोगे?" उन्होंने पूछा।

"आप क्या देंगे?" रामभरोसे ने उनका मन टटोला, "बाज़ार में मजूरी का रेट तो आपको भी मालूम ही है?"

"साठ रुपये रोज़।" सरदार जी ने कहा।

और रामभरोसे चौंक गया, "साठ रुपये दिन भर का?" अपने गाँव में उन्हें पचास रुपये से ज़्यादा मज़दूरी कभी नहीं मिलती थी। सो वह भीतर-ही-भीतर मगन था। सरदार जी ने सोचा कि रेट कम होने से शायद उसे मंजूर नहीं है। सो उसने अपनी मूँछें ऐंठते हुए कहा, "देखो जी! तुम लोगों के ठहरने के लिए भी बन्दोबस्त कर देंगे। वाहे गुरु की कृपा से खेत पर ही मेरा ट्यूबवेल है। वहाँ नहाने-धोने और बिजली की भी दिक्कत नहीं होगी।" और उसने उसको स्वीकार कर लिया। सरदार समझ गया कि पच्चीस लोगों के झुंड का अगुआ रामभरोसे ही है। सो उसने रामभरोसे को अपनी सीट के पास बिठा लिया। उसने बाकी सबको अपने ट्रैक्टर में पीछे की तरफ लगी ट्रॉली में बिठा लिया और रवाना हो गया।

सरदार जी ने अपने फार्म पर उन सबको उतार दिया। वहाँ छप्पर की बड़ी कुटिया-सी थी जिसमें पहले मजदूर रहा करते थे। बगल में सरदार जी की ट्यूबवेल थी। दूसरी ओर आम का बड़ा बगीचा था। अगले दिन से रोपनी का काम उन्हें शुरू कर देना था, इसलिए सरदार गुरिंदर सिंह ने पाँच किलो भुने हुए चने और दो किलो गुड़ लाकर उन्हें खाने को दे दिये। रामभरोसे ने सबको चना-गुड़ खाने के लिए बाँट दिया। एक घंटे के बाद सरदार जी फिर आये और एक हज़ार रुपये एडवांस दे गये। उन्होंने पास की दुकान भी उन सबको बता दी। सो वे लोग वहाँ से झटपट राशन-पानी खरीद लाये। दोपहर को गोंइठा जलाकर, आटा गूँधकर, आलू पकाकर लिट्टी-चोखा तैयार किया गया। सभी ने जमकर खाया। तब सबके शरीर में फुर्ती आयी और आँखों की रोशनी तेज हो गयी।

अगले दिन से शुरू हो गयी धनरोपनी। बच्चों को छोड़कर सभी औरत-मर्द रोपनी में जुट गये। रामभरोसे ने टुनमुनिया को रोपनी करने के बजाय एक-दो दिन आराम करने को कहा, तो वह बोली, "बिना काम के कै दिन चली? सुतले पेट भर जाय के रहित त परदेस में कोई काहे आइत। जाँगरचोर हम नइखी। हो गइल आराम। रउवा बताईं कवन काम करे के बा।" एडवांस तो पहले ही दिन पट गया, बाकी मजूरी भी तुरन्त मिल गयी। खड़ी मजूरी, चोखा काम। सारे

औरत-मर्द भीतर-बाहर से खुश रहने लगे। पन्द्रह दिनों में सरदार जी के यहाँ की धनरोपनी ख़त्म हो गयी। सो उनकी राय-सलाह से वे लोग दूसरों के यहाँ भी रोपनी आदि का काम करने लगे। फिर निकाई और सिंचाई का काम मिला। फिर खाद और कीटनाशक दवाओं के छिड़कने का काम शुरू हुआ।

कीटनाशकों से अनाज में ज़हर फैलने की बात सुनकर रामभरोसे का माथा ठनका। उसने अपने सभी साथियों को बताया, "हम लोग जो कुछ भी बाज़ार से खरीदते हैं और खाते हैं उसमें ज़हर मिला होता है। यानी हम ज़हर महँगा खरीदते हैं, ज़हर खाते भी हैं और कई रोगों को जन्म देते हैं।" मगर ज़्यादातर लोग कहने लगे, "अब ज़हर-माहुर के बात हमनी के नइखे बुझात। सोचे के बात बा कि पेट कइसे भरी। गाँवे रहला पर केतना आदमी बाँस घाट पहुँचल जाइत। बेमारी में डकटरवा देह के कपड़वो बेचवा देले रहतन स। पेट में अन मिले, देह पर कपड़ा, इहे बहुत बा।" रामभरोसे ने भी उन सबकी बात ऊपर से मान ली मगर उसके भीतर-ही-भीतर कुछ सुलगने लगा। उसकी तरफ़ सिर्फ़ टुनमुनिया थी लेकिन दूसरी ओर तेईस लोगों का बहुमत था। लोकतन्त्र में जिसके जितने हाथ, उसकी उतनी ऊँची आवाज़ होती है।

जब उनके खाने-पीने, रहने, पहनने-ओढ़ने का चौचक इन्तज़ाम हो गया, तो उन्हें अपने गाँव वालों की याद बुरी तरह सताने लगी—उनकी हर साँस में भीखनपुर होता था। सो उन लोगों ने टुनमुनिया के प्रस्ताव पर गाँव के बुज़ुर्ग बच्चू राम के नाम पर तीन हज़ार रुपये का मनीऑर्डर भेज दिया और लिख दिया कि अपने टोले के सभी घर-परिवारों में उसे बराबर-बराबर बाँट दें। उन सबने सोचा कि अपने गाँव-जवार को याद करने का यही सच्चा तरीका है, चिट्ठी से झूठमूठ की तसल्ली देने का कोई मतलब नहीं।

कुछ हफ़्तों के बाद उन लोगों ने एक रेडियो खरीद लिया और सरदार जी के ट्यूबवेल से मिले बिजली कनेक्शन से रेडियो भी चलने लगा। सो उन्हें दुनिया-जहान की घटनाएँ रोज़ाना मालूम होने लगीं। फिल्मी गीत, लोक गीत वग़ैरह भी वे आसानी से सुनने लगे। उधर दुधारू गाय-भैंसों के चारा-पानी की व्यवस्था भी उनके जिम्मे आ गयी। धीरे-धीरे सरदार जी के फार्म पर करीब पचास गाय-भैंसें हो गयीं जिन्हें चारा देना, जगह बदलना, नहलाना-धुलाना वगैरह उनकी ड्यूटी में शामिल हो गया। वे जर्सी गायें और मुर्रा भैंसें एक-एक बाल्टी दूध देती थीं। उनका दूध दुहने के लिए तीन मज़दूर अलग से थे। सरदार जी उन लोगों को रोज़ाना मुफ्त में छाछ देते थे। सरदार जी खाने-पीने के बेहद शौक़ीन थे। वह इन लोगों को भी जमकर काम करने और छककर खाने को कहते थे। सरदार जी मस्त-मौला थे।

करीब दो महीने बाद रामभरोसे ने टुनमुनिया को देखा तो उसमें गजब का

निखार आ गया था जैसे पुराने गहने साफ करने पर चमक उठते हैं। अब वह दूसरी महिलाओं के साथ पाउडर, महकउवा तेल वगैरह लगाने लगी। दुकान नज़दीक ही थी सो सिंगार के सारे सामान आसानी से उन्हें मिल जाते थे। अब झुंड की महिलाएँ पंजाबी महिलाओं से मिल-जुलकर दिल बहला लेतीं। शुरू में पंजाबी भाषा न समझने की समस्या थी मगर धीरे-धीरे प्रेम-रस ने उन मेंड़ों को काटकर एक में मिला दिया था जिससे कई दिलों के पानी का प्रवाह बढ़ गया था।

एक शाम सरदार गुरिंदर सिंह ने रामभरोसे को अकेले में बुलाया। फिर वह मुस्कराने लगे। उसने पूछा, "सरदार जी! आपकी मुस्कान का क्या राज़ है?"

"देख रामभरोसे! मेरी ज़िन्दगी तेरे भरोसे है। यार, तू चाहे तो मेरी ज़िन्दगी बना दे, ना चाहे तो बिगाड़ दे।" सरदार जी ने धीरे-धीरे कहा जिससे बगल के पेड़-पौधे भी यह गुफ़्तगू न सुन सकें। उन्हें हवा की तेज गति और पत्तों की सरसराहट पर भी शक था।

"यह कैसे हो सकता है? हम लोग तो आपके अधीन रहकर मजूरी-धतूरी करते हैं। सो आप ही हम लोगों की ज़िन्दगी बना या बिगाड़ सकते हैं। सरदार जी, कहीं किसी ने कुछ गड़बड़ तो नहीं की? यदि हमारे किसी आदमी या औरत से कोई गलती हो गयी हो, तो मैं माफ़ी माँगता हूँ।" रामभरोसे ने हाथ जोड़ते हुए कहा। सरदार जी ने उसके दोनों हाथों को अपने हाथों में थाम लिया और आहिस्ता-आहिस्ता दबाने लगे। उसकी समझ में कुछ नहीं आ रहा था। कुछ साल पहले तब पंजाब में आतंकवादियों का जलवा था, उसकी याद आते ही वह डर गया। लेकिन ऐसी विपरीत परिस्थिति में वह कुछ कर भी नहीं सकता था। अब तो बकरी का गला बाघ की पकड़ में था। डर से उसकी आवाज़ मिमियाहट में बदल गयी थी। उसे लगा कि मैं सचमुच बकरी बन गया हूँ। वह अपना माथा, कान, नाक, मुँह, पेट, हाथ, पैर वग़ैरह बार-बार छूने लगा। उसकी इस अजीब हरकत का पीछा सरदार जी की नज़रें करने लगीं। वह ताड़ गये और उसे सीने से लगाते हुए बोले, "क्यों रामभरोसे! तू मुझसे डर गया? मैं तो तेरा भाई हूँ, बड़ा भाई। अब बड़ा भाई छोटे भाई से कोई मदद माँगेगा, तो क्या वह मुकर जाएगा या डर जाएगा? देखो जी, मेरी घर वाली नहीं है। वह कुछ साल पहले बीमारी से मर गयी।" सरदार जी यह कहकर चुप हो गये मानो तेज़ गति से चल रही गाड़ी में अचानक ब्रेक लगी हो।

"तो दूसरी शादी क्यों नहीं कर लेते, सरदार जी?" रामभरोसे ने कहा।

"मगर रामभरोसे! अब कहाँ शादी! मेरे घर-परिवार में और कोई नहीं है। अब कोई औरत घर में बिठा लो। जब उसके बच्चा हो जाए तो उसे पालो-पोसो। ऐसे वह हमारी औलाद हो जाएगी। फिर वह औरत अपने घर जाये। हाँ,

इसके एवज़ में मैं उसे इतनी दौलत दे दूँगा कि वह ज़िन्दगी भर किसी चीज़ की मोहताज़ नहीं होगी। हमारे यहाँ इसे कुदेशन प्रथा कहते हैं...''

"सरदार जी! अजीब रिवाज़ है आपके यहाँ। 'महाभारत' की द्रौपदी से भी ज़्यादा बदतर हालत होगी उस बेचारी की। जोरू को प्लास्टिक की गुड़िया मानकर उसके साथ ऐशो-आराम करो और फिर फेंक दो जैसे आजकल स्याही खत्म होते ही कलम को कूड़ेदान में फेंक दिया जाता है। गजब का बाज़ार है—इस्तेमाल करो और फेंक दो। लेकिन हमारे मुल्क में ऐसा-वैसा रिवाज़ नहीं है वरना...'' उसने कहा।

"देख, रामभरोसे भाई! तू इसे मज़ाक़ में मत टाल यार, मेरा मखौल मत उड़ा। मेरी मदद कर इसमें। यह मेरी ज़िन्दगी का सवाल है।'' सरदार जी ने फिर बीरबल की तरह पहेली बुझाई। रामभरोसे हँसने लगा ठठाकर। तभी टुनमुनिया ने उसे खाने के लिए जोर से आवाज़ दी। उसने सरदार जी से कहा, "अभी चलता हूँ, सरदार जी! मैं नहीं पहुँचूँगा तो सारे लोग परोसा हुआ भोजन शुरू नहीं करेंगे। आपके मज़ाक़ से खूब मजा आया। वो क्या कहते हैं, कुदेशन कि कुलच्छन? वाह रे पंजाब! हरित क्रान्ति खेती में ही नहीं हुई, औरतों में भी हो गयी। क्यों न हो, आख़िर औरत भी तो एक खेती ही होती है! कौन उसे सम्भालकर नहीं रखता? वाह री धरती, वाह री औरत!'' और वह हाथ से इशारा करते हुए सरदार जी को विचारों के गहरे दरिया में अकेले छोड़कर अपनी रामकुटी में खाना खाने चला गया।

वहाँ जाते ही वह हँस-हँसकर सारा वाकया सभी लोगों को सुनाने लगा। सभी लोग ठठाकर हँसने लगे मगर टुनमुनिया बिलकुल गम्भीर और खामोश थी। रामभरोसे ने उसकी ओर देखा, तो वह उबल पड़ी, "ई मरद लोगन लुगाई के सिरिफ खेती समझे लें। पहिले तैयार करे लें, फिर काट ले लें। झाड़ू मारे इ रेवाज के। कवन अइसन मेहरारू होखी, जे आपन कोखी के जामल लइका के दे दी? लाखों रुपया पर हमनी के इहाँ के मेहरारू कोख न बेची। ई मरद से ऊ मरद, कवनो पतुरिये नू जा सकेले। लइका जनमावे के हमनी के कारखाना नइखी जा। मेहरारू जनावर ना होले कि बच्चा आउर दूध दुह के हटिया में केहू बेच दे।''

वह टुनमुनिया का रौद्र रूप देखकर अवाक् था। वह चकित था कि अनपढ़ टुनमुनिया इतनी गहरी समझ की बातें कैसे कर रही है? वह काफ़ी देर तक गुस्से में थी। खाना छोड़कर वह अपने बिस्तर पर लेट गयी। दूसरी औरतों के समझाने-बुझाने पर ही टुनमुनिया ने देर रात में खाना खाया। रामभरोसे रात भर सोचता रहा।

काफ़ी समय के बाद धान पक गया था। उसकी सुनहली बालें ऐसे लटक

रही थीं मानो स्त्रियों के कानों की बालियाँ हों। अगले दिन वे सभी धान की कटाई में लग गये। धूप कुछ ज़्यादा कड़ी थी मगर समूह में होने तथा औरतों के कटनी गीत गाने के आगे धूप की एक नहीं चलती थी। टुनमुनिया को दर्जनों गीत याद हैं जो बुआई-निराई-कटाई के समय गाये जाते हैं। सो वह झूम-झूमकर गा रही थी और उसी रफ़्तार से वह धान के पौधे भी काट रही थी—

मति बोलूँ सैंया, रात अँजोरिया, मति बोलूँ
सैंया/ आवे द अन्हरिया, मति बोलूँ सैंया/
अँगुरी की बिछिया है लाल आउर करिया/
आवे द अन्हरिया, मति बोलूँ सैंया।।

उसकी आवाज़ का जादू रामभरोसे के सिर पर ही नहीं, सरदार जी के सिर पर भी बोलने लगा। "इतने अच्छे-अच्छे गीत मैंने आज तक पंजाबी में नहीं सुने।"

कटनी ख़त्म होने के बाद शाम को सरदार गुरिंदर सिंह ने रामभरोसे को अपने पास इशारे से बुलाया। उनका इशारा पहले आँख से था, फिर हाथ से। वह एक थर्मस में चाय लेकर आये थे। उन्होंने खुद चाय पी और उसे भी पिलाई। गरमागरम चाय पीने से रामभरोसे की कटनी की थकावट छूमन्तर हो गयी। फिर वह बोले, "देख, रामभरोसे भाई! मेरी नैया तो यही टुनमुनिया है। तू उसकी गोद में मुझे बैठा दे, तो मैं तुम्हारा अहसान ज़िन्दगी भर नहीं भूलूँगा। कई हफ्तों से मैं उसके बारे में सोच-सोचकर पागल हो गया हूँ। उसे जितना ज़्यादा भूलने की कोशिश करता हूँ, उतना ही मैं अपने को उसके करीब पाता हूँ। मैं तेरे से पक्का वादा करता हूँ कि उसे कोई तकलीफ़ नहीं होगी। मैं उसके नाम पच्चीस हज़ार रुपये फिक्स्ड डिपॉजिट में जमा कर दूँगा और अनाज-पानी, कपड़ा वगैरह मुफ़्त देता रहूँगा। सिर्फ़ एक साल तक वह मेरे साथ ज़िन्दगी गुज़ार दे और मुझे एक औलाद दे दे।" ये शब्द रामभरोसे के दिल में बाण की तरह सीधे चुभ गये। उसका गाथ झनझना उठा। जो बात वह अपने लिए टुनमुनिया से नहीं कह सका, उसे दूसरे के लिए कैसे कह पाएगा? जो दूसरी औरत के कुदेशन बनने पर बिदक गयी हो, उसे खुद कुदेशन बनने के लिए वह कैसे कह सकेगा? भयंकर उधेड़बुन में उसके मुँह से सिर्फ़ इतना ही निकला, "सरदार जी! मैंने आपका नमक खाया है, इसलिए उसे चुकाने की कोशिश करूँगा।"

फिर वह वहाँ से चला गया। मगर उसकी समझ में नहीं आ रहा था कि टुनमुनिया से वह यह बात कैसे कहे। उसने सोचा—यह सुनते ही गरम तवे पर पड़ी पानी की बूँद की तरह वह छनछना उठेगी, शायद हाथापायी भी कर बैठे। सो उससे कहने की हिम्मत न जुटा पाने के कारण उसने रामप्यारी से पूरी बात

कह दी। उसे भी टुनमुनिया से यह सब कहने की हिम्मत नहीं हुई तो रामभरोसे ने उसकी हिम्मत बढ़ाई, ''देखो! औरत ही ऐसे मामलों में दूसरी औरत को संजीदगी से समझा सकती है। मेरे कहने से उसका दर्द दुगुना हो जाएगा। शायद वह मुझे ही इस कांड का खलनायक मान ले।''

उस रात अकेले में टुनमुनिया को ले जाकर रामप्यारी ने सरदार जी की ख़्वाहिश बता दी। यह सुनते ही फनफना गयी टुनमुनिया, ''रामप्यारी! ई का कहत बाडू तू! ऊ दढ़िजरवा, सरदरवा हमरा सामने कहले रहित त मुँहवे नोच लेती। हम हाड़ तोड़िके पेट भरे खातिर मजूरी करत बानी। ओकर रखैल बने खातिर नइखे आइल। हम एको पल एइजा ना रहब। हमनी के गरीबी में एइजा आइल बानी मजूरी खातिर, देह बेचे खातिर ना!''

सुबह होने पर रामप्यारी ने रामभरोसे को सारा वाकया सुना दिया। उसे ऐसी ही प्रतिक्रिया की आशा थी। उसने सरदार जी से अपनी असफलता बता दी। मगर वह समझने को तैयार नहीं थे। उन्होंने रामभरोसे से फिर हाथ जोड़कर चिरौरी की। वह रोज़ाना शाम को उससे इस बाबत दरयाफ़्त करते मगर रामभरोसे अभी तक एक कदम आगे, दो कदम पीछे की स्थिति में था। उसने पाया कि सरदार गुरिंदर सिंह के चेहरे पर काई लगने से चमक धीरे-धीरे गायब हो रही है। वह काम करते समय जो चुहलबाजी और हँसी-मजाक किया करते थे और चुटकुला सुनाते थे, वह अब अतीत का अध्याय था। सुदूर मुल्क से आयी मज़दूर स्त्री को न पाने का गम उन्हें घुन की तरह भीतर-ही-भीतर खाये जा रहा था। सरदार जी ने करीब एक माह बाद फिर रामभरोसे को उस काम के लिए टोका। उसने अपनी मजबूरी उनके सामने बयान कर दी। मगर सरदार जी ने बहुत दबाव दिया तो उसने एक प्रस्ताव पेश किया, ''सरदार जी! आप स्वयं टुनमुनिया से अपनी बात कहिए। ज़रूरत पड़ेगी तो मैं आपकी ओर से जोर दूँगा। आपके कहने का ज़्यादा असर होगा। शायद वह आपकी बातों से कुछ पिघल जाये!''

सरदार जी को थोड़ा जोश आया। अगले दिन सुबह-सुबह सरदार जी पूरी तैयारी के साथ फार्म पर आये। वह ट्यूबवेल पर बैठे थे कि टुनमुनिया एक बाल्टी लेकर पानी लेने आयी। सरदार जी ने उसे कौतूहल से देखा। उसने सरदार जी को घूरा और पूछ बैठी, ''का बात बा?''

''कुछ नहीं टुनमुनिया!'' सरदार जी की सीटी-सपाटी गुल हो गयी। वह पानी लेकर चली गयी। उसके चले जाने के थोड़ी देर बाद सरदार जी ने रामभरोसे से कहा, ''यार! यह औरत है कि आग का अंगार? इसकी बेधती नज़रों के आगे मेरी हिम्मत जवाब दे जाती है। मैं तो भूल ही गया था कि मुझे क्या कहना है?''

रामभरोसे को लगा कि सरदार जी भीतर से घायल हैं सो उसने कहा, ''आप

सत्रह महीने हो गये हैं। मेरी ओर से ये हैं पच्चीस हज़ार रुपये। जो कुछ तुम्हें चाहिए, मुझे बता दो।''

मगर टुनमुनिया कुछ भी कहने की स्थिति में नहीं थी। वह राजपाल के पालने की ओर टुकुर-टुकुर देख रही थी। उसे कभी अपने करार की याद आती, कभी टोली के लोगों के चेहरे सामने घूमने लगते और कभी राजपाल की किलकारियाँ याद आतीं। उस दिन उसने कोई जवाब नहीं दिया। मगर वह रात भर बेचैन रही— नींद उसकी आँखों से भागकर गुरिंदर सिंह की आँखों में समा गयी थी। और राजपाल को क्या पता था?

सुबह चाय-नाश्ता करने के बाद सरदार गुरिंदर सिंह अपने फार्म पर गये। वहाँ उन्होंने रामभरोसे को करार की याद दिलायी, तो उसने कहा, ''हाँ, हाँ, मैं तो भूल ही गया था कि आपका और टुनमुनिया का करार ख़त्म हो गया है। आपको अपनी दौलत का वह वारिस मिल गया है जिसकी आपको बरसों से तलाश थी।''

इस बातचीत के बाद दोनों सरदार जी के घर की तरफ चल पड़े। रामभरोसे को लग रहा था कि एक बार टुनमुनिया के वापस लौटने के बाद उसकी टोली आबाद हो जाएगी, उसमें पुरानी रौनक लौट आएगी। घर पहुँचने पर सरदार जी ने कहा, ''करार के मुताबिक ये रहे पच्चीस हज़ार रुपये। तेरे कपड़े-गहने वगैरह मैंने सन्दूक में रखवा दिये हैं। खाने-पीने की चीज़ें ट्रैक्टर से चली जाएँगी। मैं तेरा एहसान ज़िन्दगी भर नहीं भूलूँगा, निर्मल। कभी किसी चीज़ की ज़रूरत हो तो मुझे याद करना। बन्दा हर हालत में खुशी-खुशी हाज़िर रहेगा।''

''ई पइसा रउरे राखीं, सरदार जी! ऊ लइका के काम आई। हमनी के आपन कोख बचाए खातिर जान दे देब। हमार देह रउरा काम आ गइल, इहे जिनगी के मोल बा। हाथ जोरि के एक ही निहोरा बा कि जब तक हम एइजा रहत बानी, भर नज़र लइकवा के देखे पर रोक मत लगाइब।'' और सरदार जी ने इस पर हामी भर ली।

फिर सरदार जी ने रामभरोसे, टुनमुनिया और उसके सामान को ट्रैक्टर पर लादकर फार्म तक पहुँचा दिया। टुनमुनिया अपने लोगों से मिलकर खूब रोती रही। वह रामप्यारी और दूसरी औरतों से अपने सोलह-सत्रह महीनों के अनुभवों को देर रात तक बाँटती रही। देर रात तक आपस में बातें होती रहीं। फिर खाना बना और सबने मन भर खाया। रात में सोते समय टुनमुनिया की छाती में ढेर सारा दूध उतर आया। उसने अपनी समस्या रामप्यारी को बतायी। फिर उसने अपना दूध एक कटोरी में निकालकर छप्पर के ऊपर रख दिया। रामप्यारी ने उसके ब्लाउज

ठहरकर दुबारा कोशिश करें। हो सकता है, कुछ बदलाव आ जाये।''

और करीब पन्द्रह दिनों तक वह कठिन रिहर्सल करते रहे। उसके बाद एक शाम को, जब आकाश में बादल खेल रहे थे, सरदार जी ने टुनमुनिया से उसकी कोख किराये पर लेने का प्रस्ताव बड़ी सादगी और विनम्रता से रख दिया, ''देखो, टुनमुनिया! तुम इसे जोर-जबर्दस्ती बिलकुल मत समझो। यह हमारे यहाँ का पुराना रीति-रिवाज़ है। अपना हिन्दुस्तान तो परम्पराओं का देश है। मैं तुम्हारी बहुत इज़्ज़त करता हूँ। मैं वाहे गुरु को साक्षी मानकर वचन देता हूँ कि मेरे साथ एक साल तक मेरी औरत बनकर रह जाओ और एक बच्चा जनने के बाद मैं तुम्हें वापस आने दूँगा। मैं पच्चीस हज़ार रुपये के अलावा सारा खाना-खर्चा ज़िन्दगी भर तुम्हें देता रहूँगा।''

नागिन-सी फनफनाती हुई टुनमुनिया ने टके-सा जवाब दे दिया, ''सरदार जी! ई रेवाज राउर मुलुक में बा। हमरा मुलुक में अइसन मेहरारू से लोग नफ़रत करेलें। रंडी-रखैल केकरा के कहल जाला। हमरी जान भले जाए बाकिर अइसन काम हम ना करब। इ त अधरम बा।''

रामभरोसे ट्यूबवेल के कमरे के भीतर से सब कुछ सुन रहा था। उसे देखते ही टुनमुनिया गुस्से में बोली, ''नेता जी! सुनत बानी सरदार जी के बात। रउरा हमरा के टिकस कटा दी। अब एको पल एइजा न रहब!''

रामभरोसे बिलकुल चुप रहा। तभी सरदार जी ने घी चुपड़ते हुए कहा, ''रामभरोसे! क्या मेरी अरदास को आप दोनों ठुकरा दोगे? मैं ज़िन्दगी भर के लिए इसके भरण-पोषण का इन्तज़ाम कर दूँगा। आप दोनों मेरी नीयत पर शक मत करो। बस एक साल का सवाल है। समय बीतते देर नहीं लगती।''

रामभरोसे खामोश रहा। टुनमुनिया भी चुप रही। काफी देर तक भीतर-बाहर सन्नाटे का छन्द गूँजता रहा। सरदार जी ने रामभरोसे की ओर कातर नज़रों से देखा, फिर टुनमुनिया को निहारा। टुनमुनिया की आँखें भर आयी थीं। रामभरोसे की ओर देखते हुए उसने कहा, ''नेता जी! गाँव में भूख-पियास से मर जइती। बाढ़ में बह जइती बाकिर हम ना अइती। रउरे पर भरोसा रहे। काली माई के किरिया खाके कहत बानी, रउरा आग में जरे के कहब, त जर जाइब। नदी में डूबे के कहब, त डूब जाइब।'' यह कहते हुए वह चलने के लिए उठ खड़ी हुई। सरदार जी ने बड़ी बेसब्री और विनम्रता से फिर कहा, ''रामभरोसे! अब बचा लो मेरी ज़िन्दगी की नैया। तुम इस मौके पर मुँह नहीं मोड़ो। अय्याशी नहीं, ज़रूरत के लिए मैं इसे चाहता हूँ।''

ज़मीन से उठते हुए रामभरोसे ने दबी जुबान से कहा, ''टुनमुनिया! सरदार जी का खयाल रखो।''

''आउर हमार खयाल तोहरा के नइखे, नेता जी?'' टुनमुनिया ने घूरते हुए

ऐसा सवाल दागा कि रामभरोसे चित हो गये। वह उसका जवाब धरती से आकाश तक कहीं भी रंचमात्र नहीं ढूँढ़ पा रहा था, सो सोच की भँवर में डूबे-डूबे अचानक उसे चक्कर आ गया। टुनमुनिया अपने आँचल से उसे हवा करने लगी। वह उसके माथे पर पानी के छींटे दे चुकी थी। उसने सिसकते हुए कहा, ''नेता जी राउर जिनगी से बड़ हमार जिनगी कुछ नइखे। रउरा खातिर कुआँ में कूदे में हमनी ना हिचकब।''

और एक दिन टुनमुनिया 'कुदेशन' बन गयी। अब वह सरदार जी के हवेलीनुमा विशाल घर में रहने लगी। लोगों के वापस चले जाने के बाद टुनमुनिया इतने बड़े घर में ऊबने लगी। कभी-कभी वह खो जाती। उसे अपनी टोली की याद आती, गाँव भीखनपुर की याद आती और याद आती नेता रामभरोसे की। मगर वह इन यादों की पोटली किसी के सामने नहीं खोल सकती थी। धीरे-धीरे यादों का भारी बोझ उसके दिलो-दिमाग पर लदा रहने लगा। उसका मन करता कि वह रामप्यारी, रामभरोसे और गाँव से आयी टोली के दूसरे लोगों से मिल ले, मगर अब उसे फार्म पर जाने की मनाही थी और उसे टोली को सरदार जी के यहाँ आने की मनाही हो गयी थी। कर्फ़्यू में दूध-भात का स्वाद भी खराब लगता है। अपनों से बिछड़ने का गम उसे बुरी तरह सताने लगा। सरदार जी उसे लेकर अमृतसर चले गये। गुरुद्वारा, स्वर्ण मन्दिर, बाज़ार वग़ैरह सब जगह उसे उन्होंने घुमाया मगर उसे भीतर की शान्ति कभी नहीं मिली। उसने ज़िन्दगी में पहली बार इतनी दौलत देखी मगर अकूत दौलत की सेज पर उसे चैन की नींद नहीं आती थी।

कई हफ़्तों के बाद सारा माज़रा सरदार गुरिंदर सिंह ने समझ लिया। सो उन्होंने रामप्यारी और रामभरोसे को हर इतवार को अपने घर के बरामदे में आकर मिलने का बन्दोबस्त कर दिया। सरदार गुरिंदर सिंह के मन के छोटे-से कोने में एक बड़ा चोर छिपा था, ''यदि रामभरोसे अकेले आने का मौका पाएगा, तो दोनों की देह में एक साथ आग लग सकती है? या दोनों कहीं भाग सकते हैं या दोनों मिलकर उसकी दौलत को नुकसान पहुँचा सकते हैं या दोनों उसकी जान को ख़तरा पहुँचा सकते हैं। कहते हैं कि मोहब्बत और जंग में सब कुछ जायज़ है। इसलिए दोनों अलावों के बीच में चौकीदार की तरह हमेशा मौजूद रहेगी रामप्यारी।'' मगर टुनमुनिया तन-मन दोनों की पक्की थी। अब उसका रामभरोसे के प्रति झुकाव मोह-माया से पूरी तरह मुक्त था।

फिर यौवन की हल्की-हल्की फुहारें टुनमुनिया के तन-मन को सींचने लगीं। उसकी पुरानी खिलखिलाहट अब वापस लौट आयी। उन दोनों की ज़िन्दगी में नया अंकुर फूटने लगा। तीन माह बीतने के बाद उसे बार-बार मितली होने लगी। सरदार गुरिंदर सिंह उसे डॉक्टर हरजीत कौर के पास दिखाने ले गये। डॉक्टर

ने मुस्कराते हुए कहा, ''आपकी बीवी माँ बनने वाली है। सब कुछ च
सरदार जी! ये आयरन की गोलियाँ इसे नियमित रूप से लेनी हैं। बीच
में चेकअप करा लेना कि गर्भ में बच्चे की हालत कैसी है?''

क्लीनिक से निकलते ही सरदार जी ने टुनमुनिया को अपनी बाँहों
लिया, उसका माथा चूम लिया और उसके पेट पर हाथ फेरने लगे। और
लगे, ''तू मेरी हीर है। तू मेरे मन की रानी है। तू ही मेरा सब कुछ है

उस दिन से सरदार गुरिंदर सिंह टुनमुनिया का और ज़्यादा ख़याल
लगे। वह कहते, ''आहिस्ता, आहिस्ता! तू आराम कर। मैं सारे काम निपटा
दाई तेरी मालिश भी कर देगी, तेरे कपड़े भी धो देगी, खाना-वाना सब बना
वह। मेरे बाहर जाने पर भी तू कोई काम नहीं करना।'' मगर वह पलंग तोड़
तोड़ते थक जाती। टी.वी. देखते-देखते ऊब जाती। सो वह घर के सामने के
और फुलवारी में टहलती, फुदकती हुई तितलियाँ देखती, हवा में झूलती डालि
देखती, पत्तों की सरसराहट सुनती, फूलों की सुगन्ध सूँघती, कलियों की न
परखती, मिट्टी से निकलते नये अंकुरों को निहारती, आकाश की ऊँचाई की तर
ताकती रहती और क्षितिज के उस पार जाने का दिवास्वप्न देखती रहती। कभी
कभी वह पेट के बढ़ते आकार को देखकर खुश हो जाती कि एक दिन उसकी
अपनी सन्तान इस दुनिया में आ जाएगी और उसे माँ कहेगी।

और एक सुबह उसने एक नर्सिंग होम में बेटे को जन्म दिया। गुरिंदर सि
ने पहले से ही उसका नाम सोच लिया था—'राजपाल'। राजपाल ने आते
गुरिंदर सिंह की ज़िन्दगी के चमन में बेपनाह खुशबू चारों ओर बिखेर दी

सरदार गुरिंदर सिंह ने बेटे के जन्म पर अच्छा-खासा जलसा किया। स
लोगों ने राजपाल की बलइयाँ लीं और टुनमुनिया एवं गुरिंदर सिंह को दिल
बधाई दी। उस दिन रामभरोसे और रामप्यारी के अलावा टोली के सभी
जलसे में शरीक हुए थे और सबने छककर दावत खायी थी। दाई ने राजप
की सेवा तन-मन से करनी शुरू कर दी। वह टुनमुनिया की भी तेल-फुलेल
मालिश करती। वह राजपाल को नहलाती-धुलाती, उसका गू-मूत सब स
करती। उसके लिए एक पालना सरदार जी ने खरीद लिया था। टुनमुनिया
उसे स्तन-पान कराती, तो उसे एक विशेष प्रकार के आनन्द का अनुभव हो
उसे लगता कि काश यह सुख उसे पहले मिलता। वह राजपाल को दिन में
बार चूमती-चाटती। वह खुद भूल गयी थी कि एक दिन उसे ये सब त्याग
हैं। जब वह करीब छह माह का हो गया तो तरल खाद्य-पदार्थ खाने लगा।
एक दिन सरदार गुरिंदर सिंह ने उससे कहा, ''मैं नहीं चाहता कि तू यह
छोड़कर चली जाओ। लेकिन मेरी-तेरी करार एक साल की थी और अब सोल

ठहरकर दुबारा कोशिश करें। हो सकता है, कुछ बदलाव आ जाये।''

और करीब पन्द्रह दिनों तक वह कठिन रिहर्सल करते रहे। उसके बाद एक शाम को, जब आकाश में बादल खेल रहे थे, सरदार जी ने टुनमुनिया से उसकी कोख किराये पर लेने का प्रस्ताव बड़ी सादगी और विनम्रता से रख दिया, ''देखो, टुनमुनिया! तुम इसे जोर-जबर्दस्ती बिलकुल मत समझो। यह हमारे यहाँ का पुराना रीति-रिवाज़ है। अपना हिन्दुस्तान तो परम्पराओं का देश है। मैं तुम्हारी बहुत इज़्ज़त करता हूँ। मैं वाहे गुरु को साक्षी मानकर वचन देता हूँ कि मेरे साथ एक साल तक मेरी औरत बनकर रह जाओ और एक बच्चा जनने के बाद मैं तुम्हें वापस आने दूँगा। मैं पच्चीस हज़ार रुपये के अलावा सारा खाना-खर्चा ज़िन्दगी भर तुम्हें देता रहूँगा।''

नागिन-सी फनफनाती हुई टुनमुनिया ने टके-सा जवाब दे दिया, ''सरदार जी! ई रेवाज राउर मुलुक में बा। हमरा मुलुक में अइसन मेहरारू से लोग नफरत करेलें। रंडी-रखैल केकरा के कहल जाला। हमरी जान भले जाए बाकिर अइसन काम हम ना करब। इ त अधरम बा।''

रामभरोसे ट्यूबवेल के कमरे के भीतर से सब कुछ सुन रहा था। उसे देखते ही टुनमुनिया गुस्से में बोली, ''नेता जी! सुनत बानी सरदार जी के बात। रउरा हमरा के टिकस कटा दी। अब एको पल एइजा न रहब!''

रामभरोसे बिलकुल चुप रहा। तभी सरदार जी ने घी चुपड़ते हुए कहा, ''रामभरोसे! क्या मेरी अरदास को आप दोनों ठुकरा दोगे? मैं ज़िन्दगी भर के लिए इसके भरण-पोषण का इन्तज़ाम कर दूँगा। आप दोनों मेरी नीयत पर शक मत करो। बस एक साल का सवाल है। समय बीतते देर नहीं लगती।''

रामभरोसे खामोश रहा। टुनमुनिया भी चुप रही। काफी देर तक भीतर-बाहर सन्नाटे का छन्द गूँजता रहा। सरदार जी ने रामभरोसे की ओर कातर नज़रों से देखा, फिर टुनमुनिया को निहारा। टुनमुनिया की आँखें भर आयी थीं। रामभरोसे की ओर देखते हुए उसने कहा, ''नेता जी! गाँव में भूख-पियास से मर जइती। बाढ़ में बह जइती बाकिर हम ना अइती। रउरे पर भरोसा रहे। काली माई के किरिया खाके कहत बानी, रउरा आग में जरे के कहब, त जर जाइब। नदी में डूबे के कहब, त डूब जाइब।'' यह कहते हुए वह चलने के लिए उठ खड़ी हुई। सरदार जी ने बड़ी बेसब्री और विनम्रता से फिर कहा, ''रामभरोसे! अब बचा लो मेरी ज़िन्दगी की नैया। तुम इस मौके पर मुँह नहीं मोड़ो। अय्याशी नहीं, ज़रूरत के लिए मैं इसे चाहता हूँ।''

ज़मीन से उठते हुए रामभरोसे ने दबी जुबान से कहा, ''टुनमुनिया! सरदार जी का खयाल रखो।''

''आउर हमार खयाल तोहरा के नइखे, नेता जी?'' टुनमुनिया ने घूरते हुए

ऐसा सवाल दागा कि रामभरोसे चित हो गये। वह उसका जवाब धरती से आकाश तक कहीं भी रंचमात्र नहीं ढूँढ़ पा रहा था, सो सोच की भँवर में डूबे-डूबे अचानक उसे चक्कर आ गया। टुनमुनिया अपने आँचल से उसे हवा करने लगी। वह उसके माथे पर पानी के छींटे दे चुकी थी। उसने सिसकते हुए कहा, "नेता जी राउर जिनगी से बड़ हमार जिनगी कुछ नइखे। रउरा खातिर कुआँ में कूदे में हमनी ना हिचकब।"

और एक दिन टुनमुनिया 'कुदेशन' बन गयी। अब वह सरदार जी के हवेलीनुमा विशाल घर में रहने लगी। लोगों के वापस चले जाने के बाद टुनमुनिया इतने बड़े घर में ऊबने लगी। कभी-कभी वह खो जाती। उसे अपनी टोली की याद आती, गाँव भीखनपुर की याद आती और याद आती नेता रामभरोसे की। मगर वह इन यादों की पोटली किसी के सामने नहीं खोल सकती थी। धीरे-धीरे यादों का भारी बोझ उसके दिलो-दिमाग पर लदा रहने लगा। उसका मन करता कि वह रामप्यारी, रामभरोसे और गाँव से आयी टोली के दूसरे लोगों से मिल ले, मगर अब उसे फार्म पर जाने की मनाही थी और उसे टोली को सरदार जी के यहाँ आने की मनाही हो गयी थी। कर्फ़्यू में दूध-भात का स्वाद भी खराब लगता है। अपनों से बिछड़ने का गम उसे बुरी तरह सताने लगा। सरदार जी उसे लेकर अमृतसर चले गये। गुरुद्वारा, स्वर्ण मन्दिर, बाज़ार वग़ैरह सब जगह उसे उन्होंने घुमाया मगर उसे भीतर की शान्ति कभी नहीं मिली। उसने ज़िन्दगी में पहली बार इतनी दौलत देखी मगर अकूत दौलत की सेज पर उसे चैन की नींद नहीं आती थी।

कई हफ़्तों के बाद सारा माज़रा सरदार गुरिंदर सिंह ने समझ लिया। सो उन्होंने रामप्यारी और रामभरोसे को हर इतवार को अपने घर के बरामदे में आकर मिलने का बन्दोबस्त कर दिया। सरदार गुरिंदर सिंह के मन के छोटे-से कोने में एक बड़ा चोर छिपा था, "यदि रामभरोसे अकेले आने का मौका पाएगा, तो दोनों की देह में एक साथ आग लग सकती है? या दोनों कहीं भाग सकते हैं या दोनों मिलकर उसकी दौलत को नुकसान पहुँचा सकते हैं या दोनों उसकी जान को ख़तरा पहुँचा सकते हैं। कहते हैं कि मोहब्बत और जंग में सब कुछ जायज़ है। इसलिए दोनों अलावों के बीच में चौकीदार की तरह हमेशा मौजूद रहेगी रामप्यारी।" मगर टुनमुनिया तन-मन दोनों की पक्की थी। अब उसका रामभरोसे के प्रति झुकाव मोह-माया से पूरी तरह मुक्त था।

फिर यौवन की हल्की-हल्की फुहारें टुनमुनिया के तन-मन को सींचने लगीं। उसकी पुरानी खिलखिलाहट अब वापस लौट आयी। उन दोनों की ज़िन्दगी में नया अंकुर फूटने लगा। तीन माह बीतने के बाद उसे बार-बार मितली होने लगी। सरदार गुरिंदर सिंह उसे डॉक्टर हरजीत कौर के पास दिखाने ले गये। डॉक्टर

ने मुस्कराते हुए कहा, ''आपकी बीवी माँ बनने वाली है। सब कुछ चंगा है, सरदार जी! ये आयरन की गोलियाँ इसे नियमित रूप से लेनी हैं। बीच-बीच में चेकअप करा लेना कि गर्भ में बच्चे की हालत कैसी है?''

क्लीनिक से निकलते ही सरदार जी ने टुनमुनिया को अपनी बाँहों में भर लिया, उसका माथा चूम लिया और उसके पेट पर हाथ फेरने लगे। और कहने लगे, ''तू मेरी हीर है। तू मेरे मन की रानी है। तू ही मेरा सब कुछ है!''

उस दिन से सरदार गुरिंदर सिंह टुनमुनिया का और ज़्यादा ख़याल रखने लगे। वह कहते, ''आहिस्ता, आहिस्ता! तू आराम कर। मैं सारे काम निपटा लूँगा। दाई तेरी मालिश भी कर देगी, तेरे कपड़े भी धो देगी, खाना-वाना सब बना लेगी वह। मेरे बाहर जाने पर भी तू कोई काम नहीं करना।'' मगर वह पलंग तोड़ते-तोड़ते थक जाती। टी.वी. देखते-देखते ऊब जाती। सो वह घर के सामने के बाग और फुलवारी में टहलती, फुदकती हुई तितलियाँ देखती, हवा में झूलती डालियाँ देखती, पत्तों की सरसराहट सुनती, फूलों की सुगन्ध सूँघती, कलियों की नींद परखती, मिट्टी से निकलते नये अंकुरों को निहारती, आकाश की ऊँचाई की तरफ ताकती रहती और क्षितिज के उस पार जाने का दिवास्वप्न देखती रहती। कभी-कभी वह पेट के बढ़ते आकार को देखकर खुश हो जाती कि एक दिन उसकी अपनी सन्तान इस दुनिया में आ जाएगी और उसे माँ कहेगी।

और एक सुबह उसने एक नर्सिंग होम में बेटे को जन्म दिया। गुरिंदर सिंह ने पहले से ही उसका नाम सोच लिया था—'राजपाल'। राजपाल ने आते ही गुरिंदर सिंह की ज़िन्दगी के चमन में बेपनाह खुशबू चारों ओर बिखेर दी।

सरदार गुरिंदर सिंह ने बेटे के जन्म पर अच्छा-खासा जलसा किया। सभी लोगों ने राजपाल की बलइयाँ लीं और टुनमुनिया एवं गुरिंदर सिंह को दिल से बधाई दी। उस दिन रामभरोसे और रामप्यारी के अलावा टोली के सभी लोग जलसे में शरीक हुए थे और सबने छककर दावत खायी थी। दाई ने राजपाल की सेवा तन-मन से करनी शुरू कर दी। वह टुनमुनिया की भी तेल-फुलेल से मालिश करती। वह राजपाल को नहलाती-धुलाती, उसका गू गूत सब साफ करती। उसके लिए एक पालना सरदार जी ने खरीद लिया था। टुनमुनिया जब उसे स्तन-पान कराती, तो उसे एक विशेष प्रकार के आनन्द का अनुभव होता। उसे लगता कि काश यह सुख उसे पहले मिलता। वह राजपाल को दिन में कई बार चूमती-चाटती। वह खुद भूल गयी थी कि एक दिन उसे ये सब त्याग देने हैं। जब वह करीब छह माह का हो गया तो तरल खाद्य-पदार्थ खाने लगा। सो एक दिन सरदार गुरिंदर सिंह ने उससे कहा, ''मैं नहीं चाहता कि तू यह घर छोड़कर चली जाओ। लेकिन् मेरी-तेरी करार एक साल की थी और अब सोलह-

सत्रह महीने हो गये हैं। मेरी ओर से ये हैं पच्चीस हज़ार रुपये। जो कुछ तुम्हें चाहिए, मुझे बता दो।''

मगर टुनमुनिया कुछ भी कहने की स्थिति में नहीं थी। वह राजपाल के पालने की ओर टुकुर-टुकुर देख रही थी। उसे कभी अपने करार की याद आती, कभी टोली के लोगों के चेहरे सामने घूमने लगते और कभी राजपाल की किलकारियाँ याद आतीं। उस दिन उसने कोई जवाब नहीं दिया। मगर वह रात भर बेचैन रही—नींद उसकी आँखों से भागकर गुरिंदर सिंह की आँखों में समा गयी थी। और राजपाल को क्या पता था?

सुबह चाय-नाश्ता करने के बाद सरदार गुरिंदर सिंह अपने फार्म पर गये। वहाँ उन्होंने रामभरोसे को करार की याद दिलायी, तो उसने कहा, ''हाँ, हाँ, मैं तो भूल ही गया था कि आपका और टुनमुनिया का करार ख़त्म हो गया है। आपको अपनी दौलत का वह वारिस मिल गया है जिसकी आपको बरसों से तलाश थी।''

इस बातचीत के बाद दोनों सरदार जी के घर की तरफ चल पड़े। रामभरोसे को लग रहा था कि एक बार टुनमुनिया के वापस लौटने के बाद उसकी टोली आबाद हो जाएगी, उसमें पुरानी रौनक लौट आएगी। घर पहुँचने पर सरदार जी ने कहा, ''करार के मुताबिक ये रहे पच्चीस हज़ार रुपये। तेरे कपड़े-गहने वगैरह मैंने सन्दूक में रखवा दिये हैं। खाने-पीने की चीज़ें ट्रैक्टर से चली जाएँगी। मैं तेरा एहसान ज़िन्दगी भर नहीं भूलूँगा, निर्मल। कभी किसी चीज़ की ज़रूरत हो तो मुझे याद करना। बन्दा हर हालत में खुशी-खुशी हाज़िर रहेगा।''

''ई पइसा रउरे राखीं, सरदार जी! ऊ लइका के काम आई। हमनी के आपन कोख बचाए खातिर जान दे देब। हमार देह रउरा काम आ गइल, इहे जिनगी के मोल बा। हाथ जोरि के एक ही निहोरा बा कि जब तक हम एइजा रहत बानी, भर नज़र लइकवा के देखे पर रोक मत लगाइब।'' और सरदार जी ने इस पर हामी भर ली।

फिर सरदार जी ने रामभरोसे, टुनमुनिया और उसके सामान को ट्रैक्टर पर लादकर फार्म तक पहुँचा दिया। टुनमुनिया अपने लोगों से मिलकर खूब रोती रही। वह रामप्यारी और दूसरी औरतों से अपने सोलह-सत्रह महीनों के अनुभवों को देर रात तक बाँटती रही। देर रात तक आपस में बातें होती रहीं। फिर खाना बना और सबने मन भर खाया। रात में सोते समय टुनमुनिया की छाती में ढेर सारा दूध उतर आया। उसने अपनी समस्या रामप्यारी को बतायी। फिर उसने अपना दूध एक कटोरी में निकालकर छप्पर के ऊपर रख दिया। रामप्यारी ने उसके ब्लाउज

के भीतर एक गमछा कसकर बाँध दिया जिससे अब दूध बाहर नहीं आये। मगर टुनमुनिया राजपाल के बिना अधीर हो गयी। उसकी यादों ने उसकी नींद गायब कर दी। मगर उसके पास सिसक-सिसककर रोने के अलावा कोई चारा नहीं था।

सुबह रामप्यारी ने रामभरोसे को सारा वाकया सुना दिया। वह भी कोई उपाय नहीं ढूँढ़ पा रहा था। दिनभर रामभरोसे, रामप्यारी, टुनमुनिया तथा दूसरी औरतें बेचैन रहीं। तीसरे दिन उसकी हालत खराब होने लगी। सो शाम को रामप्यारी बिना रामभरोसे से पूछे टुनमुनिया को लेकर सरदार जी के घर चली गयी। जब दोनों पैदल चलकर वहाँ पहुँचीं तो शाम ऊँघने लगी। दाई ने राजपाल को खिला-पिलाकर पालने में सुला दिया था। शायद सरदार जी पड़ोसी के घर किसी काम से चले गये थे। सो टुनमुनिया सुखद आश्चर्य देने के लिए चुपके-चुपके घर के अन्दर घुसी और आहिस्ते-से राजपाल के पास पहुँचकर उसे टुकुर-टुकुर निहारने लगी। वह नींद में मुस्करा रहा था। शायद कोई अच्छा सपना देख रहा था। टुनमुनिया ने अपने दाहिने हाथ से उसे सहलाया और उसका माथा चूमने लगी। माँ का स्पर्श पाकर राजपाल किलकारियाँ मारने लगा। तभी अचानक दाई वहाँ आ गयी। उसे लगा कि वह राजपाल को चुराने के लिए आयी है। वह जोर से चीखी, ''दौड़ो, सरदार जी! टुनमुनिया...''

और क्षणभर में सरदार गुरिंदर सिंह हाँफते हुए घर के भीतर पहुँच गये। टुनमुनिया की ओर दाई ने इशारा किया। अँधेरे में खड़ी टुनमुनिया को देखकर सरदार जी आग बबूला हो गये, ''हरामज़ादी! तूने मेरे बेटे को चुराने की हिम्मत कैसे की? इसीलिए तूने पैसे नहीं लिए! अच्छा नाटक कर लेती है तू! देने पर रकम नहीं लेती और चुपके-से मेरा भविष्य, मेरी ज़िन्दगी चुरा ले जाना चाहती है। इसीलिए तू अपने साथ रामप्यारी को भी लायी है जिससे तुम दोनों एक कमज़ोर दाई से निपट लो। इसीलिए रात का समय चुना तूने।''

टुनमुनिया को काटो तो खून नहीं। सरदार जी गुस्से से काँप रहे थे। उन्होंने टुनमुनिया को चार-पाँच थप्पड़ जड़ दिये। वह फूट-फूटकर रोने लगी। यह आवाज़ सुनकर राजपाल जोर-जोर से रोने लगा। टुनमुनिया रोते हुए उसे उठाने के लिए बढ़ी। मगर सरदार जी ने टुनमुनिया को उसे छूने नहीं दिया। उन्होंने डाँटा, ''मत छुओ मेरे लाड़ले को। तुम्हारे छूने से उस पर पाप चढ़ जाएगा। भाग जा मेरे घर से!'' फिर उन्होंने टुनमुनिया का झोंटा पकड़कर घसीटते हुए बाहर निकाल दिया और फटाक से दरवाज़ा बन्द कर लिया। रामप्यारी ने टुनमुनिया को उठाया और सहारा देकर धीरे-धीरे चलने लगी। अब रोने में रामप्यारी उसका साथ दे रही थी। तब टुनमुनिया फूट-फूटकर रोने लगी। उसके मुँह से निकला, ''हाय भगवान!

हाय रे करम! इ कोखी के कवन दोख भइल जे इ दिन देखइला, भगवान! इ कोखिया, इ दुधवा के इहै मोल मिलल? पथरा गइल छाती फाटत काहे नइखे भगवान!'' यह कहते-कहते और आँसू पीते-पीते टुनमुनिया बेहोश हो गयी।

उसकी बेहोशी से घबरा गये सरदार जी। वे उसे ट्रैक्टर पर लादकर डॉक्टर के पास ले गये। साथ में रामप्यारी भी थी। रास्ते में उन्होंने रामभरोसे को भी साथ ले लिया। रामभरोसे पूरा वाकया सुनकर तमतमा उठा। मगर उसे पहले टुनमुनिया की जान बचानी थी। इलाज होने से सुबह तक टुनमुनिया ठीक हो गयी। अस्पताल से डेरा वापस आने के बाद पूरी टोली में सरदार जी के ख़िलाफ़ गुस्सा उबल पड़ा। रामभरोसे ने सबकी सहमति से तय कर लिया कि अब सरदार गुरिंदर सिंह के फार्म पर काम करना और रहना मुमकिन नहीं।

अगले दिन रामभरोसे ने बिना लाग-लपेट के सरदार जी को वह सामूहिक फैसला सुना दिया। सरदार जी ने उससे बहुत चिरौरी-विनती की और माफी भी माँगी मगर वह टस से मस न हुआ। टुनमुनिया की ऐसी दुर्गति के लिए वह अपने को भी गुनहगार मानने लगा। मगर बहुत सोच-विचारकर उसने सरदार जी की एक बात मान ली—विदाई की दावत लेना।

दावत के लिए सरदार जी ने नाना प्रकार के पकवान बनवाये। रामभरोसे की पूरी टोली ने छककर खाया। टुनमुनिया कुछ भी नहीं खा रही थी। सो सरदार जी ने उससे हाथ जोड़कर माफ़ी माँगी और खाने की विनती की। राजपाल को अन्तिम बार देखने की शर्त मान लिए जाने पर उसने अपना कौर उठाया। छोले-भटूरे खाने के बाद फिर सरदार जी के इशारे पर एक नवयुवक सरदार ने उसे एक कटोरी खीर खाने को दी जिसमें काजू, किसमिस, छुहारा, बादाम आदि मेवे पड़े थे। उसने बड़े चाव से वह खीर खायी।

कहने को जब भोर हुई तो टोली के सभी लोग अन्यत्र जाने की तैयारी करने लगे। मगर टुनमुनिया अभी भी सो रही थी। शायद वह राजपाल को पाने का सुखद सपना देख रही थी। रामभरोसे के बार-बार कहने पर रामप्यारी ने उसे झकझोरा। मगर टुनमुनिया की नींद नहीं खुली और राजपाल को अन्तिम बार देखने की उसकी इच्छा धरी की धरी रह गयी। पूरी टोली चिल्ला-चिल्लाकर और छाती पीट-पीटकर मातम मनाने लगी। रामभरोसे अपनी मुट्ठियाँ भींचने लगा। फिर कुछ देर बिसूरकर वह पुलिस थाना की ओर दौड़ा।

□□□

णाणं

पयासयं

2470

1944

भारतीय ज्ञानपीठ